Lecture Notes
in Economics and
Mathematical Systems
Managing Editors: M. Beckmann and H. P. Künzi

Systems Theory

131

Mathematical Systems Theory

Proceedings of the International Symposium
Udine, Italy, June 16–27, 1975

Edited by G. Marchesini and S. K. Mitter

Springer-Verlag
Berlin · Heidelberg · New York 1976

Editors

Giovanni Marchesini
Department of Electrical Engineering
University of Padua
6/A Via Gradenigo
35100 Padova/Italy

Sanjoy Kumar Mitter
Department of Electrical Engineering
and Computer Science
and Electronic Systems Laboratory
Massachusetts Institute of Technology
Cambridge, Massachusetts 02139/USA

Library of Congress Cataloging in Publication Data
Conference on Mathematical System Theory, Udine,
Italy, 1975.
Mathematical systems theory.

(Lecture notes in economics and mathematical systems ; 131)
Bibliography: p.
Includes index.
1. System analysis--Congresses. 2. Sequential machine theory--Congresses. 3. Coding theory--Congresses. I. Marchesini, Giovanni, 1936-
II. Mitter, Sahjoy K., 1933- III. Title.
IV. Series.
QA402.C58 1975 629.8'01'51 76-24815

AMS Subject Classifications (1970): 90AXX, 90CXX, 92A05, 92A15, 93AXX, 93BXX

ISBN 3-540-07798-7 Springer-Verlag Berlin · Heidelberg · New York
ISBN 0-387-07798-7 Springer-Verlag New York · Heidelberg · Berlin

Printed in Germany
Printing and binding: Beltz Offsetdruck, Hemsbach/Bergstr.

Foreword

This volume is the record of lectures delivered at the Conference on Mathematical System Theory during the summer of 1975.

The conference was held at the International Centre for Mechanical Sciences in Udine, Italy, and was supported by the Consiglio Nazionale delle Richerche of Italy and the International Centre for Mechanical Sciences. The aim of the conference was to encourage fruitful and active collaboration between researchers working in the diverse areas of system theory. It was also the hope of the organizers that mathematicians participating in the conference might become interested in the purely mathematical problems being raised by systems theory as a result of their participation. The success of the conference is to be measured by the extent to which these aims were fulfilled.

Besides the formal programme of lectures recorded in this volume, many informal seminars were held. The cafes of Udine were often the scene of rich and varied discussions of recent developments in the field amongst the participants of the conference. Last but not least, listening to the ideas exposed in the lectures of others in a creative atmosphere was an important activity.

Mathematical System Theory is a recent discipline, not more than twenty-five to thirty years old. It has many points in common with mathematical physics but also has some important differences. Its aim is to understand natural and man-made systems scientifically, help create new systems, to discover mathematical structure in models where there apparently are not any. Thus its task is often that of synthesis and not just analysis. Moreover a system theorist cannot in general rely on experimental results to guide his intuition in the same way as a physicist can. Laws of behaviour are often to be found in rigorous mathematical analysis and an axiomatic approach is generally mandatory.

Mathematical System Theory is not a coherent discipline (as yet) but has many branches - automata theory, algebraic system theory, information and communication theory, control theory to name a few. It is hoped that as the field develops and matures certain underlying unifying ideas will emerge to bind the diverse areas together.

The present volume may be divided into six parts:

Automata Theory
Finite Dimensional Linear Systems
Bilinear and Non-linear Systems
Linear Infinite Dimensional Systems
Coding Theory and Filtering for Sequential Systems
General Dynamical Systems and Categorical Approach to Systems.

Thus systems having state spaces which are finite, of finite dimension or of infinite dimension are considered. Further, systems whose dynamics are linear or non-linear are discussed. The mathematical structures exploited to analyze and synthesize each class of systems are rich and varied ranging from modern linear algebra, categorical algebra, differential geometry to the deeper aspects of Hilbert space theory and functional analysis. This interplay between mathematics and conceptual problems of systems is perhaps the most important aspect of mathematical system theory.

On behalf of the participants of the conference, the editors wish to thank the Consiglio Nazionale delle Richerche, the International Centre for Mechanical Sciences and the Provincial Government of Udine for their generous support of the conference. The fine facilities of CISM and the helpfulness of its staff contributed to the success of this meeting.

List of Participants

Mr. P. ALGOET	Belgium
Prof. M. ALVAREZ	Spain
Prof. S. BAINBRIDGE	Canada
Prof. J. BARAS	U.S.A.
Dr. M.T. BECCARI	Italy
Dr. F. BEGHELLI	Italy
Dr. K.W. BERGAN	Norway
Dr. R. BETTI	Italy
Prof. R.W. BROCKETT	U.S.A.
Dr. J. CASTI	Austria
Mr. G. CELENTANO	Italy
Dr. M. DENHAM	Great Britain
Prof. P. DEWILDE	Belgium
Prof. H. EHRIG	West Germany
Prof. S. EILENBERG	England
Dr. M. FLIESS	France
Dr. E. FORNASINI	Italy
Prof. P.A. FUHRMANN	Israel
Prof. H. FUKAWA	Japan
Dr. S. GINALI	U.S.A.
Dr. J.M. GOETHALS	Belgium
Prof. M.L.J. HAUTUS	The Netherlands
Prof. M. HAZEWINKEL	The Netherlands
Mr. A. HUBERMAN	France
Dr. G. JACOB	France
Prof. R.E. KALMAN	U.S.A.
Prof. E.W. KAMEN	U.S.A.
Mr. K.H. KELLERMAYR	Austria
Dr. H.J. KREOWSKI	West Germany
Prof. J. KULIKOWSKI	Poland
Prof. G. LONGO	Italy

Prof. E.G. MANES	U.S.A.
Prof. G. MARCHESINI	Italy
Prof. S. MARCUS	U.S.A.
Prof. A. MARZOLLO	Italy
Dr. T. MATSUO	Japan
Prof. S.K. MITTER	U.S.A.
Prof. A.S. MORSE	U.S.A.
Prof. E. MOSCA	Italy
Dr. A. OSYCZKA	Poland
Dr. E. PALKA	Poland
Dr. L. PANDOLFI	Italy
Dr. P. PARASKEVOPOULOS	Greece
Mr. A. PASCOLETTI	Italy
Dr. A. PERDON	Italy
Dr. G. PICCI	Italy
Prof. F. PICHLER	Austria
Mr. G. PIRANI	Italy
Dr. R. RIAZA	Spain
Mr. A. RICUPERO	Italy
Prof. J. RISSANEN	U.S.A.
Dr. Y. ROUCHALEAU	France
Dr. A. SCOTTI	Italy
Mr. P. SERAFINI	Italy
Dr. E.D. SONTAG	U.S.A.
Prof. G. STRAKA	Austria
Mr. M. STRATTA	Italy
Prof. H.J. SUSSMANN	U.S.A.
Dr. A.C. TSOI	Great Britain
Mr. A. VAN DER WEIDEN	England
Dr. J. VANDEWALLE	Belgium
Dr. A. VAN SWIETEN	The Netherlands
Prof. J.C. WILLEMS	The Netherlands

Prof. A.S. WILLSKY	U.S.A.
Prof. H. WIMMER	West Germany
Prof. B.F. WYMAN	U.S.A.
Dr. G. ZAPPA	Italy
Dr. G. ZILLI	Italy

Table of Contents

4. Infinite Dimensional Systems

5. Coding and Filtering for Sequential Systems

6. General Dynamical Systems and Categorical Approach to Systems

DECOMPOSITION THEOREMS

by

Samuel Eilenberg

(Columbia University)

The purpose of this lecture is to give a streamlined outline (without proofs) of two decomposition theorems. The first one is the Krohn--Rhodes decomposition theorem for sequential functions, i.e. for input--output maps of sequential machines. The second one is a similar theorem for input-output maps of linear sequential machines. By putting the two theorems side by side, certain analogies and differences become apparent. These may be helpful in suggesting the proper methodology for similar theorems in other parts of system theory.

Let Σ and Γ be finite alphabets. A sequential machine

$$\mathcal{M} : \Sigma \to \Gamma$$

is given by a finite (non-empty) set Q called the set of states, a distinguished element $i \in Q$ called the initial state, and by two functions

$$\varepsilon : Q \times \Sigma \to Q$$

$$\lambda : Q \times \Sigma \to \Gamma$$

called respectively the next state function and the output function. We usually write $q\sigma$ instead of $(q,\sigma)\varepsilon$. These functions are extended to functions

$$\varepsilon : Q \times \Sigma^{*} \to Q$$

$$\lambda : Q \times \Sigma^{*} \to \Gamma^{*}$$

where Σ^* and Γ^* are the free monoids generated by Σ and Γ as follows:

$$q1 = q \qquad q(s\sigma) = (qs)\sigma$$

$$(q,1)\lambda = 1, \quad (q,s\sigma)\lambda = (q,s)\lambda(qs,\sigma)\lambda$$

The <u>result</u> (or the <u>input-output mapping</u>) defined by the machine $\mathcal{M}$ is the function

$$f : \Sigma^* \to \Gamma^*$$

given by

$$sf = (i,s)\lambda$$

Functions obtained this way are called <u>sequential functions</u>.

The machine $\mathcal{M}$ is called <u>minimal</u> if it satisfies the following two conditions

(1) $i\,\Sigma^* = Q$

(2) $(q,s)\lambda = (q',s)\lambda$ for all $s \in \Sigma^*$ implies $q = q'$

It is a well known fact that for each sequential function f there exists a minimal sequential machine $\mathcal{M}$ with result f, and that this machine is unique up to an isomorphism.

Given a sequential machine $\mathcal{M}$ as above, each element $s \in \Sigma^*$ determines a transformation $q \mapsto qs$ of Q into itself. There results an algebraic object

$$X = (Q,S)$$

where Q is a finite non-empty set and S is a monoid of transformations $Q \rightarrow Q$ including the identity transformation. Such a structure will be called a transformation monoid (abbreviated: tm).

With each sequential function f we now associate the tm TM_f defined by the minimal sequential machine with f as result.

The smallest possible tm is $X = (Q,S)$ with card $Q = 1$; S is then necessarily reduced to the identity transformation. For TM_f to be such a unit tm, means that the minimal machine of f has a single state. This holds iff f is a morphism of monoids, i.e. iff $(st)f = (sf)(tf)$ for all $s,t \in \Sigma^*$. Thus the "size" of TM_f to some extent measures the extent by which f fails to be a morphism.

The objective is to obtain theorems which relate properties of f particularly from the point of view of decomposability with analogous properties of TM_f.

We begin by defining inequalities and products for tm's. Given tm's

$$X = (Q_X, S_X), \quad Y = (Q_Y, S_Y)$$

we shall write

$$X \prec Y$$

if there exists a partial function

$$\varphi : Q_Y \rightarrow Q_X$$

with the following two properties

(3) φ is surjective, i.e. $Q_Y \varphi = Q_X$

(4) For each $s \in S_X$ there exists $t \in S_Y$ such that $s\varphi \subset \varphi t$, or equivalently such that $q\, s\varphi = q\, \varphi\, t$ whenever $q\, s\varphi \neq \emptyset$

It is important to note that $X \prec Y$ and $Y \prec X$ hold simultaneously iff X and Y are isomorphic in the obvious sense.

We also note that each finite monoid S may be treated as a tm namely (S,S) with S acting on itself by right multiplication. If S and T are finite monoids, then one proves that the relation $S \prec T$ holds iff S is a quotient monoid of submonoid of T.

The direct product $X \times Y$ is defined as follows

$$X \times Y = (Q_X \times Q_Y\ ,\ S_X \times S_Y)$$

with

$$(p,q)(s,t) = (ps,\ qt)$$

for

$$p \in Q_X, \quad q \in Q_Y, \quad s \in S_X, \quad t \in S_Y.$$

The wreath-product $X \circ Y$ is defined as follows

$$X \circ Y = (Q_X \times Q_Y,\ S_X^{Q_Y} \times S_Y)$$

where $S_X^{Q_Y}$ consists of all functions $g\colon Q_Y \to S_X$ and where

$$(p,q)(g,t) = (p(qg), qt)$$

for $p,q \in Q_X \times Q_Y$, $(g,t) \in S_X^{Q_Y} \times S_Y$.

The inequality relation $X \prec Y$ and the two products $X \times Y$ and

$X \circ Y$ have a number of expected formal properties that we shall not bother to list.

Given sequential functions

$$f_i\colon \Sigma^* \to \Gamma_i^* \qquad (i = 1,2)$$

a new sequential function

$$f_1 \wedge f_2\colon \Sigma^* \to (\Gamma_1 \times \Gamma_2)^*$$

is defined as follows: if

$$sf_i = \gamma_{i,1} \cdots \gamma_{i,n}$$

then

$$s(f_1 \wedge f_2) = \gamma_1 \cdots \gamma_n$$

$$\gamma_i = (\gamma_{i,1}, \gamma_{i,2})$$

Let further $g\colon (\Gamma_1 \times \Gamma_2)^* \to \Gamma$ be a morphism. We then say that

$$f = (f_1 \wedge f_2)\, g \tag{5}$$

is a parallel decomposition of f.

THEOREM 1. If (5) holds, then

$$TM_f \prec TM_{f_1} \times TM_{f_2}$$

THEOREM 2. Given a sequential function f: $\Sigma^* \to \Gamma^*$ and given

$$TM_f < X_1 \times X_2$$

there exists a parallel decomposition (5) of f such that

$$TM_{f_i} < X_i$$

for r = 1,2.

From these results one can deduce similar results relating parallel decompositions

$$f = (f_1 \wedge \ldots \wedge f_n)g$$

with inequalities

$$TM_f < X_1 \times \ldots \times X_n.$$

THEOREM 3. Let

$$\Sigma^* \overset{f_1}{\to} \Omega^* \overset{f_2}{\to} \Gamma^*$$

be sequential functions and let

$$f = f_1\, f_2$$

be their composition. Then f is a sequential function and

$$TM_f < TM_{f2} \circ TM_{f1}$$

THEOREM 4. Given a sequential function $f: \Sigma^* \to \Gamma^*$ and given

$$TM_f \prec X_2 \circ X_1$$

there exists a decomposition

$$f = f_1 \, f_2$$

of f into sequential functions f_1 and f_2 such that

$$TM_{f_i} \prec X_i$$

for i = 1,2.

Again both Theorem 3 and 4 generalize to decompositions

$$f = f_1 \ldots f_n \tag{6}$$

$$TM_f \prec X_n \circ \ldots \circ X_1 \tag{7}$$

All these results have straightforward proofs and do not require much technique. The last two theorems show that obtaining a factorization (6) for f in which $f_1,\ldots,f_n$ are "elementary" (meaning that $TM_{f_1},\ldots,TM_{f_n}$ are "elementary" in some sense) is equivalent with finding decompositions (7) in which the $X_n,\ldots,X_1$ are "elementary". This brings us to the next theorem which is much less elementary.

THEOREM 5. (Krohn-Rhodes Decomposition Theorem). Each tm X admits a decomposition

$$X \prec X_n \circ \ldots \circ X_1$$

where each X_i either is a finite simple group such that $X_i \prec S_X$ or is a tm with two elements and three transformations, namely the identity transformation and two constant transformations.

This theorem combined with Theorem 4 gives a decomposition theorem for sequential functions. Note that only series composition is used. Parallel composition may be introduced if we further attempt to standardize the elements of the decomposition.

We now abandon sequential functions and sequential machines and pass to linear sequential machines. We shall arrange the exposition so as to parallel as much as possible the case discussed above, and underscore the similarities and dissimilarities.

We consider a fixed field K. In the definition of a sequential machine we assume that Σ, Γ and Q are finite dimensional vector spaces over K, that ε and λ are linear transformations and that i = 0.

It follows that

$$(q,\sigma)\,\varepsilon = qF + \sigma G$$

$$(q,\sigma)\,\lambda = qH + \sigma J$$

where

$$F: Q \to Q, \qquad G: \Sigma \to Q$$

$$H: Q \to \Gamma, \qquad J: \Sigma \to \Gamma$$

are linear transformations. This is a linear sequential machine $\mathcal{M}$ (over the field K).

We denote by Σ^* the set of all infinite sequences

$$\sigma = (\sigma_0, \sigma_1, \ldots, \sigma_n, \ldots)$$

with $\sigma_i \in \Sigma$ for $i = 0,1,\ldots$. The <u>result</u> (or the <u>input-output function</u>) defined by the machine $\mathscr{M}$ is the function

$$f: \Sigma^* \to \Gamma^*$$

defined as follows. First we define

$$q = (q_0, q_1, \ldots, q_n, \ldots) \in Q^*$$

by setting

$$q_0 = 0, \; q_{n+1} = q_n F + \sigma_n G$$

Then

$$\sigma f = (\gamma_0, \gamma_1, \ldots, \gamma_n, \ldots)$$

with

$$\gamma_n = q_n H + \sigma_n J$$

Equivalently $\gamma_0 = \sigma_0 J$ and

$$\gamma_n = \sigma_n J + \sum_{i=0}^{n-1} \sigma_i G F^{n-i-1} H$$

for $n > 0$ with F^0 interpreted as the identity transformation. There are other (equivalent) ways of formulating the input-output function, but this particular one is most suitable for our purposes here.

Functions f obtained this way are called <u>linear sequential functions</u>.

The machine $\mathcal{M}$ is called <u>minimal</u> if it satisfies the following two conditions

(8) the subspaces ΣGF^i $(i = 0,1,\ldots)$ span Q

(9) $qF^iH = 0$ for all $i = 0,1,\ldots$ implies $q = 0$.

Each linear sequential function is the result of a minimal linear sequential machine which is unique up to an isomorphism.

The vector space Q of a machine $\mathcal{M}$ together with the endomorphism $F: Q \to Q$ may be viewed as a module over the ring $R = K[z]$ of polynomials in one indeterminate z with coefficients in K. This is done by setting $qz = qF$ for all $q \in Q$. If further $\mathcal{M}$ is the minimal linear sequential machine of the linear sequential function f, then we define M_f to be the R-module Q. This is the invariant of f which replaces the tm TM_f considered earlier for sequential functions. Observe that M_f has finite dimension over K. The difference is that while the notion of a tm is to some extent new and its technique had to be developed, the notion of an R-module of finite dimension over K is well known and complete structure theorems exist in algebra. It turns out that the notion of an inequality $X \prec Y$ for R-modules, even though it is easy, is not needed. Direct products and exact sequences take the place of the direct product and the wreath product for tm's. The analogues of Theorems 1 - 4 are

THEOREM 6. Let $f: \Sigma^* \to \Gamma^*$ be a linear sequential function and let

$$M_f \approx X_1 \times X_2$$

be a direct product representation of M_f as an R-module. Then

$$f = f_1 + f_2$$

where $f_i: \Sigma^* \to \Gamma^*$ $(i = 1,2)$ are linear sequential functions such that $M_{f_i} = X_i$.

THEOREM 7. Let $f: \Sigma^* \to \Gamma^*$ be a linear sequential function and let

$$O \to X_2 \to M_f \to X_1 \to O$$

be an exact sequence of R-modules. Then f is the composition

$$\Sigma^* \overset{f_1}{\to} \Omega^* \overset{f_2}{\to} \Gamma^*$$

where Ω is a finite-dimensional vector space over K and f_i $(i = 1,2)$ are linear sequential functions such that $M_{f_i} = X_i$.

The rest is known algebra. The R-module M_f admits an essentially unique decomposition

$$M_f \approx X_1 + \ldots + X_n$$

where each R-module X_i $(1 \leqslant i \leqslant n)$ is indecomposable (in the sense of direct product). It follows that

$$f = f_1 + \ldots + f_n$$

where each $f_i: \Sigma^* \to \Gamma^*$ is a linear sequential function such that M_{f_i} is indecomposable.

Next consider the case when M_f is indecomposable. It is well known that

$$M_f \approx R/(p^n)$$

where p is an irreducible polynomial in $R = K[z]$ and (p^n) is the ideal generated by the p^n for some $n \geqslant 1$. If $n = 1$, there is nothing further to do. If $n > 1$, then we have the exact sequence

$$0 \to (p^{n-1})/(p^n) \to R/(p^n) \to R/(p^{n-1}) \to 0$$

and the isomorphism

$$(p^{n-1})/(p^n) \approx R/(p)$$

There results a decomposition

$$f = f' f_n$$

where f' and f_n are linear sequential functions with

$$M_{f'} = R/(p^{n-1}), \quad M_{f_n} = R/(p)$$

Iterating this process we obtain a decomposition

$$f = f_1 \dots f_n$$

of f into linear sequential functions such that

$$M_{f_i} = R/(p)$$

for i = 1,...,n. The R-modules R/(p) are simple and no further useful decomposition is available.

We shall now state some general suggestions that may be helpful in building a decomposition theory for other situations in the theory of systems

(A) The input-output functions should go from and to structures in the same category. Otherwise the notion of composition becomes cumbersome.

(B) Mealy type machines should be considered. The use of Moore type machines (i.e. assuming an output function $\lambda: Q \to \Gamma$ in the sequential machine case and the assumption $J = 0$ for linear sequential machines) leads to complications.

(C) The existence of minimal machines (i.e. reachable and reduced state representations) is essential. The state space together with its additional structure will then supply the analogues of TM_f and M_f, which will be the algebraic invariant of f.

The exposition given here follows that of my book Automata, Languages and Machines, Academic Press - New York, Vol. A (1973), Vol. B (1976). The material on sequential functions is exposed in Chapter I, II and VI of Vol. B, while the material on linear sequential machines is in Chapter XVI of Vol. A.

ALGEBRAIC METHOD FOR CALCULATING Z-TRANSFORM

Hiroshi Fukawa
Department of Applied Mathematics and Physics,
Faculty of Engineering, Kyoto University,
Kyoto, JAPAN

The purpose of this paper is to present the simple algebraic method for calculating the z-transform. Let G(s) be the Laplace transform of g(t) and $\hat{G}(z)$ be its z-transform. Define the linear map $\mathfrak{z}$ by

$$\hat{G}(z)=\mathfrak{z}[G(s)].$$

If G(s) is a rational function in s, $\hat{G}(z)$ becomes the same type of function in z^{-1}. In such case, G(s) and $\hat{G}(z)$ can be considered as finite dimensional vectors in the s domain and in the z domain respectively. Then there exists a matrix associated with the linear map $\mathfrak{z}$. According to this idea, the z-transform $\hat{G}(z)$ corresponding to G(s) can be calculated by the simple matrix operation. This idea is also available for calculating the modified z-transform.

1. Introduction

Let G(s) be the Laplace transform of a function g(t), namely:

$$G(s)=\int_0^\infty g(t)e^{-st}dt=\mathcal{L}[g(t)] \tag{1.1}$$

and $\hat{G}(z)$ be the z-transform of g(t), namely:

$$\hat{G}(z)=\sum_{k=0}^{\infty} g(kT)z^{-k} \tag{1.2}$$

where T is a sampling period. Then it is well-known that the relation

$$\hat{G}(z)=\frac{1}{2\pi i}\int_{Br}\frac{G(s)}{1-e^{sT}z^{-1}}ds \tag{1.3}$$

holds, where Br is the Bromwich contour. By modifying the relation (1.3), a few more practical formulae for calculating the z transform $\hat{G}(z)$ are derived from the Laplace transform G(s), and the table of the z transform $\hat{G}(z)$ has already obtained for the various forms of $G(s)$.[1,2,3] However, every calculating method involves the differential operation, which is elemental but troublesome.

In contrast, according to the algebraic method presented in this paper, the calculation can be easily carried out in a systematic manner. The same method can be applied to the calculation of the modified z-transform.

2. Fundamental Idea

Instead of the relation (1.3), we introduce the following expression

$$\hat{G}(z)=\mathfrak{Z}[G(s)] \tag{2.1}$$

where $\mathfrak{Z}$ denotes a symbol of the z-transform. As well-known, if $G(s)$ is a rational function of which partial fraction expansion is

$$G(s)=\sum_{i=1}^{r}\sum_{\nu=1}^{n_i}\frac{c_{\nu i}}{(s-\alpha_i)^{\nu}}, \tag{2.2}$$

then the corresponding z-transform is of the form

$$\hat{G}(z)=\sum_{i=1}^{r}\sum_{\nu=1}^{n_i}\frac{k_{\nu i}}{(1-d_i z^{-1})^{\nu}}, \quad d_i=e^{\alpha_i T}. \tag{2.3}$$

Here our purpose is to calculate $k_{\nu i}$ from given $c_{\nu i}$.

The fundamental idea for the calculation is as follows: we regard $G(s)$ and $\hat{G}(z)$ as finite dimensional vectors in the s domain and in the z domain with bases

$$\frac{1}{(s-\alpha_i)^{\nu}} \quad , \quad \frac{1}{(1-d_i z^{-1})^{\nu}}$$

and coordinates

$$(c_{\nu i}) \quad , \quad (k_{\nu i})$$

respectively. Then the expression (2.1) shows a linear transformation of a vector $G(s)$ to another vector $\hat{G}(z)$ by the operator $\mathfrak{Z}$. Hence we can associate $\mathfrak{Z}$ with some matrix B_i and we have

$$(k_{\nu i})=B_i(c_{\nu i}).$$

This matrix B_i is determined by the effect of $\mathfrak{Z}$ on the basis elements. Let it be

$$\mathcal{Z}[\frac{1}{(s-\alpha_i)^\nu}] = \frac{b_{1\nu}}{1-d_i z^{-1}} + \frac{b_{2\nu}}{(1-d_i z^{-1})^2} + \dots + \frac{b_{\nu\nu}}{(1-d_i z^{-1})^\nu} \tag{2.4}$$

then the νth column vector of B_i is given by $b_{\mu\nu}$, $\mu=1,2,\dots$. Thus we find our main purpose in determining the coefficients $b_{\mu\nu}$ in (2.4)

3. Summation of Series $f_\nu(x) = \sum k^{\nu-1} x^k$

By definition, the z transform of

$$\frac{1}{(s-\alpha)^\nu} = \mathcal{L}[\frac{t^{\nu-1}}{(\nu-1)!} e^{\alpha t}]$$

is given by

$$\mathcal{Z}[\frac{1}{(s-\alpha)^\nu}] = \sum_{k=0}^{\infty} \frac{(kT)^{\nu-1}}{(\nu-1)!} e^{\alpha kT} z^{-k} .$$

Hence the introduction of the series

$$f_\nu(x) = \sum_{k=0}^{\infty} k^{\nu-1} x^k , \qquad |x| < 1 \tag{3.1}$$

gives the following lemma.

Lemma 1.

$$\mathcal{Z}[\frac{1}{(s-\alpha)^\nu}] = \frac{T^{\nu-1}}{(\nu-1)!} f_\nu(dz^{-1})$$

where $d=e^{\alpha T}$.

This lemma shows that it is enough for studying the effect of $\mathcal{Z}$ on $1/(s-\alpha)^\nu$ to find the closed form of the infinite series $f_\nu(x)$. $f_\nu(x)$ has the following properties which can be easily verified.

Lemma 2.

$$f_{\nu+1}(x) = x\frac{d}{dx} f_\nu(x), \qquad f_1(x) = \frac{1}{1-x}, \qquad |x| < 1. \tag{3.2}$$

Corollary. $f_\nu(x)$ *has the form*

$$f_\nu(x) = (x\frac{d}{dx})^{\nu-1} \frac{1}{1-x} = \frac{a_{1\nu}}{1-x} + \frac{a_{2\nu}}{(1-x)^2} + \dots + \frac{a_{\nu\nu}}{(1-x)^\nu}, \quad |x| < 1 \tag{3.3}$$

In order to find the coefficients $a_{\mu\nu}$, we again consider $f_\nu(x)$ as a vector and let

$$a_\nu = \begin{pmatrix} a_{1\nu} \\ a_{2\nu} \\ \cdot \\ a_{\nu\nu} \end{pmatrix}$$

be the coordinate vector of $f_\nu(x)$ with respect to the basis $1/1-x$, $1/(1-x)^2, \ldots, 1/(1-x)^\nu$. Then the functional equation (3.2) can be rewritten into the equation in a_ν as follows:

Lemma 3. The sum of the infinite series

$$f_\nu(x) = \sum_{k=0}^{\infty} k^{\nu-1} x^k$$

is given by the following form

$$f_\nu(x) = \left[\frac{1}{1-x}, \ldots, \frac{1}{(1-x)^\nu}\right] a_\nu \tag{3.4}$$

and a_ν can be calculated successively by the relation

$$a_{\nu+1} = D_\nu a_\nu, \quad a_1 = 1 \tag{3.5}$$

where

$$D_\nu = \begin{pmatrix} -1 & & & \\ 1 & -2 & & \\ & 2 & \cdot & \\ & & \cdot & -\nu \\ & & & \nu \end{pmatrix} \tag{3.6}$$

[Proof] It is found from the fact

$$x\frac{d}{dx}\frac{1}{(1-x)^\mu} = -\frac{\mu}{(1-x)^\mu} + \frac{\mu}{(1-x)^{\mu+1}} \qquad (\mu = 1, 2, \ldots, \nu)$$

that the matrix associated with the linear map xd/dx relative to the basis $1/1-x, \ldots, 1/(1-x)^\nu$ becomes D_ν.

4. Z-transform of Rational Function G(s)

In the immediate consequence of lemma 1 and 3, we obtain the next result which is one of our main purposes as stated in section 2.

Theorem 1.

$$\mathcal{Z}\left[\frac{1}{(s-\alpha)^{\nu}}\right]=\left[\frac{1}{1-dz^{-1}},\ \ldots,\ \frac{1}{(1-dz^{-1})^{\nu}}\right]\frac{T^{\nu-1}}{(\nu-1)!}\,a_{\nu},\quad d=e^{\alpha T} \tag{4.1}$$

where a_{ν} *can be calculated by use of* (3.5) *and* (3.6).

Example 1

$$a_3=\begin{vmatrix} -1 & & \\ 1 & -2 & \\ & & 2 \end{vmatrix}\begin{vmatrix} -1 \\ 1 \\ \ \end{vmatrix}=\begin{vmatrix} 1 \\ -3 \\ 2 \end{vmatrix},$$

hence we have

$$\mathcal{Z}\left[\frac{1}{(s-\alpha)^{3}}\right]=\frac{T^2}{2}\left\{\frac{1}{1-dz^{-1}}-\frac{3}{(1-dz^{-1})^2}+\frac{2}{(1-dz^{-1})^3}\right\}.$$

In section 3, we make the function $f_{\nu}(x)$ correspond to the ν dimensional coordinate vector a_{ν}. However, sometimes it is more convenient to make $f_{\nu}(x)$ correspond to n dimensional vector $\bar{a}_{\nu}$, the dimention of which is higher than that of a_{ν}, such that

$$a_{\nu}=\begin{vmatrix} a_{\nu} \\ 0 \\ \cdot \\ \cdot \\ 0 \end{vmatrix}\left.\vphantom{\begin{matrix}0\\ \cdot\\ \cdot\\ 0\end{matrix}}\right\}\ n-\nu \tag{4.2}$$

Using this notation, instead of (3.4), we have

$$f_{\nu}(x)=\left[\frac{1}{1-x},\ \ldots,\ \frac{1}{(1-x)^{n}}\right]\bar{a}_{\nu}\ . \tag{4.3}$$

It should be noticed here that the number of basis elements $1/(1-x)^{\mu}$ is not ν but n. By this notice, n expressions: (4.1) for $\nu=1,\ \ldots,\ n$ can be represented by the single expression

$$\mathcal{Z}\left[\frac{1}{s-\alpha},\ \ldots,\ \frac{1}{(s-\alpha)^{n}}\right]=\left[\frac{1}{1-dz^{-1}},\ \ldots,\ \frac{1}{(1-dz^{-1})^{n}}\right]B_n$$

where $d=e^{\alpha T}$ and

$$B_n=\left[\bar{a}_1,\ T\bar{a}_2,\ \ldots,\ \frac{T^{n-1}}{(n-1)!}\bar{a}_n\right]. \tag{4.4}$$

Thus we obtain one of the final results:

Theorem 2. Let G(s) be a rational function and its partial fraction expansion be

$$G(s)= \sum_{i=1}^{r} \sum_{\nu=1}^{n_i} \frac{c_{\nu i}}{(s-\alpha_i)} .$$

If $d_i \neq d_j$ $(i \neq j)$ *where* $d_i = e^{\alpha_i T}$, *then the z-transform to G(s) is*

$$\hat{G}(z)= \sum_{i=1}^{r} \sum_{\nu=1}^{n_i} \frac{k_{\nu i}}{(1-d_i z^{-1})^{\nu}} .$$

where

$$\begin{pmatrix} k_{1i} \\ \cdot \\ \cdot \\ \cdot \\ k_{n_i i} \end{pmatrix} = B_{n_i} \begin{pmatrix} c_{1i} \\ \cdot \\ \cdot \\ \cdot \\ c_{n_i i} \end{pmatrix} \qquad (i=1,2, \ldots, r)$$

and B_{n_i} *is given by* (4.4). *It should be noted that the matrix is independent of the pole* α_i *except its order.*

5. Calculation of a_ν

This section is devoted to calculate the coordinate vector a_ν of $f_\nu(x)$. As stated above, we regard a μ dimensional vector a_μ $(1 \leq \mu \leq \nu)$ as a vector in the ν dimensional vector space as follows:

$$\bar{a}_\mu = \begin{pmatrix} a_\mu \\ 0 \\ \cdot \\ \cdot \\ 0 \end{pmatrix} \left.\vphantom{\begin{matrix} 0 \\ \cdot \\ \cdot \\ 0 \end{matrix}}\right\} \nu-\mu$$

Then the relation (3.5) and (3.6) can be rewritten as

$$\bar{a}_{\mu+1} = D\, \bar{a}_\mu \qquad (\mu=1,2, \ldots, \nu)$$

where

$$D = \begin{pmatrix} -1 & & & & \\ 1 & -2 & & & \\ & 2 & \cdot & & \\ & & \cdot & -(\nu-1) & \\ & & & \nu-1 & -\nu \end{pmatrix} \quad \text{and} \quad \bar{a}_1 = \begin{pmatrix} 1 \\ 0 \\ \cdot \\ \cdot \\ 0 \end{pmatrix} \; (= e) . \qquad (5.1)$$

Hereafter, we use the notation e instead of a_1. Since $a_\nu = \bar{a}_\nu$, then we have

$$a_\nu = D^{\nu-1} e. \qquad (5.2)$$

Using this relation and applying the diagonalization technique to the matrix D, we can calculate a_ν.

Theorem 3.

$$a_\nu = \begin{bmatrix} 1 & & & \\ 1 & -1 & & \\ \cdot & \cdot & \cdot & \\ \binom{\nu-1}{0} & -\binom{\nu-1}{1} & \cdot & (-1)^{\nu-1}\binom{\nu-1}{\nu-1} \end{bmatrix} \begin{bmatrix} (-1)^{\nu-1} \\ (-2)^{\nu-1} \\ \cdot \\ (-\nu)^{\nu-1} \end{bmatrix}$$

Proof. Let S be the matrix

$$S = \begin{bmatrix} 1 & & & \\ 1 & -1 & & \\ \cdot & \cdot & \cdot & \\ \binom{\nu-1}{0} & -\binom{\nu-1}{1} & \cdot & (-1)^{\nu-1}\binom{\nu-1}{\nu-1} \end{bmatrix} \qquad (5.3)$$

and q_k be the k-th column vector of S. It is easily shown that

$$Dq_k = (-k)\,q_k \qquad (k=1,2, \ldots, \nu.)$$

hence D is diagonalized by S. Therefore (5.2) can be expressed as follows:

$$a_\nu = S \begin{bmatrix} (-1)^{\nu-1} & & & \\ & (-2)^{\nu-1} & & \\ & & \cdot & \\ & & & (-\nu)^{\nu-1} \end{bmatrix} S^{-1} \begin{bmatrix} 1 \\ 0 \\ \cdot \\ 0 \end{bmatrix}$$

The simple calculation shows that $S=S^{-1}$. Thus we obtain

$$a_\nu = S \begin{bmatrix} (-1)^{\nu-1} \\ (-2)^{\nu-1} \\ \cdot \\ (-\nu)^{\nu-1} \end{bmatrix}$$

which is the result.

6. Calculation of Modified Z-transform

The calculation of the modified z-transform can be carried out in the same manner. By definition, the modified z-transform of g(t) $= \mathcal{L}^{-1}[G(s)]$ is

$$\hat{G}(z,m) = z^{-1} \sum_{k=0}^{\infty} g[(k+m)T]z^{-k}, \qquad 0 \leq m \leq 1. \tag{6.1}$$

Define $\mathfrak{z}_m$ by

$$\hat{G}(z,m) = \mathfrak{z}_m[G(s)], \tag{6.2}$$

and we have

$$\mathfrak{z}_m[\frac{1}{(s-\alpha)^{\nu}}] = z^{-1} \sum_{k=0}^{\infty} \frac{[(k+m)T]^{\nu-1}}{(\nu-1)!} e^{\alpha(k+m)T} z^{-k} .$$

The introduction of the series

$$f_{\nu}(x,m) = \sum_{k=0}^{\infty} (k+m)^{\nu-1}x^k, \qquad |x| < 1 \tag{6.3}$$

gives

Lemma 4.

$$\mathfrak{z}_m [\frac{1}{(s-\alpha)^{\nu}}] = \frac{T^{\nu-1}}{(\nu-1)!} z^{-1} d^m f_{\nu}(dz^{-1}, m), \qquad d = e^{\alpha T} .$$

The series (6.3) has the similar property to $f_{\nu}(x)$:

Lemma 5.

$$f_{\nu+1}(x, m) = (x\frac{d}{dx} + m)f_{\nu}(x, m), \qquad f_1(x, m) = \frac{1}{1-x}, \qquad |x| < 1$$

Corollary. $f_{\nu}(x, m)$ *has the form in* $|x| < 1$

$$f_{\nu}(x, m) = [\frac{1}{1-x}, \ldots, \frac{1}{(1-x)^{\nu}}]a_{\nu}(m), \qquad a_{\nu}(m) = \begin{bmatrix} a_{1\nu}(m) \\ \cdot \\ \cdot \\ \cdot \\ a_{\nu\nu}(m) \end{bmatrix}$$

The functional equation in lemma 5 can be represented by the coordinate vector $a_{\nu}(m)$ as follows:

Lemma 6.

$$a_{\nu+1}(m) = D_{\nu}(m)\ a_{\nu}(m), \qquad a_1(m) = 1$$

where

$$D_\nu(m) = \begin{bmatrix} m-1 & & & \\ 1 & m-2 & & \\ & 2 & \cdot & \\ & & \cdot & (m-\nu) \\ & & & \nu \end{bmatrix} \qquad (6.4)$$

Proof. This is the immediate result from the relation

$$(x\frac{d}{dx} + m)\frac{1}{(1-x)^\mu} = \frac{(m-\mu)}{(1-x)^\mu} + \frac{\mu}{(1-x)^{\mu+1}} .$$

From lemmas 4,5 and 6, we have

Theorem 4.

$$\mathfrak{Z}_m\left[\frac{1}{(s-\alpha)^\nu}\right] = d^m\left[\frac{z^{-1}}{1-dz^{-1}}, \ldots, \frac{z^{-1}}{(1-dz^{-1})^\nu}\right]\frac{T^{\nu-1}}{(\nu-1)!}\, a_\nu(m) \qquad (6.5)$$

Example 2

$$a_3(m) = \begin{bmatrix} m-1 & \\ 1 & m-2 \\ & 2 \end{bmatrix}\begin{bmatrix} m-1 \\ 1 \end{bmatrix} = \begin{bmatrix} (m-1)^2 \\ 2m-3 \\ 2 \end{bmatrix}$$

hence

$$\mathfrak{Z}_m\left[\frac{1}{(s-\alpha)^3}\right] = \frac{T^2}{2}\, z^{-1} d^m \left[\frac{(m-1)^2}{1-dz^{-1}} + \frac{2m-3}{(1-dz^{-1})^2} + \frac{2}{(1-dz^{-1})^3}\right] .$$

As described above, we introduce the n dimensional vector

$$\bar{a}_\nu(m) = \begin{bmatrix} a_\nu(m) \\ 0 \\ \cdot \\ \cdot \\ \cdot \\ 0 \end{bmatrix} \left.\vphantom{\begin{matrix} 0\\ \cdot \\ \cdot \\ \cdot \\ 0 \end{matrix}}\right\} n-\nu$$

and define the following matrix

$$B_n(m) = [\bar{a}_1(m), \quad T\bar{a}_2(m), \ldots, \frac{T^{n-1}}{(n-1)!}\bar{a}_n(m)] ,$$

then we can express n relations (6.5) for $\nu=1, \ldots, n$ in the single form, that is

$$\mathfrak{Z}m\left[\frac{1}{s-\alpha}, \ldots, \frac{1}{(s-\alpha)^n}\right]=\left[\frac{z^{-1}}{1-dz^{-1}}, \ldots, \frac{z^{-1}}{(1-dz^{-1})^n}\right] B_n(m)d^m . \quad (6.6)$$

The above discussions lead to the following theorem.

Theorem 5. The modified z-transform of

$$G(s) = \sum_{i=1}^{r} \sum_{\nu=1}^{n_i} \frac{c_{\nu i}}{(s-\alpha_i)^\nu}$$

is given by

$$\hat{G}(z,m)= z^{-1} \sum_{i=1}^{r} \sum_{\nu=0}^{n_i} \frac{k_{\nu i}(m)}{(1-d_i z^{-1})^\nu}, \quad d_i=e^{\alpha_i T} \quad (6.7)$$

where $d_i \neq d_j$ $(i \neq j)$ *and*

$$\begin{bmatrix} k_{1i} \\ k_{2i} \\ \cdot \\ k_{n_i i} \end{bmatrix} = B_{n_i}(m)\ d^m \begin{bmatrix} c_{1i} \\ c_{2i} \\ \cdot \\ c_{n_i i} \end{bmatrix} \qquad (i=1, 2, \ldots, r) .$$

In (6.7), $\hat{G}(z,m)$ is represented by a linear combination of $z^{-1}/(1-d_i z^{-1})^\nu$. In order to represent it by a linear combination of $1/(1-d_i z^{-1})^\nu$, we can use the next proposition which is easily verfied.

Proposition.

$$z^{-1}\left[\frac{1}{1-dz^{-1}}, \ldots, \frac{1}{(1-dz^{-1})^n}\right]=\left[1, \frac{1}{1-dz^{-1}}, \ldots, \frac{1}{(1-dz^{-1})^n}\right] I_n/d \quad (6.8)$$

where

$$I_n = \begin{bmatrix} -1 & & & \\ 1 & -1 & & \\ & 1 & \cdot & \\ & & \cdot & -1 \\ & & & 1 \end{bmatrix} .$$

Example 3.

$$\begin{bmatrix} -1 & & & \\ 1 & -1 & & \\ & 1 & & \\ & & -1 & \\ & & 1 & \end{bmatrix} \begin{bmatrix} (m-1)^2 \\ 2m-3 \\ 2 \end{bmatrix} = \begin{bmatrix} -(m-1)^2 \\ (m-2)^2 \\ 2m-5 \\ 2 \end{bmatrix}$$

hence

$$\mathfrak{Z}_m\left[\frac{1}{(s-\alpha)^3}\right] = \frac{T^2}{2}\left[-(m-1)^2 + \frac{(m-2)^2}{1-dz^{-1}} + \frac{2m-5}{(1-dz^{-1})^2} + \frac{2}{(1-dz^{-1})^3}\right]d^{m-1}$$

The general form of $a_\nu(m)$ is given by the next theorem.

Theorem 6.

$$a_\nu(m) = S \begin{bmatrix} (m-1)^{\nu-1} \\ (m-2)^{\nu-1} \\ \cdot \\ (m-\nu)^{\nu-1} \end{bmatrix}$$

where S is the matrix defined by (5.3).

Proof. The proof is similar to that of theorem 3. Define the matrix D(m) by

$$D(m) = D + mE \tag{6.9}$$

where D is the matrix shown in (5.1) and E is the unit matrix of of order n. Then we obtain

$$a_\nu(m) = [\ D(m)\ ]^{\nu-1}\ e. \tag{6.10}$$

The remaining part of the proof is the same as that of theorem 3.

7. Factorization of $B_n(m)$

The matrix $B_n^-(m)$ can be factorized into two parts: one contains m and the other does not. Let

$$b_\nu(m) = \frac{T^{\nu-1}}{(\nu-1)!}a_\nu(m), \qquad b_k = \frac{T^{k-1}}{(k-1)!}\ a_k\ , \qquad (1 \leqq k \leqq \nu).$$

From (5.2), (6.9) and (6.10), we have

$$b_\nu(m) \doteq \frac{T^{\nu-1}}{(\nu-1)!}(D+mE)^{\nu-1}e = \sum_{k=1}^{\nu} \frac{(mT)^{\nu-k}}{(\nu-k)!} b_k$$

By definition,

$$B_n(m)=[b_1(m), \ldots, b_n(m)], \qquad B_n=[b_1, \ldots, b_n].$$

Therefore we obtain

Theorem 7.

$$B_n(m)=B_n M_n$$

where

$$M_n = \begin{bmatrix} 1 & mT & \cdot & \frac{(mT)^{n-1}}{(n-1)!} \\ & 1 & \cdot & \frac{(mT)^{n-2}}{(n-2)!} \\ & & \cdot & \cdot \\ & & & 1 \end{bmatrix}.$$

(6.6), (6.8) and theorem 7, suggest that the operator $\mathfrak{Z}_m$ is composed of the three parts

$$I_n/d, \quad B_n, \quad M_n d^m .$$

8. Another Approach to Calculating Modified Z-transform

The definition

$$\mathfrak{Z}_m[G(s)] = z^{-1} \sum_{k=0}^{\infty} g[(k+m)T]z^{-k}$$

shows that $\mathfrak{Z}_m$, consists of the following three operations

1) $g(t) \rightarrow g(t+mT)$
2) z-transform of $g(t+mT)$
3) multiplication by z^{-1}

we have already shown that 2) and 3) are represented by the matrices B_n and I_n/d respectively.

Let $\mathfrak{M}$ denote the operation 1). Our problem is to find the matrix associated to $\mathfrak{M}$. Bearing in mind the relation

$$\frac{(t+mT)^{\nu-1}}{(\nu-1)!} e^{\alpha(t+mT)} = \sum_{p+q=\nu} \frac{t^{p-1}}{(p-1)!} e^{\alpha t} \frac{(mT)^q}{q!} e^{\alpha mT}$$

we define $\mathfrak{M}$ properly by

$$\mathfrak{M} [\frac{1}{(s-\alpha)^{\nu}}] = \sum_{k=1}^{\nu} \frac{1}{(s-\alpha)^k} \frac{(mT)^{\nu-k}}{(\nu-k)!} d^m, \qquad d=e^{\alpha mT} .$$

This definition is equivalent to the next one:

$$\mathfrak{M} [\frac{1}{s-\alpha}, \ldots, \frac{1}{(s-\alpha)^n}] = [\frac{1}{s-\alpha}, \ldots, \frac{1}{(s-\alpha)^n}] M_n d^m$$

where M_n is the matrix shown in theorem 7. Thus we obtain

Theorem 8. The modified z-transform consists of the following three parts

$$\mathfrak{Z}_m = z^{-1} \mathfrak{Z} \mathfrak{M}$$

and the matrices associated with these three maps: z^{-1}, $\mathfrak{Z}$ *and* $\mathfrak{M}$ *are* I_n/d, B_n *and* $M_n d^m$. *relative to appropriate bases respectively.*

Going on this line further, we can reach the whole results obtained in section 6.

9. Conclusion

The author shows that the linear algebraic technique presents a useful and lucid method for calculating the z-transform and the modified z-transform. This paper is the first attempt to make clear the mutual relations among the classic/modern theory of continuous/discrete control systems.

The author would like to express his deep appreciation for the help and encouragement he has received from Professor Yoshikazu Sawaragi.

References

1 Jury, E.I. Theory and Application of the z-Transform Method. John Wiley, 1964.

2 Tou, J.T. Digital and Sampled-Data Control Systems. McGraw-Hall, 1959.

3 Freeman, H. Discrete-time Systems, An Introduction to the Theory. John Wiley, 1965.

THE FORMAL LAPLACE TRANSFORM FOR SMOOTH LINEAR SYSTEMS

M.L.J. Hautus

Department of Mathematics
Technological University Eindhoven
The Netherlands

Abstract. A class of time invariant linear systems is introduced. For systems in this class a formal Laplace transform is defined and invertibility properties are studied using this transform. The results are related to known results in literature.

§ 0. Introduction

In this paper, a class of time invariant continuous time multivariable linear systems is considered, called *smooth systems*, which contains in particular the systems with finite dimensional state space. This class is characterized by certain smoothness conditions on the impulse response matrix (= kernel). The definition is given in section 2. It is shown that a smooth kernel can be decomposed in a canonical way into a strictly causal and an impulsive part. For strictly causal smooth systems, it has been shown in [1], that a (possibly infinite dimensional) state space representation exists, but this result will not be used here. Instead we study the behavior of the impulse response matrix in the neighborhood of $t = 0$. This investigation can be performed in an algebraic way by the *formal Laplace transform* or Λ-*transform*. The Λ-transform does not give a complete representation of the original system, but only of its asymptotic behavior near $t = 0$. Consequently, it is unsuitable for investigating stability properties of the system. However, for some problems, like invertibility, it is only the behavior at $t = 0$ that matters and for such problems, the Λ-transform is a most appropriate tool.

In section 3 the Λ-transform is introduced for single variable systems. The Λ-transform is shown to be an algebra homomorphism from the space of smooth kernels into the space of formal power series. It is shown that the invertibility of a kernel K is equivalent to the nonvanishing of $\Lambda(K)$. An application to feedback systems is given. The Λ-transform, being just a convenient way of description but not really necessary for single variable systems, turns out to be very crucial in the description of multivariable systems, as is shown in section 4. Using a formal power series analogon of the MacMillan form for rational matrices, a number of invariants of a system is defined, which completely determine its inherent integration or differentiation properties and hereby its invertibility properties. In section 5, existing invertibility results in literature ([6], [7]) are extended to smooth systems and related to our results.

§ 1. Notations

A. If $f: \mathbb{R} \to \mathbb{R}$, then $\text{supp } f := \overline{\{t \in \mathbb{R} \mid f(t) \neq 0\}}$ is the *support* of f. If $f \neq 0$, then we define

$$\lambda(f) := \inf \text{supp } f, \ \rho(f) := \sup \text{supp } f .$$

If $f = 0$ we define $\lambda(f) = \infty$, $\rho(f) = -\infty$. If $f: \mathbb{R} \to \mathbb{R}$, then $\check{f}: \mathbb{R} \to \mathbb{R}$ is defined by $\check{f}(t) := f(-t)$. If in addition $\tau \in \mathbb{R}$, then $\sigma^\tau f: \mathbb{R} \to \mathbb{R}$ is given by $\sigma^\tau f(t) = f(t+\tau)$. We have $\lambda(\sigma^\tau f) = \lambda(f) - \tau$ and $\rho(\sigma^\tau f) = \rho(f) - \tau$.

B. If X and Y are spaces, then $C(X \to Y)$, $CL(X \to Y)$, $C^\infty(X \to Y)$ denote the spaces of continuous, continuous linear, infinitely often differentiable functions $f: X \to Y$ respectively. We will use the test function spaces:

$$\mathcal{D}_- := \{\varphi \in C^\infty(\mathbb{R} \to \mathbb{R}) \mid \rho(\varphi) < \infty\} ,$$

$$\mathcal{D}_+ := \{\varphi \in C^\infty(\mathbb{R} \to \mathbb{R}) \mid \lambda(\varphi) > -\infty\} ,$$

and the distribution space:

$$\mathcal{D}'_+ := CL(\mathcal{D}_- \to \mathbb{R}) .$$

The value of $u \in \mathcal{D}'_+$ in $\varphi \in \mathcal{D}_-$ is denoted $\langle u,\varphi\rangle$. See [5, Ch. VI, § 5] for further properties of these spaces, in particular, for the topology on $\mathcal{D}_+$, $\mathcal{D}_-$, $\mathcal{D}'_+$. The space $\mathcal{D}_+$ may be (and will be) considered a subspace of $\mathcal{D}'_+$ by the identification $\langle\psi,\varphi\rangle = \int_{-\infty}^{\infty} \psi(t)\varphi(t)dt$ for $\psi \in \mathcal{D}_+$, $\varphi \in \mathcal{D}_-$. Distributions in $\mathcal{D}_+$ will be called *completely smooth.*

C. If $J \subseteq \mathbb{R}$ is an open set and $u \in \mathcal{D}'_+$, we say that $u = 0$ on J or $u|_J = 0$ if $\langle u,\varphi\rangle = 0$ for every $\varphi \in \mathcal{D}_-$ with $\text{supp } \varphi \subseteq J$. We say that $u = v$ on J or $u|_J = v|_J$ if $(u-v)|_J = 0$. The complement of the union of all open sets on which u is zero is called the *support* of u, notation supp u. We define $\lambda(u) := \inf \text{supp } u$, $\rho(u) = \sup \text{supp } u$ for $u \neq 0$, and $\lambda(0) := \infty$, $\rho(0) = -\infty$. E.g. $\lambda(\delta) = \rho(\delta) = 0$, where δ is defined by $\langle\delta,\varphi\rangle = \varphi(0)$. One can show that $\lambda(u) > -\infty$ for every $u \in \mathcal{D}'_+$.

For $u \in \mathcal{D}'_+$, we define $\dot{u}$ by $\langle\dot{u},\varphi\rangle = -\langle u,\dot{\varphi}\rangle$ and for $\tau \in \mathbb{R}$, $u \in \mathcal{D}'_+$ we define $\sigma^\tau u$ by $\langle\sigma^\tau u,\varphi\rangle = \langle u,\sigma^{-\tau}\varphi\rangle$. A distribution u is called *smooth on an open interval* J if there exists $\varphi \in \mathcal{D}_+$, such that $u = \varphi$ on J.

D. The *convolution* of $u \in \mathcal{D}'_+$, $\varphi \in \mathcal{D}_+$ is defined by

$$u * \varphi(t) := \langle u,\sigma^{-t}\check{\varphi}\rangle \qquad (t \in \mathbb{R}) .$$

The function $t \mapsto u * \varphi(t)$ is completely smooth. If $u,v \in \mathcal{D}'_+$, we define

$$\langle u * v, \varphi \rangle := \langle u, (v * \check{\varphi})^{\vee} \rangle$$

for every $\varphi \in \mathcal{D}_-$ (notice that $\check{\varphi} \in \mathcal{D}_+$ if $\varphi \in \mathcal{D}_-$). With the pointwise addition and scalar multiplication $(u,v) \mapsto u + v$, $(\lambda,u) \mapsto \lambda u$ and with the convolution as multiplication, $\mathcal{D}'_+$ is a topological commutative algebra over $\mathbb{R}$. The unit element is δ. Furthermore we have $(u * v)^{\cdot} = \dot{u} * v = u * \dot{v}$, $\text{supp}(u * v) \subseteq \text{supp } u + \text{supp } v$, $\lambda(u * v) \geq \lambda(u) + \lambda(v)$, $\rho(u * v) \leq \rho(u) + \rho(v)$.

E. We denote by $\mathbb{R}^m$ and $\mathbb{R}^{r\times m}$ the spaces of m dimensional real vectors and $r \times m$ matrices respectively. If $\mathcal{L}$ is a space of real valued functions, then $\mathcal{L}^m$ and $\mathcal{L}^{r\times m}$ denote the spaces of $\mathbb{R}^m$-values and $\mathbb{R}^{r\times m}$-valued functions with component functions in $\mathcal{L}$. The definitions of $\check{\varphi}$, $\sigma^\tau\varphi$, $\lambda(\varphi)$, $\rho(\varphi)$ for vector or matrix valued functions are analogous to the definitions in the scalar case. Similar notations will be used for distribution spaces. If $A \in \mathcal{D}'^{r\times m}_+$, $u \in \mathcal{D}'^m_+$, then $A * u$ is the element in $\mathcal{D}'^r_+$ defined by

$$(A * u)_i := \sum_{j=1}^{m} A_{ij} * u_j \quad (i = 1,\ldots,r) .$$

Similarly, if $A \in \mathcal{D}'^{r\times m}_+$, $B \in \mathcal{D}'^{m\times p}_+$, then

$$(A * B)_{ij} := \sum_{k=1}^{m} A_{ik} * B_{kj} \quad (i = 1,\ldots,r;\ j = 1,\ldots,p) .$$

§ 2. Smooth input-output maps

Let $\Omega := \mathcal{D}'^m_+$ be called the *input space* and $\Gamma := \mathcal{D}'^r_+$ the *output space*. A map $f \in CL(\Omega\to\Gamma)$ is called *time-invariant* if $f(\sigma^\tau u) = \sigma^\tau f(u)$ for all $u \in \Omega$, $\tau \in \mathbb{R}$ and *causal* if $\lambda(f(u)) \geq \lambda(u)$ for all $u \in \Omega$. A causal time-invariant $f \in CL(\Omega \to \Gamma)$ will be called an *input-output map* (i/o-map). The matrix valued distribution K defined by $(K)_{ij} := f_i(\delta e_j)$, where e_j is the jth unit vector, is called the *impulse response matrix*. It follows from the causality of f that $\lambda(K) \geq 0$. A matrix version of a well known theorem of L. Schwartz states that $f(u) = K * u$ for all $u \in \Omega$ ([5, Ch. VI, § 3], [2, Ch. 6]).

2.1. Definition. The i/o-map f and its impulse response matrix K are called *smooth* if for every open interval J we have that f(u) is smooth on J, whenever $u \in \Omega$ is smooth on J. □

Intuitively, a smooth i/o-map smoothens out irregularities of the input in the past. The pure delay map $f(u) := \sigma^{-\tau}u$ is an example of a map that is not smooth.

2.2. Theorem. An i/o-map f is smooth iff its impulse response is smooth on the interval $(0,\infty)$.

Proof. The "only if" part is immediate from the definition. In order to prove the "if" part, assume that $K(t) = H(t)$ for $0 < t < \infty$, where $H \in \mathcal{D}_+^{r\times m}$. Let $J = (a,b)$ where $-\infty \le a < b \le \infty$, be an open interval and let u be smooth on J. Then there exist $\varphi \in \mathcal{D}_+^m$, $v_1, v_2 \in \Omega$ with $\rho(v_1) \le a$, $\lambda(v_2) \ge b$, such that $u = \varphi + v_1 + v_2$. We have

$$f(u) = K * u = K * \varphi + H * v_1 + (K - H) * v_1 + K * v_2 .$$

The first and the second term are completely smooth. Furthermore:

$$\rho((K - H) * v_1) \le \rho(K - H) + \rho(v_1) \le 0 + a = a$$

and

$$\lambda(K * v_2) \ge \lambda(K) + \lambda(v_2) \ge 0 + b = b .$$

Hence f(u) is smooth on J. □

2.3. Example. Let f be the i/o-map of the linear time invariant system

$$\dot{x} = Fx + Gu ,$$

$$y = Hx + A_0 u + A_1 \dot{u} + \ldots + A_p u^{(p)} ,$$

where $x \in \mathcal{D}_+'^{n}$, $u \in \Omega$, $y \in \Gamma$, $F \in \mathbb{R}^{n\times n}$, $G \in \mathbb{R}^{n\times m}$, $H \in \mathbb{R}^{r\times n}$, $A_i \in \mathbb{R}^{m\times r}$ for $i = 0,\ldots,p$. Then for every $u \in \Omega$ we have $y = K * u$, where $K = K_{sc} + K_{imp}$. Here K_{sc} is the function defined by

$$\begin{cases} K_{sc}(t) = 0 & (t < 0) \\ K_{sc}(t) = He^{Ft}G & (t \ge 0) \end{cases}$$

and

$$K_{imp} = \sum_{i=0}^{p} A_i \delta^{(i)} .$$

It follows that K is a smooth impulse response matrix. □

The decomposition of K (and hence of f) into a *strictly causal* part K_{sc} (f_{sc}) and an impulsive part K_{imp} (f_{imp}) is also possible for general smooth i/o-maps (compare [3, Prop. 23]): If K is a smooth impulse response, there exists $H \in \mathcal{D}_+^{r\times m}$, such that $H = K$ on $(0,\infty)$. We define

$$(2.1) \qquad K_{sc}(t) := \begin{cases} 0 & (t < 0) \\ H(t) & (t \ge 0) \end{cases}$$

and $K_{imp} := K - K_{sc}$. We see that the strictly causal part of K is a function and that $\operatorname{supp} K_{imp} = \{0\}$. It follows from a well-known theorem of L. Schwartz ([5, Théorème XXXV]), that K_{imp} is of the form

$$(2.2) \qquad K_{imp} = \sum_{i=0}^{p} A_i \delta^{(i)} .$$

Hence $f_{imp}(u) = K_{imp} * u = \sum_{i=0}^{p} A_i u^{(i)}$. A smooth i/o-map f is called *strictly causal* if $f = f_{sc}$ and impulsive if $f = f_{imp}$.

§ 3. Formal Laplace transform of single variable systems

In this section, we consider the case $m = r = 1$. Smooth impulse response functions will be called *kernels*. The set of all kernels is denoted K. The space K is a subalgebra of $\mathcal{D}'_+$. Convolution of two kernels corresponds to cascade connection, addition to parallel connection of two systems. We are interested in the behavior of a kernel in the neighborhood of $t = 0$ and the relevance of this behavior to system invertibility. For that aim we introduce a formal Laplace transform, which can be used to study these problems in an algebraic way. First, we need some preliminary concepts.

3.1. Definition. $d := \delta'$; $d^k := d^{k-1} * d$ $(k = 2,3,\ldots)$; $d^0 := \delta$; $d^{-1} = U$ (the Heaviside unit step function: $U(t) = 0$ $(t < 0)$, $U(t) = 1$ $(t \geq 0)$); $d^{-k} := d^{-(k-1)} * d^{-1}$ $(k = 2,3,\ldots)$.

3.2. Proposition.

1) $d^{k+\ell} = d^k * d^\ell \qquad (k, \ell \in \mathbb{Z})$

2) $d^k = \delta^{(k)} \qquad (k = 1,2,\ldots)$

3) $d^{-k}(t) = U(t)t^{k-1}/(k-1)! \qquad (k = 1,2,\ldots)$. □

3.2. Definition. If $n \in \mathbb{Z}$, $K \in K$, then

$$K^{(n)}(0) := \lim_{t \downarrow 0} (d^n * K)(t) .$$ □

3.3. Proposition. If $K = \sum_{i=0}^{p} A_i \delta^{(i)} + K_{sc}$, where K_{sc} is strictly causal, then

$$K^{(n)}(0) = \lim_{t \downarrow 0} K_{sc}^{(n)}(t) \qquad (n = 0,1,\ldots)$$

$$K^{(n)}(0) = A_{-n-1} \qquad (n = -1,\ldots,-p-1)$$

$$K^{(n)}(0) = 0 \qquad (n < p-1) .$$ □

3.4. Definition. If $K \in K$, then the *order* of K is given by

$$\omega(K) := \min\{n \in \mathbb{Z} \mid K^{(n-1)}(0) \neq 0\}$$

and $\omega(K) := \infty$ if $K^{(n)}(0) = 0$ for all $n \in \mathbb{Z}$. □

We see that $\omega(K) = -p$ for the kernel K of proposition 3.3 (if $A_p \neq 0$). A kernel K is strictly causal iff $\omega(K) > 0$.

3.5. Proposition. Let $K \in \mathcal{K}$, $q := \omega(K)$. Then there exists for every $M \in \mathbb{Z}$ a strictly causal L, such that

$$(1) \qquad K = \sum_{n=q}^{M} K^{(n-1)}(0)d^{-n} + d^{-M}L . \qquad \square$$

Formula (1) just is the Taylor expansion of K at t = 0.

We denote by $\mathcal{F}$ the set of formal power series $\sum_{-\infty}^{\infty} a_n s^{-n}$ with the property that there exists $q \in \mathbb{Z}$ such that $a_n = 0$ for $n < q$. The set $\mathcal{F}$ is an algebra if the addition and scalar multiplication are defined termwise and the multiplication as follows:

$$\sum a_n s^{-n} \sum b_n s^{-n} = \sum c_n s^{-n}$$

where $c_n := \sum_{m=-\infty}^{\infty} a_m b_{n-m}$. (Note that this is a finite sum.)
Actually $\mathcal{F}$ is also a field. The set $\mathcal{F}_f$ of finite power series $\sum_{-M}^{N} a_n s^{-n}$, the set $\mathcal{N}$ of negative power series $\sum_{0}^{\infty} a_n s^{-n}$ and the set $\mathcal{N}_s$ of strictly negative power series $\sum_{1}^{\infty} a_n s^{-n}$ are subalgebras of $\mathcal{F}$. The field $\mathcal{F}$ is isomorphic to the quotient field of $\mathcal{N}$.

3.6. Definition. The *formal Laplace transform* or *Λ-transform* of $K \in \mathcal{K}$ is the formal power series

$$\Lambda(K) = \sum_{-\infty}^{\infty} K^{(n-1)}(0)s^{-n} . \qquad \square$$

We observe, that $\Lambda(d^n) = s^n$ for every $n \in \mathbb{Z}$. Since Λ is obviously a linear map: $\Lambda: \mathcal{K} \to \mathcal{F}$ it follows that $\Lambda(p(d)) = p(s)$ for every $p \in \mathcal{F}_f$. Finally, we note that $\Lambda(K) \in \mathcal{N}_s$ iff K is strictly causal.

3.7. Theorem.

i) $\Lambda: \mathcal{K} \to \mathcal{F}$ is an algebra homomorphism.

ii) $\ker \Lambda = \mathcal{D}_+ \cap \mathcal{K} = \mathcal{D}_{\mathbb{R}_+} := \{\varphi \in \mathcal{D}_+ \mid \lambda(\varphi) \geq 0\}$.

iii) $\operatorname{im} \Lambda = \mathcal{F}$, that is, Λ is an epimorphism.

Proof.

i) The linearity of Λ has already been noted. We have to show that

$$\Lambda(K * L) = \Lambda(K)\Lambda(L)$$

for $K,L \in \mathcal{K}$. If $K = p(d)$, $L = q(d)$, where $p,q \in F_f$, then we have

$$\Lambda(K * L) = \Lambda(p(d) * q(d)) = \Lambda((pq)(d)) = (pq)(s) = p(s)q(s) =$$
$$= \Lambda(p(d))\Lambda(q(d)) = \Lambda(K)\Lambda(L) .$$

Since $\dot{K}^{(n)}(0) = K^{(n+1)}(0)$ we have

$$(2) \qquad \Lambda(\dot{K}) = s\Lambda(K) .$$

Let $K,L \in \mathcal{K}$; $M \in \mathbb{Z}$. We write $K \equiv L \bmod d^{-M}$ if $\omega(d^M * (K - L)) \geq 0$. Similarly, if $\alpha,\beta \in F$ we write $\alpha \equiv \beta \bmod (s^{-M})$ if $s^M (\alpha - \beta) \in N$. It is clear that $\alpha = \beta$ if $\alpha \equiv \beta \pmod{s^{-M}}$ for all $M \in \mathbb{Z}$. (However, a similar statement does not hold in $\mathcal{K}$.) It follows from proposition 3.5 that for every $K \in \mathcal{K}$, $M \in Z$ there exists $p \in F_f$, such that $K \equiv p(d) \pmod{d^{-M}}$. Therefore, if $K,L \in \mathcal{K}$, $M \in \mathbb{N}$, $M_1 \in \mathbb{N}$ and $K \equiv p(d) \pmod{d^{-M_1}}$, $L \equiv q(d) \pmod{d^{-M_1}}$ then we have

$$K * L \equiv p(d) * q(d) \pmod{d^{-M}}$$

if M_1 is sufficiently large. Consequently:

$$\Lambda(K * L) \equiv \Lambda(p(d) * q(d)) = \Lambda(p(d))\Lambda(q(d)) = p(s)q(s) \equiv \Lambda(K)\Lambda(L) \pmod{s^{-M}}$$

if M_1 is large.

ii) We have $K \in \mathcal{D}_{\mathbb{R}_+}$ iff $K^{(n)}(0) = 0$ for all $n \in \mathbb{Z}$.

iii) We have to show that for every $\alpha \in F$ there exists $K \in \mathcal{K}$ such that $\Lambda(K) = \alpha$. Because of i), it is no loss of generality to assume that $\alpha \in N_s$ (consider otherwise $s^{-N} * \alpha$ for sufficiently large N). A function K is a strictly causal kernel, if there exists $H \in \mathcal{D}_+$ such that $K(t) = H(t)$ $(t > 0)$, $K(t) = 0$ $(t < 0)$. Then we have $H^{(n)}(0) = K^{(n)}(0)$ (def. 3.2). Thus, the problem reduces to the following one: Given a sequence $a_0,a_1,a_2,\ldots$, find a function $H \in \mathcal{D}_+$ such that $H^{(n)}(0) = a_n$ for $n = 0,1,2,\ldots$. Let $\varphi \in C^\infty(\mathbb{R} \to \mathbb{R})$ satisfy $\varphi(t) = 1$ for $|t| \leq \frac{1}{2}$ and $\varphi(t) = 0$ $(|t| \geq 1)$, and let $c_k := \max\{|\varphi^{(k)}(t)| \mid -1 \leq t \leq 1\}$. Consider the function

$$g(t) = \sum_{n=0}^{\infty} a_n \varphi(a_n t) t^n/n! .$$

We show that g is well defined, $g \in C^\infty(\mathbb{R} \to \mathbb{R})$ and

$$(3) \qquad g^{(m)}(t) = \sum_{n=0}^{\infty} a_n \left(\frac{d}{dt}\right)^m \{\varphi(a_n t) t^n/n!\} .$$

Indeed, for $n \geq m+1$ the general term of the right-hand side of (3) is

$$\sum_{k=0}^{m} \binom{m}{k} a_n^{m-k+1} t^{m-k+1} \varphi^{(m-k)}(a_n t) t^{n-m-1}/(n-k)! \quad ,$$

which is dominated by $C_m t^{n-m-1}/(n-m-1)!$ where $C_m := \sum_{k=0}^{m} \binom{m}{k} c_k$. This follows from the inequality $|x^p \varphi^{(q)}(x)| \leq c_q$ for all $p \geq 0$, $q = 0,1,2,\ldots$, $x \in \mathbb{R}$. We conclude that

(3) is uniformly convergent on every compact interval for every m. Hence $g \in C^{\infty}$. For every $M \in \mathbb{N}$ we have

$$g(t) = \sum_{n=0}^{M} a_n \varphi(a_n t) t^n/n! + O(t^{M+1}) = \sum_{n=0}^{M} a_n t^n/n! + O(t^{M+1}) \qquad (t \to 0) .$$

Hence $a_n = g^{(n)}(0)$ for $n = 0,1,2,\ldots$ Now the function $H(t) := \varphi(t)g(t)$ satisfies the requirements. □

3.8. Definition. The order $\omega(\alpha)$ of a formal power series $\alpha = \sum_{-\infty}^{\infty} a_n s^{-n}$ is given by

$$\omega(\alpha) := \min\{n \in \mathbb{Z} \mid a_n \neq 0\} \qquad \text{if } \alpha \neq 0 ,$$

$$\omega(\alpha) := \infty \qquad \text{if } \alpha = 0 .$$

3.9. Proposition.

i) $\omega(\alpha\beta) = \omega(\alpha) + \omega(\beta)$; $\omega(\alpha + \beta) \geq \min\{\omega(\alpha),\omega(\beta)\}$;

$\omega(\alpha + \beta) = \min\{\omega(\alpha),\omega(\beta)\}$ if $\omega(\alpha) \neq \omega(\beta)$, for $\alpha,\beta \in F$.

ii) $\omega(K) = \omega(\Lambda(K))$ for $K \in \mathcal{K}$.

iii) $\omega(K * L) = \omega(K) + \omega(L)$; $\omega(K + L) \geq \min\{\omega(K),\omega(L)\}$;

$\omega(K + L) = \min\{\omega(K),\omega(L)\}$ if $\omega(K) \neq \omega(L)$. □

The Λ-transform is not a genuine transform because it is not injective. Therefore, some information about the system is lost in the Λ-transform. However, $\Lambda(K)$ does give an almost complete picture of the asymptotic behavior of K near $t = 0$. If K has an ordinary Laplace transform $\hat{K}(s)$, then $\Lambda(K)$ is the asymptotic series of $\hat{K}(s)$ for $\mathrm{Re}\ s \to \infty$. In the case of analyticity (e.g. for finite dimensional systems) one can also state uniqueness results.

The order $\omega(K)$ indicates how many inherent integrations or differentiations are performed by the system with impulse response K. A kernel of order zero will be called a *unimodular* kernel. We can write any smooth system that is not completely smooth, as a cascade connection of a pure differentiator or integrator and a system with unimodular kernel: $K = d^{-\omega} * K_0$, where $\omega = \omega(K)$.

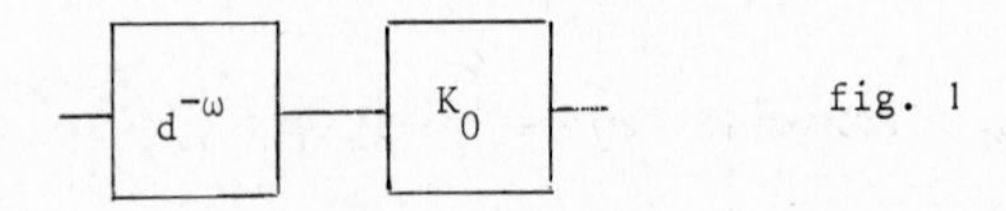

fig. 1

If $\omega > 0$, then $d^{-\omega}$ is a pure integrator (in [6], ω is called the *inherent integration* of K). If $\omega > 0$, then $d^{-\omega}$ is a pure differentiator.

3.10 Lemma. A unimodular kernel K has a unimodular inverse in K.

Proof. Without loss of generality we may assume that K is of the form $K = \delta + L$ with L strictly causal. We try to find K^{-1} of the form $K^{-1} = \delta + M$ with M strictly causal. Then M must satisfy the equation $(\delta + L) * (\delta + M) = \delta$, that is

$$M + L * M = -L \tag{4}$$

which can be written as a Volterra integral equation of the second kind:

$$M(t) + \int_0^t L(t - \tau)M(\tau)d\tau = -L(t) \qquad (t \geq 0) .$$

This equation is well known to have a C^∞ solution M on $[0,\infty)$ if L is C^∞ on $[0,\infty)$. Defining $M(t) = 0$ for $t < 0$, we see that M satisfies (4). □

3.11. Theorem. A kernel $K \in \mathcal{K}$ has an inverse in $\mathcal{K}$ iff one of the following equivalent conditions is satisfied:

i) $\omega(K) < \infty$,

ii) $K \notin \mathcal{D}_+$,

iii) $\Lambda(K) \neq 0$.

If K has an inverse, then $\omega(K^{-1}) = -\omega(K)$ and $\Lambda(K^{-1}) = (\Lambda(K))^{-1}$.

Proof. The inverse of $d^{-\omega} * K_0$ is $d^{\omega} * K_0^{-1}$. □

It follows that a strictly causal kernel never can have a strictly causal inverse.

Consider the feedback system given by figure 2.

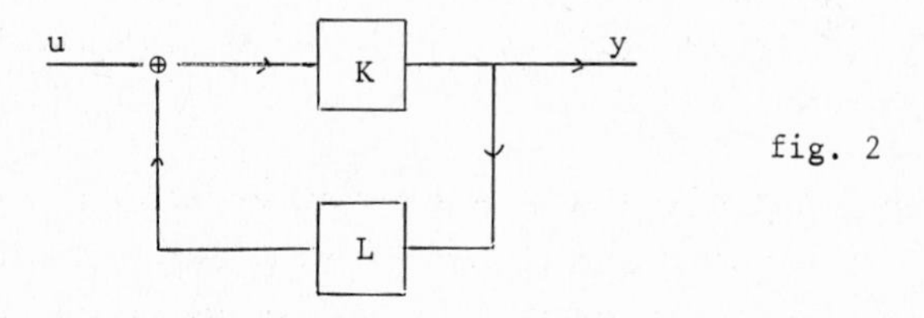

fig. 2

The i/o-behavior is given by the equation $y = K * u + K * L * y$, i.e.,

$$(\delta - K * L) * y = K * u . \tag{5}$$

3.12. Corollary. Equation (5) defines an i/o-map $y = T * u$ iff $\Lambda(K)\Lambda(L) \neq 1$.

Proof. If $\Lambda(K)\Lambda(L) \neq 1$ the kernel $\delta - K * L$ has an inverse, so that $T = (\delta - K * L)^{-1} * K$. If $\Lambda(K)\Lambda(L) = 1$ the kernel $\delta - K * L$ is completely smooth and K is not (since $\Lambda(K) \neq 0$). Hence, if $u \notin \mathcal{D}_+$, equation (5) has no solution $y \in \Gamma$. □

For the order of T we find using 3.9:

$$\omega(T) = \max\{\omega(K), -\omega(L)\} \quad \text{if } \omega(K) + \omega(L) \neq 0$$

$$\omega(T) \leq \omega(K) \quad \text{if } \omega(K) + \omega(L) = 0 .$$

If K and L are both strictly causal, T is always defined and $\omega(T) = \omega(K)$. If K and L are both unimodular kernels, it may happen that T is a differentiator (e.g., if $K = \delta$, $L = \delta - d^{-1}$, then $T = d$).

§ 4. Multivariable systems

We consider the set $K^{r\times m}$ of $r \times m$ kernels, that is, smooth impulse response matrices of dimensions $r \times m$. Similarly, we define $F^{r\times m}$ to be the set of $r \times m$ matrix values power series, or equivalently, the set of $r \times m$ matrices over F.

4.1. Definition. Let $K \in K^{r\times m}$. The *formal Laplace transform* or *Λ-transform* $\Lambda(K)$ of K is the element of $F^{r\times m}$ given by

$$(\Lambda(K))_{ij} := \Lambda(K_{ij}) \quad (i = 1,\dots,r;\ j = 1,\dots,m) .$$

4.2. Theorem.

1) $\Lambda(K+M) = \Lambda(K) + \Lambda(M)$, $\Lambda(\alpha K) = \alpha\Lambda(K)$ for $K,M \in K^{r\times m}$, $\alpha \in \mathbb{R}$.

2) $\Lambda(K * M) = \Lambda(K)\Lambda(M)$ for $K \in K^{r\times m}$, $M \in K^{m\times p}$.

Proof. This result follows immediately from theorem 3.7. □

We want to generalize the concepts of the order of a formal power series. The following is an obvious generalization of definition 3.8.

4.3. Definition. If $A(s) = \sum_{-\infty}^{\infty} A_n s^{-n} \in F^{r\times m}$, then the *principal order* of A(s) is given by

$$\omega_1(A) = \min\{n \in \mathbb{Z} \mid A_n \neq 0\} .$$ □

However, it is clear that $\omega_1(A)$ is not such a fundamental quantity as $\omega(\alpha)$ defined by (3.8). For example, if in the case $r = m$, the matrix $A \in F^{m\times m}$ has an inverse A^{-1} we do not necessarily have $\omega_1(A^{-1}) = -\omega_1(A)$. This difficulty is due to the fact that the first matrix $A_n \neq 0$ is not necessarily invertible. It turns out, that we need a sequence of order numbers: $\omega_1(A),\dots,\omega_\rho(A)$, where $\rho \leq \min(r,m)$.

4.4. Definition. A matrix $U \in F^{m\times m}$ is called *unimodular* if $U \in N^{m\times m}$ and U has an inverse $U^{-1} \in N^{m\times m}$. □

Here $N^{m\times m}$ is the set of $m \times m$ matrices with entries in the set N of negative formal power series (see section 3).

4.5. Theorem. If $U = \sum_{n=0}^{\infty} U_n s^{-n}$, then U is unimodular iff U_0 is invertible.

Proof. U is unimodular iff $\det U(s)$ is a zero order formal power series. □

Let us use the following notation: If $\alpha_1,\ldots,\alpha_\rho$ are nonzero elements of a field and $\rho \le \min(r,m)$, then $\mathrm{diag}^{r\times m}(\alpha_1,\ldots,\alpha_\rho)$ denotes the $r \times m$ matrix with elements a_{ij} where $a_{ii} = \alpha_i$ for $i = 1,\ldots,\rho$ and $a_{ij} = 0$ otherwise. Then we have the following result:

4.6. Theorem. Every $A \in F^{r\times m}$ can be decomposed as

$$A(s) = U(s)D(s)V(s)$$

where $U \in N^{r\times r}$, $V \in N^{m\times m}$ are unimodular and $D(s)$ is of the form $D(s) = \mathrm{diag}^{r\times m}(s^{-\omega_1},\ldots,s^{-\omega_\rho})$ for some $\rho \le \min(r,m)$. Here $\omega_1 \le \omega_2 \le \ldots \le \omega_\rho$. The matrix $D(s)$ is uniquely determined by $A(s)$. Furthermore $\omega_1 = \omega_1(A)$ defined in 4.3.

Proof. If $\omega_1 := \omega_1(A)$, the matrix $B(s) = s^{\omega_1}A(s)$ is in $N^{r\times m}$. Since N is a principal ideal domain, we may write $B(s) = U(s)D_0(s)V(s)$ where $U(s)$ and $V(s)$ are unimodular and $D_0(s) = \mathrm{diag}^{r\times m}(s^{-\nu_1},\ldots,s^{-\nu_\rho})$ is the Smith Canonical Form of $B(s)$ in the ring N (see [4, theorem II.9]). Since $s^{-\nu_i} \mid s^{-\nu_{i+1}}$ in N, we have $\nu_1 \le \nu_2 \le \ldots \le \nu_\rho$. It follows from $B(s) = \sum_{n=0}^{\infty} B_0 s^{-n}$ with $B_0 \neq 0$ that $\nu_1 = 0$. We have $D(s) = s^{-\omega_1} D_0(s) =$

$= \mathrm{diag}^{r\times m}(s^{-\omega_1},\ldots,s^{-\omega_\rho})$ with $\omega_i = \omega_1 + \nu_i$. The uniqueness of $D(s)$ follows from the uniqueness of the Smith form of $B(s)$. □

4.7. Definition. The numbers $\omega_1,\ldots,\omega_\rho$ are called the *order indices* of $A(s)$. The matrix $D(s)$ will be referred to as the F-MacMillan form of $A(s)$. □

The integer ρ is the *rank* of $A(s)$; $\rho = \mathrm{rank}\, A(s)$. The analogy of the F-MacMillan form with the rational MacMillan form is obvious. The situation here is much simpler than in the case of rational matrices, due to the fact that the ideals of N are of the form $s^{-\nu}N$ ($\nu = 0,1,2,\ldots$) and thus completely determined by the integer ν. In the case that $A(s)$ is the transfer matrix of a finite dimensional system, that is, if $A(s)$ is rational (see example 2.3) one might expect, that the F-MacMillan form can be obtained from the rational MacMillan form. This is not possible, as shown by the following example:

4.8. <u>Example</u>. The system with transfer matrix $\begin{bmatrix} s^{-1} & 1 \\ 0 & s^{-1} \end{bmatrix}$ has $s^{-1}I$ as MacMillan form and $\mathrm{diag}(1,s^{-2})$ as F-MacMillan form. On the other hand the system withe transfer matrix $s^{-1}I$ has the same matrix (viz. $s^{-1}I$) as MacMillan form and F-MacMillan form. □

One may say that the F-MacMillan form contains the infinite and the rational MacMillan form the finite invariant factors of A(s).

Because of the simple structure of the ring N a simple direct proof can be given of the existence of the MacMillan form. We denote the (i,j)th entry of A(s) by $a_{ij}(s) = s^{-\nu_{ij}} u_{ij}(s)$ where u_{ij} is unimodular. Determine (i,j) such that ν_{ij} is minimal, denote this exponent by ω_1 and interchange, if necessary, rows and columns such that the element $s^{-\omega_1} u_{ij}$ is at the (1,1)th position. Then, multiply the first row by u_{ij}^{-1}. Thus we obtain the entry $s^{-\omega_1}$ in the (1,1)th position, whereas for every other entry $s^{-\nu_{ij}} u_{ij}(s)$ we have $\nu_{ij} \geq \omega_1$. Then, subtract $s^{-\nu_{i1}+\omega_1} u_{i1}(s)$ times the first row from the i-th row for i = 2,...,r. After these operations, the first column is $[s^{-\omega_1},0,\ldots,0]'$. Similarly, by column operations, the first row is made equal to $[s^{-\omega_1},0,\ldots,0]$. Then the matrix has the form

$$\begin{bmatrix} s^{-\omega_1} & 0 \\ 0 & B(s) \end{bmatrix}, \quad \text{where } \omega_1(B) \geq \omega_1 .$$

A repetition of these operations yields ultimately the F-MacMillan form. And, as usual, the row and column operations performed may be viewed as left and right multiplications with unimodular matrices. Also the uniqueness of the order indices can be proved without appealing to the general Smith canonical form, as will be shown in the next section.

Now, we want to investigate the implications of the foregoing for r × m kernels. We say that a kernel K is *unimodular* if $\Lambda(K)$ is unimodular. We have the following analogue of lemma 3.10:

4.9. <u>Lemma</u>. Let K be a unimodular kernel. Then K has a unimodular inverse K^{-1} with $\Lambda(K^{-1}) = (\Lambda(K))^{-1}$.

<u>Proof</u>. K has an inverse since det K has an inverse. Note that $\Lambda(\det K) = \det \Lambda(K)$. □

Let K be an r × m kernel and let $A := \Lambda(K)$. Consider the formula of theorem 4.6: $A(s) = U(s)D(s)V(s)$. According to theorem 3.7 iii), there exists a unimodular r × r kernel $\tilde{M}$ and a unimodular m × m kernel $\tilde{N}$ such that $\Lambda(\tilde{M}) = U$, $\Lambda(\tilde{N}) = V$. Consider the kernel $\tilde{C} := \tilde{M}^{-1} * K * \tilde{N}^{-1}$ and note that $\Lambda(\tilde{C}) = D(s)$. This does not necessarily imply that $\tilde{C}$ is of the form $\mathrm{diag}^{r\times m}(d^{-\omega_1},\ldots,d^{-\omega_\rho})$, since Λ is noninjective. However, it is easily seen that by elementary row and column operations in the ring K, that is, by left and right multiplications with unimodular kernels, $\tilde{C}$ can be brought into the form

$$(1) \qquad C = \begin{bmatrix} D_0(d) & 0 \\ 0 & H \end{bmatrix}$$

where $D_0(d) = \mathrm{diag}^{\rho\times\rho}(d^{-\omega_1},\ldots,d^{-\omega_\rho})$ and H is a (possibly empty) completely smooth $(r-\rho)\times(m-\rho)$ kernel. Thus we obtain:

4.11. Theorem. Every $r \times m$ kernel K can be decomposed as

$$K = M * C * N$$

where M and N are unimodular and C is of the form (1). □

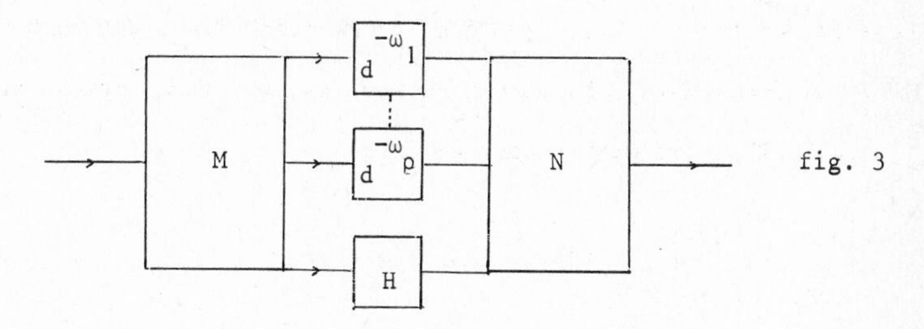

fig. 3

One may view the numbers $\omega_1,\ldots,\omega_\rho$ as a refinement of the concept of inherent integration or differentiation (see[5]). We denote these numbers by $\omega_k(K)$ $(k = 1,\ldots,\rho)$ and will call them the *order indices* of K.

4.12. Corollary. An $r \times m$ kernel K has a left inverse L (that is, an $m \times r$ kernel such that $L * K = \delta I$) iff $\rho = m$, and a right inverse iff $\rho = r$. If $r = m = \rho$, then the left and right inverse coincide and are unique.

Proof. A left inverse L of K corresponds to a left inverse E of C (of theorem 4.11) according to the formulas $E = N * L * M$ and $L = N^{-1} * E * M^{-1}$. Consequently, it suffices to investigate the invertibility of C. If $\rho = m$, then $\mathrm{diag}^{m\times r}(d^{\omega_1},\ldots,d^{\omega_m})$ is a left inverse of C. If C has a left inverse E then $\Lambda(E)\Lambda(C) = I$, hence $\Lambda(C)$ must have the full column rank, that is $\rho = m$. The statement about the right inverse is obtained by transposition. Finally, the last statement of the theorem is a well known elementary result in algebra. □

We want to determine the order indices of a left inverse. However, if $r > m$, the left inverse is not uniquely determined, and in particular, the order indices are not unique. We can state the following result:

4.13. Corollary. If the $r \times m$ kernel K has a left inverse L, then $\omega_k(L) \le -\omega_{m-k+1}(K)$. Moreover, there exists a left inverse L_0 of K with $\omega_k(L_0) = -\omega_{m-k+1}(K)$.

Proof. Again we consider left inverses of C instead of K. It is clear that $L_0 := \mathrm{diag}^{m\times r}(d^{\omega_1},\dots,d^{\omega_m})$ satisfies the equalities of the theorem. In order to prove the inequality for an arbitrary inverse L we note that L necessarily has the form $L = [L_0, L_1]$. Let us consider the Λ-transform $B(s) = [B_0(s), B_1(s)]$ of L. Then $B_0(s) = \mathrm{diag}^{m\times m}(s^{\omega_1},\dots,s^{\omega_m})$ and $B_1(s)$ is arbitrary. By suitable row and column permutations of $B(s)$ we obtain a matrix $Q(s) = [Q_0(s), Q_1(s)]$, with $Q_0(s) = \mathrm{diag}^{m\times m}(s^{\omega_m},\dots,s^{\omega_1})$. We have to show that $\omega_k(Q) \ge -\omega_{m-k+1}$. Assume that the statement has been shown for $(m-1) \times (r-1)$ matrices. If for all entries q_{1j} of the first row of Q we have $\omega(q_{1j}) \ge -\omega_m$, then we can use elementary column operations to obtain a matrix of the form

$$\begin{bmatrix} s^{\omega_m} & 0 & 0 \\ 0 & \hat{Q}_0 & \hat{Q}_1 \end{bmatrix}$$

where

$$\hat{Q}_0 = \mathrm{diag}^{(m-1)\times(m-1)}(s^{\omega_{m-1}},\dots,s^{\omega_1}) \;. \tag{2}$$

Now suppose that $\nu := \max\{\omega(q_{1j}) \mid j = 1,\dots,r-m\} < -\omega_m$ and let $\omega(q_{1\ell}) = \nu$. Then we interchange the first and the ℓ-th column and subsequently "clean" the first column and row. Thus we obtain

$$\begin{bmatrix} s^{-\nu} & 0 & 0 \\ 0 & \hat{Q}_0 & \hat{Q}_1 \end{bmatrix}$$

with $\hat{Q}_0$ given by (2). After this step we may apply the induction hypothesis. □

There is no simple relation between the order indices of A, B and AB. The only result we mention is:

$$\omega_k(AB) \ge \max\{\omega_1(A) + \omega_k(B),\ \omega_k(A) + \omega_1(B)\} \;,$$

if A and B are square nonsingular matrices, which is an immediate consequence of [4, theorem II.14]. Naturally, $\omega_k(UA) = \omega_k(A)$ is U is unimodular.

Finally, we remark that a feedback system as given by fig. 2 defines an i/o-map iff $\Lambda(K)\Lambda(L)$ does not have 1 as an eigenvalue in F.

§ 5. Relation with the invertibility conditions of Silverman and Sain-Massey

In this section we assume for notational convenience that the kernels considered have nonnegative principal order, that is, that the Λ-transforms of the kernels are elements of $N^{r\times m}$.

5.1. Definition. If $A = \sum_{n=0}^{\infty} A_n s^{-n} \in N^{r\times m}$, then

$$M_k(A) := \begin{bmatrix} A_0 & 0 & & 0 & 0 \\ A_1 & A_0 & & \vdots & \vdots \\ A_2 & A_1 & A_0 & \vdots & \vdots \\ \vdots & & & A_0 & 0 \\ A_k & & & A_1 & A_0 \end{bmatrix}$$

for $k = 0,1,2,\ldots$. □

The following results are easily verified:

5.2. Lemma.

1) The map M_k: $N^{r\times m} \to \mathbb{R}^{(k+1)r\times(k+1)m}$ is linear.

2) $M_k(AB) = M_k(A)M_k(B)$.

3) $M_k(A)$ is invertible iff A is unimodular.

The invertibility condition of Sain-Massey ([6]) was expressed in terms of the quantities $\ell_k(A) := \text{rank } M_k(A) - \text{rank } M_{k-1}(A)$ $(k = 1,2,\ldots)$, $\ell_0(A) := \text{rank } M_0(A)$. Due to lemma 5.2 we have rank $M_k(A)$ = rank $M_k(D)$ and hence $\ell_k(A) = \ell_k(D)$ where D is the F-MacMillan form of A.

Let $D(s) = D_0 + D_1 s^{-1} + \ldots$. Then $D_k = 0$ unless $k = \omega_j$ for some j, in which case D_k has a one on the diagonal. There are more diagonal elements equal to one if some ω_j's coincide. But in any case, the nonzero elements of the matrices D_k are on distinct positions on the diagonal (one can express this property by the formula: $D_k D_\ell = 0$ if $k \neq \ell$). It follows that the nonzero columns of $M_k(D)$ are independent. In particular, $\ell_k(D)$ is the number of nonzero columns in the first block column, that is $\ell_k(D)$ is the number of ω_j's satisfying $\omega_j \leq k$. Consequently,

$$\ell_k(A) = \max\{j \mid \omega_j(A) \leq k\} .$$

Conversely, we have

$$\omega_j(A) = \min\{k \mid \ell_k(A) \geq j\} .$$

The relation between ℓ_k and ω_j is clarified by a diagram:

5.3. Example.

	ω_1	ω_2	ω_3	ω_4
ℓ_0	x			
ℓ_1	x	x	x	
ℓ_2	x	x	x	
ℓ_3	x	x	x	
ℓ_4	x	x	x	
ℓ_5	x	x	x	x

$\omega_1 = 0$, $\omega_2 = 1$, $\omega_3 = 1$, $\omega_4 = 5$, $\ell_0 = 1$, $\ell_1 = \ell_2 = \ell_3 = \ell_4 = 3$, $\ell_5 = 4$. If $A \in N^{4\times 4}$, then the F-MacMillan form of A is $\mathrm{diag}^{4\times 4}(1,s^{-1},s^{-1},s^{-5})$.

We see that the numbers $\ell_k(A)$ uniquely determine the F-MacMillan form and conversely. The condition for the existence of an L-integrating left inverse of A given by Sain and Massey is $\ell_L = m$, which is obviously equivalent to $\rho = m$ and $\omega_m = L$.

In [7], Silverman has given an algorithm for determining the invertibility of an finite dimensional system. The input-output aspects of Silverman's algorithm can be generalized to the more general smooth linear systems. The algorithm consists of the construction of a finite sequence of $r \times m$ F-matrices $A^0(s), A^1(s), \ldots, A^\alpha(s)$, where $A^0(s) = A(s)$, the matrix the right invertibility of which is to be investigated, whereas either $A^\alpha(s)$ is obviously right invertible since A_0^α has full row rank or $A^\alpha(s)$ is obviously not right invertible since it is of the form $A^\alpha(s) = \begin{bmatrix} A^{-\alpha}(s) \\ 0 \end{bmatrix}$. The construction is defined as follows: We define $q_0 := \mathrm{rank}\ A_0^0$. There exists an invertible $\mathbb{R}$-matrix Q^0 such that $A_0^0 = Q^0 \begin{bmatrix} \bar{A}_0^0 \\ 0 \end{bmatrix}$, where $\bar{A}_0^0$ is a $q_0 \times m$ matrix. Then we define A^1 by

$$A(s) = A^0(s) = Q^0 D^0(s) A^1(s)$$

where

$$D^0(s) = \begin{bmatrix} I_{q_0} & 0 \\ 0 & s^{-1}_{r-q_0} \end{bmatrix} .$$

This procedure is repeated: We define $q_1 = \mathrm{rank}\ A_0^1$. Obviously $q_1 \geq q_0$. There exists an invertible matrix Q^1 of the form

$$Q^1 = \begin{bmatrix} I_{q_0} & 0 \\ S^1 & R^1 \end{bmatrix}$$

such that

$$A_0^1 = Q^1 \begin{bmatrix} \bar{A}_0^1 \\ 0 \end{bmatrix} ,$$

where A_0^1 is a $q_1 \times m$ matrix. Notice that the block decompositions of Q^1 and $\begin{bmatrix} A_0^1 \\ 0 \end{bmatrix}$ are not compatible. It is clear that ultimately (for $k = \alpha$) we have either $q_k = r$ or $A^\alpha(s) = \begin{bmatrix} \bar{A}^\alpha(s) \\ 0 \end{bmatrix}$. Although this process is finite, it is not possible to give a priori bounds on α, contrary to the finite dimensional case. The ultimate result of this calculatuon is

$$A(s) = Q^0 D^0(s) Q^1 D^1(s) \ldots Q^{\alpha-1} D^{\alpha-1}(s) A^\alpha(s) ,$$

where

$$D^k(s) = \begin{bmatrix} I_{q_k} & 0 \\ 0 & s^{-1} I_{q_k} \end{bmatrix}$$

$$Q^k = \begin{bmatrix} I_{q_{k-1}} & 0 \\ S^k & R^k \end{bmatrix}$$

$$A\ (s) = \begin{bmatrix} \bar{A}^\alpha(s) \\ 0 \end{bmatrix}$$

or $q_\alpha = r$, and $\bar{A}_0^\alpha$ has full row rank (rank $\bar{A}_0^\alpha = q_\alpha$).
After some computation we find

$$Q^0 D^0(s) \ldots Q^{\alpha-1} D^{\alpha-1}(s) = Q(s) D(s) ,$$

where $Q(s)$ is a unimodular matrix and $D_0(s) = D^0(s) \ldots D^{\alpha-1}(s)$. Furthermore, $A^\alpha(s) = \begin{bmatrix} I_{q_\alpha} & 0 \\ 0 & 0 \end{bmatrix} U(s)$, where $U(s)$ is unimodular. We see that $A(s) = Q(s)D(s)U(s)$, where $D(s) = D_0(s) \begin{bmatrix} I_{q_\alpha} & 0 \\ 0 & 0 \end{bmatrix}$, is the F-MacMillan form of $A(s)$. Calculation of $D(s)$ yields that we have $q_k = \ell_k$ and $q_\alpha = \rho$, $\alpha = \omega_\rho$. The equality of q_k and ℓ_k has also been noted by Singh [9].

5.4. <u>Example</u>. If $A(s) = \begin{bmatrix} s^{-1} & 1 \\ 0 & s^{-1} \end{bmatrix}$, then

$$M_2(A) = \left[\begin{array}{cc|cc|cc} 0 & 1 & 0 & 0 & 0 & 0 \\ 0 & 0 & 0 & 0 & 0 & 0 \\ \hline 1 & 0 & 0 & 1 & 0 & 0 \\ 0 & 1 & 0 & 0 & 0 & 0 \\ \hline 0 & 0 & 1 & 0 & 0 & 1 \\ 0 & 0 & 0 & 1 & 0 & 0 \end{array}\right] .$$

We have $\ell_0 = \ell_1 = 1$, $\ell_2 = 2$. Hence, the ω_i's are determined by the following diagram:

	ω_1	ω_2
ℓ_0	x	
ℓ_1	x	
ℓ_2	x	x

that is, $\omega_1 = 0$, $\omega_2 = 2$, so that $D(s) = \mathrm{diag}(1,s^{-2})$. Applying Silverman's algorithm we find successively:

$$A^0(s) = \begin{bmatrix} s^{-1} & 1 \\ 0 & s^{-1} \end{bmatrix},\ A_0^0 = \begin{bmatrix} 0 & 1 \\ 0 & 0 \end{bmatrix},\ q_0 = 1 ,$$

$$Q^0 = I,\ D^0(s) = \mathrm{diag}(1,s^{-1}),\ A^1(s) = \begin{bmatrix} s^{-1} & 1 \\ 0 & 1 \end{bmatrix},\ A_0^1 = \begin{bmatrix} 0 & 1 \\ 0 & 1 \end{bmatrix},\ Q^1 = \begin{bmatrix} 1 & 0 \\ 1 & 1 \end{bmatrix},$$

$$D^1(s) = \mathrm{diag}(1,s^{-1}),\ A^2(s) = \begin{bmatrix} s^{-1} & 1 \\ -1 & 0 \end{bmatrix} .$$

Hence, $\alpha = 2$, $q_0 = q_1 = 1$, $q_2 = 2$, $D(s) = D^1(s)D^0(s) = \mathrm{diag}(1,s^{-2})$. □

It should be mentioned that the emphasis in Silverman's algorithm lies on the state space aspects of invertible systems, which cannot be considered in this setting.

As is shown in [1], a strictly causal smooth system has a state space representation, with systems equation of the form $\dot{x} = Fx + Gu$, $y = Hx$. If a smooth system is not strictly causal one can find a state space representation of the form $\dot{x} = Fx + Gu$, $y = Hx + A_0u + A_1u^1 + \ldots + A_pu^{(p)}$. Here x is an element of a complete barreled locally convex Hausdorff space, F is a continuous linear operator and the differential equation is to be interpreted in the distributional sense. If

$$\Lambda(K) = \sum_{-\infty}^{\infty} A_n s^{-n} ,$$

then $A_n = HF^nG$ for $n \geq 1$. It would be interesting to see whether more detailed information about the inverse of a smooth system is obtainable from the state space representation, like in [7], [8] for the finite dimensional situation.

References

[1] Kalman, R.E. and Hautus, M.L.J., *Realization for continuous-time linear dynamical systems: Rigorous theory in the style of Schwartz*, Ordinary differential equations, 1971, NRL-MRC Conference; Acad. Press, New York, 1972.

[2] Kamen, E.W., *Representation of linear continuous time systems by spaces of distributions*, I.R.I.A. Report, INF 2713/72016, 1972.

[3] Kamen, E.W., *Module structure of infinite dimensional systems with applications to controllability*, to appear in SIAM J. on Control.

[4] Newman, M., *Integral matrices*, Acad. Press, New York, 1972.

[5] Schwartz, L., *Théorie des Distributions I, II;* Hermann, Paris, 1951.

[6] Sain, M.K. and Massey, J.L., *Invertibility of linear time-invariant dynamical systems*, IEEE Trans. Aut. Cont., AC-14, 1969, pp. 141-149.

[7] Silverman, L.M., *Inversion of multivariable linear systems*, IEEE Trans. Aut. Cont., AC-14, 1969, pp. 270-276.

[8] Silverman, L.M. and Payne, H.J., *Input-output structure of linear systems with application to the decoupling problem*, SIAM J. Control, Vol. 9, 1971, pp. 193-233.

[9] Singh, S.P., *A note on inversion of linear systems*, IEEE Trans. Aut. Cont., 1970, pp. 492-493.

ON INVARIANTS, CANONICAL FORMS AND MODULI FOR LINEAR, CONSTANT, FINITE DIMENSIONAL, DYNAMICAL SYSTEMS

Michiel Hazewinkel
Dept. Math., Econometric Inst.
Erasmus Univ. of Rotterdam
Rotterdam, The Netherlands

Rudolf E. Kalman
Center for Mathematical System Theory
Univ. of Florida
Gainesville, Fla. 32611 USA

1. INTRODUCTION AND SURVEY OF RESULTS

A linear, constant, finite dimensional dynamical system is thought of as being represented by a triple of matrices (F,G,H), where F is an n x n matrix, G an n x m matrix, and H an p x n matrix; i.e. there are m inputs, p outputs and the state space dimension is n. The dynamical system itself is

(1.1) $$\dot{x} = Fx + Gu, \quad y = Hx$$

or, if one prefers discrete time systems

(1.2) $$x_{t+1} = Fx_t + Gu_t, \quad y_t = Hx_t$$

A change of coordinates in state space changes the triple of matrices (F,G,H) into the triple (SFS^{-1}, SG, HS^{-1}). Let $\underline{DS}$ denote the space of all triples (F,G,H); i.e. $\underline{DS}$ is affine space of dimension $np + n^2 + nm$.
Then we have just defined an action of GL_n on $\underline{DS}$. This paper is concerned with the following type problems. To what extent does the quotient $\underline{DS}/GL_n$ exist ? Does the quotient have a nice geometric structure ? Do there exist globally defined algebraic continuous canonical forms for triples (F,G,H)?
Most of the paper is concerned with the input aspect only, i.e. instead of studying triples (F,G,H) under the action $(F,G,H)^S = (SFS^{-1},SG,HS^{-1})$ we study pairs (F,G) under the action $(F,G)^S = (SFS^{-1},SG)$. Let $\underline{IS}$ be the affine space of all pairs (F,G) and $\underline{IS}_{cr}$ the open subvariety of all completely reachable pairs. It turns out that the orbit space $\underline{IS}_{cr}/GL_n$ has a nice geometric structure. In fact, it is a quasi-projective algebraic variety. Moreover this variety $\underline{M}_{m,n} = \underline{IS}_{cr}/GL_n$ turns out to be a fine moduli space for algebraic families of completely reachable pairs (suitably defined). I.e. the points of $\underline{M}_{m,n}$ correspond bijectively to equivalence classes of completely reachable pairs and there exists over $\underline{M}_{m,n}$ a universal family from which every family can be obtained (uniquely) by pullback.

However, if there are two or more inputs the underlying bundle of this universal family is non trivial and this ruins all chances of finding <u>continuous</u> algebraic canonical forms for $\underline{IS}_{cr}$. This in turn also implies the nonexistence of continuous algebraic canonical forms for $\underline{IS}$, $\underline{DS}_{cr}$ and $\underline{DS}$. There exist of course (many) discontinuous canonical forms. (To keep the non existence result in proper perspective: the Jordan canonical form for square matrices is also not continuous).

In this paper we shall work over an arbitrary field k, which, for convenience, can be taken to be algebraically closed. However, all the constructions performed yield varieties defined over k itself. The category of varieties over k is denoted $\underline{Sch}_k$. Much of the material which follows is also contained in [2] in one way or another. The emphasis and presentation are different, however; here we stress the underlying ideas rather than the algebraic geometric techniques. Also this paper contains additional new material, notably subsections 3.9, 6.1, 6.3, 7.1, 7.2, 7.3, 7.4, 7.5.

2. GRASSMANN VARIETIES

Let $\underline{A}_{n,s}$ denote the affine space of all n x s matrices, where s > n; i.e. $\underline{A}_{n,s}$ is affine space of dimension ns. Let $\underline{A}^{reg}_{n,s}$ denote the (Zariski) open dense subvariety of $\underline{A}_{n,s}$ consisting of matrices of maximal rank. The group GL_n acts on $\underline{A}_{n,s}$ (and $\underline{A}^{reg}_{n,s}$) by multiplication on the left: $(S,A) \mapsto SA$.

The orbit space $\underline{A}^{reg}_{n,s}/GL_n$ has a nice geometric structure; it is a smooth projective algebraic variety of dimension n x (s-n), known as the Grassmann variety of n-planes in s-space, and denoted $\underline{G}_{n,s}$. This interpretation arises as follows. Let A be an n x s matrix of rank n. The n rows of A span an n-dimensional subspace of affine space of dimension s, and, clearly, the rows of SA span the same subspace.

The (canonical) projective embedding of $\underline{G}_{n,s}$ is obtained as follows.

A <u>selection</u> α of $\{1,\ldots,s\}$ is a subset of size n. For each selection α and n x s matrix A, let A_α be the submatrix of A consisting of those columns of A which are indexed by an element of α. Let $N = \binom{n}{s} - 1$, the number of selections minus 1. We now define a morphism from $\underline{G}_{n,s}$ to projective N-space

$$(2.1) \qquad G_{n,s} \to \underline{P}^N, \; A \to (\det(A_\alpha))_\alpha$$

where det denotes determinant. This is an embedding and exhibits $\underline{G}_{n,s}$ as a closed subvariety of $\underline{P}^N$.

Choose a selection α. The open subvariety of $\underline{G}_{n,s}$ where $\det(A_\alpha) \neq 0$ is isomorphic to affine n x (s-n) space: a point $x \in \underline{G}_{n,s}$ corresponds to the unique n x s matrix A_x for which (i) $(A_x)_\alpha = I_n$, the n x n unit matrix and (ii) the rows of

A_x span the linear subspace x.
For further details concerning $\underline{G}_{n,s}$, e.g. for a description of the equations defining $\underline{G}_{n,s}$ as a closed subvariety of $\underline{P}^N$ cf. e.g. [4]. For more details concerning $\underline{G}_{n,s}$ from the differential topological point of view cf. e.g. [3].

3. THE COARSE MODULI SPACE $\underline{M}_{m,n}$

Let $\underline{IS}$ denote the affine space of all pairs of matrices (F,G). The group GL_n acts on $\underline{IS}$ by $(F,G) \to (SFS^{-1},SG)$.

3.1. The Morphism R and Completely Reachable Pairs.

We define the morphism R from $\underline{IS}$ to $\underline{A}_{n,(n+1)m}$ by

$$R(F,G) = (G\ FG\ \dots\ F^nG) \tag{3.1.1}$$

The pair (F,G) is said to be completely reachable if R(F,G) has rank n. Let $\underline{IS}_{cr}$ denote the Zariski open subvariety of $\underline{IS}$ consisting of the completely reachable pairs. It follows that R induces a morphism

$$R: \underline{IS}_{cr} \to A^{reg}_{n,(n+1)m} \tag{3.1.2}$$

Note that R is a GL_n-invariant morphism. I.e.

$$R(SFS^{-1},SG) = SR(F,G) \tag{3.1.3}$$

3.2. Nice Selections and Successor Selections.

In section 2 we have seen that selections play on important role in the description of the quotient $\underline{A}_{n,s}/GL_n$. In view of (3.1.3) it is to be expected that they will also be important in the case of GL_n acting on $\underline{IS}$. Certain selections of the (n+1)m columns of the R(F,G) play a special role. To define them we number the (n+1)m columns by pairs of integers (lexicographically ordered) as follows

$$01,\ \dots,\ 0m;\ 11,\ \dots,\ 1m;\ \dots\ ;\ n1,\ \dots,\ nm$$

A selection α is called nice if $(i,j) \in \alpha \Rightarrow (i',j) \in \alpha$ for all $i' \leq i$.
Given a nice selection α its successor selections are obtained as follows: take any $(i,j) \in \{01,\dots,nm\}$ such that $(i,j) \notin \alpha$ but $(i',j) \in \alpha$ for all $i' < i$. Now take away from $\alpha \cup (i,j)$ any of the original elements of α. The result is a successor selection. Note that a successor selection may be nice but need not be.
Example, take m = 4, n = 6

*	x	x	x
.	x	*	x
.	x	.	*
.	*	.	.
.	.	.	.
.	.	.	.

0,1 ... 0,4

. .

. .

. .

6,1 ... 6,4

The crosses constitute a nice selection. Its successor selections are obtained by adding one of the stars and deleting one of the crosses.

3.3. Lemma.

If (F,G) is a completely reachable pair then there is a nice selection α such that $\det(R(F,G)_\alpha) \neq 0$.

3.4. Successor Indices.

Let α be a nice selection. The successor indices of α are those elements $(i,j) \in \{01,\dots,nm\}$ such that $(i',j) \in \alpha$ for all $i' < i$. I.e. in the example of subsection 3.2 the *'s mark the successor indices of the nice selection given by the x's.

We now define an algebraic morphism $\psi_\alpha : \underline{A}^{mn} \to \underline{IS}$ as follows. Let $\sigma(\alpha)$ be the set of successor indices of the nice selection α. The subset $\alpha \cup \sigma(\alpha)$ has precisely n + m elements. Give this subset the ordering induced by the (lexicographic) ordering of $\{01,\dots,nm\}$. Write an element $x \in \underline{A}^{mn}$ as an array of m columns of length n; let x_i denote the i-th column in this array. We now assign to each element (i,j) of $\alpha \cup \sigma(\alpha)$ a column c(i,j) of length n as follows: if (i,j) is the ℓ-th element of α then $c_{(i,j)} = e_\ell$, the ℓ-th unit vector. If (i,j) is the ℓ-th element of $\sigma(\alpha)$ then $c_{(i,j)} = x_\ell$, the ℓ-th column of x. Writing G_i (resp. F_i) for the i-th column of G (resp. F) we now define ψ_α by

$\psi_\alpha(x) = (F,G)$, where

G_i = column assigned to i-th element of $\alpha \cup \sigma(\alpha)$

F_i = column assigned to (m+i)-th element of $\alpha \cup \sigma(\alpha)$.

Thus in the example of subsection 3.2 we have

$$G_1 = x_1,\ G_2 = e_1,\ G_3 = e_2,\ G_4 = e_3$$

$$F_1 = e_4,\ F_2 = x_2,\ F_3 = e_5,\ F_4 = e_6,\ F_5 = x_3,\ F_6 = x_4$$

Note that if $\psi_\alpha(x) = (F,G)$, then $R(F,G)_\alpha$ = unit matrix, and if (i,j) is the ℓ-th element of $\sigma(\alpha)$ then $R(F,G)_{(i,j)}$, the (i,j)-th column of R(F,G), is equal to x_ℓ, the ℓ-th column of x. (This is easy to check; if $(i,j) \in \alpha \cup \sigma(\alpha)$ is the $(m+\ell)$-th element of $\alpha \cup \sigma(\alpha)$ then (i-1,j) is the ℓ-th element of α).

3.5. Lemma.

$R\psi_\alpha : \underline{A}^{mn} \to \underline{A}^{reg}_{n,(n+1)m}$ is an embedding which as image the subvariety of $\underline{A}^{reg}_{n,(n+1)m}$ consisting of the matrices of the form $R(F,G)$ for which $R(F,G)_\alpha = I_n$, the $n \times n$ unit matrix.

Proof. Follows from 3.4 above.

3.6. Lemma.

Let α be a nice selection. Denote with U_α the subvariety of $\underline{IS}_{cr}$ consisting of all completely reachable pairs (F,G) for which $\det(R(F,G)_\alpha) \neq 0$.
Then $U_\alpha \simeq GL_n \times \underline{A}^{nm}$.

Proof. Let $(F,G) \in U_\alpha$. There is a unique invertible matrix S such that $(S^{-1}R(F,G))_\alpha = I_n$, the $n \times n$ unit matrix. In fact $S = R(F,G)_\alpha$. Further $S^{-1}R(F,G) = R(S^{-1}FS, S^{-1}G)$. Now apply lemma 3.5.

3.7. The Coarse Moduli Space $\underline{M}_{m,n}$

It follows directly from lemma 3.6 that the quotients U_α/GL_n exist for all nice selections α. (Note also that U_α is GL_n-invariant). To construct the quotient $\underline{IS}_{cr}/GL_n$ it therefore suffices to patch the various affine pieces $V_\alpha = U_\alpha/GL_n \simeq \underline{A}^{mn}$ together. This is done as follows: let α,β be two nice selections. Let

$$V_{\alpha\beta} = \{x \in V_\alpha | \det((R\psi_\alpha(x))_\beta) \neq 0\}$$

$$V_{\beta\alpha} = \{x \in V_\beta | \det((R\psi_\beta(x))_\alpha) \neq 0\}$$

The open subvarieties $V_{\alpha\beta}$ of V_α and $V_{\beta\alpha}$ of V_β are now identified by means of the isomorphisms $\phi_{\alpha\beta} : V_{\alpha\beta} \to V_{\beta\alpha}$ defined by

$$\phi_{\alpha\beta}(x) = x' \in V_{\beta\alpha},$$

where x' is the unique point of $V_{\beta\alpha}$ such that

$$R\psi_\beta(x') = (R\psi_\alpha(x))_\beta^{-1} R\psi_\alpha(x).$$

This is a well defined isomorphism in view of lemma 3.5. Patching together all the V_α for all nice selections α gives us, in view of lemma 3.3, a prescheme $\underline{M}_{m,n}$ of which the points correspond bijectively to the orbits of GL_n in $\underline{IS}_{cr}$. This does not yet show that $\underline{M}_{m,n}$ is a variety. However, using the same general techniques it is not difficult to write down equations for $\underline{M}_{m,n}$.

More precisely: the assignment: orbit(F,G) $\to$ $(\det(R(F,G)_\gamma)_\gamma \in \underline{G}_{n,(n+1)m} \subset \underline{P}^N$, where γ runs through all selections, embeds $\underline{M}_{m,n}$ in $G_{n,(n+1)m} \subset \underline{P}^N$. One now writes down a set of homogeneous equations.

(3.7.1) $$q_{\alpha\beta}(\ldots,x_\gamma,\ldots) = 0$$

(one equation for each pair: (nice selection α, successor selection of α)). The variety $\underline{M}_{m,n}$ as a subvariety of $\underline{P}^N$ (or $\underline{G}_{n,(n+1)m}$) then consists of those points $(x_\gamma)_\gamma$ satisfying the equations (3.6.1) such that moreover for at least one nice selection α, $x_\alpha \neq 0$. Thus $\underline{M}_{m,n}$ is a quasi projective variety. Cf. [2] for more details. (Note also that the affine pieces + patching data description of $\underline{M}_{m,n}$ given above is compatible with the affine pieces + patching data description of $\underline{G}_{n,(n+1)m}$ indicated in section 2.

3.8. Example.

$\underline{M}_{2,2}$ is obtained by patching together three affine pieces $V_\alpha, V_\beta, V_\gamma$, all isomorphic to $\underline{A}^4$. Let $V_\alpha, V_\beta, V_\gamma$ be the affine pieces corresponding respectively to the nice selections $\alpha = \{01,02\}$, $\beta = \{01,11\}$, $\gamma = \{02,12\}$.
Take coordinates (a_1,a_2,a_3,a_4) for V_α, (b_1,b_2,b_3,b_4) for V_β, (c_1,c_2,c_3,c_4) for V_γ arranged in columns (a_1,a_2) and (a_3,a_4), etc....
Then we see that

$$V_{\alpha\beta} = \{a \in V_\alpha \mid a_3 \neq 0\}$$
$$V_{\beta\alpha} = \{b \in V_\beta \mid b_2 \neq 0\}$$

and the identification isomorphism is given by

$$b_1 = -a_1a_3^{-1} \qquad b_2 = a_3^{-1}$$
$$b_3 = a_2a_3 - a_1a_4 \qquad b_4 = a_1 + a_4$$

Further

$$V_{\alpha\gamma} = \{a \in V_\alpha \mid a_2 \neq 0\}$$
$$V_{\gamma\alpha} = \{c \in V_\gamma \mid c_2 \neq 0\}$$

with identifications

$$c_1 = -a_2^{-1}a_4 \qquad c_3 = a_2a_3 - a_1a_4$$
$$c_2 = a_2^{-1} \qquad c_4 = a_1 + a_4$$

And finally

$$V_{\beta\gamma} = \{b \in V_\beta \mid b_1^2 + b_1b_2b_4 - b_2^2b_3 \neq 0\}$$
$$V_{\gamma\beta} = \{c \in V_\gamma \mid c_1^2 + c_1c_2c_4 - c_2^2c_3 \neq 0\}$$

with identifications

$$c_1 = (b_1+b_2b_4)(b_1^2+b_1b_2b_4-b_2^2b_3)^{-1} \qquad c_3 = b_3$$

$$c_2 = (-b_2)(b_1^2+b_1b_2b_4-b_2^2b_3)^{-1} \qquad c_4 = b_4$$

3.9. Warning.

We have seen that $\underline{IS}_{cr}/GL_n = \underline{M}_{m,n}$. Now $\underline{DS}_{cr} = \underline{IS}_{cr} \times \underline{A}^{pn}$ and the action of GL_n on $\underline{DS}_{cr}$ is such its restriction to $\underline{IS}_{cr}$ is faithfull. It does not follow from this that $\underline{DS}_{cr}/GL_n \simeq \underline{M}_{m,n} \times \underline{A}^{pn}$ as is incorrectly claimed in subsection 4.6.A of [2]. (This would be the case if $\underline{IS}_{cr}$ were isomorphic to $\underline{M}_{m,n} \times GL_n$; this, however, is not true if $m \geq 2$). The following example may serve to illustrate the difficulty involved.

Let GL_1 act on $\underline{A}^2 \times \underline{A}^1$ as follows

$$\lambda(x_1,x_2,y) = (\lambda x_1,\lambda x_2,\lambda y)$$

Let $\underline{A}^2_{reg_1} = \{x \in \underline{A}^2 | x_1 \neq 0 \text{ or } x_2 \neq 0\}$. The quotients $\underline{A}^2_{reg_1}/GL_1$ and $(\underline{A}^2_{reg} \times \underline{A}^1)/GL_1$ both exist and are respectively equal to $\underline{P}^1$, the projective line, and $\underline{P}^2\setminus\{pt\}$, the projective plane minus the point (0,0,1). Thus we have

$$(\underline{A}^2_{reg}/GL_1) \times \underline{A}^1 = \underline{P}^1 \times \underline{A}^1$$

$$(\underline{A}^2_{reg} \times \underline{A}^1)/GL_1 = \underline{P}^2\setminus\{pt\}$$

But the algebraic varieties $\underline{P}^2\setminus\{pt\}$ and $\underline{P}^1 \times \underline{A}^1$ are not isomorphic.

Remark. It is true that the geometric quotient $\underline{DS}_{cr}/GL_n$ exists and it is a quasi-projective variety as we expect to show in a subsequent note.

4. FAMILIES OF DYNAMICAL SYSTEMS

The next topic we take up is that of a family of input pairs (F,G) parametrized by a variety S. The notion of a (locally trivial) vectorbundle is assumed to be known (cf. e.g. [1] Ch.2 for the algebraic case, or [3] for the topological version).

4.1. Families of Completely Reachable Pairs over a Variety.

As a first primitive approximation of a family of completely reachable pairs parametrized by a variety S we could define a family over S to be a morphism $S \to \underline{IS}_{cr}$. This turns out not to be suffiently general. Cf. 6.2 below. A more general concept is: a family Σ of pairs over a variety S consists of

(i) an n-vectorbundle E over S

(ii) a vectorbundle endomorphism $F: E \to E$

(iii) m sections $g_1, \ldots, g_m : S \to E$

Given a point $s \in S$ we have over s

$(i)_s$ the fibre $E(s)$ which is a vectorspace of dimension n

$(ii)_s$ a vectorspace endomorphism $F(s): E(s) \to E(s)$

$(iii)_s$ m vectors $g_1(s), \ldots, g_m(s)$ in $E(s)$

i.e. after choosing a basis in $E(s)$ we have a pair (F,G). The family Σ is said to be completely reachable if these induced pairs over the points of S are all completely reachable, i.e. if the vectors

$$F(s)^i g_j(s), \quad i = 1, \ldots, n; \; j = 1, \ldots, m$$

span all of $E(s)$ for all $s \in S$.

A family in the sense of a morphism $S \to \underline{IS}_{cr}$ corresponds to a family Σ over S for which the bundle E is isomorphic to $S \times \underline{A}^n$, the trivial n-vectorbundle over S. Two families Σ, Σ' over S are said to be isomorphic if there exists a vectorbundle isomorphism $\phi: E \to E'$ such that $\phi F = F'\phi$ and such that $\phi g_i = g_i'$.

Remark. There is another possible definition of families of input pairs; however, this other definition is not "rigid" enough for "fine moduli scheme" purposes. Cf. [2] for details.

4.2. The Functor: Isomorphism Classes of Families of Input Pairs.

Let Σ be a family of input pairs over a variety S, and let $f: T \to S$ be a morphism of varieties. Let $\Sigma = (E,F,g_1,\ldots,g_m)$. We now define an induced family $f^!\Sigma$ over T by pulling everything back along f. I.e. $f^!\Sigma = (f^!E, f^!F, f^!g_1, \ldots, f^!g_m)$, where $f^!E$ is the induced bundle over T, $f^!F$ the induced endomorphism over T and if we identify $(f^!E)(t)$ with $E(f(t))$ then $(f^!g_i)(t) = g_i(f(t))$. (The bundle $f^!E$ has as its fibre over t the fibre of E over $f(t)$; these fibres are fitted together in the obvious way). The family $f^!\Sigma$ is completely reachable if (and only if) the family Σ is completely reachable. We now define a functor $\mathcal{F}_{m,n}: \underline{Sch}_k \to \underline{Sets}$ from varieties over k to the category of sets as follows.

$\mathcal{F}_{m,n}(S)$ = set of isomorphism classes of completely reachable families of pairs with m inputs and state space dimension n

$\mathcal{F}_{m,n}(f) : \mathcal{F}_{m,n}(S) \to \mathcal{F}_{m,n}(T)$ is the mapping induced by $\Sigma \mapsto f^!\Sigma$ if $f: T \to S$ is a morphism in $\underline{Sch}_k$.

5. THE FINE MODULI SCHEME $\underline{M}_{m,n}$

5.1. $\underline{M}_{m,n}$ is a Coarse Moduli Scheme.

Let Σ be a completely reachable family of pairs over a variety S. Then for every $s \in S$ we have (after choosing a basis in $E(s)$) a completely reachable pair $(F(s),G(s))$.

The pair $(F(s),G(s))$ is unique modulo a choice of basis in $E(s)$ and hence defines a unique point of $\underline{M}_{m,n}$. Thus we find a continuous algebraic map $f_\Sigma : S \to \underline{M}_{m,n}$. This map f_Σ only depends on the isomorphism class of Σ. It turns out that we have defined a morphism of functors $\Phi: \mathfrak{F}_{m,n} \to \underline{Sch}_k(\ ,\underline{M}_{m,n})$, $\Phi(S)(\Sigma) = (f_\Sigma : S \to \underline{M}_{m,n})$. Note also that $\Phi(\mathrm{Spec}(k))$ is an isomorphism. Finally one can prove that every functor morphism $\Psi: \mathfrak{F}_{m,n} \to \underline{Sch}_k(\ ,\underline{M})$ into a representable functor factors uniquely through Φ, via a morphism $h: \underline{M}_{m,n} \to \underline{M}$. I.e. $\underline{M}_{m,n}$ is a coarse moduli scheme. (Cf. [2], [5] or [6] for a definition of this notion).

In fact, more is true: $\underline{M}_{m,n}$ is a fine moduli scheme, which by definition means that the functor morphism Φ above is an isomorphism of functors. Or in other words: there exists a universal completely reachable family Σ^u over $\underline{M}_{m,n}$ such that for every family Σ over a variety S there is a unique morphism $f: S \to \underline{M}_{m,n}$ such that $f^!\Sigma^u = \Sigma$. The next thing to do is to construct this universal family Σ^u.

5.2. Construction of the Universal Family Σ^u.

Let $V_\alpha \simeq \underline{A}^{mn}$ be the affine piece of $\underline{M}_{m,n}$ corresponding to the nice selection α. Over V_α we take the trivial bundle $E_\alpha = V_\alpha \times \underline{A}^n$. Let $\psi_\alpha : V_\alpha \to \underline{IS}_{cr}$ be the morphism defined in subsection 3.4. Write $\psi_\alpha(x) = (F_\alpha(x),G_\alpha(x))$. We now define the bundle endomorphism $F_\alpha : E_\alpha \to E_\alpha$ by the formula $F_\alpha(x,v) = (x,F_\alpha(x)v)$ and the sections $g_{1\alpha}, \ldots, g_{m\alpha} : V_\alpha \to E_\alpha$ are defined by $g_{i\alpha}(x) = (x$, i-th column of $G_\alpha(x))$.

We now construct the universal family Σ^u by patching together the partial families $(E_\alpha,F_\alpha,g_{1\alpha},\ldots,g_{m\alpha})$. This is done as follows. Let $E_{\alpha\beta} = E_\alpha|V_{\alpha\beta}$, $E_{\beta\alpha} = E_\beta|V_{\beta\alpha}$ and let $\phi_{\alpha\beta} : V_{\alpha\beta} \to V_{\beta\alpha}$ be the isomorphism constructed in 3.7 above. We now define the isomorphism $\tilde{\phi}_{\alpha\beta} : E_{\alpha\beta} \to E_{\beta\alpha}$ by the formula

$$\tilde{\phi}_{\alpha\beta}(x,v) = (\phi_{\alpha\beta}(x),\ (R\psi_\alpha(x))_\beta^{-1} v) \tag{5.2.1}$$

It is easy to check that these isomorphisms are compatible with the endomorphisms F_α, F_β and the sections $g_{i\alpha}, g_{i\beta}$, $i = 1, \ldots, m$, so that we find a family $\Sigma^u = (E^u,F^u,g_1^u,\ldots,g_m^u)$ such that $\Sigma^u|V_\alpha \simeq (E_\alpha,F_\alpha,g_{1\alpha},\ldots,g_{m\alpha})$ and hence such that the point of $\underline{M}_{m,n}$ corresponding to $\Sigma^u(s)$ is precisely s. I.e. $f_{\Sigma^u} : \underline{M}_{m,n} \to \underline{M}_{m,n}$, the morphism induced by the family Σ^u over $\underline{M}_{m,n}$ (cf. 5.1 above), is the identity morphism.

5.3. Theorem.

$\underline{M}_{m,n}$ is a fine moduli space with universal family Σ^u.

(For a proof cf. [2])

5.4. Remark.

Let E be the canonical n-bundle over the Grassmannian $\underline{G}_{n,(n+1)m}$, i.e. $E = \{(x,v) \in \underline{G}_{n,(n+1)m} \times \underline{A}^{(n+1)m} \mid v \in x\}$, where x is interpreted as an n-dimensional linear subspace of $\underline{A}^{(n+1)m}$. Let $\bar{R}$: $\underline{M}_{m,n} \to \underline{G}_{n,(n+1)m}$ be the embedding induced by the GL_n-invariant embedding R: $\underline{IS}_{cr} \to \underline{A}^{reg}_{n,(n+1)m}$. Then $\bar{R}^! E = E^u$ the underlying bundle of the universal family Σ^u over $\underline{M}_{m,n}$.

6. CANONICAL FORMS

In this section we discuss the existence and nonexistence of canonical forms.

6.1. Triviality of E^u and the Existence of Canonical Forms.

Suppose that E^u, the underlying bundle of the universal family Σ^u, were trivial; i.e. there is an isomorphism χ: $E^u \to \underline{M}_{m,n} \times \underline{A}^n$. Let e_i: $\underline{M}_{m,n} \to \underline{M}_{m,n} \times \underline{A}^n$ be the section $e_i(x) = (x,e_i)$ where e_i is the i-th unit (column) vector in $\underline{A}^n$. If there were such an isomorphism χ we would have a canonical basis, viz. $\{\chi^{-1}e_1(x),\ldots,\chi^{-1}e_n(x)\}$, in every fibre $E^u(x)$ of E^u which varies continuously with x. Let $(F_\chi(x),G_\chi(x))$ be the matrices corresponding to $\Sigma^u(x)$ with respect to this basis. Let π : $\underline{IS}_{cr} \to \underline{M}_{m,n}$ be the natural projection. Then

$$(F,G) \mapsto \pi(F,G) = x \mapsto (F_\chi(x),G_\chi(x))$$

would be a globally defined continuous algebraic canonical form on $\underline{IS}_{cr}$. Inversely, suppose there were a globally defined continuous algebraic canonical form on $\underline{IS}_{cr}$, say $(F,G) \mapsto (\bar{F},\bar{G})$. We can now define a family Σ^c over $\underline{M}_{m,n}$ as follows, $\Sigma^c = (E^c,F^c,g_1^c,\ldots,g_m^c)$, where $E^c = \underline{M}_{m,n} \times \underline{A}^n$, $F^c(x,v) = (x,\bar{F}_x v)$, $g_i^c(x) = (x,\text{i-th column of } \bar{G}_x)$ where (F_x,G_x) is any pair such that $\pi(F_x,G_x) = x$. Because Σ^u is universal there is a unique morphism f: $\underline{M}_{m,n} \to \underline{M}_{m,n}$ such that $f^!\Sigma^u = \Sigma^c$. But because $\pi(\bar{F}_x,\bar{G}_x) = x$, f is the identity morphism (cf. section 5.1), which would imply that $E^c \simeq E^u$, i.e. that E^u is trivial.
We have therefore proved

Theorem. The existence of a globally defined, continuous algebraic canonical form for $\underline{IS}_{cr}$ is equivalent to the triviality of E^u, the underlying bundle of the universal family Σ^u over $\underline{M}_{m,n}$.

6.2. Nonexistence of Canonical Forms for $\underline{IS}_{cr}$.

Let i: $\underline{G}_{n,(n+1)m} \to \underline{P}^N$ be the canonical embedding of the Grassmannian into projective space (cf. section 2). Let L be the canonical line bundle over $\underline{P}^N$, i.e. L(x) = the affine line which x represents. Let E be the canonical n-bundle over $\underline{G}_{n,(n+1)m}$. Then

$$(6.2.1) \qquad \overset{n}{\wedge} E \simeq i^{!}L$$

where $\overset{n}{\wedge}$ denotes the n-th exterior product.

Let $\bar{R}$: $\underline{M}_{m,n} \to \underline{G}_{n,(n+1)m}$ be the embedding induced by R. By 5.4 we have that $\bar{R}^{!}E = E^{u}$. Hence $\overset{n}{\wedge}E^{u} \simeq \bar{R}^{!}i^{!}L$, which is a very ample line bundle. Hence the sections of $\overset{n}{\wedge}E^{u} \to \underline{M}_{m,n}$ separate the points of $\underline{M}_{m,n}$ (cf. [1] Ch.II). Hence if E^{u} were trivial, then $\overset{n}{\wedge}E^{u}$ would be the trivial line bundle and sections of the trivial line bundle correspond bijectively to morphisms $\underline{M}_{m,n} \to \underline{A}^{1}$. I.e. if E^{u} were trivial then the morphisms $\underline{M}_{m,n} \to \underline{A}^{1}$ would separate points. It is easily seen (cf. [2] for details) from the affine pieces + patching data description of $\underline{M}_{m,n}$ that there are not enough morphisms $\underline{M}_{m,n} \to \underline{A}^{1}$ to do this when $m \geq 2$. Thus E^{u} is not trivial and there does not exist a continuous canonical form for $\underline{IS}_{cr}$ if $m \geq 2$. The nontriviality of E^{u} justifies the definition of family which we have used.

6.3. Nonexistence of Canonical Forms (continued)

There is an easier way to prove the nonexistence of canonical forms for $\underline{IS}_{cr}$. Suppose there existed a continuous canonical form fro $\underline{IS}_{cr}$, say $(F,G) \to (\bar{F},\bar{G})$ then we have n^2 morphisms a_{ij}: $\underline{M}_{m,n} \to A^{1}$ defined as follows

$a_{ij}(x)$ = (i,j)-th entry of $(\bar{F}_x,\bar{G}_x)$, where (F_x,G_x) is any pair such that $\pi(F_x,G_x) = x$. These morphisms would separate the points of $\underline{M}_{m,n}$. But this cannot be done by morphisms to $\underline{A}^{1}$ if $m \geq 2$, hence a continuous canonical form does not exist for $\underline{IS}_{cr}$ if $m \geq 2$. A fortiori there does not exist a continuous canonical form for $\underline{IS}$ if $m \geq 2$.

There is a GL_n-invariant embedding $\underline{IS}_{cr} \to \underline{DS}_{cr}$, viz. $(F,G) \to (F,G,0)$ where 0 denotes an appropriate zero matrix. Hence there also does not exist a continuous canonical form for $\underline{DS}_{cr}$ and $\underline{DS}$ if $m \geq 2$. If $m = 1$ there does exist a global continuous canonical form for $\underline{IS}_{cr}$ and $\underline{DS}_{cr}$. Summing up, we have

6.4. Theorem.

If $m = 1$, there is a globally defined continuous algebraic canonical form for $\underline{IS}_{cr}$ and $\underline{DS}_{cr}$.

If $m \geq 2$, there is no globally defined canonical form for $\underline{IS}$, $\underline{IS}_{cr}$, $\underline{DS}$, $\underline{DS}_{cr}$.

7. CONCLUDING REMARKS AND OPEN QUESTIONS.

7.1. The moduli space $\underline{M}_{m,n}$ is not complete (for all m,n); i.e. it is not a closed subvariety of $\underline{P}^{N}$ (or $\underline{G}_{n,(n+1)m}$). Let $\bar{\underline{M}}_{m,n}$ be its closure. E.g. $F_1 = t^{-1}e_1$

$F_i = ie_i$, $i = 2, \ldots, n$; $G_1 = t^n e_1 + e_2 + \ldots + e_n$, $G_i = 0$, $i = 2, \ldots, m$. Then

$$R(F,G) = \left(\begin{array}{cccc|cccc|c|cccc} t^n & 0 & \ldots & 0 & t^{n-1} & 0 & \ldots & 0 & & 1 & 0 & \ldots & 0 \\ 1 & 0 & \ldots & 0 & 2 & 0 & \ldots & 0 & & 2^n & 0 & \ldots & 0 \\ \vdots & \vdots & & \vdots & \vdots & \vdots & & \vdots & \cdots & \vdots & \vdots & & \vdots \\ 1 & 0 & \ldots & 0 & n & 0 & \ldots & 0 & & n^n & 0 & \ldots & 0 \end{array}\right)$$

which as t goes to 0 specializes to an element of $\underline{A}^{reg}_{n,(n+1)m}$ which is not of the form $R(F',G')$ for any $(F',G') \in \underline{IS}_{cr}$ (in view of lemma 3.3) and which hence gives rise to a point in $\bar{\underline{M}}_{m,n}$ which is not in $\underline{M}_{m,n}$.
The question arises whether it is possible to interpret the missing points, i.e. the points of $\bar{\underline{M}}_{m,n} \setminus \underline{M}_{m,n}$, as (generalized?) dynamical systems?

7.2. The group GL_m of basis changes in input space acts on $\underline{M}_{m,n}$. If $m < n$, then there is an open dense subset U of $\underline{M}_{m,n}$ such that the stabilizer of this action is GL_1 (diagonally embedded in GL_m) for all $x \in U$. (So what we really have is an action of PGL_{m-1} on $\underline{M}_{m,n}$). By general theorems (cf. [5]) we then know that a geometric quotient V/GL_m exists for a suitable dense open subset V of $\underline{M}_{m,n}$.
Problem: calculate the maximal V and describe the quotient V/GL_m. In particular (in view of canonical forms) one would like to know whether the points of V/GL_m can be separated by morphisms to $\underline{A}^1$.

7.3. Let V_α be the subvariety of $\underline{M}_{m,n}$ corresponding to the nice selection α. There is a global continuous algebraic canonical form for $\pi^{-1}V_\alpha$, where $\pi : \underline{IS}_{cr} \to \underline{M}_{m,n}$ is the natural projection, viz $(F,G) \to \psi_\alpha\pi(F,G)$, where ψ_α is the morphism defined in 3.4 above. The V_α are also maximal subvarieties for which a canonical form exists for $\pi^{-1}V_\alpha$. However, not every subvariety V of $\underline{M}_{m,n}$ for which a canonical form exists for $\pi^{-1}V$, is contained in one of the V_α, α a nice selection. E.g. let β be a not nice selection and $W_\beta = \{(F,G) \in \underline{IS}_{cr} \mid \det R(F,G)_\beta \neq 0\}$. Then there is a canonical form on W_β.
(NB. W_β can be nonempty as the family)

$$F_t = \begin{pmatrix} 1 & -t \\ 0 & 1 \end{pmatrix}, \quad G_t = \begin{pmatrix} 1 & t \\ 1 & 1 \end{pmatrix}$$

shows. This family also shows that W_β need not be contained in any of the V_α, α a nice selection. The following could be true, let λ be a linear form in the expressions $\det(R(F,G)_\beta)$, where β runs through all selections. Let W_λ be the subvariety of $\underline{IS}_{cr}$ where λ is $\neq 0$. If $V \subset \underline{IS}_{cr}$ is a subvariety for which a canonical form exists, then V is contained in one of the W_λ.

7.4. What kind of morphisms between the various $\underline{M}_{m,n}$ does the partial realization algorithm induce ? This could be interesting also because of 7.1.

7.5. We have seen in theorem 6.4 that there is no canonical form on $\underline{DS}_{cr}$ or $\underline{DS}$ if $m \geq 2$. Let $\underline{DS}_{cr,co}$ be the subspace of $\underline{DS}$ consisting of completely reachable and completely observable linear dynamical systems. The nonexistence of a canonical form for $\underline{IS}_{cr}$ does not imply the nonexistence of a canonical form for $\underline{DS}_{cr,co}$, and, a priori, a canonical form for $\underline{DS}_{cr,co}$ could exist also for $m \geq 2$. Indeed, such a canonical form does exist if $p = 1$ (p is the number of outputs), n and m arbitrary. The geometric quotient $\underline{DS}_{cr,co}/GL_n$ does exist, cf. also section 3.9 above, but in this case there also exists an embedding of $\underline{DS}_{cr,co}/GL_n$ in an affine space, so that the argument of 6.3 above cannot be used to prove nonexistence of canonical forms. Possibly one shall have to use results like 6.1 to decide whether $\underline{DS}_{cr,co}$ admits a global continuous algebraic canonical form or not.

REFERENCES.

1. A. Grothendieck, J. Dieudonné. Eléments de la géometrie algébrique. Ch. I, II,..., Publ. Math. I.H.E.S. 4, 8, ..., 1960, 1961,...
2. M. Hazewinkel, R.E. Kalman. Moduli and Canonical Forms for Linear Dynamical Systems. Report 7504, Econometric Inst., Erasmus University of Rotterdam, 1975.
3. D. Husemoller. Theory of Fibre Bundles. McGraw-Hill, 1966
4. S. Kleiman, D. Laksov. "Schubert Calculus". American Math. Monthly 79, 1972, 1061-1082.
5. D. Mumford. Geometric Invariant Theory. Springer 1965.
6. D. Mumford, K. Suominen. Introduction to the Theory of Moduli. In: F. Oort (ed). Algebraic Geometry. Proceedings of the Fifth Nordic Summer School in Mathematics. Noordhoff, 1972.

SYSTEM INVARIANTS UNDER FEEDBACK AND CASCADE CONTROL*

A. S. Morse
Dept. of Engineering and Applied Science
Yale University
New Haven, Connecticut 06520/USA

INTRODUCTION

In this paper we study the effects of feedback and cascade control on the structural properties of linear systems and their transfer matrices. Basic to the study is the fact that the set of all transfer functions with denominators in a prescribed set S, is a principal ideal domain $\mathbb{R}_S$ (§1). This observation enables us to immediately conclude that the orbit of a $\mathbb{R}_S$-transfer matrix H under a suitably defined group of transformations corresponding to dynamic cascade control, is uniquely characterized by the free $\mathbb{R}_S$-module generated by the columns of H(§2). A minor modification of Hermite's normal form for nonsingular matrices over a principal ideal domain [1], then yields a canonical transfer matrix representing the orbit of H. These results and the approach leading to them, clarify and extend previous work [2].

In §3 we examine some of the relationships which exist between dynamically equivalent transfer matrices (i.e., transfer matrices within the same orbit). We introduce the concept of an irreducible transfer matrix and we show that each transfer matrix H is dynamically equivalent to one which is irreducible; H itself turns out to be irreducible just in case no dynamically equivalent transfer matrix has lower McMillan degree. Two irreducible transfer matrices prove to be dynamically equivalent if and only if their respective canonical realizations are feedback equivalent in the sense of [3]. These results are used in §4 to prove that the orbit of any monic, canonical system (C,A,B) under the group of state-coordinate, state-feedback and input-coordinate transformations, is uniquely characterized by a free module determined by (C,A,B).

1. PRELIMINARIES

One version of the model following problem treated in [4] is to design a cascade control system with transfer matrix Q_{mxr}, for a linear process with transfer matrix H_{pxm} so that the transfer matrix HQ of the resulting controlled process coincides with the transfer matrix M_{pxr} of a prespecified linear model. Since

$$HQ = M \tag{1}$$

is a linear equation over the rational functions, the only problem in deciding if a

* This research was supported by the U. S. Air Force Office of Scientific Research under Grant 72-2211.

solution Q exists, stems from the requirement that Q be realizable as a dynamical system. However, if (1) is viewed as a linear equation over a suitably defined ring $\mathbb{R}_S$, then this requirement is automatically satisfied by any solution Q over $\mathbb{R}_S$. In the sequel we define $\mathbb{R}_S$ and discuss some of its properties. We return to the model following problem at the end of the section.

Let $\mathbb{R}[\lambda]$ denote the ring of polynomials over the real field $\mathbb{R}$, and write $\mathbb{R}(\lambda)$ for the field of fractions of $\mathbb{R}[\lambda]$. Rational functions $\alpha/\beta \in \mathbb{R}(\lambda)$ will always be represented in reduced form; i.e., if $\alpha \neq 0$, then α and β are coprime. By a transfer function we of course mean either the zero element in $\mathbb{R}(\lambda)$ or any nonzero element α/β with deg. $(\alpha) \leq$ deg. (β); transfer functions are then rational functions admitting dynamical realizations.

In certain applications it is useful to consider only those transfer functions with poles in some prescribed region of the complex plane. It is convenient to characterize such transfer functions in the following way. Let S be a multiplicative subset of $\mathbb{R}[\lambda]$ consisting of the polynomial 1 together with all monic polynomials of positive degree generated by a (possibly infinite) set of monic prime factors; at least one prime factor must be of degree one[†]. It is easy to verify that the set of all transfer functions α/β with $\beta \in S$, is closed under addition and multiplication, includes 1, and is therefore a subring of $\mathbb{R}(\lambda)$. This ring, denoted by $\mathbb{R}_S$, is the ring of real transfer functions with denominators in S.

Examples:

1. If $\alpha \in \mathbb{R}[\lambda]$ is a polynomial of positive degree and S is the set of all monic polynomials coprime with α, then $\mathbb{R}_S$ is the ring of transfer functions with denominators coprime with α. In this case we say that α (or any polynomial dividing α) is coprime with S.
2. If Λ is a symmetric set of complex numbers and S is the set of all monic polynomials with all roots in Λ, then $\mathbb{R}_S$ is the ring of transfer functions with all poles in Λ. If Λ is the set of points in the open left-half complex plane, then $\mathbb{R}_S$ is the ring of (continuous-time) stable transfer functions. If Λ is the set of all complex numbers, then $\mathbb{R}_S$ is the ring of all transfer functions.

We wish to describe some of the algebraic properties of $\mathbb{R}_S$. For this, let $\#: \mathbb{R}_S \to$ {non-negative integers} denote the function defined by $\#(\alpha/\beta) =$ deg. (β) - deg. (α) and $\#(0) = \infty$; $\#(\alpha/\beta)$ is called the size of α/β. Next let $\nabla: \mathbb{R}_S \to \mathbb{R}[\lambda]$ denote the function defined by $\nabla(0) = 0$ and $\nabla(\alpha/\beta) = \delta$, where δ is the unique monic polynomial of greatest degree which is coprime with S and which divides the nonzero polynomial α. It is easy to verify that $\#((\alpha/\beta)(\gamma/\delta)) = \#(\alpha/\beta) + \#(\gamma/\delta)$ and that $\nabla((\alpha/\beta)(\gamma/\delta)) = \nabla(\alpha/\beta)\nabla(\gamma/\delta)$ for all α/β, $\gamma/\delta \in \mathbb{R}_S$.

The invertibles of $\mathbb{R}_S$ can now be described as all $\alpha/\beta \in \mathbb{R}_S$ for which $\#(\alpha/\beta) = 0$ and $\nabla(\alpha/\beta) = 1$. It is straightforward to show that if α/β and γ/δ are elements of

---

† This last requirement insures that for each nonnegative integer n, there is at least one polynomial of degree n in S.

$\mathbb{R}_S$ then α/β divides γ/δ just when $\#(\alpha/\beta) \leq \#(\gamma/\delta)$ and $\nabla(\alpha/\beta)$ divides $\nabla(\gamma/\delta)$ in $\mathbb{R}[\lambda]$. From this it follows that μ/ρ is a greatest common divisor (gcd) of α/β and γ/δ just in case $\nabla(\mu/\rho) = \mathrm{GCD}(\nabla(\alpha/\beta),\nabla(\gamma/\delta))^{\dagger}$ and $\#(\mu/\rho) = \min.\ \{\#(\alpha/\beta),\#(\gamma/\delta)$.

For our purposes, the most important property of $\mathbb{R}_S$ is as follows.

Proposition 1: *$\mathbb{R}_S$ is a principal ideal domain.*

Remark: If S is the set of all monic polynomials in $\mathbb{R}[\lambda]$, then $\mathbb{R}_S$ turns out to be isomorphic to the ring of rational power series over $\mathbb{R}$[††]. Since the latter ring is known to be a principal ideal domain, the proof of Proposition 1 is immediate for this case. While it appears likely that Proposition 1 can be proved for arbitrary S in a similar manner, a direct argument seems more appropriate here.

Lemma 1: *Let α_1/β_1 and α_2/β_2 be nonzero elements of $\mathbb{R}_S$ with gcd α_3/β_3. There exist elements γ_1/δ_1 and γ_2/δ_2 in $\mathbb{R}_S$ such that*

$$(\gamma_1/\delta_1)(\alpha_1/\beta_1) + (\gamma_2/\delta_2)(\alpha_2/\beta_2) = (\alpha_3/\beta_3) \tag{2}$$

Proof: For $i \in \{1,2,3\}$, write $m_i = \#(\alpha_i/\beta_i)$, $\rho_i = \nabla(\alpha_i/\beta_i)$ and $n_i = \deg.\ (\rho_i)$. Assume that $m_1 \leq m_2$; then $m_3 = m_1$. Since S contains a degree one polynomial π, it is possible to find invertibles μ_i/σ_i, $i \in \{1,2,3\}$ such that

$$\alpha_i/\beta_i = (\sigma_i/\mu_i)(\rho_i/(\pi^{(m_i+n_i)})), \quad i \in \{1,2,3\} \tag{3}$$

Furthermore, since $\rho_3 = \mathrm{GCD}(\rho_1,\rho_2)$ there exist polynomials θ_1,θ_2 such that

$$\rho_1\theta_1 + \rho_2\theta_2 = \rho_3(\pi^{(n_1+ n_2- 2n_3+ m_2- m_1)}) \tag{4}$$

By selecting θ_2 of minimal degree modulo the polynomial ρ_1/ρ_3, it is possible to insure that $\deg.\ (\theta_2) < \deg.\ (\rho_1/\rho_3) = (n_1- n_3)$. Thus $\theta_2/(\pi^{(n_1-n_3)}) \in \mathbb{R}_S$ and if we define

$$\gamma_2/\delta_2 \equiv (\sigma_3/\mu_3)(\mu_2/\sigma_2)(\theta_2/(\pi^{(n_1- n_3)}) \tag{5}$$

then $\gamma_2/\delta_2 \in \mathbb{R}_S$ as well.

Since $\deg.\ (\theta_2) < (n_1- n_3)$, there follows $\deg.\ (\theta_2\rho_2) < n_1- n_3+ n_2 \leq n_1+ n_2- n_3 + m_2- m_1 = \deg.\ (\rho_3\pi^{(n_1+ n_2- 2n_3+ m_2- m_1)})$. This and (4) imply that $\deg.\ (\rho_1\theta_1) = \deg.\ (\rho_3\pi^{(n_1+ n_2- 2n_3+ m_2- m_1)})$ and thus that $\deg.\ (\theta_1) = (n_2- n_3+ m_2- m_1)$. Thus $\theta_1/(\pi^{(n_2- n_3+ m_2- m_1)}) \in \mathbb{R}_S$ and if we set

$$\gamma_1/\delta_1 = (\sigma_3/\mu_3)(\mu_1/\sigma_1)(\theta_1/\pi^{(n_2- n_3+ m_2- m_1)}) \tag{6}$$

then $\gamma_1/\delta_1 \in \mathbb{R}_S$ as well.

By now multiplying both sides of (4) by $(\sigma_3/\mu_3)(1/\pi^{(n_1+ n_2- n_3+ m_2)})$, rearranging terms, and making the substitutions (3),(5),(6), one obtains (2) which is the desired result. □

---

† If σ and ω are polynomials in $\mathbb{R}[\lambda]$, $\mathrm{GCD}(\sigma,\omega)$ is their unique monic greatest common divisor.

†† This observation is due to E. D. Sontag.

Proof of Proposition 1: Since $\mathbb{R}_S$ is clearly an integral domain, it's enough to show that any nonzero ideal I in $\mathbb{R}_S$ is generated by a single element. Let $x \in I$ be any nonzero element for which degree $\nabla(x)$ is as small as possible. Let C be the class of all elements $w \in I$ with degree $\nabla(w)$ = degree $\nabla(x)$. Let $g \in C$ be an element of least size. We claim that g generates I. To show this, let $y \in I$ be arbitrary and let z be a gcd of g and y; by Lemma 1, $z \in I$. Thus degree $\nabla(g) \leq$ degree $\nabla(z)$, since $g \in C$. But z divides g implying $\nabla(z)$ divides $\nabla(g)$ in $R[\lambda]$. It follows that

$$\nabla(z) = \nabla(g) \tag{7}$$

Since $z \in I$, (7) implies $z \in C$. Thus $\#(g) \leq \#(z)$; but z divides g implying $\#(z) \leq \#(g)$; therefore $\#(g) = \#(z)$. From this and (7) it follows that g divides z. But z divides y implying that g divides y which yields the desired result. □

Now consider equation (1) which appears at the beginning of this section. If the denominator of each entry in H,Q and M is an element of S or if S is the set of all monic polynomials in $\mathbb{R}[\lambda]$, then H,Q and M are $\mathbb{R}_S$-matrices and the model following problem amounts to solving (1) over a principal ideal domain. Explicit conditions for a solution Q to exist are well-known. For example, if $\mathbb{R}_S^p$ is the free $\mathbb{R}_S$-module generated by the unit p-vectors, and if Im H (resp. Im M) is the free submodule of $\mathbb{R}_S^p$ generated by the columns of H (resp. M), then (1) is solvable if and only if Im M $\subset$ Im H. Alternatively, (1) will have a solution Q just in case the Smith form [1] of the partitioned matrix [H,M] equals the Smith form of $[H, 0_{pxr}]$. Classical algorithms for computing Q, if it exists, can be found in [1].

Our objective here has not been so much to solve a model following problem as to illustrate just how easily such a problem can be solved once it is recognized that what's really involved is the solution of a linear equation over a principal ideal domain. In the next section we use a similar approach to quickly determine when two transfer matrices are equivalent under dynamic compensation.

2. DYNAMIC EQUIVALENCE

Let S remain fixed and write $\underline{C}^{pxm}$ for the class of all pxm matrices over $\mathbb{R}_S$. Generalizing slightly a definition in [2], we say that two matrices H_1 and H_2 in $\underline{C}^{pxm}$ are dynamically equivalent (mod S) if there exist $\mathbb{R}_S$ matrices Q_1 and Q_2 such that

$$\left.\begin{aligned} H_1Q_1 &= H_2 \\ H_2Q_2 &= H_1 \end{aligned}\right\} \tag{8}$$

Dynamic equivalence is clearly an equivalence relation on $\underline{C}^{pxm}$.

In the sequel we continue to view H_i as the transfer matrix of a fixed process, Q_i as the transfer matrix of a dynamic control system and H_iQ_i as the transfer matrix of the system which results when control system i is cascaded with process i. Thus H_1 and H_2 are dynamically equivalent just in case process 1 can be made to follow process 2 with cascade dynamic control, and visa versa.

Examples:

1. If S is the set of all monic polynomials in $\mathbb{R}[\lambda]$, then $\underline{C}^{p \times m}$ is the class of all $p \times m$ transfer matrices. In this case the preceding definition of dynamic equivalence coincides with the one in [2].
2. If S is the set of all stable monic polynomials in $\mathbb{R}[\lambda]$, then two stable transfer matrices are dynamically equivalent provided (8) holds with Q_1 and Q_2, the transfer matrices of stable control systems.

The existence of matrices Q_i which satisfy (8) is equivalent to the conditions $\text{Im } H_1 \subset \text{Im } H_2$ and $\text{Im } H_2 \subset \text{Im } H_1$. It is therefore clear that

Theorem 1: *H_1 and H_2 in $\underline{C}^{p \times m}$ are dynamically equivalent if and only if*

$$\text{Im } H_1 = \text{Im } H_2$$

It is also clear that

Corollary 1: *The function which assigns to each $H \in \underline{C}^{p \times m}$ its image in the class of all submodules of $\mathbb{R}_S^p$, is a complete invariant under dynamic equivalence.*

Next we briefly outline how one might construct a canonical representative for the dynamic equivalence class of a given matrix $H \in \underline{C}^{p \times m}$. For this first observe as a direct consequence of Theorem 1 that

Lemma 2: *H_1 and H_2 in $\underline{C}^{p \times m}$ are dynamically equivalent if and only if there exists an invertible $\mathbb{R}_S$-matrix Q such that $H_1 Q = H_2$.*

Lemma 2 implies that H_1 and H_2 are dynamically equivalent just in case H_1 and H_2 are right equivalent $\mathbb{R}_S$-matrices. This observation allows us to use a modified version of the Hermite normal form [1] of a nonsingular $\mathbb{R}_S$-matrix as a canonical form for $H \in \underline{C}^{p \times m}$. For this we will require

1. *A complete system of nonassociates of $\mathbb{R}_S$*: Let π be any monic polynomial of degree one in S; and write N_π for the set of all nonzero $\mathbb{R}_S$-elements of the form δ/π^n where $n \geq 0$ and δ is any monic polynomial, coprime with S. It is easy to verify that N_π has the required property; i.e., for each nonzero $\alpha/\beta \in \mathbb{R}_S$ there is exactly one element in N_π differing from α/β by at most an invertible element of $\mathbb{R}_S$.
2. *A complete residue system for each element in N_π*: Let

$$R_n = \begin{cases} \{\gamma/\pi^m : \ \gamma/\pi^m \in \mathbb{R}_S, \ m < n\} & \text{if } n > 0 \\ \{0\} & \text{if } n = 0 \end{cases}$$

The following lemma implies that R_n is a complete residue system for $\delta/\pi^n \in N_\pi$; i.e., for each nonzero element $x \in \mathbb{R}_S$ and each element $\delta/\pi^n \in N_\pi$ there exist unique elements $y \in R_n$ and z in the ideal generated by δ/π^n such that $x = y + z$.

Lemma 3: *For each $\alpha/\beta \in \mathbb{R}_S$ and each $\delta/\pi^n \in N_\pi$ there exist $\sigma_1/\sigma_2 \in \mathbb{R}_S$ and a unique element $\gamma/\pi^m \in R_n$ such that*

$$\alpha/\beta = (\sigma_1/\sigma_2)(\delta/\pi^n) + \gamma/\pi^m \tag{9}$$

Proof: The lemma is clearly true for n = 0; therefore assume n > 0. Since δ and β are necessarily coprime, there exist polynomials ρ and θ such that

$$\alpha\pi^{n-1} = \rho\beta + \delta\theta \tag{10}$$

Select θ so that deg. (θ) < deg. (β). Therefore if we define

$$\sigma_1/\sigma_2 = (\pi\theta)/\beta \tag{11}$$

then $\sigma_1/\sigma_2 \in \mathbb{R}_S$.

Since δ/π^n and α/β are in $\mathbb{R}_S$, there follows deg. $(\delta) \leqslant n$ and deg. $(\alpha) \leq$ deg. (β). Thus deg. $(\alpha\pi^{n-1}) < n +$ deg. (β) and deg. $(\delta\theta) \leqslant n +$ deg. $(\theta) < n +$ deg. (β). These relations and (10) imply deg. $(\rho\beta) < n +$ deg. (β) and therefore deg. $(\rho) < n$. It follows that if γ/π^m is the reduced form of ρ/π^{n-1}, i.e.,

$$\gamma/\pi^m = \rho/\pi^{n-1}, \tag{12}$$

then $\gamma/\pi^m \in R_n$. By dividing both sides of (10) by $\pi^{n-1}\beta$, rearranging terms and then making the substitutions (11) and (12), one obtains (9) which is the desired result.

Now suppose that

$$\alpha/\beta = (\bar{\sigma}_1/\bar{\sigma}_2)(\delta/\pi^n) + \bar{\gamma}/\pi^{\bar{m}} \tag{13}$$

for some $\bar{\sigma}_1/\bar{\sigma}_2 \in R_S$ and $\bar{\gamma}/\pi^{\bar{m}} \in R_n$. Without loss of generality, assume $\bar{m} \geqslant m$. Then subtraction of (13) from (9) yields

$$0 = (\sigma_1/\sigma_2 - \bar{\sigma}_1/\bar{\sigma}_2)(\delta/\pi^n) + \mu/\pi^{\bar{m}} \tag{14}$$

where

$$\mu = \gamma\pi^{(\bar{m} - m)} - \bar{\gamma} \tag{15}$$

Noting that $n > \bar{m}$, (14) can be rewritten as $(\sigma_1/\sigma_2 - \bar{\sigma}_1/\bar{\sigma}_2)\delta = -\pi^{(n - \bar{m})}\mu$. Since δ is coprime with $\sigma_2, \bar{\sigma}_2$ and π, δ must divide μ. Thus

$$(\sigma_1/\sigma_2 - \bar{\sigma}_1/\bar{\sigma}_2) = -\pi^{(n - \bar{m})}\eta \tag{16}$$

where η is the quotient of μ divided by δ. Since the left side of (16) is in $\mathbb{R}_S$ and since $n > \bar{m}$, (16) can be true only if $\eta = 0$. Thus $\mu = 0$ and from (15)

$$\gamma/\pi^m = \bar{\gamma}/\pi^{\bar{m}}$$

which proves that the representation is unique. □

It is now possible to define a set of representative matrices, one matrix for each dynamic equivalence class in $\underline{C}^{pxm}$. We call $H^* \in \underline{C}^{pxm}$ a canonical form under dynamic equivalence provided

i. $H^* = [\bar{H}_{pxr}, 0_{px(m-r)}]$ where $r = \text{rank } H^*$

ii. the nonsingular matrix H^{**} determined by the first r linearly independent rows of $\bar{H}$ is lower triangular with ith diagonal entry $\delta_i/\pi^{n_i} \in N_\pi$, and ijth element in R_{n_i} for $j < i$.

Let $\underline{C}_*^{pxm}$ denote the set of all matrices in $\underline{C}^{pxm}$ with these properties.

Each matrix $H \in \underline{C}^{pxm}$ determines a dynamically equivalent matrix $H \in \underline{C}_*^{pxm}$ according to the following procedure.

1. First construct an invertible $\mathbb{R}_S$-matrix Q so that
$$HQ = [\tilde{H}_{pxr}, 0_{px(m-r)}] \text{ where } r = \text{rank } H$$
2. Let $\hat{H}$ denote the nonsingular matrix determined by the first r linearly independent rows of $\tilde{H}$. Construct an invertible $\mathbb{R}_S$-matrix T so that $H^{**} \equiv \hat{H}T$ has the general structure described in ii. above. The matrix H^{**}, known as the Hermite normal form of $\hat{H}$, is uniquely determined by $\hat{H}$ and is the same for all matrices right equivalent to $\hat{H}$ [1].
3. Set $P = Q(\text{block diag. } [T, I_{(m-r)x(m-r)}])$ and define $H^* = HP$.

It is clear that the above procedure defines a matrix $H^* \in \underline{C}_*^{pxm}$. Furthermore, since P is invertible, Lemma 2 insures that H and H^* are dynamically equivalent. The actual steps involved in constructing T and Q can be found in [1].

Note that the rank r and the row indices of the first r linearly independent rows of $H \in \underline{C}^{pxm}$ are the same for all matrices which are right equivalent and therefore dynamically equivalent to H. Using this fact together with the uniqueness of the Hermite normal form, it is quite straightforward to verify that for each $H \in \underline{C}^{pxm}$ there is exactly one $H^* \in \underline{C}_*^{pxm}$, that H^* is uniquely determined by any matrix in the dynamic equivalence class of H, and thus that $\underline{C}_*^{pxm}$ is a set of canonical forms relative to dynamic equivalence.

Example 1, [2]:

If S = all monic polynomials in $\mathbb{R}[\lambda]$, $\pi = \lambda$ and
$$H = \begin{bmatrix} 1/(\lambda+1) & 1/(\lambda+2) \\ 1/(\lambda+3) & 1/(\lambda+4) \end{bmatrix} \quad \text{then } H^* = \begin{bmatrix} 1/\lambda & 0 \\ (\lambda-2)/\lambda^2 & 1/\lambda^3 \end{bmatrix}$$

Example 2:

If S and π are as above and
$$H = \begin{bmatrix} 1/(\lambda+1) & 1/(\lambda+2) \\ 1/(\lambda+3) & 1/(\lambda+5) \end{bmatrix} \quad \text{then } H^* = \begin{bmatrix} 1/\lambda & 0 \\ 1/\lambda & 1/\lambda^2 \end{bmatrix}$$
However, if S = all monic polynomials with roots in the open left-half complex plane, and $\pi = \lambda + a$, for fixed positive $a \in \mathbb{R}$, then
$$H^* = \begin{bmatrix} 1/(\lambda+a) & 0 \\ (\lambda+(a-1)/2)/(\lambda+a)^2 & (\lambda-1)/(\lambda+a)^3 \end{bmatrix}$$
Remark: If S is as in example 1, then the Smith form of an $\mathbb{R}_S$-matrix H admits the structure
$$\begin{bmatrix} D_{rxr} & 0 \\ 0 & 0 \end{bmatrix}_{pxm}$$
where r = rank H, $D = \text{diag. } [1/\pi^{n_1}, 1/\pi^{n_2}, \ldots, 1/\pi^{n_r}]$ and $n_1 \geq n_2 \geq \ldots \geq n_r$. Suppose H is a proper rational matrix with canonical realization (C,A,B); then the

n_i are all positive. In this case it can easily be shown that the n_i coincide with the controllability indices of the prime canonical subsystem which appears in the $\mathfrak{C}^*$ canonical form of (C,A,B) discussed in [3].

3. IRREDUCIBLE TRANSFER MATRICES

We now examine in greater detail some of the relationships which exist between dynamically equivalent transfer matrices. Our principal objective will be to characterize those transfer matrices within a specified dynamic equivalence class which have least McMillan degree. It will be shown that canonical (i.e., controllable and observable) realizations of such transfer matrices are related to one another by transformations of the state-feedback type.

In order to keep notation as uncluttered as possible, we deal exclusively with process transfer matrices which are proper.* All results which follow can be easily extended to arbitrary $\mathbb{R}_S$-matrices by using the concepts developed in [5].

Let $\underline{P}^{p\times m}$ denote the class of all proper transfer matrices in $\underline{C}^{p\times m}$. Since proper matrices remain proper under cascade control,** the definition of dynamic equivalence can be restricted to $\underline{P}^{p\times m}$ and all preceding results hold for this case.

Let A,B, and C denote the maps or parameter matrices of the m-input, p-output, controllable linear system $\dot{x} = Ax + Bu$, $y = Cx$; assume that the characteristic polynomial (c.p.) of A is in S, thereby insuring that the transfer matrix of (C,A,B) is in $\underline{P}^{p\times m}$ {cf., Lemma 5}. If $\underline{F}_S$ denotes the (nonempty) class of all state-feedback maps F such that c.p. $(A + BF) \in S$, then the transfer matrix of $(C, A + BF, B)$ is in $\underline{P}^{p\times m}$ for all $F \in \underline{F}_S$. For each $F \in \underline{F}_S$, let V_F denote the unobservable space of the pair $(C, A + BF)$; i.e., V_F is the largest $(A + BF)$-invariant subspace contained in kernel C.

Lemma 4: The class of all subspaces V_F, $F \in \underline{F}_S$, contains a unique largest element $V^*(C,A,B)$.

This lemma has been constructively proved in [6] for the case when S is the set of all stable, monic polynomials in $\mathbb{R}[\lambda]$. The slightly more general assertion made here can be proved in essentially the same way.

A transfer matrix $H \in \underline{P}^{p\times m}$ is called irreducible just in case it admits a controllable realization (C,A,B) for which $V^*(C,A,B) = 0$†. Irreducible transfer matrices have a number of interesting properties, some of which will now be described.

Proposition 2: Each transfer matrix in $\underline{P}^{p\times m}$ is dynamically equivalent to an irreducible transfer matrix.

---

* A transfer matrix is proper if its entries are proper rational functions; i.e., rational functions of positive size.

** The transfer matrix of a cascade control system need not be proper here, but it must be a $\mathbb{R}_S$-matrix.

† Irreducible transfer matrices should not be confused with irreducible (i.e., canonical) realizations of transfer matrices.

The constructive proof of this proposition depends on the following lemma.

Lemma 5: Let A,B,C and F be fixed $\mathbb{R}$-matrices.

i. If c.p. (A) ε S, then $C(\lambda I - A)^{-1}B$ is an $\mathbb{R}_S$-matrix

ii. If (C,A,B) is a canonical system and $C(\lambda I - A)^{-1}B$ is an $\mathbb{R}_S$-matrix, then c.p. (A) ε S.

iii. If c.p. (A) ε S and c.p. (A + BF) ε S, then $I - F(\lambda I - A)^{-1}B$ is an invertible $\mathbb{R}_S$-matrix.

Proof: The least common denominator of the entries in $C(\lambda I - A)^{-1}B$ divides the minimal polynomial of A and, if (C,A,B) is canonical, equals the minimal polynomial of A. Since the minimal polynomial of A divides c.p. (A), i is true. Assertion ii follows from the fact that c.p. (A) divides a power of the minimal polynomial of A. By i, $F(\lambda I - A)^{-1}B$ and $F(\lambda I - A - BF)^{-1}B$ are $\mathbb{R}_S$-matrices. Since $I + F(\lambda I - A - BF)^{-1}B$ is the rational function inverse of $I - F(\lambda I - A)^{-1}B$, it follows that iii is true. □

Proof of Proposition 2: Let H ε $\underline{P}^{pxm}$ be fixed with canonical realization (C,A,B); by ii/Lemma 5, c.p. (A) ε S. Write $V^* \equiv V^*(C,A,B)$ and let F be any map such that c.p. (A + BF) ε S and $(A + BF)V^* \subset V^*$. Set $H^* \equiv C(\lambda I - A - BF)^{-1}B$; i/Lemma 5 implies $H^* \varepsilon \underline{P}^{pxm}$. By the return difference formula for transfer matrices, $H^* = H(I - F(\lambda I - A)^{-1}B)^{-1}$; iii/Lemma 5 asserts that $(I - F(\lambda I - A)^{-1}B)$ is invertible over $\mathbb{R}_S$. Thus by Lemma 2, H and H^* are dynamically equivalent.

To show that H^* is irreducible, write X for the state space of (C,A,B), $P: X \to X/V^*$ for the canonical projection, $\bar{A}$ for the map induced by A + BF in X/V^*, $\bar{B}$ = PB, and $\bar{C}$ for the unique solution to $C = \bar{C}P$. Then c.p. $(\bar{A})$ ε S, and $(\bar{C},\bar{A},\bar{B})$ is a controllable realization of H^*. To show that H^* is irreducible, it is therefore enough to prove that $V^*(\bar{C},\bar{A},\bar{B}) = 0$.

Set $\bar{V}^* = V^*(\bar{C},\bar{A},\bar{B})$ and let $\bar{F}$ be any map such that c.p. $(\bar{A} + \bar{B}\bar{F})$ ε S and $(\bar{A} + \bar{B}\bar{F})\bar{V}^* \subset \bar{V}^*$. Set $F_1 = F + \bar{F}P$. Since kernel $P = V^*$, there follows $(A + BF_1)|V^* = (A + BF)|V^*$.† Thus c.p. $(A + BF_1)$ = (c.p. $(\bar{A} + \bar{B}\bar{F})$)(c.p. $((A + BF_1)|V^*)$) = (c.p. $(\bar{A} + \bar{B}\bar{F})$)(c.p. $((A + BF)|V^*)$); this implies that c.p. $(A + BF_1)$ ε S. Therefore, by Lemma 4, any $(A + BF_1)$-invariant subspace in kernel C must be a subspace of V^*. The subspace $V \equiv P^{-1}\bar{V}^*$ has these properties so $V \subset V^*$; hence $PV = 0$. But $PV = \bar{V}^*$ implying $\bar{V}^* = 0$. Thus H^* is irreducible. □

Let H ε $\underline{P}^{pxm}$ be fixed and write $\delta(H)$ for its McMillan degree; i.e., $\delta(H)$ is the dimension of a canonical realization of H. Call H minimal if its McMillan degree is no greater than the McMillan degree of any transfer matrix dynamically equivalent to H.

Suppose that H, H^*, X and V^* are as in the proof of Proposition 2. Then $\delta(H)$ = dim. (X) and $\delta(H^*) \leq$ dim. (X/V^*); clearly $\delta(H^*) \leq \delta(H)$. Since H ε $\underline{P}^{pxm}$ is arbitrary and H^* is irreducible and dynamically equivalent to H, it must be that

† $A|V$ denotes the restriction of A to the A-invariant subspace V.

Corollary 2: *If* $H \in \underline{P}^{p\times m}$ *is minimal, then* H *is irreducible.*

In the sequel it will be shown that the converse is also true.

Let us call two transfer matrices H_1 and H_2 in $\underline{P}^{p\times m}$, *feedback equivalent* (mod S) if $\delta(H_1) = \delta(H_2)$ and if for some canonical realization (C,A,B) of H_1, there exists a feedback map F and an $\mathbb{R}$-invertible map G such that $H_2 = C(\lambda I - A - BF)^{-1}BG$. It is not difficult to show that feedback equivalence as defined here is indeed an equivalence relation on $\underline{P}^{p\times m}$; however, the requirement that $\delta(H_1) = \delta(H_2)$ is essential for this to be so. The following proposition provides an important link between feedback equivalence and dynamic equivalence.

Proposition 3: *If* $H \in \underline{P}^{p\times m}$ *is irreducible and* Q *is* $\mathbb{R}_S$*-invertible, then* $\delta(HQ) \geq \delta(H)$. *If, in addition,* $\delta(HQ) = \delta(H)$, *then* H *and* HQ *are feedback equivalent transfer matrices.*

The proof of this proposition depends on the following lemma.

Lemma 6: *Let* (C,A,B) *be a controllable realization of an irreducible transfer matrix in* $\underline{P}^{p\times m}$. *Let* V *be any* (A,B)*-invariant subspace*† *in kernel* C. *Then*

$$V \cap (\text{Image } B) = 0$$

In addition, for all F *such that* $(A + BF)V \subset V$, c.p. $((A + BF)|V)$ *is fixed, independent of* F *and coprime with* S.

Since the assertions of this lemma can be readily deduced from the results of [6], Theorem 4.3 of [7] and Lemma 4.1 of [8], a proof will not be given here.

Proof of Proposition 3: Let (C,A,B) be a canonical realization of H, with state space X. Write $Q = G + \hat{Q}$ where $\hat{Q}$ is proper and G is an $\mathbb{R}$-matrix. Let $(\hat{C},\hat{A},\hat{B})$ be a canonical realization of $B\hat{Q}$ with state space $\hat{X}$. Clearly

$$\text{Image } \hat{C} \subset \text{Image } B \tag{17}$$

and

$$\bar{C} = [C,0] \quad , \quad \bar{A} = \begin{bmatrix} A & \hat{C} \\ 0 & \hat{A} \end{bmatrix} \quad , \quad \bar{B} = \begin{bmatrix} BG \\ \hat{B} \end{bmatrix}$$

is a realization of HQ with state space $\bar{X} = X \oplus \hat{X}$.

Claim: $(\bar{C},\bar{A})$ is observable.

Write V for the unobservable space of $(\bar{C},\bar{A})$ and define $A_0 : \bar{X} \to \bar{X}$, $D : \bar{X} \to \bar{X}$ and $P : \bar{X} \to X$ so that

$$A_0 = \begin{bmatrix} A & 0 \\ 0 & 0 \end{bmatrix}, \quad D = \begin{bmatrix} 0 & \hat{C} \\ 0 & \hat{A} \end{bmatrix}, \quad P = [I,0].$$

Since $\bar{A} = A_0 + D$ and V is $\bar{A}$-invariant,

$$(A_0 + D)V \subset V; \tag{18}$$

thus $A_0 V \subset \text{Image } D + V$. Since $PA_0 = AP$ and $PD = \hat{C}$, there follows $APV \subset \text{Image } \hat{C} + PV$; hence by (17), PV is (A,B)-invariant. Furthermore, $\bar{C}V = 0$ and $\bar{C} = CP$, so $PV \subset$ ker-

† Definitions and properties of (A,B)-invariant and controllability subspaces can be found in [7] and [8].

nel C. Since (C,A,B) is a controllable realization of an irreducible transfer matrix, Lemma 6 applied to $P\mathcal{V}$ yields

$$P\mathcal{V} \cap (\text{Image } B) = 0 \tag{19}$$

Note that $A_0(\hat{X}\cap\mathcal{V}) = 0$; hence from (18), $D(\hat{X}\cap\mathcal{V}) \subset \mathcal{V}$. Thus $P\mathcal{V} \supset PD(\hat{X}\cap\mathcal{V}) \supset \hat{C}(\hat{X}\cap\mathcal{V})$; this and (17) imply that $P\mathcal{V}\cap(\text{Image } B) \supset \hat{C}(\hat{X}\cap\mathcal{V})$. In view of (19), $\hat{X}\cap\mathcal{V} \subset$ kernel $\hat{C}$. Clearly $D(\hat{X}\cap\mathcal{V}) = \hat{A}(\hat{X}\cap\mathcal{V})$; but $D(\hat{X}\cap\mathcal{V}) \subset \mathcal{V}$ and $\hat{A}(\hat{X}\cap\mathcal{V}) \subset \hat{X}$, so $\hat{A}(\hat{X}\cap\mathcal{V}) \subset \hat{X}\cap\mathcal{V}$. Since $(\hat{C},\hat{A})$ is observable, $\hat{X}\cap\mathcal{V} = 0$. This implies that

$$\dim.\ P\mathcal{V} = \dim.\mathcal{V} \tag{20}$$

The relation $PD = \hat{C}$ together with (17) and (20), insure the existence of a map F such that $BFPv = PDv$ for all $v \in \mathcal{V}$. Hence for $v \in \mathcal{V}$, $(A + BF)Pv = (AP + PD)v = P(A_0 + D)v = P\bar{A}v \in P\mathcal{V}$. Therefore $P\mathcal{V}$ is $(A + BF)$-invariant and with $P|\mathcal{V}$ denoting the isomorphism $v \longmapsto Pv$, the diagram

$$\begin{array}{ccc} P\mathcal{V} & \xrightarrow{(A + BF)|P\mathcal{V}} & P\mathcal{V} \\ {\scriptstyle P|\mathcal{V}}\uparrow & & \uparrow{\scriptstyle P|\mathcal{V}} \\ \mathcal{V} & \xrightarrow[\bar{A}|\mathcal{V}]{} & \mathcal{V} \end{array}$$

commutes. Clearly c.p. $(\bar{A}|\mathcal{V})$ = c.p. $((A + BF)|P\mathcal{V})$. But Lemma 6 applied to $P\mathcal{V}$ shows that c.p. $((A + BF)|P\mathcal{V})$ is coprime with S. Therefore c.p. $(\bar{A}|\mathcal{V})$ is also coprime with S.

If M is the map induced by $\bar{A}$ in $\bar{X}/(X\cap\mathcal{V})$, then c.p. $(M|(\mathcal{V}/(X\cap\mathcal{V})))$ divides c.p. $(\bar{A}|\mathcal{V})$. It follows that

$$\text{c.p. } (M|(\mathcal{V}/(X\cap\mathcal{V}))) \text{ is coprime with S} \tag{21}$$

Let $\hat{P}:\bar{X} \to \hat{X}$ denote the projection $(x + \hat{x}) \longmapsto \hat{x}$. Clearly $\hat{P}\bar{A} = \hat{A}\hat{P}$; since $\mathcal{V}$ is $\bar{A}$-invariant, $\hat{P}\mathcal{V}$ is $\hat{A}$-invariant. Furthermore, the triangular structure of $\bar{A}$ implies that $\hat{A}|\hat{P}\mathcal{V}$ is similar to $M|(\mathcal{V}/X\cap\mathcal{V})$. It follows from (21) that

$$\text{c.p. } (\hat{A}|\hat{P}\mathcal{V}) \text{ is coprime with S} \tag{22}$$

But $(\hat{C},\hat{A},\hat{B})$ is a canonical realization of an $\mathbb{R}_S$-matrix; thus by ii/Lemma 5, c.p. $(\hat{A}|\hat{P}\mathcal{V}) \in S$. This and (22) can be true only if $\hat{P}\mathcal{V} = 0$, or equivalently, if $\mathcal{V} \subset X$. Thus $A\mathcal{V} = \bar{A}\mathcal{V} \subset \mathcal{V}$ and $C\mathcal{V} = \bar{C}\mathcal{V} = 0$. Since (C,A,B) is an observable realization, $\mathcal{V} = 0$. Thus $(\bar{C},\bar{A})$ is observable.

To show that $\delta(HQ) \geq \delta(H)$, let $\bar{R}$ denote the controllable space of $(\bar{A},\bar{B})$; then $\bar{A}\bar{R} \subset \bar{R}$ and Image $\bar{B} \subset \bar{R}$. Since $BG = P\bar{B}$ and $P\bar{A} = AP$, it follows that Image $BG \subset P\bar{R}$ and that $AP\bar{R} \subset P\bar{R}$. By definition, $G = Q - \hat{Q}$ where $\hat{Q}$ is proper and Q is $\mathbb{R}_S$-invertible; hence G must be $\mathbb{R}$-invertible and Image BG = Image B. The controllable space of (A,B), namely X, must therefore be a subspace of $P\bar{R}$, which in turn is a subspace of X. It follows that

$$P\bar{R} = X \tag{23}$$

Since $(\bar{C},\bar{A},\bar{B})$ is an observable realization of HQ, $\delta(HQ)$ = dim. $\bar{R}$. But from (23) dim. $\bar{R} \geq$ dim. X, which in turn equals $\delta(H)$. Therefore $\delta(HQ) \geq \delta(H)$.

Construction of F: If $\delta(HQ) = \delta(H)$, then dim. $\bar{R}$ = dim. X; thus from (23) dim. $\bar{R}$ = dim. $P\bar{R}$. This insures that a map F can be constructed so that $BFPr = PDr$, $\forall\, r \in \bar{R}$. It follows that $(A + BF)Pr = P\bar{A}r \in P\bar{R}$, $\forall\, r \in \bar{R}$. Since $BG = P\bar{B}$ and Im $\bar{B} \subset \bar{R}$, by induction on i, $P(A + BF)^i BG = P\bar{A}^i BG = P\bar{A}^i\bar{B}$ for $i \geq 0$; but $\bar{C} = CP$ so $C(A + BF)^i BG = \bar{C}\bar{A}^i\bar{B}$, $i \geq 0$. Thus $(C, A + BF, BG)$ realizes HQ; since G is nonsingular and $\delta(HQ) = \delta(H)$, HQ and H are feedback equivalent. □

Now let $H^* \in \underline{P}^{pxm}$ be irreducible and let $H \in \underline{P}^{pxm}$ be any minimal transfer matrix dynamically equivalent to H^*. By Lemma 2, $H = H^*Q$ for some $\mathbb{R}_S$-invertible matrix Q. It follows from Proposition 3, that $\delta(H^*) \leq \delta(H)$; but since H is minimal, $\delta(H^*) = \delta(H)$, so H^* must be minimal as well. Thus any irreducible transfer matrix in $\underline{P}^{pxm}$ is minimal. This observation combined with Corollary 2 yields

Theorem 2: A transfer matrix in $\underline{P}^{pxm}$ is minimal if and only if it is irreducible.

Let H_1 and H_2 be irreducible transfer matrices in $\underline{P}^{pxm}$. Then H_1 and H_2 are both minimal. If H_1 and H_2 are dynamically equivalent, then minimality implies $\delta(H_1) = \delta(H_2)$; also by Lemma 2, $H_1Q = H_2$ for some invertible $\mathbb{R}_S$-matrix. Thus $\delta(H_1Q) = \delta(H_1)$ so by Proposition 3, H_1 and H_2 are feedback equivalent.

Conversely, if H_1 and H_2 are feedback equivalent, then $\delta(H_1) = \delta(H_2)$ and, if (C_1, A_1, B_1) is a canonical realization of H_1, there exist F and G, with G $\mathbb{R}$-invertible, such that $C_1(\lambda I - A_1 - B_1F)^{-1}B_1G = H_2$. Since $\delta(H_2) = \delta(H_1)$, $(C_1, A_1 + B_1F, B_1G)$ is a canonical realization of H_2. By ii/Lemma 5, c.p. $(A_1) \in S$ and c.p. $(A_1 + B_1F) \in S$. Thus by iii/Lemma 5 the matrix $Q \equiv (I - F(\lambda I - A_1)^{-1}B_1)$ is $\mathbb{R}_S$-invertible, as is the matrix $Q^{-1}G$. Since H_1 and H_2 are related by the return difference formula $H_1Q^{-1}G = H_2$, it follows from Lemma 2 that H_1 and H_2 are dynamically equivalent. We summarize:

Theorem 3: Two irreducible matrices in $\underline{P}^{pxm}$ are dynamically equivalent if and only if they are feedback equivalent.

4. FEEDBACK EQUIVALENT SYSTEMS

We now come to the main purpose of this paper which is to classify linear systems up to transformations of the state-feedback type. The results which follow are applicable to monic linear systems.

A canonical linear system (C,A,B) is monic if its transfer matrix $C(\lambda I - A)^{-1}B$ is the matrix of a monomorphism of vector spaces over $\mathbb{R}(\lambda)$. Monic systems are completely characterized by Theorem 5 of [9]. The theorem states that if (C,A,B) is monic, then

$$V \cap (\text{Image } B) = 0, \qquad (24)$$

where V is the largest (A,B)-invariant subspace in kernel C. Equation (24), Theorem 4.3 of [7] and Lemma 4.1 of [8] imply that if F is any map such that $(A + BF)V \subset V$, and α is the minimal polynomial of $(A + BF)|V$, then α is uniquely determined by

(C,A,B); α is the (C,A,B)-transmission polynomial of greatest degree [3]. If S_α is the set of all monic polynomials in $\mathbb{R}[\lambda]$, which are coprime with α, then $\mathbb{R}_{S_\alpha}$ is uniquely determined by (C,A,B). Using (24) and the aforementioned theorem and lemma in [6] and [8] respectively, it is easy to verify that for all F such that c.p. $(A + BF) \in S_\alpha$, $(C, A + BF)$ is observable and $C(\lambda I - A - BF)^{-1}B$ is an irreducible $\mathbb{R}_{S_\alpha}$-matrix.

Let $\Sigma^{(p,n,m)}$ denote the class of all p-output, n-dimensional, m-input, monic, canonical systems. Call two systems (C,A,B) and $(\bar{C},\bar{A},\bar{B})$ in $\Sigma^{(p,n,m)}$ <u>feedback equivalent</u> if there exist $\mathbb{R}$-invertible matrices T and G and a feedback matrix F such that

$$CT^{-1} = \bar{C}\ , \quad T(A + BF)T^{-1} = \bar{A}\ , \quad TBG = \bar{B}$$

Feedback equivalence is clearly an equivalence relation on $\Sigma^{(p,n,m)}$.

Let $(C,A,B) \in \Sigma^{(p,n,m)}$ be fixed and let α denote its transmission polynomial of greatest degree. Let $\underline{F}$ be the nonempty class of maps F such that c.p. $(A + BF) \in S_\alpha$. Then for each $F \in \underline{F}$, $H_F \equiv C(\lambda I - A - BF)^{-1}B$ is an irreducible $\mathbb{R}_{S_\alpha}$-matrix. If $F_1, F_2 \in \underline{F}$ are arbitrary, then H_{F_1} and H_{F_2} are related by the return difference formula $H_{F_1}Q^{-1} = H_{F_2}$ where $Q = (I - (F_2 - F_1)(\lambda I - A - BF_1)^{-1}B)$. Since F_1 and F_2 are in $\underline{F}$, iii/Lemma 5 implies that Q is an invertible $\mathbb{R}_{S_\alpha}$ matrix. It follows that H_{F_1} and H_{F_2} are dynamically equivalent (mod S_α) and thus by Theorem 1, that Im H_{F_1} = Im H_{F_2}. This means that Im H_{F_1} is uniquely determined by (C,A,B), independent of the choice of $F_1 \in \underline{F}$. Henceforth we write $\mathcal{H}(C,A,B)$ for Im H_{F_1} and $\alpha(C,A,B)$ for α.

The main result of this paper is

<u>Theorem</u> 4: <u>Two systems</u> (C,A,B) <u>and</u> $(\bar{C},\bar{A},\bar{B})$ <u>in</u> $\Sigma^{(p,n,m)}$ <u>are feedback equivalent if and only if</u>

$$\alpha(C,A,B) = \alpha(\bar{C},\bar{A},\bar{B}) \tag{25}$$

<u>and</u>

$$\mathcal{H}(C,A,B) = \mathcal{H}(\bar{C},\bar{A},\bar{B}) \tag{26}$$

The theorem implies that any system $(C,A,B) \in \Sigma^{(p,n,m)}$ is uniquely determined up to similarity, feedback and input-coordinate transformations by its transmission polynomial $\alpha(C,A,B)$ together with the free $\mathbb{R}_{S_{\alpha(C,A,B)}}$-module $\mathcal{H}(C,A,B)$ generated by the columns of $C(\lambda I - A - BF)^{-1}B$, F being selected so that c.p. $(A + BF) \in S_{\alpha(C,A,B)}$.

<u>Proof of Theorem</u> 4: Let (C,A,B) and $\bar{C},\bar{A},\bar{B}) \in \Sigma^{(p,n,m)}$ be fixed. If (C,A,B) and $(\bar{C},\bar{A},\bar{B})$ are feedback equivalent, then $\alpha(C,A,B) = \alpha(\bar{C},\bar{A},\bar{B})$. {cf., [3]}. Set $\alpha = \alpha(C,A,B)$ and let F_1;and F_2 be chosen so that c.p. $(A + BF_1) \in S_\alpha$ and c.p. $(\bar{A} + \bar{B}F_2) \in S_\alpha$. Since feedback equivalence is transitive, $(C, A + BF_1, B)$ and $(\bar{C}, \bar{A} + \bar{B}F_2, \bar{B})$ are feedback equivalent. Thus there exist maps T,F and G, with T and G nonsingular, such that $CT^{-1} = \bar{C}$, $T(A + BF_1 + BF)T^{-1} = \bar{A} + \bar{B}F_2$ and $TBG = \bar{B}$. Since $(A + BF_1 + BF)$ is similar to $\bar{A} + \bar{B}F_2$, c.p. $(A + BF_1 + BF) \in S_\alpha$. Thus by Lemma 5, the transfer matrices

$H \equiv C(\lambda I - A - BF_1)^{-1}B$, $\bar{H} = \bar{C}(\lambda I - \bar{A} - \bar{B}F_2)^{-1}\bar{B}$ and $Q = (I - F(\lambda I - A - BF_1)^{-1}B)$ are $\mathbb{R}_{S_\alpha}$ matrices and Q is $\mathbb{R}_{S_\alpha}$-invertible. The return difference formula $HQ^{-1}G = \bar{H}$ and Lemma 2 imply that H and $\bar{H}$ are dynamically equivalent (mod S_α). It follows from Theorem 1 and the definition of $H(\cdot)$, that (26) is true.

Now suppose that (25) and (26) hold. Let α, F_1, F_2, H and $\bar{H}$ be as above; (25) and (26) imply that H and $\bar{H}$ are dynamically equivalent (mod S_α). Since H and $\bar{H}$ are irreducible $\mathbb{R}_{S_\alpha}$-matrices, by Theorem 3, H and $\bar{H}$ are feedback equivalent. Thus there exist maps F and G, with G nonsingular, such that $\bar{H} = C(\lambda I - A - BF_1 - BF)^{-1}BG$. Since $(\bar{C},\bar{A} + \bar{B}F_2,\bar{B})$ and $(C,A + BF_1 + BF,BG)$ both canonically realize $\bar{H}$ it follows that there exists a map T such that

$$CT^{-1} = \bar{C} \quad , \quad T(A + BF_1 + BF)T^{-1} = \bar{A} + \bar{B}F_2 \quad , \; TBG = \bar{B}$$

In addition, if $\tilde{F} \equiv F_1 + F - GF_2T$, then $T(A + B\tilde{F})T^{-1} = \bar{A}$. Thus (C,A,B) and $(\bar{C},\bar{A},\bar{B})$ are feedback equivalent. □

CONCLUDING REMARKS

As stated, Theorem 4 only classifies those canonical systems which are monic. It would be interesting to develop a more general result (along similar lines) applicable to all canonical linear systems. For an alternative approach to the classification problem treated here, see [10].

REFERENCES

[1] MacDuffee, C. C., The Theory of Matrices, Chelsea, New York.

[2] Wolovich, W. A. and Falb, P. L., "Invariants and Canonical Forms Under Dynamic Compensation," Brown University Technical Report, 1975.

[3] Morse, A. S., Structural invariants of linear multivariable systems, SIAM J. Control, 11 (3), 446, 1973.

[4] __________, Structure and design of linear model following systems, IEEE Trans. Auto. Control, AC-18 (4), 346, 1973.

[5] __________, Output controllability and system synthesis, SIAM J. Control, 9 (2), 143, 1971.

[6] Wonham, W. M., Algebraic methods in linear multivariable control, System Structure, IEEE Special Publication No. 71C61-CSS, 89, 1971.

[7] Wonham, W. M. and Morse, A. S., Decoupling and pole assignment in linear multivariable systems: a geometric approach, SIAM J. Control, 8 (1), 1, 1970.

[8] Morse, A. S. and Wonham, W. M., Decoupling and pole assignment by dynamic compensation, ibid, 8 (3), 317, 1970.

[9] __________, Status of noninteracting control, IEEE Trans. Auto. Control, AC-16 (6), 568, 1971.

[10] Wang, S. H. and Davison, E. J., "Canonical Forms of Linear Multivariable Systems," University of Toronto Control Systems Report No. 7203, 1972.

LINEAR DIFFERENCE SYSTEMS ON PARTIALLY ORDERED SETS

Bostwick F. Wyman
Mathematics Department
The Ohio State University
Columbus, Ohio 43210, USA

ABSTRACT

A theory of time-varying linear discrete-time difference systems is presented, for which the time set may be any locally finite partially ordered set. A subring of the incidence algebra of the time set acts on the state-module, and the resulting input and output functors lead to a realization theory for time-varying difference equations. The resulting canonical realizations can be computed explicitly and generalize the classical theory of weighting patterns.

1. INTRODUCTION

The subject of this study is the problem of modeling or representing an observed stationary stochastic process with p components, $y(0), y(1), \cdots$, by a model of the following familiar type:

$$\begin{aligned} x(t+1) &= Ax(t) + Be(t) \\ y(t) &= Cx(t) + e(t);\ x(0) = 0\ . \end{aligned} \tag{1}$$

Here, $x(\cdot)$ is an intermediate state-process, and $e(\cdot)$ is what remains when a chosen system described by the matrix triple (A,B,C) has been "matched" to the observed data according to (1). Often, an observed input process is added to the first of the equations in (1), but since this adds nothing essential to the problem, it will be excluded here.

It has been customary to regard the residual process $e(\cdot)$ as an error, and to determine the model parameters (A,B,C) so as to minimize the negative of the likelihood function ($e(\cdot)$ is then assumed to be an independent gaussian process):

$$L_N = \frac{N}{2}\,[\log \det R + p(1+\log 2\pi)] \tag{2}$$

subject to the constraint (1), where

$$R = \frac{1}{N-1} \sum_{t=1}^{N} e(t)e'(t). \tag{3}$$

Moreover, importantly, the structure of the model is then regarded as known a priori.

While this classical estimation technique has a certain merit in obtaining the input-output behavior of the model (1) as a representation of the process, it can provide at best only a partial solution to the presently considered modeling problem. In our modeling problem we not only are after a model which has a good input-output behavior, but one that also has the structure parameters well estimated.

We then consider the structure parameters themselves to be important objects of estimation, for we regard them to represent valuable information about the hidden links tying together the variables y(t). We say "hidden" because this information is accessible to us only indirectly through the measurement y(t).

In the literature the estimation of the structure parameters has generally been regarded as a difficult and cumbersome problem for which more or less intuitive but still ad hoc solutions have been suggested.[1,2,3,15] Of these the last reference, dealing exclusively with the order estimation of scalar processes, appears to be nearest to the procedure studied here.

This paper is a continuation of the study of the entropy criterion introduced for the model estimation problem in Ref. 4. This criterion expresses the total entropy of the estimated variables, and, therefore, both the structure and the real-valued system parameters can be estimated by minimizing one single criterion.

After a discussion of a suitable family of canonical forms in Section 2, we shall rederive the criterion in Ref. 4 by informal but quite informative arguments in Section 3. In Section 4, then, we first prove that the maximum likelihood criterion cannot make a distinction between the different structures. We further outline a proof of a theorem which states that by minimizing the entropy criterion for increasingly long sequences of observations, generated by a system of the considered type, the minimizing structure will be a correct one, and that the associated minimizing parameters will be asymptotically equivalent to the maximum likelihood estimates. This we regard as a strong indication of that the introduced criterion captures the essential features of the estimation problem considered here.

2. PARAMETRIZATION OF LINEAR SYSTEMS

In this section we shall describe briefly how the set of linear systems of order n with p inputs and equally as many outputs can be represented by a set of structure and system parameters. This material is based on well known facts about canonical forms and the associated parameters.[3,5,6,7] We shall, however, introduce a crucial variant which plays an important role in the results in Section 4.

The output of a system (1) can be written as:

$$y(t) = \sum_{j=0}^{\infty} H_j \, e(t-j), \tag{4}$$

where the p x p - matrices $H_0, H_1, \cdots$ define the impulse response of the system. If we consider (1) as a representation of the process $y(\cdot)$ such that $e(t) = y(t) - y(t/t-1)$ is the prediction error when the components of $y(t)$ have been projected orthogonally on the space Y_{t-1}, spanned by the components of $y(t-j)$ for $j=1,2,\cdots$, then we may write the projection of $y(t+r)$ on Y_t as:

$$y(t+r/t) = H_r \, e(t) + H_{r+1} \, e(t-1) + \cdots . \tag{5}$$

This follows at once from (4) and the facts that the components of $e(t), e(t-1), \cdots$ also span Y_t while $e_i(t)$ is orthogonal to Y_{t-1} for $i=1,\cdots,p$. We assume throughout that the y-process is of full rank; i.e., that $E\, e(t)e'(t) > 0$.

Denote by H^r, $r \leq \infty$, the blockwise Hankel matrix,

$$H^r = \begin{bmatrix} H_1 & H_2 & \cdots \\ H_2 & H_3 & \cdots \\ \cdot & & \\ \cdot & & \\ \cdot & & \\ H_r & H_{r+1} & \cdots \end{bmatrix} . \tag{6}$$

Then we can write from (5):

$$u^r \triangleq \begin{bmatrix} y(t+1/t \\ y(t+2/t) \\ \cdot \\ \cdot \\ \cdot \\ y(t+r/t) \end{bmatrix} = H^r \begin{bmatrix} e(t) \\ e(t+1) \\ \cdot \\ \cdot \\ \cdot \\ \\ \end{bmatrix} \tag{7}$$

As the $e(\cdot)$-process is of full rank, too, there are just as many linearly independent random variables in the list u^r as there are such rows in H^r. Moreover, if the i_1'th, i_2'th,$\cdots$ rows in H^r are linearly independent, then so are the corresponding components of u^r.

Now, the ranks of the matrices H^r have a maximum, n, called the order of the system (1), which is the dimensionality of the linear space spanned by the rows of H^∞ or, equivalently, by the components of u^∞. There is a well known canonical construction of a basis from the rows of a Hankel matrix H^∞ of rank n. This is the so-called lexicographic basis, which was perhaps first described by Luenberger[5]. This basis includes the first p rows, since we assumed $y(\cdot)$ to be a full rank process. The next basis element is the first of the following rows which is not a linear combination of the previously selected ones, etc. Clearly, each lexicographic basis uniquely determines the entire Hankel matrix and hence the corresponding system. Let $\mathscr{H}_n$ denote the set of all Hankel matrices of pxp-blocks with rank n in which the first p rows are linearly independent. The set $\mathscr{H}_n$, then, is in 1-1 correspondence with the set $\mathscr{L}_n$ of the lexicographic bases constructed from the members of $\mathscr{H}_n$.

It follows from the fact that H^∞ is a block Hankel matrix that if the k'th row h_k is not a basis element, neither is h_{k+p}. This further implies that each of the initial consecutive block rows must include at least one basis element, and hence the rank of H^n must be n, the rank of H^∞.

To each lexicographic basis there then corresponds the set of indices $I = \{i(1),\cdots,i(n)\}$ of its rows with the property that i(k) in this set implies that i(k-p) is also in the same set for k>p; and, conversely, each such set defines a lexicographic basis of some Hankel matrix generated by an n'th order system. Observe that $i(j)=j$ for $j\le p$. How many distinct such index sets can there be with n elements? If m(n,p) denotes this number we can derive the recursion:

$$m(n,p) = \sum_{k=1}^{p} \binom{p}{k} m(n-p,k);\ m(n,1) = 1. \tag{8}$$

For instance, $m(n,2) = n-1$, and for $n=10$ and $p=3$ there are 36 index sets.

The set $\mathscr{L}_n$ then gets partitioned in $m(n,p)$ blocks, two bases belonging to one block when they have the same index set or structure. We shall also let $m(n,p)$ denote the set of the index sets or structures I.

Example. Consider the case with $n=4$ and $p=2$. There are the three index sets:

$$I_1 = \{1,2,3,4\},\ I_2 = \{1,2,3,5\},\ I_3 = \{1,2,4,6\} .$$

We shall return to this example several times below.

We shall now proceed to describing two different sets of matrix triple representations based on the above defined index sets. In the first the system matrix A is determined by

$$\begin{bmatrix} h_{i(1)+p} \\ \cdot \\ \cdot \\ \cdot \\ h_{i(k)+p} \\ \cdot \\ \cdot \\ \cdot \\ h_{i(n)+p} \end{bmatrix} = A \begin{bmatrix} h_{i(1)} \\ \cdot \\ \cdot \\ \cdot \\ h_{i(k)} \\ \cdot \\ \cdot \\ \cdot \\ h_{i(n)} \end{bmatrix} \quad \text{for } \{h_{i(1)},\cdots,h_{i(n)}\} \text{ a lexicographic basis.} \tag{9}$$

B consists of the p first columns in the right-most matrix in (9), and $C = (I\ 0)$.

In this what we may call lexicographic canonical form the defining property of the index set is taken advantage of to minimize the number of the free parameters in A. Indeed, if $i(k)+p$ does not belong to the index set, we know that $h_{i(k)+p}$ lies in the span of the rows h_j for those indices j in I which are smaller than $i(k)+p$. Consequently, certain coefficients in the k'th row of A will be zero. We shall illustrate this with an example.

Example (continued). For the index set I_1 the matrix A has the following structure:

$$A_1 = \begin{bmatrix} 0 & 0 & 1 & 0 \\ 0 & 0 & 0 & 1 \\ x & x & x & x \\ x & x & x & x \end{bmatrix}, \qquad (10a)$$

where "x" denotes the free parameters (to be estimated). Similarly, for I_2 and I_3 we have

$$A_2 = \begin{bmatrix} 0 & 0 & 1 & 0 \\ x & x & x & 0 \\ 0 & 0 & 0 & 1 \\ x & x & x & x \end{bmatrix} \text{ and } A_3 = \begin{bmatrix} x & x & 0 & 0 \\ 0 & 0 & 1 & 0 \\ 0 & 0 & 0 & 1 \\ x & x & x & x \end{bmatrix}. \qquad (10b,c)$$

It should be noticed that a given system of order 4 with 2 outputs and inputs admits one and only one of the representations (10a), (10b) or (10c) with $C = (I\ 0)$ in each. This follows from the fact that we have described above a function which takes a system to its lexicographic canonical form.

To obtain the second set of <u>representations</u> or <u>forms</u>, suppose that we pick one of the m(n,p) index sets for a Hankel matrix of rank n. The corresponding rows may or may not be a basis for the rows of this matrix. If the picked rows indeed are a basis, let A, B, and C be defined as in (9). Observe in particular that this basis need not be a lexicographic one. This means that a row not belonging to the basis need not be in the linear span of those above it (although, naturally, it is in the span of all the basis elements). Accordingly, we can no longer insert the zeros in A corresponding to the lexicographic structure, and therefore the matrices A in the example for the three bases must be parametrized as follows:

<u>Example</u> (continued). If the index set I_j for $j=1,2,3$ is known to define a basis, although not necessarily a lexicographic one, the corresponding structures of the A-matrices are given by:

$$A_1 = \begin{bmatrix} 0 & 0 & 1 & 0 \\ 0 & 0 & 0 & 1 \\ x & x & x & x \\ x & x & x & x \end{bmatrix}, \ A_2 = \begin{bmatrix} 0 & 0 & 1 & 0 \\ x & x & x & x \\ 0 & 0 & 0 & 1 \\ x & x & x & x \end{bmatrix}, \ A_3 = \begin{bmatrix} x & x & x & x \\ 0 & 0 & 1 & 0 \\ 0 & 0 & 0 & 1 \\ x & x & x & x \end{bmatrix}. \qquad (11a,b,c)$$

The B- and the C-matrices for each are just as above.

A given fourth order system with two outputs can be represented by at least one of these forms, but this time it may also be represented by two or by all of the three forms. In other words, a given Hankel matrix may admit bases of several of the m(n,p) types, and the sets of systems defined by the corresponding representations do not partition the set of all systems of order n with p outputs and inputs, which contrasts the situation with the lexicographic type of canonical forms. Observe too, that in this second class of representations, which are used throughout the paper, A has the same number of np parameters, while it, in the lexicographic form, may have fewer but not more than np.

We shall use the phrase "correct" or "true" form or parametrization for those forms in which the system generating the data is representable. The estimation or identification problem, then, is to find one of these correct forms and the corresponding values of its parameters. The possible lexicographic structure of the basis may then be detected by the parameter values in that certain parameters are nearly zero (cf., e.g., (10b) and (11b)). If one desires to test statistically for this, it is numerically a much better posed problem to test if a given parameter has the value zero than to decide if a given row of the Hankel matrix lies in the linear span of those above it.

The lack of uniqueness of a representation, as with the canonical forms of the second type, may be a disadvantage from a strictly theoretical viewpoint, but it is a valuable feature for the purpose of identification, as becomes clearer in Section 4. Basically, the reasons for this are that we have a wider selection of bases available, and in principle it is possible to start with any of these and "move" to the next by an equivalence transformation until an optimum is found. This would not be possible if we use the lexicographic parametrization. Further, it is an advantage to have the same number of parameters in each canonical form. And finally, as can be shown, the associated parameters in each form are still complete and independent in the sense of [7], and hence they can be estimated from the observations.

3. ESTIMATION CRITERION

To recall from the introduction the estimation problem we are concerned with is to try to match a model of the type (1) to the observed data. Another way to think of this is to regard the problem as one of finding the most concise, i.e., non-redundant, way of describing the data sequence $y(0),\cdots,y(N)$ in the form $(A,B,C,e(0),\cdots,e(N))$. We already know what the non-redundant descriptions of the system (A,B,C) are: the canonical forms in which no parameter can be expressed in terms of the others and hence eliminated. But what might be the non-redundant descriptions of the sequence $e(0),\cdots,e(N)$? Once we settle this we shall have a way to derive the estimation criterion in a direct albeit informal way. A more formal derivation was given in Ref. 4, with Jaynes' Principle of Minimal Prejudice as the starting point[12].

A vector random variable x of k components with zero mean and covariance R has its maximum information content or entropy when its distribution is normal[8]. This entropy is given by:

$$H(x) = \frac{1}{2}\log\det R + \frac{1}{2}k(1 + \log 2\pi). \qquad (12)$$

It then follows immediately that from two sequences of p-component random variables, $e(1),\cdots,e(N)$ and $f(1),\cdots,f(N)$ with the same "amplitude", $E\, e'(i)e(i) = E\, f'(i)f(i)$, the uncorrelated one has the larger entropy. This also agrees with intuition: such a sequence is less redundant; to put it another way, correlation means redundancy, which when removed by a linear rotational transformation which leaves the entropy unchanged, results in an uncorrelated sequence with reduced amplitudes.

These considerations imply that the most concise description $(A,B,C,e(0),\cdots,e(N))$ is one in which (A,B,C) is in a canonical form and $e(0),\cdots,e(N)$ is a sequence with maximum entropy, which in particular implies that this sequence is independent. This leads to the estimation method in which we should minimize the entropy of the sequence $\underline{e} = (e(0),\cdots,e(N))$, regarded as uncorrelated, or:

$$H(\underline{e}) = \frac{N}{2} \left(\log \det R + p(1+\log 2\pi) \right) \tag{13}$$

with respect to the parameters A and B, (C is constant). For R we may take,

$$R = \frac{1}{N-1} \sum_{t=1}^{N} e(t)e'(t). \tag{14}$$

Moreover, e(t) is what remains from the data according to the equations:

$$\begin{aligned} x(t+1) &= A\,x(t) + B\,e(t), \quad x(0) = 0 \\ y(t) &= C\,x(t) + e(t)\,. \end{aligned} \tag{15}$$

This criterion (13)-(15) agrees with the maximum likelihood criterion (1)-(3) for stationary processes.

In the preceding (A,B,C) has to be in a canonical form, but which one? We shall show in the next section that this question cannot be settled by just finding the minimum of $H(\underline{e})$, since for almost all finite sets of data $y(0),\cdots,y(N)$, the same minimum is obtained regardless of which canonical form is selected. Here we keep n, the order of the model, as fixed.

Pursuing our general search for the most concise description, we should now pick the least redundant of the canonical forms (A^i,B^i,C) of order n, where i denotes one of the m(n,p) structures. Each canonical form (A^i,B^i,C), in which the parameters $\theta^i = (A^i,B^i)$ are determined by some estimation method, is itself a random variable. If we denote the mean of θ^i by m^i and its covariance about the mean by Q(i), we have the maximum entropy of θ^i given by:

$$H(\theta^i) = \frac{1}{2} \log \det Q(i) + \frac{1}{2} k(1+\log 2\pi), \tag{16}$$

where k denotes the number of parameters in the canonical form i. The least redundant θ^i would be one where the components are uncorrelated, and we should accordingly pick the canonical form which minimizes the expression:

$$\frac{1}{2} \sum_{j=1}^{k} \log q_{jj}(i) + \frac{1}{2} k(1+\log 2\pi), \tag{17}$$

where $q_{jj}(i)$ denotes the j'th diagonal element of Q(i); i.e., the variance of the j'th component of θ^i.

Another approach to choosing that canonical form i which makes θ^i "least redundant" is to minimize the complexity of θ^i, as defined by van Emden[13]:

$$\log \operatorname{tr} Q(i) - \log \det Q(I) - k\log k. \tag{18}$$

The criterion (18) might have the advantage over (17) that it is unaffected by rescaling of the parameter vector. On the other hand, since we confine ourselves to comparisons of the finite set of i in m(n,p), the criteria (17) and (18) will be virtually equivalent, and in addition, (17) has a more direct interpretation in terms of entropy, analogously with that of e.

Combining all this we arrive at the total criterion, which is to be minimized over the structures i in m(n,p) and the parameter θ^i:

$$V(\theta^i) = \log\det R + p(1+\log 2\pi) + \frac{1}{N}\sum_{j=1}^{k'} \log q_{jj}(i) + \frac{k}{N}(1+\log 2\pi) \tag{19}$$

where R is given by (14)-(15) and $Q(i) = \{q_{rs}(i)\}$ may be taken as the following matrix:

$$Q(i) = \{q_{rs}(i)\} = \frac{2}{N}\left\{\frac{\partial^2 \log \det R}{\partial\theta_r^i \, \partial\theta_s^i}\right\}^{-1} \tag{20}$$

as derived in Ref. 4. Hence for large N, Q(i) is approximately proportional to the inverse of the Hessian matrix of V, and the third term in V(•) is a measure of how sensitive V is with respect to changes in its arguments. Q(i) may be calculated from certain recurrence equations as indicated in Ref. 4.

4. THE QUESTION OF CONSISTENCY

The important question now is whether the minimizing structure i and the corresponding parameter vector θ^i are consistent estimates of the system that generates the data y(0),y(1),•••. Here, of course, we now imagine that some such system does generate the data. The order n is also unknown, but again we imagine that (19) will be repeatedly minimized for $n=1,2,\cdots$ until the last term, which grows with n ($k=2np$), and the third term become so large that the decrease in the first term is overcome and the minimum of V starts growing again. We should mention that a criterion consisting of the first term in (19) and the fourth term in the form k/N was earlier derived by Akaike[9], also with information theoretic arguments. Such a criterion, then, will be useful in cases where there is essentially just one structure.

Since the criterion (19) as a function of θ differs from the maximum likelihood criterion only by a term of size $O(1/N)$, and the maximum likelihood estimates are proved to be consistent[10,11], it is certainly plausible that so are the estimates resulting from (19) for a correct structure. A full proof of this can be done along the lines in References 10 and 14 by showing that for any correct $i, V(\theta^i)$ converges uniformly in θ^i to the expected value of the log likelihood function almost surely (a.s.) as N tends to infinity; we shall not go into further details here.

Let $S(n,i)$ denote the set of all systems (1) of order n having the (non-lexicographic) structure i of the $m(n,p)$ possibles ones as discussed in Section 2. $S(n,i)$ can be shown to correspond to a dense subset of R^{2np}, where R is the real line.

<u>Theorem 1.</u> For every i and j and each system (A^i,B^i,C) in a dense subset of $S(n,i)$, there corresponds an equivalent system (A^j,B^j,C) in $S(n,j)$; i.e., $A^i = T A^j T^{-1}$, $B^i = T B^j$, and $C = C T$ for some nonsingular T.

Proof. Let $\{i(1),\cdots,i(n)\}$ denote the indices of the basis rows x_k of the Hankel matrix of a system (A^i,B^i,C) in $S(n,i)$. Then

$$\begin{bmatrix} x_{i(1)+p} \\ \cdot \\ \cdot \\ \cdot \\ x_{i(n)+p} \end{bmatrix} = A^i \begin{bmatrix} x_{i(1)} \\ \cdot \\ \cdot \\ \cdot \\ x_{i(n)} \end{bmatrix} \qquad (21)$$

and

$$\begin{bmatrix} x_1 \\ \cdot \\ \cdot \\ \cdot \\ x_{p+1+(n-p)p} \end{bmatrix} = E \begin{bmatrix} x_{i(1)} \\ \cdot \\ \cdot \\ \cdot \\ x_{i(n)} \end{bmatrix} \tag{22}$$

for a unique matrix E. The dense set of S(n,i) in question consists of all systems in S(n,i) for which all the n x n - submatrices of E are non-singular.

Let S(n,j) have the index set {j(1),···,j(n)} for the basis rows. If a system (A^j,B^j,C) is to be equivalent with (A^i,B^i,C) then they must have the same Hankel matrix. From (22) we then have:

$$\begin{bmatrix} x_{j(1)} \\ \cdot \\ \cdot \\ \cdot \\ x_{j(n)} \end{bmatrix} = T \begin{bmatrix} x_{i(1)} \\ \cdot \\ \cdot \\ \cdot \\ x_{i(n)} \end{bmatrix}$$

for a non-singular n x n - submatrix T of E. This transformation is seen to produce the system (A^j,B^j,C) which is equivalent with (A^i,B^i,C). The proof is complete.

Suppose now that the data y(0),···,y(N) has been generated by a system (A^i,B^i,C) with the structure i; denote by I_S all the structures in which the true system can be represented. Clearly I_S contains at least the structure corresponding to i. The maximum likelihood estimate $(\hat{A}^i,\hat{B}^i,C)$, for almost all sets of data, belongs to the dense set of S(n,i) defined in the proof of Theorem 1. Accordingly, in any other structure as well, there is a system which is equivalent to $(\hat{A}^i,\hat{B}^i,C)$, and therefore produces the same minimum for the two first terms in V, Eq. (19). This is what we claimed in Section 3; namely, that these terms in V and hence the likelihood function cannot distinguish between correct and incorrect structures.

Let $(A_N^j,B_N^j) = \theta_N^j$ denote the estimate which minimizes (19) (over a compactum) in structure j for a sequence y(0),···,y(N), generated by a system (A^i,B^i,C); i.e., by θ^i, in the lexicographic form i, and let j(N) denote the minimizing structure. In the following theorem we shall speak about convergence of the structure parameter j(N). We then simply interpret j(N) as the ordinal number of the structure j(N) in the set m(n,p) of all structures which is ordered in some fashion. These numbers in turn are regarded as real numbers with their metric. A convergence then means that for some N_0, $j(N)=j$ for all N larger than N_0, and a.s. (almost sure) convergence is interpreted in the usual fashion.

<u>Theorem 2.</u> For almost all sequences $y(0),y(1),\cdots$ the structure parameter $j(N)$ converges to the set I_S; i.e., from a number N_0 on the parameter values $j(N)$ remain in this set.

Outline of proof. Consider first the structure i and θ_N^i. According to the discussion in the second paragraph of this section, $\theta_N^i \to \theta^i$ (the true value) a.s. Furthermore, R as defined in (14)-(15) converges to, say, $\bar{R}$ for this structure i, and we also have

$$\lim_{N\to\infty} V(\theta_N^i) = \bar{V} = \log\det\bar{R} + p(1+\log 2\pi), \tag{23}$$

since the last two terms in (19) vanish as $N\to\infty$. $\bar{V}$ is the global minimum of the limit of the criterion function, which follows from the fact that e(t) in (14) is the optimal orthogonal prediction error. As described above, (A_N^i,B_N^i,C) belongs a.s. to a dense subset of S(n,i). By Theorem 1 for each j there exists a T_N and $(\bar{A}_N^j,\bar{B}_N^j,C)$ such that

$$\bar{A}_N^j = T_N A_N^i T_N^{-1} \text{ and } \bar{B}_N^j = T_N B_N^i . \tag{24}$$

Further, as $A_N^i \to A^i$, the sequence T_N converges a.s. to, say T, which follows from (22). Denote the associated parameter vector by $\bar{\theta}_N^j$, which, of course, may differ from the minimizing parameter vector θ_N^j in structure j.

Let j be a limit point of the minimizing sequence j(N); i.e., j(M)=j for an infinite set of numbers M. Since $V(\theta_M^j) \le V(\theta_M^i)$ we have:

$$V(\theta_M^j) \to \bar{V} \text{ as } M\to\infty .$$

This means that R, evaluated at θ_M^j, converges to $\bar{R}$, and since $\bar{\theta}_M^j$ and θ_M^i give by (24) the same value for log det R, we conclude by (23) that both θ_M^j and $\bar{\theta}_M^j$ give the same limiting value for log det R. But this can only happen when the difference of the impulse responses of (A_M^j, B_M^j, C) and $(\bar{A}_M^j, \bar{B}_M^j, C)$ also converges to zero in, say, the norm

$$||H|| = \sum_{i=0}^{\infty} ||H_i|| .$$

Since in each structure j, the function $H \mapsto (A,B,C)$ is continuous, the difference $\theta_M^j - \bar{\theta}_M^j \to 0$.

Suppose now that $j \notin I_S$. As there is no system in S(n,j) which is equivalent with (A^i, B^i, C), we conclude from (24) that $\bar{A}_M^j$ diverges, and hence so does θ_M^j. Along the diverging path, the value of log det R tends to log det $\bar{R}$. But this means by (20) that M Q(j) grows beyond any bound while M Q(i) remains bounded. Hence the difference $V(\theta_M^j) - V(\theta_M^i)$ is always strictly positive, and so the structure $j \notin I_S$ is not a minimizing one, which is a contradiction. Therefore, $j \in I_S$. This completes the proof.

5. CONCLUDING COMMENTS

The determination of state space models for stochastic processes from measured data y(•) consists of two equally important parts: In the first, a suitable structure (canonical form) for representation of the process has to be found, and in the second the associated parameters have to be evaluated. A common approach to this problem is to separate these two steps by first estimating the observability indices and then determining the parameters in the corresponding canonical form by, say, the maximum likelihood method.[1,3,7]

We have studied here a procedure which combines the structure and the parameter estimation steps in one by introducing a criterion which is sensitive to both these effects. The parameter-dependent part of the criterion is asymptotically equivalent to the negative of the log likelihood function, and therefore, the parameter estimates still have all the valuable asymptotic properties of the maximum likelihood estimates. Moreover, as Theorem 2 shows, the minimization of the criterion guarantees, in a statistical sense, a correct structure for the process when the amount of the observed data is sufficiently large.

What we have said so far would have been true for any canonical form, in particular for the lexicographic one. However, the suggested second type of structure has the further merit that it allows for a free change of structures during the numerical minimization of the criterion. Indeed, it follows from Theorem 1 that at any stage of the minimization procedure, the current model in the current structure is a.s. equivalent to a model in any other structure. Therefore, it is possible by a non-singular transformation of the parameters to switch over to another structure without altering the value of the first, parameter-dependent, part of the criterion. Moreover, because of this transformation, it is possible to evaluate the effect of the structure change on the third, structure-dependent, term of the criterion without having to re-evaluate the criterion.

Such structure changes during the minimization seem to us to be a valuable feature of the representations of the second type. Indeed, they make it possible to treat the optimum structure and the parameter estimation problems in a similar way by a stepwise minimization of the criterion, where both quantities are updated in a single procedure. We leave the details of this to another context.

As the final remark, we may view our second type of non-canonical representations as an attempt to provide a workable substitute for a global continuous canonical form whose non-existence has recently been proved by Hazewinkel and Kalman.[16]

REFERENCES

1. Tse, E. and Weinert, H., "Structure Determination and Parameter Identification for Multivariable Stochastic Linear Systems," Proc. 1973 JACC, Columbus, Ohio.
2. Mayne, D. Q., "A Canonical Model for Identification of Multivariable Linear Systems," IEEE Trans Autom. Control, AC-17, 728, 1972.
3. Ljung, L. and Rissanen, J., "On Canonical Forms, Parameter Identifiability and the Concept of Complexity," (submitted to the 4'th IFAC Symposium on Identification and System Parameter Estimation, Moscow, 1976).
4. Rissanen, J., "Minmax Entropy Estimation of Models for Vector Processes," to appear in System Identification: Advances and Case Studies, (D. G. Lainiotis and R. K. Mehra, Eds.), Marcel Dekker, Inc., New York.
5. Luenberger, D. G., "Canonical Forms for Linear Multivariable Systems," IEEE Trans. Autom. Control, AC-12, 290, 1967.
6. Akaike, H., "Markovian Representation of Stochastic Processes by Canonical Variables," SIAM J. Control, 13, 1975.
7. Rissanen, J., "Basis of Invariants and Canonical Forms for Linear Dynamic Systems, Automatica," 10, 175, 1974.
8. Shannon, C. E., "A Mathematical Theory of Communication," Bell System Technical J., 27, 623, 1948.
9. Akaike, H., "A New Look at the Statistical Model Identification," IEEE Trans. Autom. Control, AC-19, 716, 1974.
10. Rissanen, J. and Caines, P., "Consistency of Maximum Likelihood Estimators for ARMA Processes," Control Systems Report No. 7424, Univ. of Toronto, Dec 1974.
11. Ljung, L., "On the Consistency of Prediction Error Identification Methods," to appear in System Identification: Advances and Case Studies, (D. G. Lainiotis and R. K. Mehra, Eds.), Marcel Dekker, Inc., New York.
12. Jaynes, E. T., "Information Theory and Statistical Mechanics," The Physical Review, 106, 620, 1957 and 108, 171, 1957.
13. Van Emden, M. H., "An analysis of Complexity, Mathematical Centre Tracts," 35, Mathematisch Centrum, Amsterdam, 1971.
14. Ljung, L., "On Consistency and Identifiability," to appear in Proc. Conf. on Stochastic Systems, Lexington, Kentucky, June 1975, North Holland Publishing Co.
15. Chan, C. W., Harris, C. J., and Wellstead, P. E., "An Order Testing Criterion for Mixed Autoregressive Moving Average Processes," Control Systems Centre, Report No: 229, Univ. of Manchester Inst. of Sc. and Technology, Dec. 1973.
16. Hazewinkel, M. and Kalman, R., "Moduli and Canonical Forms for Linear Dynamical Systems," Report 7504/M, Erasmus Universiteit Rotterdam Econometrisch Institut, 1974.

ESTIMATION OF OPTIMUM STRUCTURES AND PARAMETERS FOR LINEAR SYSTEMS

J. Rissanen

IBM Research Laboratory
San Jose, California 95193/USA

L. Ljung*

Division of Automatic Control
Lund Institute of Technology
Lund, Sweden

ABSTRACT: The problem of determining a linear state space model for a stochastic process from measured output data contains, as an important part, the problem of choosing of a suitable structure for the model. A criterion is studied which measures the fit between the model and the data as well as the validity of a structure assumption. Hence the parameter and structure estimates can be obtained by minimization of a single criterion. The minimizing model is shown to converge to one with a correct structure and parameter values in this structure.

* At present with Information Systems Laboratory, Stanford University, Stanford, California 94305/USA.

SECTION 1. INTRODUCTION

This paper is a preliminary report on a theory of linear systems on locally-finite partially-ordered time sets. It should be viewed as a generalization of the theory of time-varying discrete-time linear system theory, and it is directly inspired by E. Kamen's algebraic approach to that theory [1-2].

The methodology of this report is adapted from my San Francisco abstract [3], which is reviewed briefly in Section 2. There, the state space of a linear system is considered as a module over a ring of "generalized difference operators," and input/output behavior is described in terms of two pairs of adjoint functors. This procedure is followed exactly in the present treatment, using a new kind of incidence algebra for the ring of operators. The abstract input/output behavior is related to time-varying difference equations, and the structure theory of time varying canonical realizations is discussed.

Although we refer constantly to "partially-ordered time," in many examples the relevant index set should be thought of as "space," or even "space-time." Examples of this type are mentioned in [3-7] and [13].

An attempt will be made to motivate the various abstract constructions used, starting with the examples in the next section. Proofs of technical results will be given in a later publication.

Acknowledgments. I would like to thank E. Kamen for his constant encouragement and for sharing an early version of his results. D. Elliott tried to persuade me long ago that system-theorists should care about incidence algebras, and J. Ferrar has helped me a great deal with the ring theory aspects of these problems.

SECTION 2. SOME PRELIMINARY EXAMPLES

This tutorial section presents some of the main motivating examples. Other examples, with a more detailed treatment, appear in Sections 7 and 8.

2.1. Stationary discrete-time linear systems.

Consider the "classical" systems $\Sigma = (X, U, Y; F, G, H)$, where X, U, and Y are modules over a commutative ring (or field) k, and $F: X \to X$, $G: U \to X$, $H: X \to Y$ are k-module maps. R. E. Kalman, [8], and countless later authors, consider X as a module ${}_F X$ over the polynomial ring $k[z]$, defining $zx = Fx$, for all $x \in X$.

For each U, there is a $k[z]$-module ΩU of "input strings," a $k[z]$-module mapping $\tilde{G}: \Omega U \to X$, and a commutative "state-input diagram" of k-modules

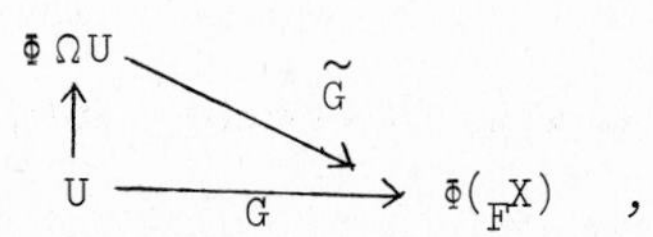

,

where Φ is the forgetful functor from the category ((k[z]-mod)) of k[z]-modules to ((k-mod)).

The map $\widetilde{G}$ is uniquely determined by these conditions: more precisely, Ω is the left-adjoint to Φ. Of course, $\Omega U \cong U[z] = U \underset{k}{\otimes} k[z]$ is the familiar module introduced by Kalman.

The forgetful functor Φ also has a right-adjoint Γ. For every k-module Y there is an "output-string" module ΓY over k[z], and a map $H: X \to Y$ gives a unique $\widetilde{H}: {}_F X \to \Gamma Y$ such that

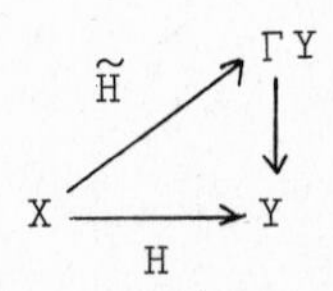

commutes. In this case, $\Gamma Y \cong Y((z^{-1}))/Y[z]$, Kalman's module of left-truncated Laurent series. Alternatively, $\Gamma Y \cong \mathrm{Hom}_k(k[z],Y)$, which is a general construction for right adjoints in this context.

A system Σ gives a k[z]-module map $f_\Sigma: \Omega U \to \Gamma Y$ by $f_\Sigma = \widetilde{H}\widetilde{G}$. Conversely, any k[z]-module map can be factored through a unique minimal realization:

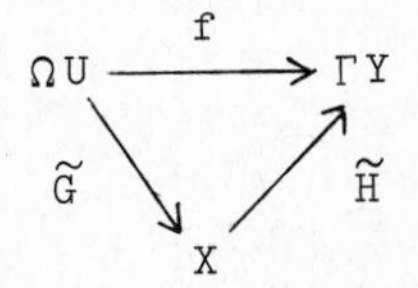

with $\widetilde{G}$ surjective and $\widetilde{H}$ injective, giving a system (X, U, Y; F, G, H). The map f can be thought of as a sequence $\{A_t\}$, $t \geq 0$, $A_i: U \to Y$, and the diagram gives $A_t = H F^{t-1} G$, $t \geq 1$.

All this is very well known. See [3] for the categorical formulation. The San Francisco volume [4] contains a superb bibliography.

2.2. Time-varying systems on $\mathbb{Z}$.

This example is adapted from [1] with some small changes. Let T be the ordinary integers with the usual order. Consider a system $\Sigma = (X, U, Y; F(t), G(t), H(t))$, where X, U, and Y are modules over a communtative ring k, and for each $t \in T$, $F(t): X \to X$, $G(t): U \to X$, and $H(t): X \to Y$ are k-module mappings. In this example, we write mappings on the right: $x \mapsto x\,F(t)$.

The "instantaneous" state, input, and output modules are replaced by the corresponding "trajectory-modules" as follows.

$\mathcal{X} = X^T = \{$all functions $x: T \to X\}$, and similarly, $\mathcal{U} = U^T$, $\mathcal{Y} = Y^T$. Also $\mathcal{K} = k^T$, which is a commutative ring with coordinatewise operations $((a + b)(t) = a(t) + b(t)$, etc.).

The trajectory modules $\mathcal{X}$, $\mathcal{U}$, and $\mathcal{Y}$ are right $\mathcal{K}$-modules: $(x\,a)(t) = x(t)a(t)$. Define right $\mathcal{K}$-module maps $G: \mathcal{U} \to \mathcal{X}$ and $H: \mathcal{X} \to \mathcal{Y}$ by $(uG)(t) = u(t)G(t)$ and $(xH)(t) = x(t)H(t)$, respectively.

The next ingredient is an appropriate module structure on $\mathcal{X}$. Motivated by the vector difference equation

$$x(t) = x(t - 1)F(t) ,$$

Kamen introduces a new variable z and an action $z: \mathcal{X} \to \mathcal{X}$ by

$$(xz)(t) = x(t - 1)F(t) .$$

An apparent difficulty arises, since this action does not commute with the scalar functions in $\mathcal{K}$.

For x in $\mathcal{X}$ and $a \in \mathcal{K}$, we have

$$((xa)z)(t) = x(t - 1)a(t - 1)F(t) = x(t - 1)F(t)a(t - 1)$$

$$((xz)a)(t) = x(t - 1)F(t)a(t) .$$

This "difficulty" is the key to the theory! First, define a shift-operator $\sigma: \mathcal{K} \to \mathcal{K}$ by $a^{\sigma}(t) = a(t - 1)$. Then, consider the non-commutative ring

$$\mathcal{K}_{\sigma}[z] = \{z^n a_n + z^{n-1} a_{n-1} + \dots + z a_1 + a_0 : a_i \in \mathcal{K} , n \geq 0\}$$

with multiplication defined as for polynomials, except for the crucial relation

$$az = za^{\sigma} \quad \text{for all} \quad a \in \mathcal{K} .$$

The formula $(xz)(t) = x(t - 1)F(t)$ gives $\mathcal{X}$ a right $\mathcal{K}_{\sigma}[z]$-module structure. The forgetful functor $((\text{mod-}\mathcal{K}_{\sigma}[z])) \to ((\text{mod-}\mathcal{K}))$ has left and right adjoints Ω and Γ, and we get an i/o-diagram

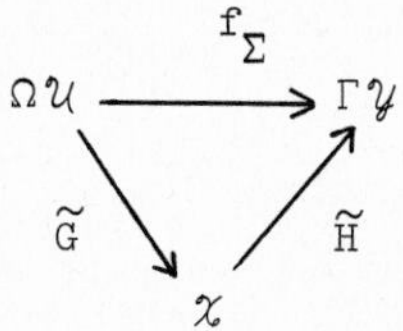

This has been developed by Kamen without using categorical concepts, and it has been applied by him and his student K. Hafiz, mainly to state-input problems such as control-canonical form, [1-2]. Later we discuss Ω and Γ in detail and describe explicit consequences of a realization theory.

2.3. Partially ordered time sets.

Here we give a rough introduction to the general theory. Precise definitions and details will occupy most of the paper.

We start with a partially ordered set ("poset"), $(T, \leq)$, which is <u>locally-finite</u> in the sense that for $s \leq t$ in T, the <u>segment</u> $[s,t] = \{r: s \leq r \leq t\}$

is a finite set. The integers $(\mathbb{Z},\leq)$ is locally finite, as are $(\mathbb{Z}^n,\leq)$, where $(a_1,a_2,\ldots,a_n) \leq (b_1,b_2,\ldots,b_n)$ when $a_i \leq b_i$ for each i. Trees are locally finite posets, but the real numbers $(\mathbb{R},\leq)$ are not.

We are also given a commutative ring k of scalars and three right k-modules X, U, and Y. For each t in T, suppose given k-module mappings $G(t)\colon U \to X$, and $H(t)\colon X \to Y$. Equivalently, we consider $\mathcal{K}$-module maps $G\colon \mathcal{U} \to \mathcal{X}$, $H\colon \mathcal{X} \to \mathcal{Y}$, where $\mathcal{K} = k^T$, $\mathcal{X} = X^T$, $\mathcal{U} = U^T$, and $\mathcal{Y} = Y^T$ as before.

Next, we need a <u>dynamical structure</u> on $\mathcal{X}$, which we take to be given as follows: for each pair s, t in T with $s \leq t$, we are given a k-endomorphism $\Phi(s,t)\colon X \to X$ (which will be written on the right). The system theoretic intuition is that if $x(s)$ is a state at time s, then $x(s)\Phi(s,t)$ is the state at time t resulting from the free evolution of the system. In other words, $\Phi(s,t)$ is a "weighting pattern." See, e.g. [9, p. 91]. This motivates the following two axioms:

(1) $\Phi(t,t) = I$, the identity for all $t \in T$.

(2) $\Phi(s,r)\Phi(r,t) = \Phi(s,t)$ for all $s \leq r \leq t$ in T.

Suppose given, then, a "system" $\Sigma = (X, U, Y; \Phi, G(t), H(t))$, or $\Sigma = (\mathcal{X}, \mathcal{U}, \mathcal{Y}\colon \Phi, G, H)$. How can we use Φ to describe a module structure on $\mathcal{X}$? More fundamentally, what ring will we use as a ring of generalized difference operators? The ring proposed here is a slight modification of the <u>incidence algebra of a poset</u>, which has been studied extensively in the combinatorics literature by G. C. Rota and others; see especially [10] and references cited there.

If T is a locally finite poset, then the incidence algebra $\hat{R}$ of T over a ring k consists of functions $f\colon T \times T \to k$, with $f(s,t)$ defined for $s \leq t$. Addition comes from addition in k, and multiplication is given by "convolution":

$$(\alpha\beta)(s,t) = \sum_{s \leq r \leq t} \alpha(s,r)\beta(r,t) .$$

(The sum is finite since T is locally finite.) This ring $\hat{R}$ is an associative, non-commutative ring with identity δ given by $\delta(s,t) = 1$ if $s = t$, otherwise 0. For the system-theoretic interpretation of $\hat{R}$, we think of each α in $\hat{R}$ as assigning a "weight" $\alpha(s,t)$ to each segment $[s,t]$. The state trajectory set $\mathcal{X}$ is (almost) an $\hat{R}$-module, with action

$$(x\alpha)(t) = \sum_{s \leq t} x(s)\Phi(s,t)\alpha(s,t) .$$

Verbally: $x\alpha$ is a new state trajectory obtained by assigning to time t that state $(x\alpha)(t)$ resulting from all preceding states $x(s)$, $s \leq t$, by free evolution and superposition, except that the contribution from $x(s)$ is assigned the weight $\alpha(s,t)$.

This method breaks down, however, since the right hand sum may not be finite. Some finiteness assumption must be added, either to the definition of $\mathcal{X}$ or to the functions $f \in \hat{R}$. We choose to define a subalgebra $R \subset \hat{R}$ of incidence functions with "finite memory." That is, for each $t \in T$, $f(s,t) \neq 0$ only for finitely many $s \leq t$. The above formula does in fact define a right R-module structure on $\mathcal{X}$.

Since the general philosophy of linear system theory advocated here requires two rings, we take $\mathcal{K} = k^T$ as base ring, where $\mathcal{K}$ is imbedded in R as follows: let $a: T \to k$ be in $\mathcal{K}$. Consider a as a function of two variables by $a(s,t) = a(t)$ if $s = t$; $a(s,t) = 0$ otherwise. The inclusion $\mathcal{K} \subset R$ induces the usual adjoint functors and allows us to set up a complete realization theory in Section 4. In particular, we have an input functor $\Omega\mathcal{U} = \mathcal{U} \otimes_{\mathcal{K}} R$ and an output functor $\Gamma\mathcal{Y} = \mathrm{Hom}_{\mathcal{K}}(R,\mathcal{Y})$.

We conclude this introduction with a brief discussion of difference equations and a concrete statement of the fundamental realization result. Consider modules $\mathcal{U}$ and $\mathcal{Y}$ as above, but also supply a right R-module structure by

$$(u\,\theta)(t) = \sum_{s \leq t} u(s)\theta(s,t)$$

for all $\theta \in R$. A <u>difference equation</u> is an equation of the form $y\alpha = u\beta$, or, written out

$$\sum_{s \leq t} y(s)\alpha(s,t) = \sum_{s \leq t} u(s)\beta(s,t) \quad \text{for all } t \in T .$$

(For example, to put $y(t) - y(t-1)a(t) = u(t)$ in this form, simply define

$$\alpha(s,t) = \begin{cases} 1 & \text{if } s = t \\ a(t) & \text{if } s = t-1, \\ 0 & \text{otherwise} \end{cases}$$

and let $\beta = \delta$; i.e. $\beta(s,t) = 1$ if and only if $s = t$.) Difference equations over posets were introduced in [11].

Returning to the general equation $y\alpha = u\beta$, assume $\alpha(t,t)$ is a unit in k. Then α is invertible in $\hat{R}$, but not necessarily in R. We are tempted to assert that $y = u\beta\alpha^{-1}$ is "the solution" to the difference equation. However, this expression may be meaningless since α^{-1} need not have finite memory. We use the trick (familiar by now) of using a finite memory function to insure finite sums.

Given $u \in \mathcal{U}$, we define a function $uf: R \to \mathcal{K}$ by

$$(uf)(\theta)(t) = \sum_{s \leq t} u(s)\theta(s,t)(\beta\alpha^{-1})(s,t) .$$

The sum is finite, since θ has finite memory. Also if $(uf)(\zeta)$ exists (which cannot be expected if the zeta function ζ is not in R), then $(uf)(\zeta) = \sum_{s \leq t} u(s)(\beta\alpha^{-1})(s,t) = u\beta\alpha^{-1}$ is "the solution" as before.

In any case, the map $u \to uf$ gives $f: \mathcal{U} \to \mathrm{Hom}_{\mathcal{K}}(R,\mathcal{Y}) = \Gamma\mathcal{Y}$, which extends to an abstract i/o map $f: \Omega\mathcal{U} \to \Gamma\mathcal{Y}$. We can try to find a <u>canonical realization</u>

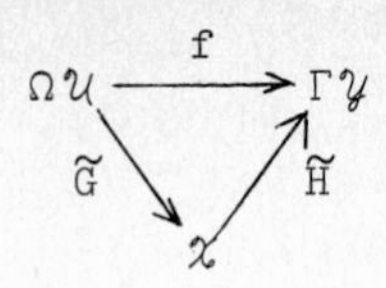

and investigate the implications of this representation. A precise structure theorem for $\mathcal{X}$ will be stated later: roughly, these modules come from "weighting patterns" as above. We conclude this introduction with an approximate statement of the result: THEOREM. Suppose given a difference equation $y\alpha = u\beta$ on the poset T, as above. Then there exists a family $\{X_t : t \in T\}$ of k-modules, a family $\Phi(s,t): X_s \to X_t$, $s \leq t$, of k-linear maps, and families $\{G(s): s \in T\}$, $G(s): U \to X_s$, and $\{H(t): t \in T\}$, $H(t): X_t \to Y$, such that $\beta\alpha^{-1}(s,t) = G(s)\Phi(s,t)H(t)$ for all $s \leq t$. This representation can be chosen to be "canonical" in a certain sense, and if this is done, it is essentially unique.

SECTION 3. GENERALIZED DIFFERENCE SYSTEMS

In this section, we recall the definition of a "generalized difference system," which was called an "S/k-system" in [3], and state the expected realization theorem. We assume a reasonable acquaintance with adjoint functors.

Throughout, we consider a base ring $\mathcal{K}$ which is assumed commutative with identity. We denote various right $\mathcal{K}$-modules as $\mathcal{X}$, $\mathcal{U}$, and $\mathcal{Y}$. Our first definition gives a suggestive name to a standard object.

3.1. Definition.

An <u>algebra of generalized difference operators</u> over a commutative ring $\mathcal{K}$ with identity 1 is a ring R and a ring homomorphism $\varphi: \mathcal{K} \to R$ such that $\varphi(1)$ is the identity in R.

Note that R need not be commutative, and the image of $\mathcal{K}$ need not be central. We consider R as a $\mathcal{K}$-$\mathcal{K}$ bimodule by $ar = \varphi(a)r$ and $rb = r\varphi(b)$ for all $r \in R$, $a,b \in \mathcal{K}$. The map φ will usually be omitted from the notation.

We denote by $((\text{mod} - R))$ and $((\text{mod} - \mathcal{K}))$ the categories of right R- and right $\mathcal{K}$-modules. There is an unnamed "forgetful functor": $((\text{mod} - R)) \to ((\text{mod} - \mathcal{K}))$, which has a left adjoint Ω and a right adjoint Γ. See, for example [12, p. 28]. We call Ω and Γ the <u>input</u> - and <u>output</u> - functors for $\varphi: \mathcal{K} \to R$.

Explicitly, we can take $\Omega(-) = - \otimes_{\mathcal{K}} R$ and $\Gamma(-) = \text{Hom}_{\mathcal{K}}(R, -)$, where $\text{Hom}_{\mathcal{K}}$ denotes the set of all right $\mathcal{K}$-linear maps. The right R-module structure on $\Omega(\mathcal{U})$ is familiar, and with our conventions, the right R-structure on $\text{Hom}_{\mathcal{K}}(R, \mathcal{Y})$ is given by $(h\alpha)(\beta) = h(\alpha\beta)$ for $h \in \text{Hom}_{\mathcal{K}}(R, \mathcal{Y})$ and $\alpha, \beta \in R$.

If $G: \mathcal{U} \to \mathcal{X}$ is a right $\mathcal{K}$-module map, then the R-linear extension $\tilde{G}: \Omega\mathcal{U} \to \mathcal{X}$ given by the adjunction has the explicit formula $(u \otimes \alpha)\tilde{G} = uG \cdot \alpha$ for simple tensors. If $H: \mathcal{X} \to \mathcal{Y}$ is a right $\mathcal{K}$-module map, the corresponding R-linear map

$\widetilde{H}: \mathcal{X} \to \Gamma\mathcal{Y}$ is given by $(x\widetilde{H})(\alpha) = (x\alpha)H$ for $x \in \mathcal{X}$ and $\alpha \in R$. We often use the unit and counit mappings $u \to u \otimes 1: \mathcal{U} \to \Omega\mathcal{U}$ and $h \to h(1): \Gamma\mathcal{Y} \to \mathcal{Y}$ without comments or names.

Now we are ready to define systems and their corresponding input/output (i/o) maps in this general context.

3.2. Definition.

Let $\mathcal{K}$ be a commutative ring and $\varphi: \mathcal{K} \to R$ an algebra of generalized difference operators. Then a generalized difference system over $\mathcal{K}$ with algebra R, or an $R/\mathcal{K}$-system for short, is a quintuple $\Sigma = (\mathcal{X}, \mathcal{U}, \mathcal{Y}; G, H)$, where $\mathcal{X}$ is a right R-module, $\mathcal{U}$ and $\mathcal{Y}$ are right $\mathcal{K}$-modules, and $G: \mathcal{U} \to \mathcal{X}$, $H: \mathcal{X} \to \mathcal{Y}$ are $\mathcal{K}$-module maps.

As above, we consider the R-module maps $\widetilde{G}: \Omega\mathcal{U} \to \mathcal{X}$ and $\widetilde{H}: \mathcal{X} \to \Gamma\mathcal{Y}$. Then the i/o-diagram of Σ is the commutative R-module diagram

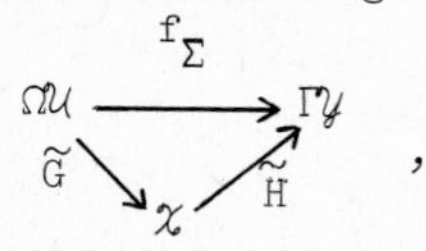

where, by definition, $f_\Sigma = \widetilde{G} \circ \widetilde{H}$, and f_Σ is called the i/o-map of Σ.

The system Σ is called reachable if $\widetilde{G}$ is surjective, observable if $\widetilde{H}$ is injective, and canonical if it is both reachable and observable.

We can also present the notion of an abstract i/o-map and define realizations. The usual existence and uniqueness results will follow.

3.3. Definition.

Suppose given an algebra of generalized difference operators $\varphi: \mathcal{K} \to R$ with input functor Ω and output functor Γ. An i/o-map is an R-module homomorphism $f: \Omega\mathcal{U} \to \Gamma\mathcal{Y}$ for any $\mathcal{K}$-modules $\mathcal{U}$ and $\mathcal{Y}$.

An $R/\mathcal{K}$ system $\Sigma = (\mathcal{X}, \mathcal{U}, \mathcal{Y}; G, H)$ is a realization of f if $f = f_\Sigma$. Equivalently, the R-module diagram

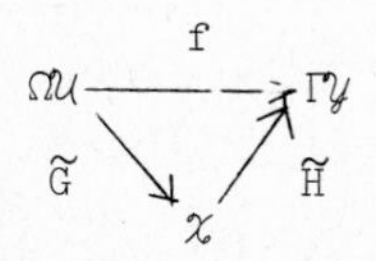

commutes. The system Σ is called a canonical realization of it is canonical as a system.

Note that no finiteness conditions are required for realizations. In fact, the appropriate "finiteness" conditions in this general context are quite elusive. From our point of view, every i/o map has at least one realization, for example: $\mathcal{X} = \Omega\mathcal{U}$, $\widetilde{G}$ = identity, $\widetilde{H} = f$. Moreover, the following familiar result holds.

3.4. Theorem.

If $f: \Omega\mathcal{U} \to \Gamma\mathcal{Y}$ is an i/o-map, then f has a canonical realization $\Sigma = (\mathcal{X}, \mathcal{U}, \mathcal{Y}; G, H)$. Furthermore Σ is unique: if $\Sigma_1 = (\mathcal{X}_1, \mathcal{U}_1, \mathcal{Y}_1; G_1, H_1)$ is another canonical realization, there is a unique R-module isomorphism $T: \mathcal{X} \to \mathcal{X}_1$ such that $GT = G_1$ and $TH_1 = H$.

Proof. This follows by arguments similar to those found in the literature; see [14] for example. One approach to uniqueness is to show that a canonical realization is terminal in the category of reachable realizations, properly defined.

Finally, we point out that adjoint functors have been used extensively in system theory, and various canonical realization results have been proved. The interested reader should consult [4], especially the introductory surveys by Arbib-Manes and Anderson in that volume. Sontag [6] contains a different approach altogether.

SECTION 4. POSETS, INCIDENCE ALGEBRAS, AND DYNAMICAL MODULES

In this section we present the "time-set" T, which will be any locally finite poset. A commutative ring k will be fixed for the discussion, and $\hat{R}$ will denote the incidence algebra of T over k, as defined in the introduction. The reader should consult [10] and references cited there for more information about incidence algebras. We define a subring of $\hat{R}$ which is a more suitable application for system theory.

4.1. Definition.

Let T be a locally finite poset. An incidence function $f \in \hat{R}$ is said to have finite memory if for every $t \in T$, $\{s: s \leq t \text{ and } f(s,t) \neq 0\}$ is finite. The set of all functions with finite memory is denoted R and is called the finite memory algebra of T.

The reader should check that R is really a subring of $\hat{R}$. Next we give some particular elements of R which will be important. For fixed $s \leq t \in T$, let

$$\delta_{s,t}(u,v) = \begin{cases} 1 & \text{if } u = s \text{ and } v = t \\ 0 & \text{otherwise} \end{cases},$$

and for $t \in T$, let $e_t = \delta_{t,t}$. Some useful identities involving these functions can be found in [10, p. 271]. We observe that the e_t are orthogonal idempotents: $e_t^2 = e_t$ and $e_s e_t = 0$ if $s \neq t$. We write $\mathcal{K} = k^T$, and consider $\mathcal{K}$ as the subring of "diagonal functions,"

$$a(s,t) = \begin{cases} a(t) & \text{if } s = t \\ 0 & \text{otherwise} \end{cases}.$$

If k is a field, then $\mathcal{K}$ is the "semi-simple part" of R ; cf. [10, Prop. 3.3].

A few remarks about the finite memory algebra may be helpful. The ring R is "large" in the sense that it is dense in $\hat{R}$ with respect to the standard topology [10, p. 271]. Also, R is large enough to determine T uniquely; cf. "Stanley's Theorem" [10, Th. 3.2], which holds with $\hat{R}$ replaced with R. Of course, $R = \hat{R}$ if T is finite, or if each $t \in T$ has only a finite number of precedessors. If this last condition fails for T, then the important zeta function ($\zeta(s,t) = 1$ whenever $s \leq t$) is not in R. On the other hand, the Möbius function $\mu = \zeta^{-1}$ does lie in R for many posets. This illustrates the philosophy that R has fewer units, and hence a richer ideal structure, than $\hat{R}$. All this will make more sense in the specific cases considered later. Now, in preparation for the study of $R/\mathcal{K}$-system, we discuss the structure of right R-modules.

Let $\mathcal{X}$ be any right R-module. Then for each $t \in T$ we define

$$\mathcal{X}_t = \mathcal{X} e_t = \{xe_t : x \in \mathcal{X}\} .$$

If $a \in \mathcal{K} \subset R$, then $ae_t = e_t a$ for all t, so that each X_t is a right $\mathcal{K}$-submodule of $\mathcal{X}$. We denote xe_t by $x(t)$ and call it the "t-component" of x. The collection of mappings $x \to x(t)$ give a $\mathcal{K}$-module map

$$(*) \qquad \mathcal{X} \to \prod_{t \in T} X_t .$$

It is easy to see that the map $(*)$ is an isomorphism of right $\mathcal{K}$-modules if T is a finite poset: write $\delta = \Sigma e_t$ in R. If T is infinite, $(*)$ is conjectirally an isomorphism (J. Ferrar), but the proof of this seems to depend heavily on set theory and ultrafilters! We dodge the problem with a definition.

4.2. Definition.

Let T be a poset, and let R be the finite memory algebra of T. A right R-module $\mathcal{X}$ is called dynamical if the map in $(*)$ above is a $\mathcal{K}$-module isomorphism.

Note that the modules of the form X^T discussed in the introduction are dynamical, and it can be proved that canonical realizations of reasonable i/o-maps are dynamical, justifying the definition. (cf. the next section). On the other hand, we cannot in general expect the "t-coordinate" modules X_t to be finitely generated over the base ring k.

Dynamical R-modules can be given an attractive explicit description.

4.3. Proposition.

Suppose given a poset T, and a commutative ring k of scalars, and let R be the finite memory algebra of T over k. Suppose $\mathcal{X} \cong \prod_{t \in T} X_t$ is a dynamical R-module. Then:

(a) Each X_t is a right k-module.

(b) For each $s \leq t$ in T, there exists a k-linear map $\Phi(s, t): X_s \to X_t$, such that for all $\alpha \in R$:

$$(x\alpha)(t) = \sum_{s \leq t} x(s)\Phi(s,t)\alpha(s,t) \quad .$$

(c) The maps $\Phi(s,t)$ satisfy the following "coherence" assumptions.

1. $\Phi(t,t) = I$, the identity on X_t for all $t \in T$
2. $\Phi(s,r) \cdot \Phi(r,t) = \Phi(s,t)$ for all $s \leq r \leq t$ in T.

Conversely, a collection $\{X_t\}$ of k-modules together with k-linear mappings $\Phi(s,t): X_s \to X_t$ satisfying (c) given rise to a dynamical R-module.

Proof. We limit our comments here to the fact that $\delta_{s,t} = e_s \cdot \delta_{s,t} \cdot e_t$ in R, so that right multiplication by $\delta_{s,t}$ on $\mathcal{X}$ maps X_s to X_t whenever $s \leq t$. This action gives the $\Phi(s,t)$.

A system-theoretic interpretation of this description will be given in the next section.

SECTION 5. LINEAR SYSTEMS ON LOCALLY FINITE POSETS

Suppose given a locally finite poset T and a ring k of scalars, giving rise to a ring $\mathcal{K} = k^T$ of k-valued time functions and a finite memory algebra R. As usual, there is a natural imbedding $\mathcal{K} \subset R$, and our first inclination is to define a <u>linear dynamical system on</u> T to be nothing but an $R/\mathcal{K}$ system in the sense of Section 3. This is essentially what we will do, but we put some restriction on the modules involved.

5.1. Definition.

Suppose given T, k, $\mathcal{K}$, and R as above. Let $\{U_t\}$, $\{Y_t\}$, $t \in T$, be two families of k-modules, and define $\mathcal{U} = \prod_{t \in T} U_t$, $\mathcal{Y} = \prod_{t \in T} Y_t$. Furthermore, let $\mathcal{X} \cong \prod_{t \in T} X_t$ be a dynamical R-module, and for each $t \in T$ assume given $G(t): U_t \to X_t$ and $H(t): X_t \to Y_t$. Define $\mathcal{K}$-module maps $G: \mathcal{U} \to \mathcal{X}$ by $G = \prod_{t \in T} G(t)$ and $H: \mathcal{X} \to \mathcal{Y}$ by $H = \prod_{t \in T} H(t)$. Then the $R/\mathcal{K}$ system $\Sigma = (\mathcal{X}, \mathcal{U}, \mathcal{Y}; G, H)$ will be called a <u>linear dynamical system on the poset</u> T.

The system-theoretic intuition goes as follows: we are given families of time-varying input spaces $\{U_t\}$ and output spaces $\{Y_t\}$. The state-module $\mathcal{X}$ is dynamical, so that it is described by a family of "instantaneous" state-modules X_t together with "state-transition maps" $\Phi(s,t): X_s \to X_t$, when $s \leq t$, as described in the introduction. The maps $G(t)$ and $H(t)$ are "instantaneous", or in Kamen's terminology, "local in time." They are pieced together to form G and H which are "global in time."

Just as in the abstract case, a system on T gives rise to an i/o-diagram of right R-modules

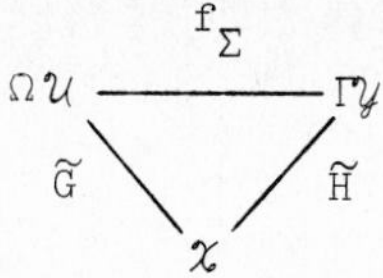

which we proceed to interpret in suggestive language.

We consider $\Omega\mathcal{U} = \mathcal{U} \otimes_{\mathcal{K}} R$ first. The map $G\colon \mathcal{U} \to \mathcal{X}$ is given by $(uG)(t) = u(t)G(t)$ for all $t \in T$. A simple tensor $u \otimes \alpha$ is thought of as an input function u together with a "weight" α, and

$$(u \otimes \alpha)G(t) = (uG)(t) = \sum_{s \leq t} u(s)G(s)\alpha(s,t)\Phi(s,t) .$$

Verbally, the input $u(s)$ at time s is multiplied by the weight $\alpha(s,t)$ and allowed to evolve freely to time t. The resulting "superposition" is finite, since α has finite memory.

The output module is given by $\Gamma\mathcal{Y} = \mathrm{Hom}_{\mathcal{K}}(R,\mathcal{Y})$. Suppose $x \in X$ and $\beta \in R$; then

$$(x\tilde{H})(\beta)(t) = (x\beta)H(t) = \sum_{s \leq t} x(s)\Phi(s,t)\beta(s,t)H(t) .$$

That is, the value of $x\tilde{H}$ at β depends on a weighted sum (determined by β) over part of the history of x. We return to this interpretation in the next section.

In this context we can also speak of "abstract" input/output maps: these will be right R-module mappings $f\colon \Omega\mathcal{U} \to \Gamma\mathcal{Y}$. According to Section 3, each such i/o map has a unique canonical realization Σ, which is an $R/\mathcal{K}$ - system. However, our "systems on the poset T" are not completely general $R/\mathcal{K}$ - systems, so some further discussion is necessary. The next theorem is the optimal result, but has only been proved under some restrictive hypotheses.

5.2. Theorem.

Suppose $\mathcal{U} = \prod U_t$ and $\mathcal{Y} = \prod Y_t$ as in Definition 5.1. If $f\colon \Omega\mathcal{U} \to \Gamma\mathcal{Y}$ is an i/o map, then the canonical realization of f as an $R/\mathcal{K}$ - system is a linear dynamical system on the poset T. In particular, the canonical state-module $\mathcal{X}$ is dynamical.

Proof. This theorem has been proved completely only under the hypothesis that the base ring k is noetherian and the (U_t) form a constant finitely generated family. That is, all $U_t \cong U$ for some fixed f.g. k-module U. (These hypotheses insure $\mathcal{U} \cong U \otimes_k \mathcal{K}$, which is a technical requirement in my proof.) I conjecture that these hypotheses are not really necessary.

The explicit ingredients in the canonical realization are given in Definition 5.1. It is easy to see that an explicit form of the uniqueness result goes as follows:

5.3. Theorem.

Suppose $f\colon \Omega\mathcal{U} \to \Gamma\mathcal{Y}$ Is an i/o map with canonical realization $\Sigma = (\mathcal{X}, \mathcal{U}, \mathcal{Y}; G, H)$ and $\Sigma' = (\mathcal{X}', \mathcal{U}', \mathcal{Y}': G', H')$. Then for each $t \in T$, there exists a k-linear isomorphism $P(t)\colon X_t \to X'_t$ such that $\Phi(s,t)P(t) = P(s)\Phi'(s,t)$ whenever $s \leq t$; and $G(s)P(s) = G'(s)$ and $H'(t)P(t)$ for all $s,t \in T$.

These results will be applied to difference equations in the next section.

SECTION 6. DIFFERENCE EQUATIONS AND REALIZATIONS

In this section we define difference equations on posets, in a way suggested by work of Elliott and Mullans [11]. The same idea occurs in a different context in Hafiz [2, p. 25ff.]. These difference equations are used to construct i/o-maps, and a fairly concrete view of realization theory is presented. The reader should consult the next two sections for motivating examples.

Suppose given a locally finite poset T, scalar ring k, $\mathcal{K} = k^T$, and finite-memory algebra R as usual. In this section we fix k-modules U and Y and set $\mathcal{U} = U^T$ and $\mathcal{Y} = Y^T$. These are right $\mathcal{K}$-modules, and we can make them into right R-modules by setting, for instance

$$(u\alpha)(t) = \sum_{s \leq t} u(s)\alpha(s,t) ,$$

with $u \in \mathcal{U}$ and $\alpha \in R$.

Suppose, in addition, that $U \cong Y \cong k$, so that $\mathcal{U} \cong \mathcal{Y} \cong \mathcal{K}$. In this context a <u>scalar difference equation</u> is given by $y\alpha = u\beta$, with $u \in \mathcal{U}$, $y \in \mathcal{Y}$, and $\alpha, \beta \in R$. We always assume that α is invertible in the full incidence algebra $\hat{R}$, and set $A(s,t) = (\beta\alpha^{-1})(s,t)$. In this case, just as in the introduction, the <u>i/o-map of the equation</u> is $f\colon \Omega U \to \Gamma Y$ determined by

$$(uf)(\theta)(t) = \sum_{s \leq t} u(s)\theta(s,t)A(s,t) . \qquad (*)$$

More generally, suppose given any two k-modules U and Y, and for each $s \leq t$ a k-linear map $A(s,t)\colon U \to Y$. Then the equation $(*)$ can still be used to define an i/o-map. Problems of this sort occur in [1] and [11].

6.1. Definition.

Given a poset T, commutative base-ring k, and families of k-modules U_t and Y_t, $t \in T$, a <u>set of input/output data</u> is a collection of k-linear maps $A(s,t)\colon U_s \to Y_t$, $s \leq t$.

Define $\mathcal{U} = \prod_{t \in T} U_t$ and $\mathcal{Y} = \prod_{t \in T} Y_t$. Then the i/o-map corresponding to these data is the R-linear map $f: \Omega\mathcal{U} \to \Gamma\mathcal{Y}$ given by

$$(uf)(\theta)(t) = \sum_{s \leq t} u(s)\theta(s,t)A(s,t) \quad .$$

According to Theorem 5.2 (which has, however, not been proved in complete generality), an i/o-map has a unique canonical realization which is a system on the poset T . (Recall that this requires a special form for $\mathcal{X}$). Translating this result, we get:

6.2. Theorem

Suppose given i/o-data $A(s,t): U_s \to Y_t$ as above. Then there exists a family $\{X_t : t \in T\}$, k-linear maps $\Phi(s,t): X_s \to X_t$ such that $\Phi(t,t) = I$ on X_t and $\Phi(s,r)\Phi(r,t) = \Phi(s,t)$ when $s \leq r \leq t$, together with $G(s): U_s \to X_s$ and $H(t): X_t \to Y_t$ such that for every $s,t,\ s \leq t$:

$$A(s,t) = G(s)\Phi(s,t)H(t) \ .$$

These ingredients can be chosen to form a canonical $R/\mathcal{K}$-system, and if this is done they are unique up to isomorphism: see Theorem 5.3.

Proof. We only show how to verify the formula. Suppose

$$\begin{array}{ccc} \Omega\mathcal{U} & \xrightarrow{\quad f \quad} & \Gamma\mathcal{Y} \\ & {\scriptstyle \widetilde{G}} \searrow \quad \nearrow {\scriptstyle \widetilde{H}} & \\ & \mathcal{X} & \end{array}$$

is the realization. We have $f = \widetilde{G}\,\widetilde{H}$; so that for any $\alpha \in R$ and $t \in T$;

$$(uf)(\alpha)(t) = u\widetilde{G}\,\widetilde{H}(\alpha)$$

$$\sum_{s \leq t} u(s)A(s,t)\alpha(s,t) = \sum_{s \leq t} u(s)G(s)\Phi(s,t)\alpha(s,t)H(t)$$

where $\Phi(s,t)$ is the state-transition map for $\mathcal{X}$. Since this is true for any $\alpha \in R$, choosing $\alpha = \delta_{s,t}$ yields $A(s) = G(s)\Phi(s,t)H(t)$ as required.

SECTION 7. TIME-VARYING DIFFERENCE EQUATIONS ON $\mathbb{Z}$

In this section we present the theory of time-varying difference equations on $\mathbb{Z}$ as an example of the general theory. Many of the results are due to Kamen and Hafiz [1-2] , who have carried out a much more general and detailed analysis of systems on $\mathbb{Z}$. The use of the incidence algebra and the explicit treatment of realizations are new.

First we describe the relationship between Kamen's ring $\mathcal{K}_\sigma[z]$ (cf. 2.2 in the Introduction) and the incidence algebra of $\mathbb{Z}$. In this section we set $T = \mathbb{Z}$. Let $\hat{R}$ be the full incidence algebra of T, and define a special function z in $\hat{R}$ by

$$z(s,t) = \begin{cases} 1 & \text{if } s = t-1 \\ 0 & \text{otherwise} \end{cases} .$$

As usual $\mathcal{K} = k^T \subset \hat{R}$. If $a \in \mathcal{K}$, then an easy calculation shows that $az = za^\sigma$, where $a^\sigma(t) = t-1$. This is Kamen's fundamental relation, and exhibits $\mathcal{K}_\sigma[z]$ as a subring of $\hat{R}$. Actually we get

$$\mathcal{K}_\sigma[z] \subset R \subset \hat{R}$$

where R is the finite memory algebra. Both inclusions are proper, and $\mathcal{K}_\sigma[z]$ may be regarded as "incidence functions with uniformly bounded memory." The next lemma gives an interesting new description of $\hat{R}$.

7.1. Lemma

If T is the integers, then $\hat{R} \cong \mathcal{K}_\sigma[[z]]$.

Proof. Here $\mathcal{K}_\sigma[[z]]$ denotes formal power series in z subject to the relation $az = za^\sigma$. The proof of the isomorphism is straightforward, the map $\hat{R} \to \mathcal{K}_\sigma[[z]]$ being

$$\theta \to \sum_{i=1}^{\infty} z^i \theta_{-i}$$

where for $\theta \in \hat{R}$, $\theta_{-i}(t) = \theta(t-i,t)$.

We consider scalar difference equations of the form

$$y(t) - y(t-1)a_{n-1}(t) - \dots - y(t-n)a_o(t) = u(t) ,$$

or, in another form,

$$y(1 - za_{n-1} - \dots - z^n a_o) = u .$$

Here $\mathcal{U}$ and $\mathcal{Y}$ are both isomorphic to $\mathcal{K}$. The functors are given by

$$\Omega\mathcal{U} = \mathcal{U} \underset{\mathcal{K}}{\otimes} R \cong R$$

$$\Gamma\mathcal{Y} = \mathrm{Hom}_{\mathcal{K}}(R,\mathcal{Y}) \cong \mathrm{Hom}_{\mathcal{K}}(R,\mathcal{K}) .$$

Writing $\alpha = 1 - za_{n-1} - \dots - z^n a_o$, we see that α is invertible in $\hat{R}$. Following (6.1) and identifying $\Omega\mathcal{U}$ and $\Gamma\mathcal{Y}$ as above, we define an i/o map

$$f \colon R \to \mathrm{Hom}_{\mathcal{K}}(R,\mathcal{K})$$

$$(\delta f)(\theta)(t) = \sum_{s \le t} (\delta\, \theta \alpha^{-1})(s) .$$

A rather involved calculation establishes the basic result:

7.2. Proposition

With notation as above, define $\alpha^* = z^n - a_{n-1}z^{n-1} - a_{n-2}z^{n-2} - \ldots - a_o$. Then $\alpha^* R \subset \ker f$.

This means we should be able to use α^* to make a realization. An approximation argument shows that $R/\alpha^* R \cong \mathcal{K}_\sigma[z]/\alpha^*\mathcal{K}_\sigma[z]$, and this last module has been studied extensively in [2]. Applying that work with some changes in notation, we get:

7.3. Proposition

Consider the difference equation

$$y(t) - y(t-1)a_{n-1}(t) - \ldots - y(t-n)a_o(t) = u(t) .$$

This equation has a canonical realization $\Sigma = (\mathcal{X}, \mathcal{U}, \mathcal{Y}; \Phi(s,t), G, H)$ where

$$\mathcal{U} = \mathcal{K} \;,\; \mathcal{Y} = \mathcal{K} \;,\; \mathcal{X} = \mathcal{K}^n = (k^n)^T$$

$$G(s) \equiv 1 \;;\; H(t) \equiv 1$$

$\Phi(s,t) = F(s+1) \ldots F(t)$, where

$$F(t) = \begin{pmatrix} 0 & 0 & 0 & a_o(t) \\ 1 & 0 & 0 & a_1(t) \\ 0 & 1 & 0 & . \\ . & . & . & . \\ . & . & . & . \\ . & . & . & . \\ 0 & 0 & 1 & a_{n-1}(t) \end{pmatrix}$$

Note that Proposition 7.2 only implies that Σ is reachable, but one can check directly that Σ ib observable. This shows a <u>posteriori</u> that $\ker f = a^* R$ in (7.2).

I would like to close this section by admitting that time-varying realization theory on $\mathbb{Z}$ needs a lot more work. I do not know a reliable way of computing the kernels of input/output maps, a task which is usually previous to the application of [2]. For instance, if $\beta \neq \delta$, the equation $y\alpha = u\beta$ is substantially more complicated than $y\alpha = u$!

SECTION 8. THREE EXAMPLES

In this section we present briefly three explicit examples which indicate the breadth and difficulties of the theory.

8.1. EXAMPLE: Feed-through lines and the Jacobson Radical.

Let T be any locally finite poset. Let $U = Y$ be a k-module, and let $\mathcal{U} = U^T$. Consider the "feed-through" equation

$$y(t) = u(t) .$$

This leads to the i/o-data $A(s,t) = \delta(s,t)I$, that is, $A(s,t)$ is the identity: $U \to Y$ if $s = t$, and zero otherwise. The corresponding i/o-map $f: \Omega\mathcal{U} \to \Gamma\mathcal{Y}$ is given by

$$(u \otimes \alpha)f(\theta)(t) = u(s)\alpha(t,t)\theta(t,t)$$

Let $J = \{\alpha \in R: \alpha(t,t) = 0 \text{ for all } t \in T\}$. (If k is a field, J is the Jacobson radical of R .) We see that $\Omega\mathcal{U} \cdot J \subseteq \ker(f)$, and since $\Omega\mathcal{U}/\Omega\mathcal{U}\cdot J \cong \mathcal{U}$, this leads to a canonical realization $\Sigma = (\mathcal{X}, \mathcal{U}, \mathcal{Y} ; \Phi(s,t), G, H)$ with $\mathcal{X} \cong \mathcal{U}$, $G(s) \equiv 1$, $H(t) \equiv 1$, $\Phi(s,t) = A(s,t) = \delta(s,t)I$, as expected. This shows that the theory should be thought of as a "Moore-theory" rather than a "Mealy-theory" since we need enough states to accomodate the input space.

8.2. EXAMPLE. Systems on a Binary Tree.

Let T be a binary tree, which we write sideways:

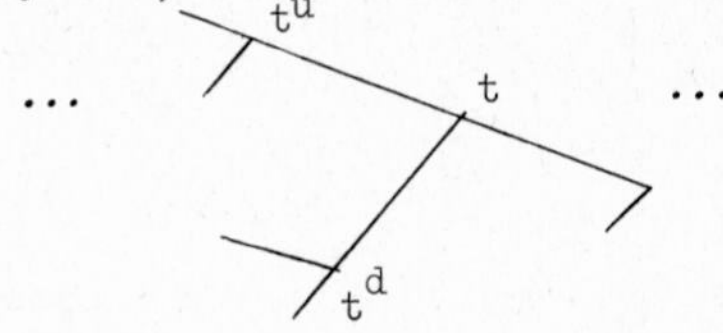

Each $t \in T$ has an up-predecessor t^u and a down-predecessor t^d . We introduce two special incidence functions z_u and z_d , where, for instance

$$z_u(s,t) = \begin{cases} 1 & \text{if } s = t^u \\ 0 & \text{otherwise} \end{cases} .$$

Define shift operators u and d on $K = k^T$ by $a^u(t) = a(t^u)$ and $a^d(t) = a(t^d)$. The communtation relation $az_u = z_u a^u$ and $az_d = z_d a^d$ are easily verified, leading to

$$\mathcal{K}\{z_u, z_d\} \subset R \subset \hat{R} = \mathcal{K}\{\{z_u, z_d\}\}$$

where the braces indicate that z_u and z_d do not commute with each other; the commutation relations of variables with coefficients are given above.

Suppose given the scalar difference equation:

$$y(t) - y(t^u)a(t) - y(t^d)b(t) = u(t)$$

or:

$$y(\delta - z_u a - z_d b) = u .$$

In this scalar case, we can identify $\Omega\mathcal{U} = R$ and $\Gamma\mathcal{Y} = \mathrm{Hom}_{\mathcal{K}}(R,\mathcal{K})$. Just as in the last section we get an i/o-map

$$f : R \to \mathrm{Hom}_{\mathcal{K}}(R,\mathcal{K})$$

$$(\gamma f)(\theta) = \sum_{s \leq t} (\gamma\theta)(s,t)A(s,t)$$

where $A(s,t) = (\delta - z_u a - z_d b)^{-1}(s,t)$.

Although the theory is not well enough developed to compute $\ker f$ reliably, a little experimentation suggests at least that $I = (z_u - a,\ z_d - b)R \subset \ker f$. This leads to a realization attempt with $\mathcal{X} = R/I = \mathcal{K} = k^T$, so that the local state-space X_t is one-dimensional for each t . The weighting pattern $\Phi(s,t)$ is computed as follows: $\Phi(t^u,t) = a(t)$, $\Phi(t^d,t) = b(t)$. If $s \leq t$, there is a unique path $s = s_1 < s_2 < \ldots < s_n = t$ such that for each i either $s^u_{i+1} = s_i$ or $s^d_{i+1} = s_i$: define $\Phi(s,t) = \Phi(s_1,s_2)\ \Phi(s_2,s_3)\ \ldots\ \Phi(s_{n-1},s_n)$. Finally, take $G(s) \equiv 1$ and $H(t) \equiv 1$. One can show by brute force that $A(s,t) = G(s)\ \Phi(s,t)\ H(t)$, but this state of affairs is very unsatisfactory and a coherent algebraic technique is needed.

8.3. EXAMPLE. Systems on the discrete-plane.

In this example $T = \mathbb{Z} \oplus \mathbb{Z}$. Introduce the unit vectors $e_1 = (1,0)$ and $e_2 = (0,1)$, and two functions z_1, z_2 by

$$z_i(s,t) = \begin{cases} 1 & \text{if } s = t - e_i \\ 0 & \text{otherwise} \end{cases}$$

The functions z_1 and z_2 commute with each other, but shift functions $a \in \mathcal{K} = k^T$ in a predictable way. One sees that

$$\mathcal{K}[z_1,z_2] \subset R \subset \hat{R} = \mathcal{K}[[z_1,z_2]] \ .$$

We consider the scalar equation

$$y(t) - y(t-e_1) - y(t-e_2) = u(t) \ , \quad \text{or}$$

$$y(\delta - z_1 - z_2) = u$$

The corresponding i/o map $f : R \to \mathrm{Hom}_{\mathcal{K}}(R,\mathcal{K})$ has $\ker f = (z_1 z_2 - z_1 - z_2)R$, leading to a canonical state-module

$$\mathcal{X} \cong \left(\frac{k[z_1,z_2]}{(z_1 z_2 - z_1 - z_2)} \right)^T \ .$$

It $s \leq t$, write $t - s = (i,\ j)$, and set $\Phi(s,t) = z_1^i\ z_2^j : X_s \to X_t$. This example is particularly interesting because the local state-spaces X_t are not finite-dimensional over k .

REFERENCES.

1. Kamen, E. W., A New Algebraic Approach to Linear Time-Varying Systems. Preprint: March, 1974. To appear in J. Comp. Sys. Sci.

2. Hafiz, K. M., New Results on Discrete-Time, Time-Varying Linear Systems. Thesis: Georgia Institute of Technology, March, 1975.

3. Wyman, B. F., Linear Systems over Rings of Operators, in [4, pp. 218-223].

4. Manes, E. G., ed., Category Theory Applied to Computation and Control, Lecture Notes in Computer Science, Vol. 25, Springer-Verlag, 1975

5. Johnston, R., Linear Systems over Various Rings, M.I.T. Dissertation, Report ESL-R-497, 1973.

6. Sontag, E.D., On Linear Systems and Non-commutative Rings. Pre-print, 1974-75. To appear in J. Math. Systems Theory.

7. Roesser, R. P., A Discrete State-Space Model for Linear Image Processing, IEEE Trans. Auto. Control, Vol. AC-20, 1975, 1-10.

8. Kalman, R. E., Algebraic Structure of Linear Dynamical Systems I. The Module of Σ. Proc. Nat. Acad. Sci. (USA), 54, 1965, 1503-1508.

9. Brockett, R. W., Finite Dimensional Linear Systems, Wiley, 1970.

10. Doubilet, P., G.-C. Rota, and R. Stanley, On the Foundations of Combinatorial Theory VI: The Idea of Generating Function, Proc. Sixth Berkeley Symp. Math. Stat. and Prob. Vol. II, 267-318. U. Calif. Press, 1972.

11. Elliott, D. and R. Mullans, Linear Systems on Partially Ordered Time Sets, IEEE Decision and Control Conference, Dec. 1973.

12. Cartan, H., and S. Eilenberg, Homological Algebra, Princeton, 1956.

13. Ho, Y. C., and K. C. Chu, Team Decision Theory and Information Structures in Optimal Control Problems, Part I. IEEE Trans. Auto. Control, Vol. AC-17 (1972), 15-22.

14. Goguen, J., Minimal Realization of Machines in Closed Categories, Bull. A.M.S., 78, 1972, 777-783.

FUNCTIONAL EXPANSIONS AND HIGHER ORDER NECESSARY CONDITIONS IN OPTIMAL CONTROL

Roger W. Brockett
Harvard University
Division of Engineering and Applied Physics
Cambridge, Massachusetts 02138
U.S.A.

ABSTRACT

The recently developed Volterra Series expansion formulas are extended here to include differential equations whose right hand sides have a nonlinear u dependence. They are also worked out in the case where only a finite number of derivatives exist. This gives a theory which provides a very natural and conceptually clear approach to the problem of finding necessary conditions of arbitrarily high order for optimal controls.

1. INTRODUCTION

In reference [1] it is shown that for

$$\dot{x}(t) = f[x(t)]+u(t)g[x(t)]; \quad y(t) = h[x(t)]; \quad x(0) = \text{given}$$

there exists a Volterra series

$$y(t) = y_o(t) + \sum_{k=1}^{\infty} \int_0^{\sigma_1} \cdots \int_0^{\sigma_k} w_k(t,\sigma_1,\sigma_2,\ldots\sigma_k)$$

$$u(\sigma_1)u(\sigma_2)\ldots u(\sigma_k)d\sigma_1 d\sigma_2 \ldots d\sigma_k$$

which represents the input-output map defined by the given differential equation on any finite interval $[0,T]$ provided $|u(t)|$ is not too large and f, g and h are analytic. In [2] there is described a type of summability with respect to which the Volterra series converges for all piecewise continuous, bounded u. If, instead of analyticity, we assume that f and h are merely k-times differentiable and that g is k-1 times differentiable then the methods used in [1] allow one to compute the first k kernels in a Volterra series expansion and to conclude that the actual y(t) and the approximation produced by the first k kernels differ by terms which are of order higher than k.

This paper was written while the author held a Guggenheim Fellowship, partial support from the U.S. Office of Naval Research under the Joint Services Electronics Program by Contract N00014-75-C-0648, Division of Engineering and Applied Physics, Harvard University, Cambridge, Mass. 02138 is also acknowledged.

If we have m inputs, $u_1,u_2,\ldots u_m$ instead of just one, then the results are the same but we have more indices. We look for a series of the form

$$y(t) = \sum_{k=1}^{\infty} \int_0^t \int_0^{\sigma_1} \cdots \int_0^{\sigma_k} \sum w^k_{\ell\ldots p}(\sigma_1,\sigma_2,\ldots\sigma_k) \cdot$$

$$u_\ell(\sigma_1)\ldots u_p(\sigma_k)d\sigma_1 d\sigma_2\ldots d\sigma_k$$

where the inner sum is over all possible k-tuples with entries 1,2,...m.

Consider a new situation now. Suppose there is only one input but that it enters nonlinearly. For example, consider

$$\dot{x}(t) = f[x(t)]+g_1[x(t)]u(t)+g_2[x(t)]u^2(t)+\ldots+g_m[x(t)]u^m(t)$$

Treating $u^1,u^2,\ldots u^m$ as distinct inputs we get an expansion of the form

$$y(t) = \int_0^t \int_0^{\sigma_1} \cdots \int_0^{\sigma_k} w_k(t,\sigma_1,\sigma_2\ldots\sigma_k)u(\sigma_1)u(\sigma_2)\ldots u(\sigma_k)d\sigma_1 d\sigma_2\ldots d\sigma_k+y_o(t)$$

but now in general the kernels are no longer functions but must be regarded as distributions in the sense of Schwartz. Again, if we have only k-derivatives then a kth order approximation exists.

The relevance of this for optimal control problems may be explained as follows. Suppose that we have

$$\dot{x}(t) = f[x(t),u(t)]; \quad x(0) = x_o; \quad u(t) \in \Omega$$

$$y(t_1) = \int_0^{t_1} L[x(t),u(t)]dt$$

Assume the final end point on x is free and that we wish to minimize $y(t_1)$. We consider then the input-output system

$$\frac{d}{dt}\tilde{x} = \begin{bmatrix}\dot{x}_o\\ \dot{x}\end{bmatrix} = \begin{bmatrix}L[x(t),u(t)]\\ f[x(t),u(t)]\end{bmatrix} = f[\tilde{x}(t),u(t)]; \quad y(t) = x_o(t)$$

If the functions f and L have a certain degree of differentiability we may expand this input-output map in a Volterra series up to some order

$$y(t) = y_o(t) + \int_0^t w_1(t,\sigma_1)u(\sigma_1)d\sigma_1 +$$

$$\int_0^t \int_0^{\sigma_1} w_2(t,\sigma_1,\sigma_2)u(\sigma_1)u(\sigma_2)d\sigma_1 d\sigma_2+\ldots$$

This is, of course, nothing but a Taylor series expansion for y(t) in terms of u. If 0 is an interior point of Ω and if the control $u(\cdot) = 0$ is to be optimal then it is more or less clear, but we will enlarge upon these points later, that the first kernel must be zero, the second nonnegative definite, the third zero on those functions which the second annihilates, etc. Regardless of whether the second kernel is symmetrized with respect to σ_1 and σ_2 or not, a necessary condition for the nonnegativity of the second term is $w_2(t_1,\sigma,\sigma) \geq 0$ for all $0 \leq \sigma \leq t_1$. This gives the second order necessary condition first discussed by Gabasov [3] and Jacobson [4]. However, there are additional conditions on w_2 which must be satisfied. For example, if w_2 is symmetrized with respect to σ_1 and σ_2 then we will show that

$$w(t_1,\sigma_1,\sigma_1)w(t_1,\sigma_2,\sigma_2)-[w(t_1,\sigma_1,\sigma_2)]^2 \geq 0; \quad 0 \leq \sigma_1,\sigma_2 < t_1$$

is a necessary condition. Taking the limit as σ_1 approaches σ_2 we get Kelly's necessary condition [5]. These conditions can be extended to the vector input case and have higher order analogs as well.

In this paper we carry out some of the details substantiating the claims made above and their extensions. Our main point is that in optimal control, as in ordinary function minimization, the most natural approach toward getting necessary conditions utilizes the Taylor series. By taking this point of view we get a conceptually straightforward and computationally tractable method of discovering and organizing results on singular control. We draw freely on the developments of [1] in order to avoid reproducing the analysis given there. The reader is urged to consult that paper for background material since the arguments given here are somewhat sketchy in places.

2. LINEARIZATION OF MAPS

We describe here in the context of maps from $\mathbb{R}^n$ into $\mathbb{R}^m$ the basic idea behind the Volterra series in the analytic case and the C^k case. Both the idea and the notation will be useful later.

Let x be an n-tuple with components $x_1,x_2,...x_n$. By $x^{[p]}$ we understand an $\binom{n+p-1}{p}$-tuple (binomial coefficient) whose components are the linearly independent symmetric forms which are homogeneous of degree p in $x_1,x_2,...x_n$. That is $x^{[p]}$ has components $\alpha_1x_1^p$, $\alpha_2x_1^{p-1}x_2$, $\alpha_3x_1^{p-1}x_3,...$ where the coefficients α_i are chosen in such a way as to have

$$||x||^{2p} = ||x^{[p]}||^2$$

with $||\cdot||$ being the Euclidean norm. Such a choice of coefficients is possible since we need only equate the coefficients of like terms in

$$(x_1^2+x_2^2+...x_n^2)^p = (\alpha_1x_1^p)^2 + (\alpha_2x_1^{p-1}x_2)^2 + ...$$

and since all coefficients on the left are positive the square roots necessary to express the α_i will be real.

Suppose $f : \mathbb{R}^n \to \mathbb{R}^n$ has k continuous derivatives. We can express f by means of a Taylor series about x = 0 of a Taylor series about $x = 0$

$$f(x) = F_o + F_1 x + F_2 x^{[2]} + \dots F_k x^{[k]} + \tilde{f}(x)$$

where

$$\lim_{||x|| \to 0} ||\tilde{f}(x)||/||x||^k = 0 .$$

Let $y = f(x)$. Then $y^{[1]}, y^{[2]} \dots y^{[k]}$ can also be computed to a certain degree of accuracy. That is,

$$\begin{bmatrix} y^{[1]} \\ y^{[2]} \\ \vdots \\ y^{[k]} \end{bmatrix} = \begin{bmatrix} F_o \\ F_o^{[2]} \\ \vdots \\ F_o^{[k]} \end{bmatrix} + \begin{bmatrix} F_{11} & F_{12} & \dots & F_{1k} \\ F_{21} & F_{22} & \dots & F_{2k} \\ \dots & \dots & \dots & \dots \\ 0 & 0 & \dots & F_{kk} \end{bmatrix} \begin{bmatrix} x^{[1]} \\ x^{[2]} \\ \vdots \\ x^{[k]} \end{bmatrix} + \tilde{f}_k(x)$$

The kth-approximation of F is embedded in the <u>first</u> approximation of this new mapping. In fact, the kth approximation of the original map determines the first approximation completely in view of the following property of the F_{ij}. If $y = Ax$ is linear we define $A^{[p]}$ as the map which gives $y^{[p]} = A^{[p]} x^{[p]}$. A little thought will convince the reader that in terms of this notation

$$F_{ij} = F_{1,j-1}^{[i]} \ ; \quad j \geqslant i$$

Of course $F_{ij} = 0$ for $j < i-1$.

3. LINEARIZATION OF DIFFERENTIAL EQUATIONS

Now consider the differential equation

$$\dot{x}(t) = f[x(t), u(t), t]; \quad x(0) = x_o \qquad (*)$$

If f has a continuous first derivative with respect to $[x,u]$ on some open set S then we can linearize (*) about the free solution passing through $x_o \in S$ which corresponds to a particular choice of u to get

$$\delta\dot{x}(t) = (\frac{\partial f}{\partial x})\delta x(t) + (\frac{\partial f}{\partial u})\delta u(t)$$

It is easy to show that in S the true solution of (*) is given by

$$x(t) = x_o(t) + \delta x(t) + n(t)$$

where on $[0,T]$

$$\lim_{\sup|\delta u(t)| \to 0} ||n(t)||/(\sup_{t \in [0,T]} |\delta u(t)|) = 0$$

The development of the previous section suggests then that in order to compute a higher order approximation to x it would be appropriate to form a differential equation for $x^{[1]}, x^{[2]}, \dots x^{[p]}$ and to linearize it. Suppose that we have (from now on we do not display t dependence to simplify the notation)

$$\dot{x} = F_o + F_1 \begin{bmatrix} x \\ u \end{bmatrix} + F_2 \begin{bmatrix} x \\ u \end{bmatrix}^{[2]} + \ldots + F_k \begin{bmatrix} x \\ u \end{bmatrix}^{[k]} + \tilde{f}(x,u)$$

This can be extended, using an obvious notation, to

$$\frac{d}{dt} x^{[2]} = F_{21}x^{[1]} + F_{22}x^{[2]} + \ldots F_{2k}x^{[k]} + G^2_{20}$$
$$+ uG^2_{21}x^{[1]} + uG^2_{22}x^{[2]} + \ldots uG^2_{2k}x^{[k]}$$
$$\cdot \quad \cdot \quad \cdot \quad \cdot \quad \cdot \quad \cdot \quad \cdot \quad \cdot \quad \cdot$$
$$+ u^k G^k_{21}x^{[1]} + u^k G^k_{22}x^{[k]} + \ldots u^k G^k_{2k}x^{[k]} + \tilde{f}(x,u)$$

Carrying out a similar expansion for higher powers results in a set of equations of the general form

$$\dot{x}_p = Fx_p(t) + G^o + uG^1x_p + u^2G^2x_p + \ldots + u^kG^kx_p + \tilde{\tilde{f}}(x,u)$$

where x_p denotes a vector consisting of $x, x^{[2]}, \ldots x^{[p]}$ and $\tilde{\tilde{f}}$ is of order p or higher in (x,u). This is sometimes called Carleman linearization. Krener [6] works out the case where u is linear and this case plays an important role in [1].

4. VOLTERRA SERIES: SPECIAL CASE

Consider the question of computing the Volterra series for systems which are linear in x, but nonlinear in u. Suppose

$$\dot{x}(t) = A(t)x(t) + \sum_{i=1}^{m} b_i[u(t)]F_i(t)x(t); \quad y(t) = h(x)$$

where the b_i are analytic functions of the scalar u and h is an analytic function of x. Let $\Phi(\cdot,\cdot)$ be the fundamental solution of $\dot{x}(t) = A(t)x(t)$ and define z via $z(t) = \Phi(0,t)x(t)$. Then

$$\dot{z}(t) = \sum_{i=1}^{m} b_i[u(t)]\Phi(0,t)F_i(t)\Phi(t,0)z(t)$$

But this has a solution given by the Peano-Baker series

$$z(t) = z_o + \int_0^t \sum_{i=1}^{m} b_i[u(\sigma)]F_i(\sigma)d\sigma +$$
$$\int_0^t \int_0^{\sigma_1} \sum_{i=1}^{m} b_i[u(\sigma_1)]F_i(\sigma_1) \sum_{i=1}^{m} b_i[u(\sigma_2)F_i(\sigma_2)d\sigma_2 d\sigma_1 + \ldots$$

Moreover by expanding the $b_i[\ ]$ in their power series we obtain a development of the form

$$z(t) = z_o + \sum_{n=1}^{\infty} \int_0^t \cdots \int_0^{\sigma_{n-1}} w_n(t,\sigma_1,\sigma_2,\ldots\sigma_n)$$
$$u(\sigma_1)u(\sigma_2)\ldots u(\sigma_n)d\sigma_1 d\sigma_2 \ldots d\sigma_n$$

However, since the expansion for the b_i will in general yield terms such as, $u^2(\sigma_1)$, etc. it is necessary to permit the w_n to contain delta functions such as $\delta(\sigma_i-\sigma_{i-1})$. (These factors can be eliminated, at the price of a more elaborate notation in which the number of integrals and the degree of the u dependence are allowed to differ.)

To complete the computation of the Volterra series we simply transform back to x and expand h to get

$$y(t) = h[\Phi(t,0)z_o + \sum_{n=1}^{\infty} \int_0^t \cdots \int_0^{\sigma_{n-1}} (t,0)w_n(t,\sigma_1,\sigma_2\ldots\sigma_n)$$
$$u(\sigma_1)u(\sigma_2)\ldots u(\sigma_n)d\sigma_1 d\sigma_2\ldots d\sigma_n]$$

and finally

$$y(t) = y_o(t) + \sum_{n=1}^{\infty} \int_0^t \cdots \int_0^{\sigma_{n-1}} w_n(t,\sigma_1,\sigma_2\ldots\sigma_n)$$
$$u(\sigma_1)u(\sigma_2)\ldots u(\sigma_n)d\sigma_1 d\sigma_2\ldots d\sigma_n$$

In this last step it is not necessary to introduce additional delta functions since the products of kernels can be expressed as single kernels.

With respect to convergence, we can only claim convergence on $[0,T]$ if $|u(\sigma)| \leqslant \varepsilon$ with sufficiently small, since the expansions of the b_i and the expansion of h need not be globally convergent. Moreover, it is clear that if the b_i are only k times differentiable we can compute k kernels and the error between the true solution and the approximation based on these k kernels will be $o(\varepsilon^k)$.

We state a theorem which describes the situation in the scalar input case. This theorem is of interest in its own right and as a lemma in the proof of the more general theorem 2 which appears later.

<u>Theorem 1</u>: Suppose that $A(\cdot)$ and $F(\cdot)$ are n by n matrix valued functions of time which are bounded and piecewise continuous on $[0,\infty)$. If $b_i: \mathbb{R}^1 \to \mathbb{R}^1$ has k continuous derivatives in some neighborhood of 0 and if

$$\dot{x}(t) = A(t)x(t) + \sum_{i=1}^{p} b_i[u(t)]F_i(t)x(t);$$

$$y(t) = c(t)x(t); \quad x_o = x(0)$$

then there exists (distribution valued) kernels $w_1, w_2, \ldots w_k$ such that for any given T and any $\varepsilon > 0$ there exists $\delta > 0$ with

$$\sup_{0\leqslant t\leqslant T} \Big| y(t)-y_o(t) - \int_0^t w_1(t-\sigma)u(\sigma)d\sigma - \ldots$$
$$\int_0^t \int_0^{\sigma_{k-1}} w_k(t,\sigma_1,\sigma_2\ldots\sigma_k)u(\sigma_1)u(\sigma_2)\ldots u(\sigma_k)d\sigma_1 d\sigma_2\ldots d\sigma_k \Big| \leqslant \varepsilon \sup_{0\leqslant t\leqslant T}|u(t)|^k$$

for u( ) piecewise continuous and $|u(t)| \leq \delta$ for $0 \leq t \leq T$.

5. VOLTERRA SERIES: GENERAL CASE

In section 3 we saw that we could associate with the equation

$$\dot{x}(t) = f[x(t),u(t)]; \quad y(t) = h[x(t)]; \quad x(0) = x_o$$

a higher order "linearization" if f has a certain degree of differentiability. Specifically if it has k derivatives with respect to x and u in a neighborhood of the free response trajectory we can write

$$\dot{x}_k(t) = F_k(t)x_k(t)+G_k^o(t)+G_k^1u(t)+\ldots+G_k^ku^k(t)+\tilde{f}(x,u)$$

where x_k is $(x^{[1]},x^{[2]},\ldots,x^{[k]})$ and $||\tilde{f}||/(||x_k||+||u_k||)$ goes to zero as the denominator does. Droping the $\tilde{f}(x,u)$ term we can compute the first order Volterra approximation for x_k. This then gives a Volterra expansion for x accurate up to terms of order k. The error caused by omission of the $\tilde{f}$ term is of higher order so that this expansion is in fact correct to order k. If h has k continuous derivatives in a neighborhood of the free response trajectory then it too can be expanded in a Taylor series. The composition of these two expansions gives the first k terms in a Volterra series for the input-output map. The proofs of these remarks are not difficult and, in any case parallel developments in [1] and [6]. We summarize with the following theorem.

<u>Theorem 2</u>: Suppose that $f : \mathbb{R}^1 \times \mathbb{R}^n \to \mathbb{R}^n$ and $h : \mathbb{R}^n \to \mathbb{R}^p$ have k continuous derivatives in some open set $S \subset (-a,a) \times \mathbb{R}^n$. Suppose that

$$\dot{x}(t) = f[u(t),x(t)]; \quad y(t) = h[x(t),t]; \quad x(0) = x_o$$

and that the free response (corresponding to $u(\cdot) = 0$) belongs to S. Then there exists kernels $w_1,w_2,\ldots w_k$, such that for any given $T > 0$ and any $\varepsilon > 0$ there is $\delta > 0$ with

$$\sup_{0\leq t\leq T} \Big|y(t)-y_o(t)-\int_0^t w_1(t,\sigma_1)u(\sigma_1)d\sigma_1 + \ldots \int_0^t\int_0^{\sigma_1}\ldots\int_0^{\sigma_k} w_k(t,\sigma_1,\sigma_2\ldots\sigma_k)u(\sigma_1)u(\sigma_2)\ldots u(\sigma_k) d\sigma_1 d\sigma_2\ldots d\sigma_k\Big| \leq \varepsilon \sup_{0\leq t\leq T}|u(t)|^k$$

for u( ) piecewise continuous and $|u(t)| \leq \delta$ for $0 \leq t \leq T$.

<u>Remark</u>: There is an alternative procedure which may also prove useful in some cases. Consider a differential system

$$\dot{x}(t) = f[x(t),u(t)]; \quad y(t) = h[x(t)]; \quad x(0) = x_o$$

where f is analytic in (x,u) and h is analytic in x. Suppose that we set

$u(t) = \int_0^t v(t)dt$ for some piecewise continuous function v. Then of course

$$\begin{aligned} \dot{x}(t) &= f(x,u) \\ \dot{u}(t) &= v(t) \end{aligned} \quad ; \qquad g(t) = h[x(t)]$$

Now the results of [1] apply, because the input enters linearly, and we can write

$$y(t) = y_o(t) + \int_0^t w_1(t,\sigma)v(\sigma)d\sigma +$$

$$\int_0^t \int_0^{\sigma_1} w_2(t,\sigma_1,\sigma_2)v(\sigma_1)v(\sigma_2)d\sigma_1 d\sigma_2 + \ldots$$

If the kernel functions are such that we may integrate by parts, then it is possible to trade v for u. For example,

$$y(t) = y_o(t)+w_1(t,\sigma)u(\sigma)\Big|_0^t - \int_0^t \frac{\partial}{\partial\sigma} w_1(t,\sigma)u(\sigma)d\sigma + \ldots$$

(In this case $w(t,t) = w(0,0) = 0$ so only the integral term makes a contribution.)

6. NECESSARY CONDITIONS IN OPTIMAL CONTROL

Now consider the problem of maximizing $\phi[x(t)]$ for a free end point problem with

$$\dot{x}(t) = f[x(t),u(t)]; \quad x(0) = x_o; \quad u(t) \in \Omega \tag{E}$$

We will work out in detail some of the necessary conditions under the assumption that f is twice differentiable with respect to x and u and ϕ is twice differentiable with respect to x. Suppose that $u_o(\cdot)$ is a candidate for an optimal control. Let $\delta u = u-u_o$ and let $\tilde{x}_o$ be the solution of (E) corresponding to u_o. Expanding this about (x_o,u_o) we have

$$\begin{aligned} \delta\dot{x} &= f[x(t),u(t)]-f[\tilde{x}_o(t),u_o(t)] \\ &= A_{1,1}(t)\delta x+A_{1,2}[\delta x]^{[2]}+\delta u b^1(t)+\delta u B^1_{11}(t)\delta x+(\delta u)^2 b^2(t)+e_1 \\ \phi[x] &= \phi[x_o]+H_1\delta x+H_2[\delta x]^{[2]} + e_2 \end{aligned}$$

where e_1 and e_2 are terms of order higher than 2 in $(\delta x,\delta u)$. Applying theorem 2 we get

$$\phi(t) = \phi_o(t) + \int_0^t w_1(t,\sigma)\delta u(\sigma)d\sigma + \int_0^t \int_0^{\sigma} w_2(t,\sigma_1,\sigma_2)\delta u(\sigma_1)\delta u(\sigma_2)d\sigma_1 d\sigma_2+\ldots$$

If $u_o(t)$ belongs to the interior of Ω then for u_o to be a minimizing control on $[0,T]$ it must happen that the linear kernel vanishes and that the second kernel is nonnegative definite. (If w_1 is nonzero then we can make a first order change in $\phi(t)$ in either direction by matching u to w_1; thus w_1 must be zero. If w_1 is zero and w_2 is not positive definite then we can make a second order correction to ϕ which will reduce it; hence u_o could not be optimal.)

In the usual approach to these problems one would look at the state-costate pair

$$\dot{x}(t) = f[x(t),u(t)]$$

$$\dot{p}(t) = -\left(\frac{\partial f}{\partial x}\right)^T p(t)$$

and form the Hamiltonian

$$H[x(t),p(t),u(t)] = \langle p(t),f[x(t),u(t)]\rangle$$

The condition for singular control is that the process of minimizing H with respect to u does not define u. Since for $u = u_o$ the coefficient of $u-u_o = \delta u$ is

$$\langle p(t),b_1(\sigma)\delta u(\sigma)\rangle = \langle p(0),\Phi_{\frac{\partial f}{\partial x}}(t,\sigma)b_1(\sigma)\rangle u(\sigma)$$

The final condition on p which corresponds to a free end condition on x is such that the vanishing of the first kernel $w_1(t,\sigma)$ sets up the singular situation.

Remark: A necessary condition for the second order kernel to be nonnegative definite is that for all $t_1,t_2,\ldots t_k$ the matrix

$$M = \begin{bmatrix} w(t_1,t_1) & w(t_1,t_2) & \ldots & w(t_1,t_n) \\ w(t_2,t_1) & w(t_2,t_2) & \ldots & w(t_2,t_n) \\ \cdots & \cdots & \cdots & \cdots \\ w(t_n,t_1) & w(t_n,t_2) & \ldots & w(t_n,t_n) \end{bmatrix}$$

should be nonnegative definite. In particular, regardless of whether M is symmetric or not, it is necessary that $w(t_i,t_i)$ be nonnegative. If $M = M'$ then it is necessary that the upper 2 by 2 block should be nonnegative definite. If we let $t_1 = t$ and $t_2 = t+\delta$ this condition gives

$$\begin{bmatrix} w(t,t) & w(t,t+\delta) \\ w(t+\delta,t) & w(t+\delta,t+\delta) \end{bmatrix} \geq 0$$

Hence if $w(\cdot,\cdot)$ is continuous and if the indicated derivatives exist except possibly at (t,t) then we get on letting δ go to zero the necessary condition

$$\frac{\delta w(t,\sigma)}{\partial t} - \frac{\partial w(t,\sigma)}{\partial \sigma} \geq 0 . \qquad \text{(K)}$$

This is then (a rather far from sufficient) necessary condition on w. It does have the merit that it can be checked easily.

Example: We now take a familiar problem and illustrate how conceptually straight forward the Volterra series method is. Consider the standard linear quadratic problem with vector input

$$\dot{x}(t) = A(t)x(t) + B(t)u(t); \quad x(0) = 0$$

$$J(t_1) = \int_0^{t_1} x'(t)Q(t)x(t)+u'(t)C(t)x(t)+u'(t)R(t)u(t)dt$$

We can express $J(t_1)$ as

$$J(t_1) = \int_0^{t_1} u'(t)R(t)u(t)dt + \int_0^{t_1} u'(t)C(t) \int_0^{t} \Phi(t,\sigma)B(\sigma)u(\sigma)d\sigma dt$$

$$+ \int_0^{t_1}\left(\int_0^{t} \Phi(t,\sigma)B(\sigma)u(\sigma)d\sigma\right)'Q(t)\left(\int_0^{t} \Phi(t,\sigma)B(\sigma)u(\sigma)d\sigma\right)dt$$

In order to bring this into standard Volterra form we must arrange matters so as to display a kernel function

$$J(t_1) = \int_0^{t_1}\int_0^{t} u'(t)W(t_1,t,\sigma)u(\sigma)d\sigma$$

The first term can be put in this form by inserting a delta function. The last term can be integrated by parts to bring it into standard form. Define

$$M(t) = \int_0^{t} \Phi'(t,0)Q(t)\Phi(t,0)dt$$

Then

$$J(t_1) = \int_0^{t_1}\int_0^{t_1} u'(t)[R(t)\delta(t-\sigma)+C(t)\Phi(t,\sigma)B(\sigma)]u(\sigma)d\sigma dt$$

$$+ \int_0^{t_1}\int_0^{t} u'(\sigma)B'(\sigma)\Phi'(0,\sigma)M(t_1)\Phi(0,\phi)u(\rho)d\sigma d\rho$$

$$- \int_0^{t_1}\int_0^{t} u'(\sigma)B'(\sigma)\Phi'(0,\sigma)M(\sigma)\Phi(0,\rho)B(\rho)d\rho d\sigma$$

In order for $u \equiv 0$ to be minimizing this kernel must be nonnegative definite. Thus we need $R(t) \geq 0$; the Legendre-Clebsch condition. If $R \equiv 0$ then we must have

$$C(t)B(t)+B(t)\Phi'(0,t)[M(t_1)-M(t)]\Phi(0,t)B(t) \geq 0$$

a condition discovered by Gabasov [3] and Jacobson [4].

Condition (K) above gives the Kelly condition [5] that

$$-\dot{C}B+C\dot{B}-2CAB+B'QB \geq 0$$

The main interest in this approach to necessary conditions does not stem from this example, however, which is well understood from other points of view. Rather one can use this method to derive higher order necessary conditions such as would be appropriate for treating problems in which the first nonzero kernel is of order 2p. In this case we see easily that

$$w(t_1,\sigma,\sigma,\ldots\sigma) \geq 0$$

is a necessary condition analogous to the Gabasov-Jacobson condition.

The main limitation on this method as developed here is that it applied only to the free endpoint case. It appears that this limitation can be overcome at the price of some additional complexity in notation. It should be enlightening to compare this with Krener's work [7] on the constrained endpoint problems.

REFERENCES

1. R.W. Brockett,"Volterra Series and Geometric Control Theory," Automatica, Vol. 12, March, 1976.
2. R.W. Brockett, "Nonlinear Systems and Differential Geometry," IEEE Proceedings, Volume 64, January 1976.
3. R. Gabasov, "Necessary Conditions for Optimality of Singular Controls," Engineering Cybernetics, 1968.
4. D.H. Jacobson, "A New Necessary Condition of Optimality for Singular Control Problems," SIAM J. on Control, Vol. 7, 1969, pp. 578-595.
5. H.J. Kelly, "A Second Variation Test for Singular Extremals," J. AIAA, Vol. 2, 1964, pp. 1380-1382.
6. A.J. Krener, "Linearization and Bilinearization of Control Systems," Proc. 1974 Allerton Conference on Circuit and System Theory, Univ. of Illinois, Urbana, Illinois, 1974.
7. A.J. Krener, "The High Order Maximum Principle," in Geometric Methods in System Theory, D.Q. Mayne and R.W. Brockett eds., Reidel, Dordrecht, Holland, 1973.

UN OUTIL ALGEBRIQUE : LES SERIES FORMELLES NON COMMUTATIVES

Michel Fliess *
Université Paris VIII
et
IRIA - LABORIA

Table des matières :

----

* Adresse Postale : 38, rue Godefroy Cavaignac 75011 PARIS, France

Introduction

La transformation de Laplace permet, par l'introduction des fonctions de transfert, un calcul symbolique pour les asservissements linéaires et stationnaires, dont l'utilité n'est plus à défendre. On a recherché des moyens analogues dans les cas non linéaires et non stationnaires. A la suite de Wiener |41|, on a introduit les séries de Volterra (cf. Barrett |3|) de la forme :

$$(1)\quad y(t) = h_0(t) + \int_0^t h_1(t,\tau_1)u(\tau_1)d\tau_1 + \int_0^t\int_0^t h_2(t,\tau_2,\tau_1)u(\tau_2)u(\tau_1)d\tau_2 d\tau_1 + \ldots,$$

où $u,y : [0,\infty[\to \underline{R}$ sont l'entrée et la sortie. En dépit de nombreux travaux, la faiblesse de cette approche réside dans la détermination effective des noyaux h_0, h_1, h_2,....

Lorsque ces noyaux sont des fonctions analytiques, on montre que (1) peut être représenté par une série formelle en indéterminées non commutatives. Si celle-ci est rationnelle, le système peut recevoir la représentation par espace d'état suivante :

$$(2)\quad \begin{cases} \dot{q}(t) = (A_0 + u(t)A_1)q(t) \\ y(t) = \lambda q(t) , \end{cases} \qquad (t \in [0,\infty[\ )$$

où le vecteur d'état q appartient à un espace vectoriel Q de dimension finie, A_0, $A_1 : Q \to Q$, $\lambda : Q \to \underline{R}$ sont des applications linéaires. Ces systèmes, appelés dans la littérature "bilinéaires", ont, en raison d'une terminologie due à Volterra (cf. |29|, |39|), reçu le nom de "réguliers". Dans un domaine compact donné, tout asservissement continu peut être uniformément approché par des sytèmes de type (2), qui devraient ainsi satisfaire à un grand nombre de problèmes en physique, mécanique et ingéniérie. Ce d'autant plus que la réalisation de (2) et, en temps discret, de

$$\begin{cases} q(t+1) = (A_0 + u(t)A_1)q(t) \\ y(t) = \lambda q(t), \end{cases} \qquad (t = 0,1,2,\ldots)$$

est justifiable des méthodes de calcul des séries rationnelles, comme les matrices de Hankel (cf. |19|).

L'emploi des séries formelles non commutatives en rapport avec des équations différentielles linéaires, ne doit pas surprendre si l'on rappelle qu'elles ont été définies en 1906 par F. Hausdorff pour la démonstration de la formule dite de Baker-Hausdorff. Notre outil privilégié, les séries rationnelles non commutatives, a été introduit en 1959 par Schützenberger (cf. |11|) en liaison avec la théorie des automates et des langages formels. Il y a là un lien entre asservissements et automates, qui nous semble bien plus convaincant que ceux recherchés par d'autres auteurs (Kalman, Falb, Arbib |27| , Brockett, Willsky |6|,...). D'ailleurs, notre méthodologie permet une théorie complète des asservissements réguliers, non autonomes, à temps discret, qui s'applique au cas linéaire, et qui est l'analogue strict du traitement des automates finis variables (cf. Dauscha, Nürnberg, Starke, Winkler |12|).

L'intrusion de méthodes nées en informatique et linguistique mathématiques peut être mise en parallèle avec le développement, dû notamment à Thom |35|,|36|, de la topologie différentielle comme moyen d'explication en biologie, linguistique, ... Au-delà de toute question de validité, ces deux démarches sont difficilement comparables, car

car l'une, la topologique, est globale et qualitative, alors que la nôtre est locale et quantitative. Nous donnons approximativement à "global" et "local" les sens qu'ils ont en Analyse. Nous ne savons en effet tenir compte de possibles singularités, qui sont à la base de la théorie des "catastrophes" de Thom. Par opposition à une description qualitative (|35|, p. 20), nous entendons par quantitatif le fait que l'ingénieur doit pouvoir construire un appareillage modélisé par un asservissement régulier. Pour conclure cette disgression, nous suggérons un rapprochement de global et local avec les concepts dégagés par F. de Saussure dans son *Cours de Linguistique Générale*, de "diachronie" et de "synchronie". En effet, Chomsky |10|, dont le formalisme est, grâce aux travaux de Schützenberger (cf. |11|), à la base de nos méthodes, dit explicitement viser un modèle synchronique, tandis que le but de Thom |36|, chap. XIII, est la compréhension de l'évolution diachronique.

La plupart des résultats de cet article-ci ont été présentés dans quatres Notes |16, 17, 20, 21| à l'Académie des Sciences de Paris.

Remerciements : Nous exprimons notre plus vive reconnaissance au Professeur R. E. Kalman qui, en nous invitant au Center for Mathematical System Theory de l'Université de Floride à Gainesville, nous a fait découvrir la théorie mathématique des asservissements et nous a suggéré un possible lien avec automates et langages.

I. - Rappels sur les séries formelles non commutatives

a) Définitions et premières propriétés

Soient X un ensemble non vide supposé, dans ce travail, fini, appelé alphabet, X^* le monoïde libre qu'il engendre, dont les éléments sont les mots, et où l'élément neutre, le mot vide, est noté "1". K étant un semi-anneau (1) commutatif unitaire, soient $K<X>$ et $K<<X>>$ les semi- anneaux des polynômes et des séries formels, à coefficients dans K, en les indéterminées associatives (non commutatives si card $X \geqslant 2$) $x \in X$. Une série $s \in K<<X>>$ est notée

$$s = \sum\{(s,w)w \mid w \in X^*\} \quad , \text{ où } (s,w) \in K.$$

s est quasi-inversible (ou propre) ssi son terme constant $(s,1)$ est nul. Le quasi-inverse

$$\tilde{s} = \sum \{s^n \mid n > 1\}$$

est déterminé univoquement par la relation

$$s + s\tilde{s} = \tilde{s} = s + \tilde{s}s.$$

---

(1) Un semi-anneau K est un ensemble muni de deux lois de composition, l'addition et la multiplication satisfaisant aux axiomes : (i) pour l'addition, K est un monoïde commutatif ; (ii) la multiplication est associative et distributive par rapport à l'addition ; (iii) 0 désignant l'élément neutre de l'addition, pour tout $a \in K$, on a : $a0 = 0a = 0$.

Un sous-semi-anneau R de $K\langle\langle X\rangle\rangle$ est dit <u>rationnellement clos</u> ssi le quasi-inverse de tout élément quasi-inversible de R, appartient encore à R [(1)]. Le semi-anneau $K\langle(X)\rangle$ des séries <u>rationnelles</u> (Schützenberger |34|) est le plus petit sous-semi-anneau rationnellement clos de $K\langle\langle X\rangle\rangle$ qui contienne $K\langle X\rangle$.

Soit $K^{N\times M}$ l'ensemble des matrices, à coefficients dans K, à N lignes et M colonnes. Une <u>représentation</u> (<u>linéaire</u>) $\mu : X^* \to K^{N\times N}$ est un homomorphisme de X^* dans le monoïde multiplicatif des matrices carrées d'ordre N. Le résultat suivant est connu sous le nom de théorème de Kleene-Schützenberger (cf. |34|).

<u>Théorème 1.1.</u> - *Une série* $r\in K\langle\langle X\rangle\rangle$ *est rationnelle si et seulement s'il existe un entier* $N\geq 1$, *une représentation* $\mu : X^* \to K^{N\times N}$, *une matrice* $p\in K^{N\times N}$ *tels que*

$$r = \sum \{ (\mathrm{Tr}\, p\, \mu w) w \mid w \in X^* \} ,$$

où $\mathrm{Tr}\, p\, \mu w$ *désigne la trace de la matrice* $p\mu w$.

<u>Corollaire 1.2.</u> (cf. |19|) - *Une série* $r\in K\langle\langle X\rangle\rangle$ *est rationnelle si et seulement s'il existe un entier* $N\geq 1$, *une représentation* $\mu : X^* \to K^{N\times N}$, *des matrices ligne* $\lambda\in K^{1\times N}$ *et colonne* $\gamma\in K^{N\times 1}$ *tels que*

(I.1) $$r = \sum \{ (\lambda \mu w \gamma) w \mid w\in X^* \} .$$

<u>Remarques</u>.- (i) Lorsque K est un corps et lorsque X est réduit à une seule lettre x, les séries rationnelles ne sont que le développement de Taylor à l'origine des fonctions rationnelles P/Q, où $P,Q\in K[x]$, $Q(0)\neq 0$.

(ii) Lorsque K est le semi-anneau de Boole $B = \{0,1\}$, où $1+1=1$, les séries de $B\langle\langle X\rangle\rangle$ correspondent aux langages formels habituels, c'est-à-dire aux parties de X^*, et les séries rationnelles aux langages de même nom (cf. |19|). Alors le théorème de Kleene-Schützenberger exprime, avec la formule (I.1), qu'un langage est rationnel ss'il est accepté par un automate fini : c'est le théorème de Kleene habituel.

b) <u>Matrices de Hankel et modules sériels</u>

Dans ce paragraphe, K est un corps.

A toute série $s\in K\langle\langle X\rangle\rangle$, on associe un tableau infini, appelé <u>matrice de Hankel</u>, et noté $H(s)$, dont lignes et colonnes sont indexés par X^* et tel que le coefficient d'indice $(w_1,w_2)\in X^*\times X^*$ soit $(s, w_1 w_2)$. Le rang de $H(s)$, qui, par définition, est celui de s, est immédiat à introduire. On peut énoncer (cf. |19|) :

<u>Théorème 1.3.</u> - K *étant un corps, une condition nécessaire et suffisante pour qu'une série* $r\in K\langle\langle X\rangle\rangle$ *soit rationnelle est qu'elle soit de rang fini* $\overline{N}$. *Il existe alors une représentation* $\overline{\mu} : X^* \to K^{\overline{N}\times\overline{N}}$, *des matrices ligne* $\overline{\lambda}\in K^{1\times\overline{N}}$ *et colonne* $\overline{\gamma}\in K^{\overline{N}\times 1}$, *telles que*

$$r = \sum \{ (\overline{\lambda}\, \overline{\mu}\, w\, \overline{\gamma}) w \mid w\in X^* \} .$$

---

(1) Lorsque K est un anneau, on peut remplacer la quasi-inversion par l'<u>inversion</u>. s est inversible ss'il existe s^{-1} telle que $ss^{-1} = s^{-1}s = 1$; elle l'est ssi $(s,1)$ est inversible dans K.

On peut déterminer :

- *deux ensembles de* $\overline{N}$ *mots* $\{d_j\}_{j=1}^{\overline{N}}$, $\{g_i\}_{i=1}^{\overline{N}}$;

- *une application* $\chi : X^* \to K^{\overline{N}x\overline{N}}$ *définie, pour tout* $w \varepsilon X^*$, *par* $(\chi w)_{ij} = (r, g_i w d_j)$ $(i,j = 1, \dots, \overline{N})$, *telle que* $\chi 1$ *soit inversible et vérifiant* $\chi w = \chi 1 \overline{\mu} w$;

- $\overline{N}^2$ *matrices* $m_{ij} \varepsilon K^{\overline{N}x\overline{N}}$ $(i,j = 1, \dots, \overline{N})$ *vérifiant, pour tout* $w \varepsilon X^*$, $\overline{\mu} w = \sum_{ij} (r, g_i w d_j) m_{ij}$

Soit un entier $N \geqslant 1$, *une représentation* $\mu : X^* \to K^{NxN}$, *des matrices lignes* $\lambda \varepsilon K^{1xN}$ *et colonne* $\gamma \varepsilon K^{Nx1}$ *tels que* $r = \sum \{(\lambda \mu w \gamma) w | w \varepsilon X^*\}$

Alors, $N \geqslant \overline{N}$. *Si* $N = \overline{N}$, *il existe une matrice inversible* $P \varepsilon K^{\overline{N}x\overline{N}}$ *telle que* $P \mu w P^{-1} = \overline{\mu} w$, $\overline{\lambda} P = \lambda$, $P\gamma = \overline{\gamma}$ *(les représentations* $\overline{\mu}$ *et* μ *sont donc semblables).*

La démonstration se fait de manière synthétique à partir de la notion de K<X> -module sériel droit ou gauche, empruntée à A. Heller, et qui généralise les modules de Kalman |27|, chap. 10. Un K<X>-module sériel gauche est un triple (E,c,f) où E est un K<X>-module gauche, c un élément de E, $f : E \to K$ une application K-linéaire. A cet objet correspond la série $s = \sum \{(fwc) w | w \varepsilon X^*\}$.

Réciproquement, à une série $s \varepsilon K\langle\langle X \rangle\rangle$, correspond un module sériel gauche (E,c,f) où E n'est autre que K<X> considéré comme K<X>-module gauche, c le polynôme unité, f l'application qui à $p \varepsilon K\langle X \rangle$ associe $\sum \{(p,w)(s,w) | w \varepsilon X^*\}$. Un morphisme $\phi : (E,c,f) \to (E',c',f')$ de K<X>-modules sériels gauches est un morphisme du K<X>-module E dans le K<X>-module E', tel que $\phi c = c'$, ${}^t\phi f' = f$ (${}^t\phi$ désigne l'application duale de ϕ, considéré comme morphisme de K-espaces vectoriels). On a isomorphisme si ϕ est un isomorphisme entre E et E'.

(E,c,f) est dit réduit (ou minimal) ssi $E = K\langle X \rangle c$ et $\{n | n \varepsilon E, f(K\langle X \rangle n) = 0\} = \{0\}$.

Soient $N = \{n | n \varepsilon K\langle X \rangle c, f(K\langle X \rangle n) = 0\}$ un sous-K<X>-module de K<X>c, $R = K\langle X \rangle c / N$, $\alpha : K\langle X \rangle c \to R$ l'épimorphisme canonique. Il existe une application K-linéaire $f_0 : R \to K$ telle que la restriction de f à K<X>c soit égale à $f_0 \alpha$. (R, c_0, f_0), où $c_0 = \alpha c$ est clairement un module sériel réduit. Il est d'autre part évident que deux modules sériels réduits, associés à la même série, sont isomorphes.

On peut doter le K-espace vectoriel engendré par les colonnes de $H(s)$ d'une structure de K<X>-module gauche de la manière suivante : tout mot $w \varepsilon X^*$ opérant sur la colonne d'indice $u \varepsilon X^*$ lui fait correspondre la colonne d'indice wu. R est isomorphe à ce module, ce qui en donne une interprétaion remarquable.

Définitions et propriétés analogues pour les modules sériels droits et les lignes de la matrice de Hankel.

Soit $r \varepsilon K\langle(X)\rangle$ donnée comme en (I.1). A μ, λ, γ, on peut associer le module sériel gauche (E,c,f) où :

- $E = K^{Nx1}$, K-espace vectoriel sur lequel opèrent canoniquement les matrices μw et que l'on peut considérer comme un K<X>-module gauche ;
- $c = \gamma$;

- f est l'application repérée par λ, c'est-à-dire telle que $fwc = \lambda\mu w\gamma$.
Les assertions du théorème concernant les liens entre $\bar{\mu}$, $\bar{\lambda}$, $\bar{\gamma}$, et μ, λ, γ apparaissent alors comme la traduction en langage matriciel des propriétés des modules sériels réduits.
La représentation $\bar{\mu}$, qui est définie à une similitude près, est appelée représentation (linéaire) réduite (ou minimale).
Remarques. - (i) Schützenberger |34| et Isidori |25| donnent des approches plus constructives.
(ii) Le résultat précédent généralise celui, bien connu (cf. Gantmacher |23|, p.201), sur les matrices de Hankel des fonctions rationnelles en une indéterminée.
(iii) La réduction de l'automate fini acceptant un langage rationnel donné est analogue (cf. Eilenberg |13|, chap. III), dans son principe, à celle des représentations.

c) Polynômes

Une représentation $\mu : X^* \to K^{N \times N}$ est dite nilpotente ssi, pour tout mot non vide w, la matrice μw l'est. D'après un résultat dû à Levitzki (cf. Kaplansky |28|, p. 135), la représentation peut être triangularisée, en ce sens que toutes les matrices μw peuvent l'être simultanément.
Proposition 1.4. (cf. |15|) - *Une série rationnelle de* $K\langle\langle X\rangle\rangle$, *où* K *est un corps, est un polynôme si et seulement si elle peut être produite par une représentation nilpotente. Alors, la représentation réduite est nilpotente.*
Preuve. - D'après le théorème de Levitski, la condition de nilpotence est suffisante. La formule $\bar{\mu}w = \sum m_{ij}(r, g_i w d_j)$ du théorème (1.3.) montre la nécessité et ce pour la représentation réduite.

II. - Asservissements réguliers à temps discret

a) Description interne et indiscernabilité

K étant un corps commutatif, un asservissement régulier (ou bilinéaire) à temps discret (ou échantillonné)peut être décrit par

$$\text{(II.1)} \quad \begin{cases} q(t+1) = (A_0 + \sum_{i=1}^{n} u_i(t) A_i) q(t) \\ y(t) = \lambda q(t) \end{cases} \quad (t = 0, 1, 2, \dots)$$

où q(t) appartient à l'espace vectoriel d'état Q (q(0) est donné), $A_0, A_1, \dots, A_n$: $Q \to Q$, $\lambda : Q \to K$ sont des applications K-linéaires, $u_1, \dots, u_n : \underline{N} \to K$ sont les entrées (ou commandes, ou gouvernes, ou contrôles), $y : \underline{N} \to K$ la sortie.
Deux asservissements sont dits indiscernables ssi, pour les mêmes entrées, on obtient les mêmes sorties.

Proposition 2.1. - *Un asservissement de la forme*

$$\left|\begin{array}{l} q(t+1) = (A_0 + \sum_{i=1}^{n} u_i(t)A_i)q(t) + \sum_{i=1}^{n} u_i(t)b_i + b_0 \\ y(t) = \lambda q(t) + y_0 \end{array}\right.$$

($b_0, b_1, \dots, b_n \varepsilon Q$, $y_0 \varepsilon K$) *est indiscernable d'un asservissement régulier.*

Preuve. - Elle est inspirée de Brockett |5|.

(i) Soit d'abord $\left|\begin{array}{l} q(t+1) = (A_0 + \sum u_i(t)A_i)q(t) + \sum u_i(t)b_i + b_0 \\ y(t) = \lambda q(t) . \end{array}\right.$

Posons $Q' = Q \oplus K$, $A'_j : Q' \to Q'$ $(j = 0, 1, \dots, n)$, $\lambda' : Q' \to K$ applications K-linéaires définies par :

$A'_0|Q = A_0$, $A'_0|K(1) = (b_0, 1)$; $A'_i|Q = A_i$, $A'_i|K(1) = (b_i, 0)$ $(i = 1, \dots, n)$;
$\lambda'|Q = \lambda$, $\lambda'|K = 0$ (la barre | indique la restriction de l'application).

Il y a indiscernabilité avec l'asservissement régulier d'espace d'état Q' défini par $A'_0, A'_1, \dots, A'_n, \lambda'$, de vecteur d'état initial $(q(0), 1)$.

(ii) Soit maintenant $\left|\begin{array}{l} q(t+1) = A_0 + \sum u_i(t)A_i)q(t) \\ y(t) = \lambda q(t) + y_0 \end{array}\right.$

Posons $Q'' = Q \oplus K$, $A''_j : Q'' \to Q''$, $\lambda'' : Q'' \to K$ applications K-linéaires définies par :

$A''_0|Q = A_0$, $A''_0|K(1) = (0,1)$; $A''_i|Q = A''_i$, $A''_i|K = 0$ $(i = 1, \dots, n)$; $\lambda''|Q = \lambda$
$\lambda''|K(1) = y_0$.

Il y a indiscernabilité avec l'asservissement régulier défini par Q'', $A''_0, A''_1, \dots, A''_n$, λ'', où $q''(0) = (q(0), 1)$

Corollaire 2.2. - *La famille des asservissements réguliers contient strictement celle des asservissements linéaires de la forme*

$$\left|\begin{array}{l} \eta(t+1) = F\eta(t) + \sum_{i=1}^{n} u_i(t)g_i \\ y(t) = H\eta(t) \end{array}\right.$$

d'espace d'état N $[\eta(0) = 0]$, *où* $g_1, \dots, g_n \varepsilon N$, *et où* $F : N \to N$, $H : N \to K$ *sont des applications K-linéaires.*

Remarque.- Il n'y aurait, bien entendu, aucune difficulté à généraliser à des sorties multidimensionnelles, c'est à dire prises dans des espaces vectoriels de dimension finie.

b) Séries formelles génératrices

(i) Entrées homogènes

Un asservissement régulier est dit à entrée homogène (1) ss' il est de la forme

---

(1) On dit aussi *homogeneous in the state.*

$$(\text{II}.2)\quad \begin{cases} q(t+1) = (\sum_{i=1}^{n} u_i(t)A_i)q(t) \\ y(t) = \lambda q(t) \end{cases} .$$

Associons-lui l'alphabet $X = \{x_1, \ldots, x_n\}$, où la lettre x_i correspond à l'entrée u_i. A tout mot non vide $x_{i_\nu} \ldots x_{i_1} x_{i_0} \in XX^*$ correspond la sortie $y(\nu+1)$ pour les entrées ainsi définies :

$$u_i(t) = \begin{cases} 1 & \text{si} \quad i_t = i \\ 0 & \text{sinon} . \end{cases}$$

Prenant cette sortie pour coefficient du mot, celui du mot vide étant $\lambda q(0)$, on définit une série $\underline{G} \in K\langle\langle X\rangle\rangle$, dite <u>série génératrice</u> de (II.2). Il est évident que deux asservissements indiscernables ont même série génératrice.

Réciproquement, soit $\underline{G} \in K\langle\langle X\rangle\rangle$. D'après le paragraphe I.b, il lui correspond un $K\langle X\rangle$-module sériel gauche (E,c,f) et l'asservissement (II.2) d'espace d'état Q=E $[q(0)=c]$, où λ=f et où A_i (i=1,...,n) est l'opérateur induit par x_i.

<u>Théorème 2.3.</u>- *Il y a bijection canonique entre les asservissements réguliers à entrées homogènes, définis à une indiscernabilité près, et les séries formelles non commutatives.*

(ii) Cas général

A (II.1) associons l'asservissement à entrée homogène

$$(\text{II}.3)\quad \begin{cases} q(t+1) = (\sum_{j=0}^{n} u_j(t)A_j)q(t) \\ y(t) = \lambda q(t), \end{cases}$$

où $u_0 : \underline{N} \to K$ est une nouvelle entrée. Soient $X = \{x_0, x_1, \ldots, x_n\}$ et $\underline{G} \in K\langle\langle X\rangle\rangle$ la série génératrice de (II.3). Par définition, $\underline{G}$ est la série génératrice de (II.1).
La sortie $y(\nu+1)$ se calcule ainsi : on ne garde dans $\underline{G}$ que les monômes de degré $\nu+1$, on remplace dans chacun d'eux x_{k_1} par 1 si $x_{k_1}=x_0$, par $u_i(1)$ si $x_{k_1}=x_i$ (i=1...1), $y(\nu+1)$ est la somme ainsi obtenue.

<u>Théorème 2.4.</u> - *Il y a bijection canonique entre les asservissements réguliers à entrées non homogènes, définis à une indiscernabilité près, et les séries formelles non commutatives.*

<u>Remarque.</u> - Si les sorties des asservissements étaient multidimensionnelles, il faudrait prendre un vecteur de séries génératrices.

c) Réalisabilité et réduction

Un asservissement régulier est <u>réalisable</u> ss'il est indiscernable d'un asservissement régulier d'espace d'état de dimension finie. Le théorème 1.1. de Kleene-Schützenberger permet d'énoncer :

<u>Théorème 2.5.</u> - *Un asservissement régulier est réalisable si et seulement si sa série génératrice est rationnelle.*

La matrice de Hankel de la série génératrice et son rang sont, par définition, ceux du système. Le théorème 1.3 peut être réénoncé en termes de réduction du système.

Théorème 2.6. - *L'asservissement régulier* (Σ)

$$\left|\begin{array}{l} q(t+1) = (A_0 + \sum_{i=1}^{n} u_i(t)A_i)q(t) \\ y(t) = \lambda q(t) \end{array}\right.$$

est réalisable si et seulement s'il est de rang fini $\overline{N}$. *Il est alors indiscernable de* $(\overline{\Sigma})$

$$\left|\begin{array}{l} \overline{q}(t+1) = (\overline{A}_0 + \sum u_i(t)\overline{A}_i)\overline{q}(t) \\ y(t) = \lambda\overline{q}(t) \end{array}\right. ,$$

d'espace d'état Q *de dimension* $\overline{N}$. *Tout asservissement* (Σ')

$$\left|\begin{array}{l} q'(t+1) = A'_0 + \sum u_i(t)A'_i)q'(t) \\ y(t) = \lambda' q'(t) \end{array}\right. ,$$

indiscernable de (Σ), *a un espace d'état* Q' *de dimension* $N'\geq\overline{N}$. *Si* $N'=N$, *il existe un isomorphisme* $P:Q'\to\overline{Q}$, *tel que :*

$$PA'_jP^{-1} = \overline{A}_j \quad (j = 0,\dots,n), \quad \overline{\lambda}P = \lambda', \quad Pq'(0) = \overline{q}(0).$$

$(\overline{\Sigma})$ est dit réduit (ou minimal): il est défini à un isomorphisme près.

Remarque. - Afin de ne pas allonger par trop cet article, nous ne ferons que mentionner les problèmes de commandabilité et d'observabilité. Les propriétés du module sériel réduit (cf. par. I.b) montre qu'un asservissement régulier est réduit ss'il est observable et si son espace d'état est sous-tendu par l'ensemble des vecteurs accessibles à partir de l'état initial. On ne peut exiger la propriété plus forte dont jouissent les systèmes linéaires, à savoir la commandabilité complète (cf. Kalman, Falb, Arbib|27|, chap. 10). Cela est prouvé par l'exemple suivant : $y(t+1) = u(t)y(t)$ $[y(0) = 1]$ a un espace d'état isomorphe à K où aucun élément non nul n'est accessible à partir de zéro.

Nous sommes maintenant en mesure de consacrer quelques lignes aux liens avec automates et langages. Un automate déterministe $\underline{A} = (X,Q,q_0,Q_F,\delta)$ est un quintuple où :

- X est l'alphabet d'entrée,
- Q est l'ensemble des états, non nécessairement fini, où q_0 est l'état initial et Q_F l'ensemble des états finals,
- $\delta:Q\times X^*\to Q$ est la fonction de transition définie par récurrence sur la longueur des mots :

$$\forall_Q q \qquad \delta(q,1) = q$$
$$\forall_{X^*} w \quad \forall_X x \qquad \delta(q,wx) = \delta(\delta(q,w),x).$$

Le langage (formel), c'est-à-dire la partie de X^*, acceptée (ou reconnue) par $\underline{A}$ est, par définition,

$$L_{\underline{A}} = \{w \mid \delta(q_0,w)\in Q_F\}.$$

Un tel automate peut être considéré comme lié à l'action du monoïde libre X^* sur Q. On a ainsi une structure algébrique tout à fait semblable aux $K\langle X\rangle$-modules, et qu'Eilenberg |13| a utilisée pour la minimisation des automates sous le nom de

X-modules. Le langage accepté et la série génératrice jouent des rôles analogues.

Remarque. - Les liens entre asservissements et automates ont déjà été soupçonnés par de nombreux auteurs (cf. Kalman, Falb, Arbib |27|, Brockett, Willsky |6|,etc)

d) Asservissements réguliers non autonomes

Un asservissement régulier

$$(II.4) \quad \begin{cases} q(t+1) = (A_0(t) + \sum_{i=1}^{n} u_i(t)A_i(t))q(t) \\ y(t) = \lambda(t)q(t) \end{cases}$$

est dit non autonome ssi l'un au moins des opérateurs $A_0, A_1, \ldots, A_n, \lambda$ dépend du temps t. Sinon, il est dit autonome.

Comme précédemment, on associe à (II.4) l'asservissement à entrée homogène

$$(II.5) \quad \begin{cases} q(t+1) = (\sum_{j=0}^{n} u_j(t)A_j(t))q(t) \\ y(t) = \lambda(t)q(t) \end{cases}$$

où u_0 est une nouvelle entrée. Soient $X = \{x_0, x_1, \ldots, x_n\}$ et $\underline{G} \in K\langle\langle X\rangle\rangle$ la série génératrice de (II.4) et (II.5), où le coefficient de tout mot est défini comme au paragraphe II.b.

Les théorèmes 2.3 et 2.4 conduisent à énoncer :

Proposition 2.7. - *Il existe une bijection canonique entre les asservissements réguliers non autonomes (à entrées homogènes ou non), définis à une indiscenabilité près, et les séries formelles non commutatives. Par conséquent, tout asservissement régulier non autonome est indiscernable d'un asservissement régulier autonome.*

Exemple.- Soit le système $y(t+1) = \binom{2t}{t} u(t)y(t)$ $[y(0)=1]$. Il a pour série génératrice

$$\underline{G} = \sum_{t \geq 0} \binom{2t}{t} x^t$$

qui est algébrique car développement de Taylor à l'origine de la fonction algébrique $1/\sqrt{1-4x}$. On peut aussi donner la représentation autonome suivante

$$\begin{cases} q(t+1) = Au(t)q(t) \\ y(t) = \lambda q(t) \end{cases}$$

où l'espace vectoriel d'état Q est de dimension infinie dénombrable, avec une base indexée par l'ensemble $\underline{Z}$ des entiers relatifs, de sorte que :

$$A = B^2 \text{ ou } B_{ij} = \begin{cases} 1 \text{ si } i=j+1 \text{ ou } j=i+1 \\ 0 \text{ sinon} \end{cases}, \quad q(0)_j = \begin{cases} 1 \text{ si } j=0 \\ 0 \text{ sinon} \end{cases}, \quad \lambda_i = \begin{cases} 1 \text{ si } i=0 \\ 0 \text{ sinon} \end{cases}$$

Un asservissement régulier est dit réalisable de manière non nécessairement autonome ss'il est indiscernable d'un asservissement non nécessairement autonome d'espace d'état de dimension finie.

Soit ρ_k la dimension du K-espace vectoriel sous-tendu par les colonnes d'indices les mots de longueur k de la matrice de Hankel $H(s)$ d'une série $s \in K\langle\langle X\rangle\rangle$.

$\rho = \sup_k \rho_k$ est, par définition, le <u>rang colonne homogène</u> de $H(s)$, de s et de tout asservissement ayant s pour série génératrice. Il est évidemment inférieur ou égal au rang.

<u>Théorème 2.8.</u> - *Un asservissement* (Σ) *régulier non nécessairement autonome est réalisable en tant que tel si et seulement si son rang colonne homogène est fini, égal à* $\overline{N}$. *Il est alors indiscernable d'un asservissement régulier non nécessairement autonome d'espace d'état de dimension* $\overline{N}$. *Tout asservissement régulier indiscernable de* (Σ) *a un espace d'état de dimension supérieure ou égale à* $\overline{N}$.

<u>Preuve.</u> - (i) Lorsque (Σ) est réalisable de manière non nécessairement autonome, la finitude du rang colonne homogène se montre comme celle du rang d'une série rationnelle (cf. |19|).

(ii) Supposons (Σ), de série génératrice $\underline{G}$, de rang colonne homogène fini $\overline{N}$. Soient $\overline{Q}$ un K-espace vectoriel de dimension $\overline{N}$, C (resp. C_t) le K-espace vectoriel sous-tendu par les colonnes d'indices les mots de longueur t) de la matrice de Hankel $H(\Sigma)$ de (Σ), $\phi_t : C_t \to \overline{Q}$ une injection K-linéaire, $\overline{q}(0)$ l'image par ϕ_0 de la colonne d'indice 1, $\psi_{t,j} : C_t \to C_{t+1}$ l'application K-linéaire induite par x_j $(j=0,1,\ldots,n)$, $f : C \to K$ l'application K-linéaire qui, à la colonne d'indice $w \in X^*$, associe le coefficient d'indice 1, c'est-à-dire $(\underline{G},w)$. Il existe des applications K-linéaires $\overline{A}_j(t) : \overline{Q} \to \overline{Q}$ $(j = 0,1,\ldots,n)$ telles que $\overline{A}_j(t)\phi_t = \phi_{t+1}\psi_{t,j}$, $\overline{\lambda}(t) : \overline{Q} \to K$ telle que la restriction $f|C_t$ soit égale à $\lambda(t)\phi_t$. L'asservissement défini par $\overline{Q}$, $\overline{q}(0)$, $\overline{A}_j(t)$, $\overline{\lambda}(t)$ répond au problème.

(iii) Un raisonnement trivial d'algèbre linéaire prouve que la dimension de l'espace d'état est nécessairement supérieure ou égale au rang colonne homogène.

<u>Application.</u> - Tout système à entrée homogène scalaire

$$\left| \begin{array}{l} q(t+1) = u(t)A(t)q(t) \\ y(t) = \lambda(t)q(t) \end{array} \right.$$

est de rang colonne homogène 1. Si $\sum_{t>0} a_t x^t$ est sa série génératrice, il est indiscernable de

$$\left| \begin{array}{l} \overline{q}(t+1) = u(t)\overline{q}(t) \\ y(t) = a_t\overline{q}(t), \end{array} \right.$$

d'espace d'état de dimension 1 $[\overline{q}(0) = 1]$.

<u>Remarques.</u> - (i) A l'opposé du cas autonome (par. II.c), il n'y a pas en général ici unicité de l'asservissement minimal à une similitude près. Cela est dû à l'arbitraire du choix, dans la preuve, de $\overline{Q}$, $\overline{A}_j(t)$, $\overline{\lambda}(t)$.

(ii) Les résultats de ce paragraphe sont tout à faits semblables à ceux obtenus pour les automates finis variables avec le temps. Le théorème 2.8 correspond à la caractérisation par congruences des langages acceptés par de tels automates, caractérisation due à Agasandjan |1| . L'application et la proposition 2.7 sont analogues à des résultats de Dauscha, Nürnberg, Starke et Winkler |12|, article contenant une revue assez complète des automates finis variables.

e) Asservissements réguliers linéaires

Au paragraphe II.a, nous avons vu que la famille des asservissements réguliers contient celle des linéaires de la forme

$$\begin{cases} \eta(t+1) = F\eta(t) + \sum_{i=1}^{n} u_i(t) g_i \\ y(t) = H\eta(t) \end{cases}$$

Plus généralement, un asservissement régulier est dit linéaire ssi la sortie dépend linéairement de l'entrée, ou, en d'autres termes, si le principe de superposition s'applique. La définition même des séries génératrices permet d'écrire :

Proposition 2.9 - *Un asservissement régulier de série génératrice dans* $K\langle\langle X\rangle\rangle$ *, où* $X = \{x_0, x_1, \ldots, x_n\}$ *, est linéaire si et seulement si tout mot de* X^* *de coefficient non nul, contient une et une seule occurence de l'une des lettres* $x_1,\ldots,x_n$*.*

Remarque. - Le support de la série génératrice $\underline{G}$ est, par définition, la partie de X^* supp $\underline{G} = \{ w \mid (\underline{G},w) \neq 0 \}$. Le résultat précédent s'énonce ainsi : il y a linéarité ssi

$$\text{supp } \underline{G} \subseteq \bigcup_{i=1}^{n} \{x_0\}^* x_i \{x_0\}^* .$$

Corollaire 2.10. - *Il existe un algorithme à un nombre fini de pas permettant de décider si l'asservissement régulier*

$$\begin{cases} q(t+1) = (A_0 + \sum_{i=1}^{n} u_i(t) A_i) q(t) \\ y(t) = \lambda q(t), \end{cases}$$

d'espace d'état de dimension finie N*, est linéaire ou non.*

Preuve.- Il existe (cf. |19|) un algorithme à un nombre fini de pas pour mettre l'asservissement sous forme réduite, définie par $(\overline{q}(0), \overline{A}_j, \overline{\lambda})$. Un argument simple d'algèbre linéaire montre que, si $i,i' = 1,\ldots,n$, $k = 0,1,\ldots$, on doit avoir $\overline{A}_i \overline{A}_0^k \overline{A}_{i'} = 0$.

Vérifications effectuables en un nombre fini de pas.

Proposition 2.11. *Lorsqu'un asservissement régulier est linéaire, il peut être réalisé par un asservissement linéaire, non nécessairement autonome, de la forme*

$$\text{(II.4)} \quad \begin{cases} \eta(t+1) = F\eta(t) + \sum_{i=1}^{n} G_i F^t n u_i(t) \\ y(t) = H\eta(t) , \end{cases}$$

où $\eta(t)$ *appartient à l'espace d'état* $N\,[\eta(0) = 0]$, $F: N \to N$, $G_i: N \to N$ $(i=1,\ldots,n)$, $H: N \to K$ *sont des applications* K*-linéaires.*

Preuve. - Supposons l'asservissement régulier sous forme (II.1). Il est clair que l'on peut prendre $F=A_0$, $G_i=A_i$, $n=q(0)$, $H=\lambda$.

Remarques. - (i) (II.4) est stationnaire si $Fn=n$.

(ii) La série génératrice d'un asservissement régulier linéaire est de la forme

$$\sum_{i=1}^{n} b_i(x_0) x_i a_i(x_0) ,$$

où a_i, b_i sont des séries formelles en l'indéterminée x_0. Il y a stationnarité ssi, pour tout i, $a_i(x_0) = \sum_{k \geq 0} x_0^k = (1-x_0)^{-1}$. Le n-uple $(b_1(x_0), \ldots, b_n(x_0))$ est alors

le vecteur de fonctions de transfert du système linéaire stationnaire (à condition, pour respecter les conventions habituelles, de changer x_0^k en $1/z^{k+1}$).
La réalisation des asservissements réguliers non autonomes (cf. par. II.d) conduit à celle des linéaires non autonomes.

Proposition 2.12. - *Un asservissement linéaire*

$$
\text{(II.6)} \quad \begin{cases} \eta(t+1) = F(t)\eta(t) + \sum_{i=1}^{n} u_i(t)g_i(t) \\ y(t) = H(t)\eta(t) \end{cases}
$$

est indiscernable d'un asservissement linéaire non nécessairement autonome d'espace d'état de dimension finie N *si et seulement s'il est indiscernable d'un asservissement régulier non nécessairement autonome d'espace d'état de dimension* N.

Preuve. - Supposons (II.6) d'espace d'état de dimension N. De même que pour la proposition 1.1,(II.6) est indiscernable de (II.5), d'espace d'état de dimension N+1, où :

$$
A_0(t) = \begin{pmatrix} F(t) & 0 \\ 0 & 1 \end{pmatrix} \quad , \quad A_i(t) = \begin{pmatrix} 0 & g_i(t) \\ 0 & 1 \end{pmatrix} (i=1,\dots,n)
$$

$$
q(0) = \begin{pmatrix} 0 \\ \vdots \\ 0 \\ 1 \end{pmatrix} \quad , \lambda(t) = (H(t),0).
$$

Comme la dernière composante de q(t) est toujours égale à 1, le rang colonne homogène est au plus N, d'où la possibilité de prendre un espace d'état de dimension N (cf. par. II.d).

Réciproquement,soit un asservissement régulier non autonome (II.4) d'espace d'état de dimension N. Il est indiscernable d'un asservissement linéaire (II.6) où :
$F(t) = A_0(t)$, $g_i(0) = q(0)$, $g_i(t) = A_i(t)A_0(t-1) \dots A_0(0)q(0)$ $(t\geq 1,\ i=1,\dots,n)$, $H(t) = \lambda(t)$.

Un système linéaire non autonome est dit réalisable en tant que tel ss'il est indiscernable d'un système linéaire non nécessairement autonome d'espace d'état de dimension finie. Le rang colonne homogène d'un système linéaire non nécessairement autonome est celui qu'on obtient en le considérant comme asservissement régulier. Les propositions 2.8 et 2.9 permettent d'énoncer :

Proposition 2.13. - *Un asservissement linéaire non autonome* (Λ) *est réalisable en tant que tel si et seulement s'il est de rang colonne homogène fini* $\overline{N}$. *Il peut alors être mis sous la forme :*

$$
\begin{cases} \eta(t+1) = F(t)\eta(t) + \sum_{i=1}^{n} u_i(t)g_i(t) \\ y(t) = H(t)\eta(t), \end{cases}
$$

où l'espace d'état est de dimension $\overline{N}$. *L'espace d'état de tout asservissement linéaire indiscernable de* (Λ) *est de dimension supérieure ou égale à* $\overline{N}$.

Remarque. - Weiss |40| étudie l'observabilité et la commandabilité des systèmes des systèmes linéaires non autonomes.

III. - Asservissements réguliers à temps continu

Désormais, et ce jusqu'à la fin, K est soit le corps $\underline{R}$ des réels, soit celui $\underline{C}$ des complexes.

a) Description interne et indiscernabilité

Un asservissement régulier (ou bilinéaire), à temps continu, réalisable et autonome, est décrit par

$$(III.1)\quad \begin{cases} \dot{q}(t) = (dq/dt) = (A_0 + \sum_{i=1}^{n} u_i(t)A_i)q(t) \\ y(t) = \lambda q(t) \end{cases}$$

où $q(t)$ ($t\in\underline{R}_+ = [0,\infty[$) appartient à l'espace d'état Q, qui est un K-espace vectoriel de dimension finie [$q(0)$ est donné], $u_1,\dots,u_n:\underline{R}_+\to K$ sont les entrées (ou commandes, ou gouvernes, ou contrôles) que l'on suppose localement intégrables par rapport à la mesure de Lebesgue, $A_0,\dots,A_n:Q\to Q$, $\lambda:Q\to K$ sont des applications K-linéaires.

Deux asservissements sont dits indiscernables ssi, pour les mêmes entrées, on obtient les mêmes sorties. Le résultat suivant est dû à Brockett |5|.

Proposition 3.1. - *Tout asservissement de la forme*

$$\begin{cases} \dot{q}(t) = (A_0 + \sum_{i=1}^{n} u_i(t)A_i)q(t) + \sum_{i=1}^{n} u_i(t)b_i + b_0 \\ y(t) = \lambda_0 + \lambda_1 q(t) + \lambda_2 q(t)\otimes q(t) + \dots + \lambda_k q(t)\otimes\dots\otimes q(t), \end{cases}$$

où $b_0,b_1,\dots b_n$ *appartiennent au K-espace vectoriel d'état Q, de dimension finie,* $\lambda_0\in K$ *et* $\lambda_1:Q\to K, \lambda_2:Q\otimes Q\to K,\dots,\lambda_k:\underbrace{Q\otimes\dots\otimes Q}_{k\text{ fois}}\to K$ *sont des applications K-linéaires du produit tensoriel de copies de Q, est indiscernable d'un asservissement régulier réalisable.*

Preuve. - Le fait que

$$\begin{cases} \dot{q}(t) = (A_0 + \sum u_i(t)A_i)q(t) + \sum u_i(t)b_i + b_0 \\ y(t) = \lambda_0 + \lambda_1 q(t) \end{cases}$$

soit indiscernable d'un asservissement régulier réalisable se démontre à peu près comme en temps discret (cf. proposition 2.1).

Considérons donc

$$\begin{cases} \dot{q}(t) = (A_0 + \sum u_i(t)A_i)q(t) \\ y(t) = \lambda_2 q(t)\otimes q(t). \end{cases}$$

Or $d(q(t)\otimes q(t)/dt = dq(t)/dt\otimes q(t) + q(t)\otimes dq(t)/dt =$

$$= ((A_0\otimes 1_Q + 1_Q\otimes A_0) + \sum u_i(t)(A_i\otimes 1_Q + 1_Q\otimes A_i))q(t)\otimes q(t)$$

où $1_Q:Q\to Q$ désigne l'application identité. Ce qui prouve l'indiscernabilité demandée.

Généralisation immédiate aux produits tensoriels quelconques.

Corollaire 3.2. - *La famille des asservissements réguliers contient strictement celle des asservissements linéaires de la forme*

$$\begin{cases} \dot{\eta}(t) = F\eta(t) + \sum_{i=1}^{n} u_i(t) g_i \\ y(t) = H\eta(t) \quad , \end{cases}$$

d'espace d'état N $\eta(0)=0$, *où* $g_1,\ldots,g_n \in N$, *et où* $F:N\to N$, $H:N\to K$ *sont des applications* K-*linéaires*.

Remarques. - (i) il n'y aurait, bien entendu, aucune difficulté à généraliser à des sorties multidimensionnelles.

(ii) On vérifie aisément que la généralisation aux sorties non linéaires de la proposition 3.1 n'est pas valable en temps discret (cf. par. II.a).

b) Séries formelles génératrices

(i) Entrées homogènes

Un asservissement régulier réalisable est dit à entrée homogène ss'il est de la forme

$$\text{(III.2)} \qquad \begin{cases} \dot{q}(t) = (\sum_{i=1}^{n} u_i(t) A_i) q(t) \\ y(t) = \lambda q(t) \end{cases}$$

où l'espace d'état est de dimension finie N.

La série génératrice de (III.2) est par définition la série rationnelle $\underline{G}$ de $K\langle\langle X\rangle\rangle$, où $X = \{x_1, \ldots, x_n\}$ (x_i correspond à l'entrée u_i), ainsi obtenue : $\mu: X^* \to K^{N\times N}$ est la représentation donnée par $\mu x_i = A_i$, de sorte que

$$\underline{G} = \sum \{ (\lambda \mu w q(0)) w \mid w \in X^* \} .$$

Réciproquement, à $\underline{G} \in K\langle(X)\rangle$, définie par $\mu: X^* \to K^{N\times N}$, $\lambda \in K^{1\times N}$, $\gamma \in K^{N\times 1}$, on associe l'asservissement

$$\begin{cases} \dot{q}(t) = (\sum_{i=1}^{n} u_i(t) \mu x_i) q(t) \\ y(t) = \lambda q(t) \quad , \end{cases}$$

d'espace d'état $K^{N\times 1}$ $[q(0) = \gamma]$.

La formule de Peano-Baker (cf. Gantmacher |23|, p. 121) permet d'écrire la sortie sous forme d'une série infinie, liée à la série génératrice

$$\text{(III.3)} \qquad y(t) = \lambda\Big[1 + \sum_i A_i \int_0^t u_i(\tau) d\tau + \sum_{i_1, i_2} A_{i_2} A_{i_1} \int_0^t u_{i_2}(\tau_2) d\tau_2 \int_0^{\tau_2} u_{i_1}(\tau_1) d\tau_1 + \ldots\Big] q(0).$$

Il reste à prouver que deux asservissements sont indiscernables ss'ils ont même série génératrice. Soient donc $\underline{G}_1$, $\underline{G}_2 \in K\langle\langle X\rangle\rangle$ séries génératrices de deux systèmes indiscernables. Nécessairement, leurs termes constants $(G_k, 1)$ sont égaux . En vertu de (III.3), il vient, si l'on suppose les entrées continues :

$$dy_k/dt = \lambda\Big[\sum_i u_i(t) A_i^k \Big(1 + \sum_j A_j^k \int_0^t u_j(\tau) d\tau + \ldots\Big] q(0) \qquad (k = 1,2)$$

où, nécessairement,

$$\lambda A_i^k[1 + \sum A_j^k \int_0^t u_j(\tau)d\tau + \ldots]q(0)$$

sont égaux pour tout choix des entrées. D'où

$$\lambda A_i^1 q(0) = \lambda A_i^2 q(0) \qquad (i = 1, \ldots, n)$$

ce qui prouve l'identité des monômes du premier degré dans $\underline{G}_1$ et $\underline{G}_2$. Pour les termes de degré supérieur, on opère pas à pas de manière identique.

Théorème 3.3. - *Il y a bijection canonique entre les asservissements réalisables, à entrées homogènes, définis à une indiscernabilité près, et les séries formelles rationnelles non commutatives.*

(ii) Cas général

A (III.1), associons l'asservissement à entrée homogène

$$\text{(III.4)} \qquad \left\{ \begin{array}{l} \dot{q}(t) = (\sum_{j=0}^{n} u_j(t)A_j)q(t) \\ y(t) = \lambda q(t) \end{array} \right. ,$$

où $u_0 : \underline{R}_+ \to K$ est une nouvelle entrée. Soient $X = \{x_0, x_1, \ldots, x_n\}$ et $\underline{G} \in K\langle\langle X\rangle\rangle$ la série génératrice de (III.4). Par définition, $\underline{G}$ est la série génératrice de (III.1). Une démonstration analogue à la précédente permet d'énoncer :

Théorème 3.4. - *Il y a bijection canonique entre les asservissements réguliers réalisables, à entrées non homogènes, définis à une indiscernabilité près, et les séries formelles rationnelles non commutatives.*

Remarque. - Si les sorties des asservissements étaient multidimensionnelles, il faudrait prendre un vecteur de séries génératrices.

c) Réduction

La matrice de Hankel de la série génératrice et son rang sont, par définition, ceux de l'asservissement. Les théorèmes 3.3 et 3.4 permettent d'appliquer le théorème 1.3 pour obtenir la forme réduite d'un asservissement régulier réalisable, résultat également obtenu, à l'aide d'autres moyens par Brockett |5| (algèbres de Lie), Bruni, Di Pillo, Koch |7| et d'Alessandro, Isidori, Ruberti |2| (séries de Volterra).

Théorème 3.5. - *Soit un asservissement régulier réalisable* $(\sum)$ *de rang fini* $\overline{N}$. *Il est indiscernable de* $(\overline{\sum})$.

$$\left\{ \begin{array}{l} \dot{\overline{q}}(t) = (\overline{A}_0 + \sum_{i=1}^{n} u_i(t)\overline{A}_i)\overline{q}(t) \\ y(t) = \overline{\lambda}\overline{q}(t) \end{array} \right. ,$$

d'espace d'état $\overline{Q}$ *de dimension* $\overline{N}$. *Tout asservissement réalisable*

$$\left\{ \begin{array}{l} \dot{q}'(t) = (A'_0 + \sum_{i=1}^{n} u_i(t)A'_i)q'(t) \\ y(t) = \lambda' q'(t) \end{array} \right. ,$$

indiscernable de $(\sum)$, *a un espace d'état* Q' *de dimension* $N' \geq \overline{N}$. *Si* $N' = \overline{N}$, *il existe un isomorphisme* $P : Q' \to \overline{Q}$, *tel que :*

$$PA'_jP^{-1} = \overline{A}_j \quad (j = 0, \ldots, n) \quad , \quad \overline{\lambda}P = \lambda' \quad , \quad Pq'(0) = \overline{q}(0) .$$

($\overline{\Sigma}$) est dit réduit (ou minimal) : il est défini à un isomorphisme près.

Remarque. - La commandabilité et l'observabilité, examinées par d'Alessandro, Isidori, Ruberti |2|, donnent des résultats identiques à ceux du temps discret: un asservissement est réduit ss'il est observable et si l'espace d'état est sous-tendu par l'ensemble des vecteurs d'état accessibles à partir de l'état initial. Une étude plus poussée exige l'emploi des groupes de Lie.

d) Asservissements réguliers complètement intégrables

En |19|, nous avons suggéré un parallèle entre langages et séries rationnels, parallèle que nous avons illustré par l'examen des séries et langages échangeables. Il est tentant d'examiner si à des classes particulières de séries rationnelles correspondent des asservissements réguliers aux propriétés remarquables. Cette démarche est justifiée pour les séries génératrices échangeables.

Soit $\alpha : X^* \to X^+$ l'épimorphisme canonique de X^* sur le monoïde commutatif libre X^+, engendré par X. Une série $s \in K\langle\langle X \rangle\rangle$ est dite échangeable ssi deux mots w,w' de X^* ayant même image commutative, c'est-à-dire tels que $\alpha w = \alpha w'$, ont même coefficient : $(s,w)=(s,w')$. De même, un langage $L \subseteq X^*$ est dit échangeable ssi $L = \alpha^{-1}\alpha L$. La structure des séries et langages rationnels échangeables est donnée en |19|.

A l'asservissement (III.1), associons le système aux différentielles totales

$$\begin{cases} dq = \left(\sum_{j=0}^{n} A_j d\xi_j\right) q \\ y = \lambda q \end{cases},$$

où $\xi_0 = t$, $d\xi_i = u_i(t)dt$, $\xi_i(t) = \int_0^t u_i d\tau$ $(i=1,\ldots,n)$. Le système et l'asservissement sont dits complètement intégrables ssi y(t) peut s'exprimer comme fonction de t, $\xi_1(t),\ldots,\xi_n(t)$.

Proposition 3.6. - *Un asservissement régulier réalisable est complètement intégrable si et seulement si sa série génératrice, qui est rationnelle, est échangeable.*

Preuve. - (i) Supposons la série génératrice échangeable. D'après |19|, il existe une réalisation (III.1) où les opérateurs $A_0, A_1,\ldots,A_n$ commutent deux à deux. Il vient alors

$$y(t) = \lambda\left[\exp A_0 t + \sum_i A_i \xi_i(t)\right] q(0)$$

(ii) Supposons l'asservissement complètement intégrable. Une intégration par parties conduit à :

$$\int_0^t d\xi_{j_k} \int_0^{\tau_{k-1}} \cdots \int_0^{\tau_2} d\xi_{j_2} \int_0^{\tau_1} d\xi_{j_1} = \xi_{j_k}(t) \cdots \xi_{j_1}(t) - \sum_{\sigma \in \underline{S}_k^+} \int_0^t d\xi_{j_{\sigma k}} \cdots \int_0^{\tau_1} d\xi_{j_{\sigma 1}}$$

où $\underline{S}^+_k$ désigne le groupe symétrique sur $\{1,\ldots,k\}$, privé de l'identité. Cette formule jointe à celle (III.3) de Peano-Baker donne le résultat cherché.

Remarque. - Les asservissements réguliers complètement intégrables ont, sous une terminologie différente et pour le filtrage statistique, été aussi rencontrés par Lo et Willsky |30|.

e) Asservissements réguliers nilpotents

Un asservissement régulier est dit nilpotent ssi sa série génératrice est un polynôme. En vertu de la proposition 1.4, il vient :

Proposition 3.7. - *Un asservissement régulier est nilpotent si et seulement s'il est indiscernable de l'asservissement*

$$\begin{cases} \dot{q}(t) = (A_0 + \sum u_i(t)A_i)q(t) \\ y(t) = \lambda q(t) \end{cases},$$

où le semi-groupe engendré par les matrices $A_0, A_1, \ldots, A_n$ *est nilpotent.*

Par conséquent, en vertu du théorème de Levitzki, $A_0, A_1, \ldots, A_n$ peuvent être simultanément triangularisées.

Remarque. - De même, le monoïde syntactique (cf. Eilenberg |13|, p. 62) d'un langage fini est nilpotent.

f) Asservissements réguliers linéaires

Au paragraphe III.a, nous avons vu que la famille des asservissements réguliers réalisables contient celle des linéaires de la forme

$$\begin{cases} \dot{\eta}(t) = F\eta(t) + \sum_{i=1}^{n} u_i(t) g_i \\ y(t) = H\eta(t) \end{cases},$$

d'espace d'état de dimension finie.

Plus généralement, un asservissement régulier est dit linéaire ssi la sortie dépend linéairement de l'entrée.

Proposition. - *Un asservissement régulier réalisable de série génératrice dans* $K\langle\langle X\rangle\rangle$, *où* $X = \{x_0, x_1, \ldots, x_n\}$, *est linéaire si et seulement si tout mot de* X^* *de coefficient non nul, contient une et une seule occurence de l'une des lettres* $x_1, \ldots, x_n$.

Preuve. - La condition étant évidemment suffisante, montrons en la nécessité. Soient $\underline{G} \in K\langle\langle X\rangle\rangle$ la série génératrice et $k \geq 0$ l'entier maximal tel que tout mot $w \in X^*$ du support de $\underline{G}$ soit tel que $w = x_0^k w'$, où $w' \in X^*$. Il existe donc au moins un mot $w_0 \in \mathrm{supp}\, \underline{G}$ tel que $w_0 = x_0^k x_i w_1$, où $i \in \{1, \ldots, n\}$, $w_1 \in X^*$. Il vient :

$$\underline{G} = x_0^k \left(\sum_{i=1}^{n} x_i \underline{G}_i + \underline{G}' \right)$$

($\underline{G}_i$, $\underline{G}' \in K\langle\langle X\rangle\rangle$). Si l'on montre l'absence d'occurence de $x_1, \ldots, x_n$ dans les $\underline{G}_i$, il serait aisé, suivant la même technique, de prouver la condition pour $\underline{G}'$, ce qui achèverait la démonstration. Soit $y_i(t)$ la sortie de l'asservissement ayant pour série génératrice $x_0^k x_i \underline{G}_i$. dy_i^{k+1}/dt^{k+1} est égal au produit de $u_i(t)$ par la sortie de l'asservissement ayant pour série génératrice $\underline{G}_i$. Il est clair qu'il y a linéarité ssi aucune occurence de $x_1, \ldots, x_n$ n'apparaît dans les mots de supp $\underline{G}_i$.

Remarques. - (i) Il y a linéarité ssi $\mathrm{supp}\, \underline{G} \subseteq \bigcup_i \{x_0\}^* x_i \{x_0\}^*$.

(ii) Le corollaire 2.10 reste valable sans modification de la démonstration.

Proposition 3.9.- *Lorsqu'un asservissement régulier réalisable est linéaire, il peut être réalisé par un asservissement linéaire, non nécessairement autonome, de la forme*

$$\text{(III.5)}\quad \begin{cases} \dot{\eta}(t) = F\eta(t) + \sum_{i=1}^{n} G_i(\exp Ft)nu_i(t) \\ y(t) = H\eta(t), \end{cases}$$

où $\eta(t)$ *appartient à l'espace d'état* $N[\eta(0)=0]$, *de dimension finie,* $F:N\to N, G_i:N\to N$ $(i=1,\dots,n), H:N\to K$ *sont des applications K-linéaires.*

Preuve. - Supposons l'asservissement régulier sous forme (III.1). Il est clair que l'on peut prendre $F=A_0$, $G_i=A_i$, $n=q(0)$, $H=\lambda$.

Remarque. - Il y a stationnarité ssi supp $\underline{G}\subseteq \bigcup_i \{x_0\}^* x_i$. Alors, si $\underline{G} = \sum_i a_i(x_0)x_i$, le n-uple $(a_1(x_0),\dots,a_n(x_0))$ est le vecteur de fonctions de transfert (à condition, pour respecter les conditions habituelles, de changer x_0^k en $1/z^{k+1}$). Comparer avec le paragraphe II.e.

IV. - Fonctionnelles et séries formelles non commutatives

a) Fonctionnelles représentées par séries formelles non commutatives

Soit un espace affine sur K, de dimension n, où chaque point est repéré par ses coordonnées $(\xi_1,\dots,\xi_n)$. Un chemin (C) consiste en la donnée de n fonctions continues $\{\xi_i(\sigma)\,|\,0\leq\sigma\leq s\}$. Toute série formelle $s\in K\langle\langle X\rangle\rangle$, où $X = \{x_1,\dots,x_n\}$, conduit, par analogie avec la formule de Peano-Baker (cf. Gantmacher |23|, p.121), à attacher à (C) une valeur numérique donnée par (1)

$$\text{(IV.1)}\qquad (s,1) + \sum_i (s,x_i)\int_0^s d\xi_i(\sigma) + \sum_{i_1 i_2} (s,x_{i_2}x_{i_1})\int_0^s d\xi_{i_2}\,d\xi_{i_1} + \dots ,$$

où

$$\int_0^s d\xi_i(\sigma) = \xi_i(s)-\xi_i(0),\dots,\int_0^s d\xi_{i_k}\dots d\xi_{i_1} = \int_0^s d\xi_{i_k}(\sigma)\int_0^\sigma d\xi_{i_{k-1}}\dots d\xi_{i_1}.$$

Cette valeur est intrinsèque en ce sens qu'elle est, comme on le vérifie aisément, indépendante du choix du paramétrage de (C). Deux chemins ne différant que par une translation donnent la même valeur. Une fonctionnelle, ou, selon l'heureuse expression originale de Volterra |38|, une fonction de lignes, est dite analytique ssi dans un domaine donné elle peut être, d'après (IV.1), représentée par une série non commutative, qui, ici aussi, sera dite série génératrice. Une fonctionnelle analytique est dite rationnelle ou polynômiale ssi sa série génératrice l'est. Pour la consistance de ces définitions, il reste à prouver un résultat d'unicité.

---

(1) Pour alléger l'exposé, nous ne nous occuperons pas ici de questions de convergence. Cependant, lorsque la série est rationnelle, on voit, en se ramenant à une équation différentielle linéaire, (IV,1) est toujours définie.

Proposition 4.1.- *Deux fonctionnelles analytiques sont égales si et seulement si elles ont même série génératrice.*

Preuve. - Elle est identique à celle du paragraphe III.b donnant l'unicité des séries génératrices des asservissements réguliers réalisables.

Remarques. - (i) La qualification d'analytique pour les fonctionnelles a déjà été employée avec d'autres significations (cf. Volterra et Pérès |39|,p.68, F.Pellegrino in |29|, 4ème partie).

(ii) Notre définition des fonctionnelles analytiques est à rapprocher des travaux de Chen |9|, où, au chemin (C), on associe la série de K<<X>>donnée par

$$1 + \sum_i \left(\int_0^s d\xi_i\right) x_i + \sum_{i_1 i_2} \left(\int_0^s d\xi_{i_2} d\xi_{i_1}\right) x_{i_2} x_{i_1} + \ldots .$$

b) Multiplication des fonctionnelles

La multiplication de deux fonctionnelles donne la fonctionnelle qui, à un chemin, fait correspondre le produit des valeurs prises par les deux fonctionnelles.

Proposition 4.2. - *La fonctionnelle multiple de deux fonctionnelles analytiques, est aussi analytique, de série génératrice le produit de Hurwitz* [(1)] *des deux séries génératrices.*

Preuve.- Elle est identique à celle, fort simple, donnée par Ree 33 à propos de travaux de K.-T. Chen. Une intégration par parties conduit à écrire

$$\left(\int_0^s d\xi_{i_k} \ldots d\xi_{i_1}\right)\left(\int_0^s d\xi_{i'_{k'}} \ldots d\xi_{i'_1}\right) = \int_0^s d\xi_{i_k} \left[\int_0^{} d\xi_{i_{k-1}} \ldots d\xi_{i_1}\right)\left(\int_0^s d\xi_{i'_{k'}} \ldots d\xi_{i'_1}\right)\right]$$
$$+ \left[\left(\int_0^s d\xi_{i_k} \ldots d\xi_{i_1}\right)\left(\int_0^s d\xi_{i'_{k'-1}} \ldots d\xi_{i'_1}\right)\right] \int_0^t d\xi_{i'_k} .$$

Ce qui redonne la définition par récurrence du produit de Hurwitz.

Corollaire 4.3. - *La fonctionnelle multiple de deux fonctionnelles analytiques rationnelles (resp. polynômiales) est de même nature.*

c) Approximations des fonctionnelles

Soit *C* l'espace des chemins continus muni de la topologie de la convergence compacte : une suite de chemins est dite converger vers un chemin ssi elle le fait dans tout domaine compact de l'espace affine. Compte tenu de la clôture par addition et multiplication,il faut, pour appliquer le théorème d'approximations de Weierstrass-Stone (cf. Bourbaki |4|), vérifier la propriété de séparabilité suivante : il existe une fonctionnelle (polynômiale) prenant des valeurs distinctes pour deux chemins

---

(1) Le produit de Hurwitz (*shuffle product* dans la littérature américaine) associe à deux mots un polunôme homogène, défini par récurrence sur la longueur : $1 ш 1 = 1$, $\forall_X x$ $1 ш x = x ш 1 = x$, $\forall_X x,y$ $\forall_{X^*} u,v$ $ux ш vy = (u ш vy)x + (ux ш v)y$.
Par linéarité, on prolonge à K<<X>>:$s ш s' = \sum\{(s,u)(s',v) u ш v | u,v \in X^*\}$. On montre (cf. |18|) que le produit de Hurwitz de deux séries rationnelles est une série rationnelle.

distincts C et C' (c'est-à-dire non déductibles l'un de l'autre par translation) de $\mathcal{C}$. Or, d'après un résultat dû à Chen |8|, il existe au moins un mot $w_0 = x_{i_k} \dots x_{i_1} \in X^*$ tel que :

$$\int_C d\xi_{i_k} \dots d\xi_{i_1} \neq \int_{C'} d\xi_{i_k} \dots d\xi_{i_1}$$

Théorème 4.4. - *Dans tout domaine compact de l'espace des chemins continus, toute fonctionnelle continue peut être uniformément approchée par des fonctionnelles analytiques, que l'on peut choisir rationnelles ou polynômiales.*

d) Séries de Volterra et séries formelles non commutatives

Utilisant les travaux de Volterra (cf. Volterra et Pérès |39|, Lévy |29|), Wiener, en 1942, a introduit (cf. Wiener |41|, Barrett|3|), pour représenter les asservissements non linéaires, ce qu'on appelle aujourd'hui séries de Volterra et qui a la forme suivante :

$$\text{(IV.2)}\quad y(t)=h_0(t)+\int_0^t h_1(t,\tau_1)u(\tau_1)d\tau_1+\int_0^t\int_0^t h_2(t,\tau_2,\tau_1)u(\tau_2)u(\tau_1)d\tau_2 d\tau_1+\dots ,$$

où, pour simplifier, on suppose l'entrée $u:\underline{R}_+\to K$ scalaire.

Parallèlement à nos considérations sur les fonctionnelles, un asservissement, ici encore à entrée et sortie scalaires, est dit analytique ss'il admet une série génératrice $\underline{G}\in K\langle\langle x_0,x_1\rangle\rangle$ de sorte que (cf. par. III.3)

$$\text{(IV.3)}\qquad y(t) = (\underline{G},1) + \sum_{j=0,1} (\underline{G},x_j)\int_0^t d\xi_j + \sum_{j_1 j_2} (\underline{G},x_{j_2}x_{j_1})\int_0^t d\xi_{j_2} d\xi_{j_1} +\dots$$

où $\xi_0 = t$, $\quad \xi_1(t) = \int_0^t u(\tau)d\tau$.

Théorème 4.5. - *Un asservissement donné par une série de Volterra est analytique si et seulement si l'on peut prendre pour les noyaux des fonctions analytiques.*(1)

Preuve. - La formule (IV.2) peut être précisée en "triangularisant" les noyaux: pour tout $n\geq 1$, on peut remplacer $h_n(t,\tau_n,\dots,\tau_1)$ par $h'_n(t,\tau_n,\dots,\tau_1)$ qui est nul si l'on n'a pas $t>\tau_n>\dots>\tau_1$. On égale (IV.2) et (IV.3) :

$$h_0(t) = \sum_{k>0} (\underline{G},x_0^k)t^k/k!$$

....

$$h'_n(t,\tau_n,\dots,\tau_1) = \sum_{k_0,\dots k_n\geq 0} \frac{1}{k_0!\dots k_n!} (\underline{G},x_0^{k_n}x_1\dots x_1 x_0^{k_0})(t-\tau_n)^{k_n}\dots(\tau_2-\tau_1)^{k_1}\tau_1^{k_0}.$$

Remarques. - (i) Pour les asservissements analytiques rationnels, en vertu de résultats dus à Bruni, Di Pillo, Koch |7| et d'Alessandro, Isidori, Ruberti |2|, il vient :

$$h'_n(t,\tau_n,\dots,\tau_1) = \lambda[e^{A_0(t-\tau_n)}A_1\dots e^{A_0(\tau_2-\tau_1)}A_1 e^{A_0\tau_1}]q(0)$$

---

(1) Comme nous ignorons les questions de convergence, nous entendons par analytique le fait pour les noyaux $h_0,h_1,h_2,\dots$ d'être représentés par des séries entières en les variables.

où A_0, A_1, ,$q(0)$ ont même signification qu'en (III.1).

(ii) Comme pour la remarque finale du paragraphe III.f, il y a stationnarité ssi le support de la série génératrice $\underline{G}$ vérifie :

$$\text{supp } \underline{G} \subseteq \{x_0, x_1\}^* x_1$$

(iii) Une représentation par un asservissement régulier autonome de la forme (III.1) d'un asservissement analytique irrationnel exigerait un espace d'état de dimention infinie. Question que nous n'aborderons pas, d'autant plus qu'elle est loin d'être éclaircie dans le cas classique des systèmes linéaires stationnaires à fonctions de transfert irrationnelles.

(iv) On dispose ainsi d'un outil algébrique pour les asservissements non linéaires et (ou) non stationnaires, qui devrait jouer, du moins nous l'espérons, un rôle d'une importance égale à celui des fonctions de transfert dans le cas linéaire et stationnaire.

La multiplication de deux asservissements d'entrée u et de sortie y_1, y_2 donne l'asservissement d'entrée u, ayant pour sortie, à l'instant t, $y_1(t)y_2(t)$. D'après proposition et corollaire 4.2. et 4.3., il vient :

Proposition 4.6. - *L'asservissement multiple de deux asservissements analytiques est aussi analytique, de série génératrice le produit de Hurwitz des deux séries génératrices.*

Corollaire 4.7. - *L'asservissement multiple de deux asservissements analytiques rationnels (resp. polynômiaux) est de même nature.*

Remarque. - Deux asservissements analytiques en parallèle donnent un asservissement analytique de série génératrice la somme des deux séries génératrices.

e) Approximations des asservissements.

Pour les fonctionnelles, nous sommes restés dans le champ des fonctions continues, bien trop restreint ici, ne serait-ce que pour les applications comme le filtrage statistique. Soit donc $L^\alpha (\alpha \geq 1)$ l'espace des fonctions de $\underline{R}_+$ dans K de puissance α localement intégrable et muni de la topologie de la convergence sur tout compact. L^∞ est l'ensemble des fonctions continues.

Un asservissement à entrée et sortie scalaire peut être considéré comme une application de $\underline{R}_+ \times L^\alpha$ dans K, qui obéit au principe de causalité : la sortie y(t) ne dépend de l'entrée $u(\tau)$ que pour $0 \leq \tau \leq t$. Il est dit continu si c'est une application continue de $\underline{R}_+ \times L^\alpha$ dans K .

Théorème 4.8.- *Dans tout domaine compact de $\underline{R}_+ \times L^\alpha$ tout asservissement continu peut être uniformément approché par des asservissements analytiques, que l'on peut choisir rationnels ou polynômiaux.*

Preuve. - Compte tenu de la cloture par addition et multiplication, il faut, pour appliquer le théorème de Weierstrass-Stone, vérifier la propriété de séparabilité suivante : il existe un asservissement (polynômial) prenant des valeurs distinctes pour des entrées distinctes $u_i : [0,t_i] \to K$ $(i = 1,2)$

Si $t_1 \neq t_2$, il suffit de prendre l'asservissement de série génératrice x_0.

Si $t_1 = t_2 = t$, il existe,en vertu de la densité des polynômes, au moins un entier k tel que :

$$\int_0^t u_1(\tau)\tau^k d\tau \neq \int_0^t u_2(\tau)\tau^k d\tau_k$$

On prend l'asservissement de série génératrice $x_1 x_0^k$.

Corollaire 4.9. - *Dans tout domaine compact de* $\underline{R}_+ \times L^\alpha$, *tout asservissement continu peut être uniformément approché par des asservissements réguliers réalisables, que l'on peut choisir nilpotents.*

Remarques. - (i) Ce résultat nous paraît fournir un cadre du plus grand intérêt pour l'étude effective, c'est-à-dire algorithmique et numérique, des systèmes non linéaires et non stationnaires. On sait, en effet, que c'est la généralisation, due à Fréchet (cf. Volterra et Pérès |39|, p. 61, Lévy |29|,p.78) du théorème de Weiertrass aux fonctionnelles, qui a justifié l'introduction des séries de Volterra (cf. Barrett |3|). Le calcul des noyaux des séries de Volterra, même tronquées, a toujours posé un redoutable problème, même si, dans le cas stationnaire, la transformation de Laplace multidimensionnelle a constitué un outil d'une certaine efficacité (cf. George |24|, Lubbock, Bansal |31|). Qu'il y ait ou non stationnarité des séries formelles non commutatives permettent un calcul beaucoup plus simple qui, en |22|, sera appliqué aux équations différentielles non linéaires.

(ii) L'approximation dans L^2 conduit à considérer les problèmes de filtrage statistique non linéaire. Les travaux de Marcus |44| montrent, en substance,que le filtrage des asservissements polynômiaux nilpotents est relativement aisé, ce qui fournit un cadre nouveau, peut être de la plus haute importance.

(iii) Le théorème d'approximation a été publié en |21| pour la topologie compacte des fonctions continues. Peu après, et indépendamment, il a été retrouvé par Sussmann |42| pour la topologie de la convergence vague des fonctions continues, ce qui implique le résultat de |21|. La méthode de Sussmann différente de la nôtre s'inspire de la théorie de la représentation des groupes de Lie, et plus précisèment du théorème classique de Peter-Weyl. Krener |43| montre que tout système de la forme

$$\begin{cases} \dot{q}(t) = A_0(q) + \sum_{i=1}^{n} u_i(t)\ B_i(q) \\ y(t) = C(q(t)) \end{cases}$$

où la commande apparaît linéairement, peut être approximé par des asservissements réguliers réalisables [(1)].

----

(1) L'auteur a eu connaissance des résultats de Sussmann et Krener lors du présent symposium.

Terminons en justifiant notre remplacement par le terme "régulier" de ce qui par ailleurs est appelé asservissement "bilinéaire". Ce dernier adjectif, qui veut caractériser le produit du vecteur d'état et de l'entrée, nous semble mal venu car l'application entrée-sortie définie par le système n'est pas nécessairement bilinéaire. De plus, l'appellation de système "bilinéaire" a été employée par divers auteurs (R.E. Kalman, M.A. Arbib, G. Marchesini) lorsque cette application est précisément bilinéaire. Volterra (cf. Volterra et Pérès |39|, chap. III) appelle "régulières" les fonctionnelles homogènes intervenant dans la formule (IV.2). Le terme "régulier" a aussi l'avantage d'insister sur l'absence de singularités.

V. - Remarques sur la notion d'espace d'état

Nous ne consacrerons à ces remarques que quelques lignes, car le faire de manière approfondie nous mènerait par trop loin.
C'est Kalman |26| qui, à partir de 1960, a surtout contribué à populariser la notion d'espace d'état pour les systèmes linéaires, et les concepts y afférents de commandabilité et d'observabilité. Il faut constater que les propriétés très remarquables du cas linéaire, notamment stationnaire, ont induit bien des gens à essayer de les généraliser sans examen à d'autres cas. Ainsi, on a souvent voulu définir le fait pour un espace d'état d'être réduit (ou minimal) par la propriété d'être complètement commandable et observable. C'est faux pour les asservissements réguliers. On utilise souvent de manière indistincte les termes de "non stationnaire" et "non autonome" (ou *time-varying*) . Il n'y a pas de mal dans le cas linéaire, mais, comme le montre la proposition 2.7., c'est inexact en général. Seule la stationnarité peut être définie de manière intrinsèque comme invariance par translation temporelle.
Il nous semble anormal que les discussions sur l'espace d'état, parues dans la littérature américaine, ne fassent, à notre connaissance, aucune allusion aux polémiques fort vives qui ont eu lieu au début du siècle à propos des phénomènes héréditaires (1) et dont on trouve un écho dans un livre |37| récent de Vogel, avec qui nous citerons le passage suivant de Picard |32| :

"... Dans cette étude, les lois exprimant nos idées sur le mouvement se sont trouvées condensées dans les équations différentielles, c'est-à-dire des relations entre les variables et leurs dérivées. Il ne faut pas oublier que nous avons en définitive formulé un principe de non-hérédité, en supposant que l'avenir d'un système ne dépend à un moment donné que de son état actuel, ou d'une manière plus générale (si on regarde les forces comme pouvant aussi dépendre des vitesses) que

----

(1) C'est pour étudier mathématiquement ces phénomènes que V. Volterra a introduit son calcul fonctionnel (voir, en particulier, le dernier chapitre du livre |38|)

que cet avenir dépend de l'état actuel et de l'état infiniment voisin qui précède. C'est une hypothèse restrictive et que, en apparence du moins, bien des faits contredisent. Les exemples sont nombreux, où l'avenir d'un système semble dépendre des états antérieurs : il y a hérédité. Dans des cas aussi complexes, on se dit qu'il faudra peut-être abandonner les équations différentielles, et envisager les équations fonctionnelles, où figureront des intégrales prises depuis un temps très long jusqu'au temps actuel, intégrales qui seront la part de cette hérédité. Les tenants de la mécanique classique pourront cependant prétendre que l'hérédité n'est qu'apparente et qu'elle tient à ce que nous portons notre attention sur un trop petit nombre de variables ...".
Comme l'écrivent Volterra et Pérès |39|,p. 165, la notion d'équation fonctionnelle "représente ... une extension, pour l'ordre infini, de la notion d'équation différentielle d'ordre fini" (voir aussi Vogel |37|, Fargue |14|. En terme philosophique, l'espace d'état (ou de phase) doit être introduit pour soi (*für sich*) et non considéré en soi (*an sich*).

Bibliographie

1 . - AGASANDJAN (G.A.) - Automata with a variable structure (en russe avec résumé en anglais). Dokl. Akad Nauk. SSSR, 1974, 1967, p.529-530.

2 . - ALESSANDRO (P. d'), ISIDORI (A.) et RUBERTI (A.) - Realization and structure theory of bilinear dynamical systems. SIAM J. Control,15,1963, p.567-615.

3 . - BARRETT(J.F.) - The use of functionnals in the analysis of non-linear physical models. J. Electronics Control, 15,1963, p. 567-615.

4 . - BOURBAKI (N.) - Topologie Générale (chap. 5 à 10). Hermann, Paris, 1974.

5 . - BROCKETT (R.W.) - On the algebraic structure of bilinear systems, in "Theory and Applications of Variable Structure Systems" (R.R. Mohler et A. Ruberti,éd.),p. 153-168. Academic Press, New York, 1972.

6 . - BROCKETT (R.W.) et WILLSKY (A.S.) - Some structural properties of automata defined on groups, in"Category Theory Applied to Computation and Control" (E.G. Manes, éd.) p.112-118. Lect. Notes Comput. Sci. 25 Springer-Verlag, Berlin, 1975.

7 . - BRUNI (C.), DI PILLO (G.) et KOCH (G.) - Bilinear systems : an appealing class of "nearly linear" systems in theory and applications. IEEE Trans. Autom. Contr., 19, 1974, p. 334-348.

8 . - CHEN (K.T.) - Integration of paths, a faithful representation of paths by non-commutative formal power series. Trans. Amer. Math. Soc. 89, 1958, p. 395-407.

9 . - CHEN (K.T.) - Algebraic paths, J. Algebra, 10, 1968, p. 8-36.

10 . - CHOMSKY (N.) - Aspects of the theory of syntax. M.I.T. Press, Cambridge (Mass.), 1965.

11 . - CHOMSKY (N.) et SCHÜTZENBERGER (M.P.) - The algebraic theory of context-free languages, in "Computer Programming and Formal Systems" (P. Braffort et D. Hirschberg, éd.), p. 118-161. North-Holland. Amsterdam. 1963.

12 . - DAUSCHA (W.), NÜRNBERG (G.), STARKE (P.H.) et WINKLER (K.D.) - Theorie der determinierten zeitvariabeln Automaten. Elektron. Informationsverarbeit. Kybern., 9, 1973, p. 455-511.

13 . - EILENBERG (S.)- Automata, languages and machines, vol. A. Academic Press, New York, 1974.

14 . - FARGUE (D.) - Réductibilité des systèmes héréditaires. Internat. J. Non-Linear Mechanics, 9 , 1974, p. 331-338.

15 . - FLIESS (M.) - Deux applications de la représentation matricielle d'une série rationnelle non commutative. J. Algebra, 19, 1971, p. 344-353.

16 . - FLIESS (M.) - Sur la réalisation des systèmes dynamiques bilinéaires. C.R. Acad. Sc. Paris, A-277, 1973, p. 923-926.

17 . - FLIESS (M.) - Sur les systèmes dynamiques bilinéaires qui sont linéaires. C. R. Acad. Sc. Paris, A-278, 1974, p. 1147-1149.

18 . - FLIESS (M.) - Sur divers produits de séries formelles. Bull. Soc. Math. France, 102, 1974, p. 181-191.

19 . - FLIESS (M.) - Matrices de Hankel. J. Math. Pures Appl., 53, 1974, p. 197-222.

20 . - FLIESS (M.) Sur la réalisation des systèmes dynamiques bilinéaires, non autonomes, à temps discret ; application aux systèmes linéaires. C. R. Acad. Sc. Paris, A-279, 1974, p. 243-246.

21 . - FLIESS (M.) - Séries de Volterra et séries formelles non commutatives. C. R. Acad. Sc. Paris, A-280, 1975, p. 965-967.

22 . - FLIESS (M.) - Un calcul symbolique non commutatif pour les asservissements non linéaires et non stationnaires. 7e Conf. IFIP Techn. Optim., Nice (1975) à paraître aux Lect. Notes Comput. Sci., Springer-Verlag. Berlin.

23 . - GANTMACHER (F.R.) - Théorie des matrices (traduit du russe), t. 2. Dunod, Paris, 1966.

24 . - GEORGE (D.A.) - Continuous non-linear systems. Techn. Rep., Research Lab. Electronics, M.I.T. , Cambridge (Mass.),1959.

25 . - ISIDORI (A.) - Direct contruction of minimal bilinear realizations from nonlinear input/output maps. IEEE Trans.Autom. Contr., 18, 1973,p.626-631.

26 . - KALMAN (R.E.) - On the general theory of control systems. Proc. Ist IFAC Congress, Moscow(1960), p. 481-492, Butterworths,London, 1960.

27 . - KALMAN (R.E.), FALB (P.L.) et ARBIB (M.A.) - Topics in mathematical system theory. McGraw-Hill, New-York, 1969.

28 . - KAPLANSKY (I.) - Fields and rings. University of Chicago Press, Chicago, 1969.

29 . - LÉVY (P.) - Problèmes concrets d'analyse fonctionnelle. Gauthier-Villars, Paris, 1951.

30 . - LO (J.T.H.) et WILLSKY (A.S.) - Estimation for rotational processes with one degree of freedom-Part I : Introduction and continuous-time processes. IEEE Trans. Contr., 20, 1975, p. 10-21

31 . - LUBBOCK (J.K.) et BANSAL (U.S.) - Multidimensional Laplace transforms for solution of nonlinear equations. Proc. IEE, 116, 1969, p. 2075-2082.

32 . - PICARD (E.)-La mécanique classique et ses approximations successives. Rivista Scienza, 1, 1907, p. 4-15

33 . - REE (R.) - Lie elements and an algebra associated with shuffles. Annals Math., 68, 1958, p. 210-220.

34 . - SCHÜTZENBERGER (M.P.)-On the definition of a family of automata. Inform. Contr., 4, 1961, p. 245-270.

35 . - THOM (R.) - Stabilité structurelle et morphogénèse. Benjamin, Reading (Mass.) 1972.

36 . - THOM (R.) - Modèles mathématiques de la morphogénèse.Union Générale d'Editions, Paris, 1974.

37 . - VOGEL (T.) - Pour une théorie mécaniste renouvelée. Gauthier-Villars, Paris, 1973.

38 . - VOLTERRA (V.) - Leçons sur les fonctions de lignes. Gauthier-Villars, Paris, 1913.

39 . - VOLTERRA (V.) et PÉRÈS (J.) - Théorie générale des fonctionnelles. t.1, Gauthier-Villars, Paris, 1936.

40 . - WEISS (L.) - Controllability, realization and stability of discrets-time systems. SIAM J. Contr., 10, 1972, p. 230-251.

41 . - WIENER (N.) - Nonlinear problems in random theory. M.I.T. Press, Cambridge (Mass.), 1958.

42 . - SUSSMANN (H.J.)-Semigroup representation, bilinear approximation of input/output maps, and generalized inputs. Dans ce volume.

43 . - KRENER (A.J.) - Bilinear and nonlinear realizations of input/output maps. SIAM J. Contr., 13, 1975, p. 827-834.

44 . - MARCUS (S.I.) - Estimation and analysis of nonlinear stochastic systems. Ph. D. Thesis, Dpt. of Electrical Engineering, M.I.T., Cambridge (Mass.), 1975. Voir aussi l'article de S.I. MARCUS et A.S. WILLSKY (Algebraic structure and finite dimensional nonlinear estimation) dans ce volume.

A FORMAL POWER SERIES APPROACH TO CANONICAL REALIZATION OF BILINEAR INPUT-OUTPUT MAPS

by

Ettore Fornasini and Giovanni Marchesini*
Dept. of Electrical Engineering
University of Padua, Italy

INTRODUCTION

In a previous paper [1] we presented a contribution to the realization theory of bilinear discrete-time stationary input-output maps. In particular we shown that for bilinear i/o maps characterized by realizable series the set of Nerode states (or canonical states) can be embedded in a finite dimensional vector space. The dynamics of the state is then described by recursive equations of "bilinear structure".

This paper is a continuation of [1]. We extend the notions of reachability, controllability and observability in bounded time to canonical realizations of bilinear i/o maps and we prove that each one of these conditions is equivalent to assume that the i/o map is represented by a realizable series.

1. DEFINITIONS

Let K be a field and let U_1, U_2 and Y denote the following spaces:

(a) $U_1 = U_2 = \{ u \in K^{\mathbb{Z}-}$ with compact support$\}$

(b) $Y = \{ y \in K^{\mathbb{N} - \{0\}} \}$

---

* This work was supported by CNR-GNAS under NSF-CNR joint research programme.

$U_1 \times U_2$ is termed the <u>input space</u> and Y the <u>output space</u>.

DEFINITION. A map $f: U_1 \times U_2 \to Y$ is a <u>bilinear discrete time, stationary i/o map</u> if it satisfies the following conditions:

(i) <u>bilinearity</u>

$$f(ku_1,u_2) = kf(u_1,u_2); \quad f(u_1,ku_2) = kf(u_1,u_2)$$

$$f(u_1 + v_1,u_2) = f(u_1,u_2) + f(v_1,u_2)$$

$$f(u_1,u_2 + v_2) = f(u_1,u_2) + f(u_1,v_2)$$

for any $k \in K$, $u_1, v_1 \in U_1$, $u_2, v_2 \in U_2$

(ii) <u>stationarity</u>

the map f is invariant under translation with respect to time in the following sense: the diagram

$$(1.1) \qquad \begin{array}{ccc} U_1 \times U_2 & \xrightarrow{f} & Y \\ \sigma \downarrow & & \downarrow \sigma_* \\ U_1 \times U_2 & \xrightarrow{f} & Y \end{array}$$

commutes with respect to the shift operators σ and σ_* defined as

$$\sigma: ((\ldots,u_1(-1), u_1(0)), (\ldots,u_2(-1), u_2(0))) \to$$

$$((\ldots,u_1(-1), u_1(0),0), (\ldots,u_2(-1), u_2(0),0))$$

$$\sigma_*: (y(1), y(2),\ldots) \to (y(2), y(3),\ldots)$$

REMARK. The space $K[z_1^{-1}] \times K[z_2^{-1}]$ is endowed with a $K[\sigma]$ - module structure via the operation

$$\sigma: K[z_1^{-1}] \times K[z_2^{-1}] \to K[z_1^{-1}] \times K[z_2^{-1}]: (p_1,p_2) \to (z_1^{-1}p_1,z_2^{-1}p_2);$$

in this way $U_1 \times U_2$ can be identified to the $K[\sigma]$-module $K[z_1^{-1}] \times K[z_2^{-1}]$.

It is also obvious in what sense the ring of formal power series in one indeterminate z $K[[z]]$ ("causal" power series), denoted by $K_c[[z]]$, is a $K[z^{-1}]$-module and is identified to the $K[\sigma_*]$-module Y.

Polynomials and power series are then alternative ways of viewing the elements of U_1, U_2 and Y; thus we shall not distinguish between the two representations.

REMARK. The polynomial notation is particularly useful in representing i/o relations. In fact, let $(z_1 z_2)K[[z_1,z_2]]$ be the ring (without identity) of "causal" power series, denoted by $K_c[[z_1,z_2]]$. Consider $s \in K_c[[z_1,z_2]]$, $s = \Sigma s_{ij} \; z_1^i z_2^j$ and assume that the coefficient s_{ij} represents the output at time 1 for inputs $u_1 = z_1^{-i}$, $u_2 = z_2^{-j}$. Then for any input pair $u_1 \in K[z_1^{-1}]$, $u_2 \in K[z_2^{-1}]$ the output can be represented by the formal power series in $(z_1 z_2)$ given by:

$$y = f(u_1,u_2) = \sum_{1}^{\infty}{}_r \; y(r)(z_1 z_2)^r = s \; u_1(z_1^{-1})u_2(z_2^{-1}) \odot \Sigma_{1}{}_r (z_1 z_2)^r$$

where $\odot$ denotes the Hadamard product:

$$\Sigma_{ij} a_{ij} z_1^i z_2^j \odot \Sigma_{ij} b_{ij} z_1^i z_2^j = \Sigma_{ij} a_{ij} b_{ij} z_1^i z_2^j$$

We have thus introduced a biunique correspondence between i/o maps and "causal" power series in two indeterminates.

The Nerode equivalence relation $\underset{N}{\sim}$ [2] is naturally defined in $U_1 \times U_2$: two input pairs (u_1,u_2) and (v_1,v_2) are Nerode equivalent iff the output sequences $f(u_1,u_2)$ and $f(v_1,v_2)$ are the same and remain the same whenever both (u_1,u_2) and (v_1,v_2) are followed by an arbitrary input pair $(w_1,w_2) \in U_1 \times U_2$. More precisely:

$$(u_1,u_2) \underset{N}{\sim} (v_1,v_2) \quad \text{iff} \quad f(\sigma^k(u_1,u_2) + (w_1,w_2)) = $$
$$= f(\sigma^k(v_1,v_2) + (w_1,w_2))$$

$\forall k \in \mathbb{N}$, $\forall (w_1,w_2) \in U_1 \times U_2$ with length $(w_1,w_2) \triangleq (\max(\deg w_1, \deg w_2))+1 \leqslant k$.

We denote the Nerode equivalence classes by $[u_1,u_2]$:

$$[u_1,u_2] = \{(v_1,v_2) \in U_1 \times U_2 : (v_1,v_2) \underset{N}{\sim} (u_1,u_2)\}.$$

The map f is then factorized as in the following commutative diagram:

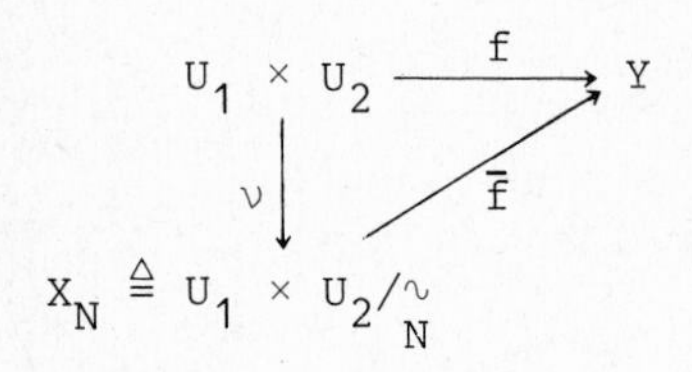

The set X_N defined by

$$X_N = \{[u_1,u_2] : (u_1,u_2) \in U_1 \times U_2\} = (U_1 \times U_2)/\underset{N}{\sim}$$

is called the <u>Nerode</u> (<u>or canonical</u>) <u>state space</u>.

2. EMBEDDING OF X_N IN A FINITE DIMENSIONAL VECTOR SPACE

The following three equivalence relations defined on U_1, U_2 and $U_1 \times U_2$ respectively play an essential role in our study (see [3,4]):

(1) $\quad u_1 \underset{1}{\sim} v_1 \quad \text{iff} \quad f(\sigma^k u_1, w_2) = f(\sigma^k v_1, w_2)$, $\forall k$, $\forall w_2 \in U_2$, $\deg w_2 < k$

(2) $\quad u_2 \underset{2}{\sim} v_2 \quad \text{iff} \quad f(w_1, \sigma^k u_2) = f(w_1, \sigma^k v_2)$, $\forall k$, $\forall w_1 \in U_1$, $\deg w_1 < k$

(3) $\quad (u_1,u_2) \underset{3}{\sim} (v_1,v_2)$ iff $f(u_1,u_2) = f(v_1,v_2)$.

In [4] it has been proved that

$$(u_1,u_2) \underset{N}{\sim} (v_1,v_2) \text{ iff } u_1 \underset{1}{\sim} v_1,\ u_2 \underset{2}{\sim} v_2, (u_1,u_2) \underset{3}{\sim} (v_1,v_2)$$

The quotient spaces $X_1 = U_1/\underset{1}{\sim}$ and $X_2 = U_2/\underset{2}{\sim}$ are naturally endowed with the structure of a linear space. In general the set $(U_1 \times U_2)/\underset{3}{\sim}$ does not admit such a structure but a standard algebraic construction allows the embedding of $U_1 \times U_2$ (i.e. $K[z_1^{-1}] \times K[z_2^{-1}]$) in the tensor space $U_1 \otimes U_2$ (i.e. $K[z_1^{-1}, z_2^{-1}]$). It follows that there exists a linear map $f_\otimes$ making the following diagram commutative

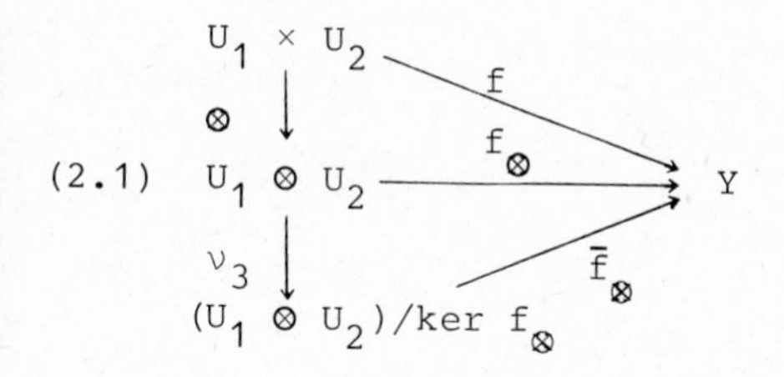

where ν_3 is onto and $\bar{f}_\otimes$ is one-to-one.

The map $f_\otimes$ induces an equivalence relation in $U_1 \otimes U_2$ and it can be verified immediately $(u_1,u_2) \underset{3}{\sim} (v_1,v_2)$ iff $u_1 \otimes u_2 = v_1 \otimes v_2 \pmod{f_\otimes}$. Thus the equivalence classes under $\underset{3}{\sim}$ are naturally embedded in the linear space $X_3 = (U_1 \otimes U_2)/\ker f_\otimes$ and the linear space $X_1 \oplus X_2 \oplus X_3$ furnishes a natural embedding for the canonical space X_N.

In [1] we related the finite dimensionality of $X_1 \oplus X_2 \oplus X_3$ to some properties of the series characterizing the i/o map f. Since some results presented there are necessary to a better comprehension of what follows, we devote the remaining of this section to recall them.

Denote by $K[(z)]$ the ring of rational power series in the indeterminate z and by $K[(z_1,z_2)]$ the ring of rational power series in z_1 and z_2.

The subring of $K[(z_1,z_2)]$ generated by $K[(z_1)]$, $K[(z_2)]$ and $K[(z_1 z_2)]$ is denoted by $K^{real}[(z_1,z_2)]$ (the ring of "realizable" power series). The elements in $K^{real}[(z_1,z_2)]$ are power series expansions of the rational functions whose denominators can be factored in the form $p_1(z_1^{-1})p_2(z_1^{-1})p(z_1^{-1}z_2^{-1})$. A further characterization of realizable series is given by the following theorem:

THEOREM 2.1. *Let* $s \in K[[z_1,z_2]]$, $s = \Sigma_{ij}\, s_{ij} z_1^i z_2^j$ *and define the following three families of formal power series* r_i, c_j *and* d_{ij} *in one indeterminate*

$$r_i = \sum_{o}^{\infty}{}_k\, s_{i,i+k}\, z^k, \qquad i = 1,2,\ldots \qquad \text{"row series"}$$

$$c_j = \sum_{o}^{\infty}{}_k\, s_{j+k,j}\, z^k, \qquad j = 1,2,\ldots \qquad \text{"column series"}$$

$$d_{ij} = \sum_{o}^{\infty} s_{i+k,j+k} z^k, \qquad i,j = 0,1,2,\ldots \qquad \text{"diagonal series"}$$

Then $s \in K^{real}[(z_1,z_2)]$ *if and only if* r_i, c_j, d_{ij} *are power series expansions of rational functions in one indeterminate having common denominator.*

The connections among dimensions of X_1, X_2 and X_3 and the structure of the formal power series s are clarified by Lemmas 2.1, 2.2 and Theorem 2.2 below.

LEMMA 2.1. *Let* $s \in K_c[[z_1,z_2]]$ *represent a bilinear i/o map* f: $U_1 \times U_2 \to Y$. *Then* X_3 *is finite dimensional iff the diagonal series* d_{ij}, $i,j = 1,2,\ldots$ *are power series expansions of rational functions having common denominator.*

LEMMA 2.2. *Let* $s \in K_c[[z_1,z_2]]$ *represent a bilinear* i/o *map* f:

$U_1 \times U_2 \to Y$. Then the space $X_1 (X_2)$ is finite dimensional iff the column series c_j, $j = 1,2,\ldots$ (row series r_i, $i = 1,2,\ldots$) are power series expansions of rational functions having common denominator.

THEOREM 2.2. Let $s \in K_c[[z_1,z_2]]$ represent a bilinear i/o map f: $U_1 \times U_2 \to Y$. Then $X_1 \oplus X_2 \oplus X_3$ is finite dimensional iff s is a realizable power series.

3. REACHABILITY AND CONTROLLABILITY OF X_N IN BOUNDED TIME

The canonical state space X_N is intrinsically reachable by definition. Thus it is worth while to investigate if X_N is also controllable and, further, if there is an upper bound for the lengths of inputs needed to reach (control) reachable (controllable) states. To be more precise we introduce the following definitions.

DEFINITION 3.1. X_N is reachable in time m if each Nerode equivalence class $[u_1,u_2] \in X_N$ contains at least one input of length less than $m + 1$. X_N is reachable in bounded time if it is reachable in time m for some m.

DEFINITION 3.2. X_N is controllable (to zero state) in time k if for each Nerode equivalence class $[u_1,u_2]$ there exists at least one input (w_1,w_2) of length less that $k+1$ such that $(\sigma^k(u_1,u_2) + (w_1,w_2)) \in [0,0]$. X_N is controllable in bounded time if it is controllable in time k for some k.

Reachability and controllability in bounded time are characteristic properties of the canonical state space of bilinear i/o maps represented by realizable series. This fact is stated in the following

THEOREM 3.1. (reachability and controllability in b.t.). Let $s \in K_c[[z_1,z_2]]$ represent a bilinear i/o map f. Then the following con-

ditions are equivalent:

(i) $s \in K_c^{real}[[z_1,z_2]]$

(ii) X_N *is reachable in bounded time*

(iii) X_N *is controllable in bounded time*

The equivalence (i) ⇔ (ii) is proved in [1]. Before proceeding to prove (i) ⇔ (iii), we shall derive two technical lemmas:

LEMMA 3.1. *Let* $f: U_1 \times U_2 \to Y$ *be a bilinear i/o map. Then there exist polynomials* $\omega_j \in K[z_j^{-1}]$, $j = 1,2$, *such that*

(i) X_j *is a* $K[z_j^{-1}]$*-module isomorphic to* $U_j/(\omega_j)$, $j = 1,2$

(ii) X_j *is finite dimensional over* K *iff* $\omega_j \neq 0$, $j = 1,2$

PROOF. Define the map $f_1: U_1 \to K[[z]]^{1 \times \infty}$ by the assignment

$$f_1(u_1) = (f(z_1^{-1}u_1,1),\ f(z_1^{-2}u_1,1)\ldots).$$

The linear space $K[[z]]^{1 \times \infty}$ admits the structure of a $K[z^{-1}]$ - module with scalar multiplication $z_1^{-1}(s_1,s_2,\ldots) = (s_2,s_3,\ldots)$. Hence the map f_1 is a $K[z_1^{-1}]$-morphism. For by definition of module scalar multiplication:

$$f_1(z_1^{-1}u_1) = (f(z_1^{-2}u_1,1),f(z_1^{-3}u_1,1),\ldots) =$$

$$= z_1^{-1}(f(z_1^{-1}u_1,1),f(z_1^{-2}u_1,1),\ldots).$$

We therefore have the following commutative diagram

$$\begin{array}{ccc} K[z_1^{-1}] \simeq U_1 & \xrightarrow{f_1} & K[[z]]^{1\times\infty} \\ \nu_1 \downarrow & \nearrow \bar{f}_1 & \\ U_1/\ker f_1 & & \end{array}$$

where $U_1/\ker f_1$ is naturally endowed with $K[z^{-1}]$-module structure and $\ker f_1$ is an ideal in $K[z_1^{-1}]$.

A similar argument is used to prove the case j = 2. It is immediately verified that $u_1 \underset{1}{\sim} v_1$ if and only if $u_1 - v_1 \in \ker f_1$. Hence $X_1 \simeq U_1/\ker f_1$ is finite dimensional if and only if $\ker f_1 = (\omega_1) \neq (0)$. Moreover $\dim X_1 = \deg \omega_1$.

LEMMA 3.2. <u>Let</u> $f: U_1 \times U_2 \to Y$ <u>be a bilinear i/o map</u>, <u>let</u> X_i, $i = 1,2$ <u>be finite dimensional and let</u> ω_i, $i = 1,2$, <u>be as in Lemma 3.1. Then the image under</u> $f_\otimes$ <u>of the ideal</u> $\omega_1\omega_2\, K[z_1^{-1}, z_2^{-1}]$ <u>is infinite dimensional iff</u> $\dim X_3 = \infty$.

PROOF. Take any monomial $z_1^{-k} z_2^{-h}$ and make the following decompositions:

$$\begin{aligned} z_1^{-k} &= \omega_1 q_1 + r_1, && \deg r_1 < \deg \omega_1 \\ z_2^{-h} &= \omega_2 q_2 + r_2, && \deg r_2 < \deg \omega_2 \\ q_i &= z_i^{-a} p_i + g_i, && \deg g_i < a \qquad i = 1,2 \end{aligned}$$

with $a = \max(\deg \omega_1, \deg \omega_2)$.

Hence

$$f_\otimes(z_1^{-k} z_2^{-h}) = f_\otimes((\omega_1 z_1^{-a} p_1 + \omega_1 g_1 + r_1)(\omega_2 z_2^{-a} p_2 + \omega_2 g_2 + r_2)) =$$

$$= f_\otimes(\omega_1\omega_2 z_1^{-a} z_2^{-a} p_1 p_2) + f_\otimes(\omega_1 g_1 r_2) + f_\otimes(\omega_2 g_2 r_1) + f_\otimes(r_1 r_2) \quad .$$

Since the degrees of r_i and g_i are less than a,

$f_\otimes(z_1^{-k}z_2^{-h} - \omega_1\omega_2 z_1^{-a}z_2^{-a}p_1p_2)$, $h,k = 0,1,\ldots$ span a finite dimensional vector space.

PROOF of (i) $\Leftrightarrow$ (iii) (controllability in b.t)

(i) $\Rightarrow$ (iii). To prove this implication, we shall show that there exists an integer k such that for each input (u_1,u_2) we can find a pair (w_1,w_2) of length less than $k+1$ satisfying

$$(3.1) \quad \sigma^k(u_1u_2) + (w_1,w_2) \in [0,0]$$

Since by Lemma 3.1 ω_1 and ω_2 are non zero polynomials, (3.1) is equivalent to $z_1^{-k}u_1 + w_1 \in (\omega_1)$, $z_2^{-k}u_2 + w_2 \in (\omega_2)$, $f(z_1^{-k}u_1 + w_1, z_2^{-k}u_2 + w_2) = 0$.

Set $\max(\deg\omega_1, \deg\omega_2) = a$ and denote by $\rho_i \in K[z_i^{-1}]$, $i = 1,2$, a pair of polynomials satisfying $u_1z_1^{-a} + \rho_1 \in (\omega_1)$, $\deg\rho_1 < a$, $u_2z_2^{-a} + \rho_2 \in (\omega_2)$, $\deg\rho_2 < a$. Observe now that reachability in time m implies the existence of polynomials $v_i \in (\omega_i)$, $\deg v_i < m$, $i = 1,2$, such that:

$$(z_1^{-m}(u_1z_1^{-a} + \rho_1), z_2^{-m}(u_2z_2^{-a} + \rho_2)) \underset{N}{\sim} (v_1,v_2).$$

Hence setting $k = m + a$ and $(w_1,w_2) = (\rho_1z_1^{-m} + v_1, \rho_2z_2^{-m} - v_2)$, we see that $\sigma^k(u_1,u_2) + (w_1,w_2) \in [0,0]$. Consequently X_N is controllable (to zero state) in time k.

(iii) $\Rightarrow$ (i). Assume that X_N is controllable in time k. Obviously this implies controllability in time k of X_1 and X_2. Hence by Lemma 3.1, ω_i are non zero and $\dim X_i = \deg\omega_i = n_i < \infty$, $i = 1,2$.

By Lemma 3.2, X_3 is finite dimensional if $\dim f_\otimes((\omega_1\omega_2)) < \infty$, i.e. if $f_\otimes(\omega_1\omega_2z_1^{-i}z_2^{-j})$, $i,j = 0,1,\ldots$ span a finite dimensional vector space. Since X_N is controllable to zero in time k, for each monomial $z_1^{-i}z_2^{-j}$, there exist polynomials $w_i \in (\omega_i)$, $\deg w_i < k$, $i = 1,2$, such that

$$0 = f_{\otimes}((z_1z_2)^{-k}(z_1^{-i}z_2^{-j}\omega_1\omega_2) + w_1w_2) =$$

$$= \sigma_*^k f_{\otimes}(z_1^{-i}z_2^{-j}) + f_{\otimes}(w_1w_2).$$

Hence $\sigma_*^k\ f_{\otimes}(z_1^{-i}z_2^{-j}\omega_1\omega_2)$, $i,j = 0,1,\ldots$ span a finite dimensional vector space an so do $f_{\otimes}(z_1^{-i}z_2^{-j}\omega_1\omega_2)$, $i,j = 0,1,\ldots$

Thus $X_1 \oplus X_1 \oplus X_3$ is finite dimensional and by Theorem 2.2 s is a realizable series.

COROLLARY. X_N <u>is connected in bounded time</u>.

4. OBSERVABILITY OF X_N IN BOUNDED TIME

Let $\pi\colon K_c[[z]] \to K$, $\pi\colon \sum_{i=1}^{\infty} a_i z^i \to a_1$ and introduce the map $f_\pi = \pi \circ f$.

DEFINITION 4.1. Two states $[u_1,u_2]$ and $[v_1,v_2]$ in X_N are <u>distinguishable in time m</u> if there exist an integer k and an input (w_1,w_2) such that

$$\text{lenght } (w_1,w_2) \leqslant k \leqslant m$$

$$f_\pi(\sigma^k(u_1,u_2) + (w_1,w_2)) \neq f_\pi(\sigma^k(v_1,v_2) + (w_1,w_2))$$

The space X_N is called <u>observable in time m</u> when any two states are distinguishable in time m. X_N is <u>observable in bounded</u> time if it is observable in time m for some m.

A natural continuation of the programme of describing X_N in terms of its system theoretic properties is the description of the relationship between observability in bounded time and the structure of the i/o map. To be more precise we shall prove the following Theorem.

THEOREM 4.1. (observability in b.t.). *Let* $s \in K_c[[z_1,z_2]]$ *represent a bilinear i/o map* f. *A necessary and sufficient condition that* s *be realizable is that* X_N *is observable in bounded time*.

Necessity. Let s be realizable. Then by Theorem 2.2, $X_1 \oplus X_2 \oplus X_3$ is finite dimensional with $\dim X_i = n_i$, $i = 1,2,3$. Assume $[u_1,u_2] \neq$ $\neq [v_1,v_2]$. We therefore have three cases to consider.

If $(u_1,u_2) \underset{3}{\not\sim} (v_1,v_2)$, then $f(u_1,u_2) \neq f(v_1,v_2)$. Since Im $f_\otimes$ is a σ_*-invariant subspace of $K[[z]]$ having dimension n_3, it is easy to verify that

$$(4.1) \quad \sum_1^{n3} {}_i \, (z_1 z_2)^i \odot (f(u_1,u_2) - f(v_1,v_2)) \neq 0$$

This implies that $[u_1,u_2]$ and $[v_1,v_2]$ are distinguishable in time n_3.

If $(u_1,u_2) \underset{3}{\sim} (v_1,v_2)$, assume $u_1 \underset{1}{\not\sim} v_1$. Hence $f_1(u_1) \neq f_1(v_1)$ implying $f(u_1 z_1^{-k},1) \neq f(v_1 z_1^{-k},1)$ for some positive integer $k \leq n_1$. Thus

$$f(\sigma^k \, (u_1,u_2) + (0,1)) \neq f(\sigma^k \, (v_1,v_2) + (0,1))$$

Recalling (4.1) we have

$$f_\pi(\sigma^{k+h} \, (u_1,u_2) + \sigma^h \, (0,1)) \neq f_\pi(\sigma^{k+h} \, (v_1,v_2) + \sigma^h \, (0,1))$$

for some non negative integer $h \leq n_1$. Hence $[u_1,u_2]$ and $[v_1,v_2]$ are distinguishable in time $n_3 + n_1$.

If $(u_1,u_2) \underset{3}{\sim} (v_1,v_2)$ and $u_2 \underset{2}{\not\sim} v_2$, we can use analogous arguments to show that $[u_1,u_2]$ and $[v_1,v_2]$ are distinguishable in time $n_3 + n_2$. It follows that $[u_1,u_2]$ and $[v_1,v_2]$ are distinguishable in time

$m = \max(n_3 + n_1, n_3 + n_2)$.

As a noticeable consequence of the proof above, there exists a finite set of experiments sufficient to distinguish two Nerode states in bounded time.

Sufficiency. Assume that X_N is observable in time m. We shall prove that $X_1 \oplus X_2 \oplus X_3$ is finite dimensional.

Suppose X_1 is infinite dimensional. Lets define the map $g: U_1 \to K^{m \times m}$, $g(\omega_1) = (k_{ij})_{i,j=1,\ldots m}$, $k_{ij} = f_\pi(u_1 z_1^{-i-j+1}, z_2^{-j+1})$. Since $f_1(u_1) = 0$ implies $g(u_1) = 0$, then $\ker f_1 \subseteq \ker g$ and the quotient $(U_1/\ker f_1)/(\ker g/\ker f_1)$ is canonically isomorphic to $U_1/\ker g$.

This, together with the assumption $\dim X_1 = \dim(U_1/\ker f_1) = \infty$ gives that $\ker g/\ker f_1$ is infinite dimensional. Consequently we can find $u_1 \in U_1$ such that $g(u_1) \neq 0$ and $f_1(u_1) \neq 0$.

Then $[u_1, 0]$ and $[0,0]$ are indistinguishable in time m. For assuming $k \leqslant m$ we have

$$f_\pi(\sigma^k(u_1,0) + (\sum_0^{k-1} {}_i a_i z_1^{-i}, \sum_0^{k-1} {}_i b_i z_2^{-i})) -$$

$$-f_\pi(\sigma^k(0,0) + (\sum_0^{k-1} {}_i a_i z_1^{-i}, \sum_0^{k-1} {}_i b_i z_2^{-i})) =$$

$$= \sum_0^{k-1} {}_i b_i f_\pi(u_1 z_1^{-k}, z_2^{-i}) = 0$$

This contradicts the assumption. A similar result can be proved for X_2.

Assume now $\dim X_i = n_i \neq \infty$, $i = 1,2$ and $\dim X_3 = \infty$. Hence recalling Lemma 3.2, $f_\otimes((\omega_1 \omega_2))$ is infinite dimensional. Let introduce the following linear maps

$$\bar{f}: K[z_1^{-1}, z_2^{-1}] \to K_c[[z]] \;, \qquad \bar{f}(p) = f_\otimes(\omega_1 \omega_2 p)$$

$$\mu: K_c[[z]] \to K^m \;, \qquad \mu(\sum_{i=1}^{\infty} a_i z^i) = (a_1, a_2, \ldots a_m)$$

Now consider an (infinite) set $I \subseteq \mathbb{N}\times\mathbb{N}$ such that $\{\bar{f}(z_1^{-i}z_2^{-j}) : (i,j) \in I\}$ is a Hamel basis for $\bar{f}(K[z_1^{-1}, z_2^{-1}])$. Since the series s can be written in the form [5]:

$$s = (N + s^*)/\omega_1\omega_2$$

where $s^* = (z_1 z_2)^{-1} \sum_{|h-k| \leq a} s'_{h,k} z_1^{-h} z_2^{-k}$, $a = \max(\deg \omega_1, \deg \omega_2)$, $N \in K[z_1^{-1}, z_2^{-1}]$, it follows that $f_\otimes(\omega_1\omega_2 z_1^{-i} z_2^{-j}) = 0$ when $|i-j| > a$. Hence each pair (i,j) in I satisfies the condition $|i-j| \leq a$.

Choose in I m+1 pairs $(i_1,j_2),(i_2,j_2),\dots,(i_{m+1},j_{m+1})$ so that

$$\begin{aligned} &i_2,\ j_2 > \max(i_1,j_1) + a \\ &i_3,\ j_3 > \max(i_2,j_2) + a \\ (4.2)\qquad &\dots\dots\dots\dots\dots\dots \\ &i_{m+1}, j_{m+1} > \max(i_m,j_m) + a \end{aligned}$$

Since the range of μ is m-dimensional we have that $\pi \circ \bar{f}(\sum_{k=1}^{m+1} \alpha_k z_1^{-i_k} z_2^{-j_k}) \neq 0$ for some list of scalars $\alpha_1, \alpha_2, \dots \alpha_{m+1}$ not all zero. On the other hand $\bar{f}(\sum_{k=1}^{m+1} \alpha_k z_1^{-i_k} z_2^{-j_k}) \neq 0$.

We shall now use the sets of indices and of scalars introduced above to construct an input pair $(\omega_1 v_1, \omega_2 v_2)$ which does not belong to $[0,0]$ and is indistinguishable from (0,0) in time m.

In fact consider the polynomials

$$v_1 = \sum_{k=1}^{m+1} z_1^{-i_k}, \qquad v_2 = \sum_{h=1}^{m+1} \alpha_h z_2^{-j_h}$$

By (4.2), $v_1 v_2 - \sum_{k=1}^{m+1} \alpha_k z_1^{-i_k} z_2^{-j_k}$ is an element of $\ker \bar{f}$ and hence

$$0 \neq \bar{f}\left(\sum_{k=1}^{m+1} \alpha_k z_1^{-i_k} z_2^{-i_k}\right) = \bar{f}(v_1 v_2) = f(\omega_1 v_1, \omega_2 v_2)$$

$$0 = \mu \circ \bar{f}\left(\sum_{k=1}^{m+1} \alpha_k z_1^{-i_k} z_2^{-i_k}\right) = \mu \circ f(\omega_1 v_1, \omega_2 v_2)$$

We therefore see that $(\omega_1 v_1, \omega_2 v_2)$ and $(0,0)$ are not equivalent under Nerode equivalence. However they are indistinguishable in time m; for

$$f_\pi(\sigma^k(\omega_1 v_1, \omega_2 v_2) + (w_1, w_2)) = f_\pi(w_1, w_2)$$

if $0 \leqslant k \leqslant m$ and length $(w_1, w_2) < k$.

This contradicts the assumption.

REMARK 1. A finite number of experiments is sufficient to observe X_N in bounded time. In fact, as proved in the necessity part of Theorem 4.1, these experiments correspond to apply a family of inputs $(0,0)$, $(0,z_2), \ldots, (0,z_2^{n_1}), (z_1,0), \ldots, (z_1^{n_2},0)$ and then to find the outputs in an interval of length n_3.

REMARK 2. A similar finite procedure can be adopted to characterize the i/o maps represented by realizable series. For, by Theorem 2.1 a realizable series in the indeterminates z_1 and z_2 can be computed from the first $2n_3$ coefficients of the series in one indeterminate $f(z_1^{-i},1)$, $i = 0,1,\ldots, 2n_1$, and $f(1,z_2^{-i})$, $i = 1,\ldots, 2n_2$;

The integers n_1, n_2, n_3 are the degrees of the lowest recurrence polynomials for "column", "row" and "diagonal" series respectively.

5. CONCLUSIONS

We have proved in this paper that the following conditions are equivalent:

(i) the i/o map $f: U_1 \times U_2 \to Y$ is represented by a realizable series

(ii) the canonical realization X_N is reachable in bounded time

(iii) the canonical realization X_N is controllable in bounded time

(iv) the canonical realization X_N is observable in bounded time.

REFERENCES

1. E. FORNASINI, G. MARCHESINI - "Algebraic Realization Theory of Bilinear Discrete Time Input/Output Maps", to appear in Journal of the Franklin Inst.; special issue on Recent Trends in System Theory, Jan. 1976.

2. A. NERODE - "Linear Automaton Transformations" Proc. Am. Math. Soc., Vol. 9, pp. 541 - 544, 1958.

3. R.E. KALMAN - "Pattern Recognition Properties of Multilinear Machines" IFAC Symp., Yerevan, Armenian, SSR, 1968.

4. M.A. ARBIB - "A Characterization of Multilinear Systems" IEEE T-AC, Vol. AC 14, Dec. 1969.

5. E. FORNASINI, G. MARCHESINI - "On the Internal Structure of Bilinear Input/output Maps" in Geometric Methods for Nonlinear Systems, R. Brockett D. Mayne eds., Reidel, Dordrecht, 1973.

HIGH ORDER ALGEBRAIC CONDITIONS FOR CONTROLLABILITY

Henry Hermes†
Department of Mathematics
University of Colorado
Boulder, Colorado 80302, USA

Introduction.

Let $X, Y^2, \ldots, Y^m$ be analytic vector fields on an analytic, n-dimensional, manifold M, and $\mathcal{S}$ the control system

$$\dot{x} = X(x) + \sum_{i=2}^{m} u_i(t) Y^i(x), \quad x(0) = p \in M, \quad (\dot{x} = dx/dt) \tag{1}$$

where, unless stated otherwise, an admissible control is a Lebesgue measurable function u with components $|u_i(t)| \leq 1$. $\mathcal{A}(t,p,\mathcal{S})$ will denote the set of all points attainable at time $t \geq 0$ by solutions of $\mathcal{S}$ corresponding to all admissible controls; $T^X(\cdot)p$ will denote the solution of $\mathcal{S}$ corresponding to $u \equiv 0$. Our problem is to determine necessary and sufficient conditions that $T^X(t)p \in \text{int.}\, \mathcal{A}(t,p,\mathcal{S})$ $\forall t > 0$.

Let TM_p denote the tangent space to M at p; $V(M)$ the real linear space of analytic vector fields on M considered as a real Lie algebra with Lie product $[X,Y]$; for $\mathcal{C} \subset V(M)$, $\mathcal{C}(p) = \{V(p) \in TM_p : V \in \mathcal{C}\}$ while $L(\mathcal{C})$ will denote the subalgebra generated by $\mathcal{C}$. For notational ease we let $(\text{ad}X, Y) = [X,Y]$ and inductively $(\text{ad}^k X, Y) = [X, (\text{ad}^{k-1} X, Y)]$.

First order theory, for the problem considered, proceeds by checking controllability of the variational equation associated with $\mathcal{S}$, taken about the reference solution $T^X(\cdot)p$. Using the test in [1, §19], and defining

$$\mathcal{S}^1 = \{\text{ad}^j X, Y^i) : j \geq 0, \; i = 2, \ldots, m\} \tag{2}$$

we easily obtain

PROPOSITION 1. (See [2, prop. 3]) A necessary and sufficient condition that the variational equation along $T^X(\cdot)p$ be controllable on every interval $[0,t]$, $t > 0$ (hence a sufficient condition that $T^X(t)p \in \text{int}\, \mathcal{A}(t,p,\mathcal{S})$ $\forall t > 0$) is that rank $\mathcal{S}^1(p) = n$.

A specialization of a theorem of Sussmann and Jurdjevic, [3, Th. 3.2], to our system $\mathcal{S}$ yields

PROPOSITION 2. ([4, prop. 1.6]) A necessary and sufficient condition that $\text{int}\, \mathcal{A}(t,p,\mathcal{S}) \neq \emptyset$ $\forall t > 0$ is that $\dim L(\mathcal{S}^1)(p) = n$.

†This research was supported by the National Science Foundation under grant GP 27957.

If control is "unlimited", i.e., u_i belongs to the Lebesgue space $\mathcal{L}_1[0,1]$ and $\dim L\{Y^2,\ldots,Y^m\}(p) = n$ the influence of $Y^2,\ldots,Y^n$ can "override" that of X and we have

PROPOSITION 3. ([5, Th. 1.1]) Let $\mathcal{D}^\infty$ denote the system $\mathcal{D}$ with control components $u_i \in \mathcal{L}_1[0,1]$ admissible. Then $\dim L\{Y^2,\ldots,Y^m\}(p) = n$ is a sufficient condition that $T^X(t)p \in \operatorname{int} \mathcal{Q}(t,p,\mathcal{D})$ $\forall t > 0$.

If $\dim L\{Y^2,\ldots,Y^n\}(p) = n$ and control values are bounded, the precise values of these bounds are essential in determining if $T^X(t)p$ belongs to the boundary, or interior, of $\mathcal{Q}(t,p,\mathcal{D})$. If $\dim L\{Y^2,\ldots,Y^m\}(p) = k < n$, a theorem of Nagano, [6], shows $L\{Y^2,\ldots,Y^m\}$ has a k-dimensional integral manifold thru p and Y^i influence motion only in this manifold. Geometrically, the easiest case is where

(a-1) $\dim L\{Y^2,\ldots,Y^m\}(q) = n-1$ for q in a nbd. of p

(a-2) $X(p) \notin L\{Y^2,\ldots,Y^m\}(p)$.

For the moment we assume $m = n$; $Y^2,\ldots,Y^n$ are involutive while $X(p), Y^2(p),\ldots,Y^n(p)$ are linearly independent so both (a-1), (a-2) are satisfied. Linear theory easily shows that for $\tau > 0$, $\mathcal{Q}(\tau,p,\mathcal{D})$ contains an $(n-1)$ dimensional manifold transverse to $T^X(\cdot)p$ having $T^X(\tau)p$ as a (relative) interior point. Let $t_1 > 0$ be given and $p^1 = T^X(t_1)p$. If for some $\varepsilon > 0$, $p^1 \in \mathcal{Q}(t_1+\varepsilon, p, \mathcal{D})$ and $p^1 \in \mathcal{Q}(t_1-\varepsilon, p, \mathcal{D})$ it easily follows that a segment $\{T^X(t_1+\sigma): -\varepsilon < \sigma < \varepsilon\} \subset \mathcal{Q}(t_1,p,\mathcal{D})$ and therefore $p^1 \in \operatorname{int} \mathcal{Q}(t_1,p,\mathcal{D})$. Let Z be a one-form such that $\langle Z(x), X(x)\rangle \equiv 1$, $\langle Z(x), Y^i(x)\rangle \equiv 0$, $i = 2,\ldots,n$ and x in a nbd. of p. We can use Z to measure time along comparison solutions of $\mathcal{D}$ which join p to p^1. This yields the following result. Let

$\nu = (\nu_2,\ldots,\nu_n)$ with ν_i a non-negative integer,

$$|\nu| = \sum_2^n \nu_i, \quad \nu! = \nu_2!\,\nu_3!\cdots\nu_n!,$$

$$a(\nu,j) = \langle Z(p), (\operatorname{ad}^j X, (\operatorname{ad}^{\nu_n} Y^n, (\ldots(\operatorname{ad}^{\nu_2} Y^2, X)\ldots)(p)\rangle$$

$$\varphi_{rj}(s) = \sum_{|\nu|=r} (1/\nu!)(-s_2)^{\nu_2}\cdots(-s_n)^{\nu_n} a(\nu,j).$$

THEOREM 1. ([4, Th. 1]) Assume $Y^2,\ldots,Y^n$ are involutive and $X(p), Y^2(p),\ldots,Y^n(p)$ are linearly independent. A necessary and sufficient condition that $\mathcal{Q}(t,p,\mathcal{D})$ have nonempty interior $\forall t > 0$ is that some r-form $\varphi_{rj}(s) \not\equiv 0$, $r \geq 1$, $j \geq 0$. If r^* is the smallest r such that $\varphi_{r^*j}(s) \not\equiv 0$ for some $j \geq 0$ and j^* the smallest j for which this happens, a necessary and sufficient condition that $T^X(t)p \in \operatorname{int} \mathcal{Q}(t,p,\mathcal{D})$ $\forall t > 0$ is that $\varphi_{r^*j^*}(s)$ is not definite (i.e. assumes both positive and negative values in every nbd. of $0 \in \mathbb{R}^{n-1}$).

If $\dim L\{Y^2,\ldots,Y^m\}(q) = k < n$ for q in a nbd. of p and $X(p) \notin L\{Y^2,\ldots,Y^m\}(p)$, we may choose $V^2,\ldots,V^{k+1} \in L\{Y^2,\ldots,Y^m\}$ and $V^{k+2},\ldots,V^n \in V(M)$ such that $V^2,\ldots,V^n$ are involutive and $X(p),V^2(p),\ldots,V^n(p)$ are linearly independent. Define the $a(\nu,j)$ and $\varphi_{rj}(s)$ as above but with Y^i replaced by V^i, $i = 2,\ldots,n$. One may show, [5, §2], that if $\dim L(\mathscr{S}^1)(p) = n$ (i.e. $\operatorname{int} \mathcal{A}(t,p,\mathscr{D}) \neq \emptyset \ \forall t > 0$) then for any choice of $V^2,\ldots,V^n$ as above, $\varphi_{rj}(s) \not\equiv 0$ for some $r \geq 1$, $j \geq 0$. Furthermore, with r^*, j^* chosen as in Theorem 1, and some $V^2,\ldots,V^n$ as above, $\varphi_{r^*j^*}(s)$ definite is a sufficient condition that for some $\varepsilon > 0$, $T^X(t)p \in \partial \mathcal{A}(t,p,\mathscr{D})$ for $0 \leq t \leq \varepsilon$. (Here ∂ denotes boundary.) Examples can be found in [4] and [5].

CONJECTURE. Assume $\dim L\{Y^2,\ldots,Y^m(q) = n-1$ for q in a nbd. of p and $X(p) \notin L\{Y^2,\ldots,Y^m\}(p)$. Choose $V^2,\ldots,V^n \in L\{Y^2,\ldots,Y^m\}$ such that $X(p),V^2(p),\ldots,V^n(p)$ are linearly independent and let $\varphi_{rj}(s)$, r^*, j^* be as in Theorem 1. Then a necessary and sufficient condition that $T^X(t)p \in \operatorname{int} \mathcal{A}(t,p,\mathscr{D})$ $\forall t > 0$ is that $\varphi_{r^*j^*}(s)$ not be definite.

In the above results, the assumption $X(p) \neq 0$ was fundamental. In many applications, in particular controlled stability about a rest solution, one desires $X(p) = 0$. For the remainder of this paper we assume this and study the problem: Determine necessary and sufficient conditions that $T^X(t)p \equiv p \in \operatorname{int} \mathcal{A}(t,p,\mathscr{D})$ $\forall t > 0$.

§1. High Order Controlled Stability.

The sufficient conditions given in propositions 1 and 3 are valid for the case $X(p) = 0$. We will be interested in higher order conditions which are independent of the control bounds, i.e. should be valid if u admissible implies $|u_i(t)| \leq \alpha_i$ where $\alpha_i > 0$ may be arbitrary. Thus our main concern will be the case when rank $\mathscr{S}^1(p) < n$ and $\dim L\{Y^2,\ldots,Y^m\}(p) < n$. Specifically, in this section we consider the system $\mathscr{D}$ with $m = n$, $X(p) = 0$ and assume $Y^2,\ldots,Y^n$ are involutive and $Y^2(p),\ldots,Y^n(p)$ are linearly independent. Then $L\{Y^2,\ldots,Y^n\}$ has an $(n-1)$ dimensional integral manifold thru each point q, which we denote $M^{n-1}(q)$. Let V be any analytic vector field with $V(p),Y^2(p),\ldots,Y^n(p)$ independent, and Z a one-form such that $\langle Z(x),V(x)\rangle \equiv 1$, $\langle Z(x),Y^i(x)\rangle \equiv 0$ for $i = 2,\ldots,n$ and x in a nbd. of p. For $s = (s_2,\ldots,s_n) \in \mathbb{R}^{n-1}$ define $q(s,p) = T^{Y^2}(s_2) \circ \ldots \circ T^{Y^n}(s_n)p$ (where $T^{Y^i}(\cdot)q$ is the solution of $\dot{x} = Y^i(x), x(0) = q$) and

$$H(s) = \langle Z(q(s,p)),X(q(s,p))\rangle .$$

Lemma 1. A necessary and sufficient condition that $T^X(t)p \in \operatorname{int} \mathcal{A}(t,p,\mathscr{D})$ $\forall t > 0$ is that $H(s)$ change sign, as a function of s, in every nbd. of $0 \in \mathbb{R}^{n-1}$.

Outline of proof. If $H(s) \equiv 0$ in a nbd. of zero, X is tangent to $M^{n-1}(p)$ and $\operatorname{int} \mathcal{Q}(t,p,\mathcal{D}) = \emptyset$ $\forall t > 0$. If $H(s) \geq 0$ (or ≤ 0) for all s in a nbd. of $0 \in \mathbb{R}^{n-1}$ clearly for sufficiently small $t > 0$, $\mathcal{Q}(t,p,\mathcal{D})$ lies on "one side" of $M^{n-1}(p)$ hence $p \in \partial \mathcal{Q}(t,p,\mathcal{D})$. Thus H changing sign is necessary.

For sufficiency, we assume H changes sign in every nbd. of $0 \in \mathbb{R}^{n-1}$ and will show that given any $t_1 > 0$ $\exists$ a $\delta > 0$ such that $T^V(\tau)p \in \mathcal{Q}(2t_1,p,\mathcal{D})$ if $|\tau| < \delta$. From linear theory, for each such τ there exists an $(n-1)$ manifold, $N^{n-1}(T^V(\tau)p)$, contained in $\mathcal{Q}(2t_1,p,\mathcal{D})$, transverse to $T^V(\cdot)p$; having $T^V(\tau)p$ as a (relative) interior point, and tangent space at $T^V(\tau)p$ spanned by $Y^2(T^V(\tau)p),\ldots,Y^n(T^V(\tau)p)$. Thus we obtain a tubular nbd. of $\{T^V(\tau)p: |\tau| < \delta\}$ belonging to $\mathcal{Q}(2t_1,p,\mathcal{D})$ hence $p \in \operatorname{int} \mathcal{Q}(2t_1,p,\mathcal{D})$.

Let $t_1 > 0$ be given. We first will show there exists $\varepsilon > 0$ such that if $0 \leq \tau \leq \varepsilon$, $T^V(\tau)p \in \mathcal{Q}(2t_1,p,\mathcal{D})$. The argument is "symmetric" i.e. also yields the existence of an $\varepsilon_1 < 0$ such that $T^V(\tau)p \in \mathcal{Q}(2t_1,p,\mathcal{D})$ if $\varepsilon_1 < \tau \leq 0$ and we can choose the desired $\delta = \min\{\varepsilon,-\varepsilon_1\}$. First note that since p is a rest point of the system $\mathcal{D}$ when $u \equiv 0$, we have $\bigcup_{0 \leq t \leq t_1} \mathcal{Q}(t,p,\mathcal{D}) = \mathcal{Q}(t_1,p,\mathcal{D})$ hence it suffices to show $\exists$ $\varepsilon > 0$ such that if $0 \leq \tau \leq \varepsilon$

$$\Big(\bigcup_{0 \leq t \leq t_1} \mathcal{Q}(t,p,\mathcal{D})\Big) \cap \Big(\bigcup_{0 \leq t \leq t_1} \mathcal{Q}(-t,T^V(\tau)p,\mathcal{D})\Big) \neq \emptyset . \tag{*}$$

Let $\mathcal{N}$ be a compact nbd. of p such that $|X(x)| \leq (1/10) \min\{|Y^2(x)|,\ldots,|Y^n(x)|\}$ for $x \in \mathcal{N}$. This means components of X in the directions of the Y^i can be "overridden" by the Y^i in $\mathcal{N}$. Let

$$E^+ = \{x \in \mathcal{N}: \langle Z(x),X(x)\rangle > 0\} .$$

Since H changes sign in every nbd. of $0 \in \mathbb{R}^{n-1}$, E^+ has nonempty interior and p belongs to the boundary of E^+.

First pick $\varepsilon > 0$ small enough so that for $0 \leq \tau \leq \varepsilon$, $E^+ \cap M^{n-1}(T^V(\tau)p) \neq \emptyset$ and for each such τ, the distance from $T^V(\tau)p$ to this intersection is sufficiently small so that a trajectory, ψ, of the system $-\mathcal{D}$ (i.e. $\dot{x} = -X(x) + \sum_2^n u_i Y^i(x)$) initiating from $T^V(\tau)p$ enters E^+ in a time τ_1 small compared to t_1. This is possible since the Y^i dominate X. Once ψ has entered E^+, the control can be chosen so that $\psi(t) \in E^+$ for $\tau_1 \leq t \leq t_1$. For $\varepsilon > 0$ sufficiently small, we can thus obtain a solution ψ of $-\mathcal{D}$ such that $\psi(0) = T^V(\tau)p$ and $\psi(t_1)$ is in E^+ and is "to the left" (side determined by $-V$) of $M^{n-1}(T^V(\tau)p)$.

Similarly, since the Y^i dominate, we may choose a control u for the system $\mathcal{D}$ so that the corresponding solution φ satisfies $\varphi(0) = p$ and $\varphi(t) \in E^+$ for $0 < t \leq t_1$ hence for such t, $\varphi(t)$ is to the right of $M^{n-1}(p)$, (the side determined by $+V$). Furthermore for $\varepsilon > 0$ sufficiently small, one can obtain $\varphi(t') = \psi(t_1)$ for some $t' \in (0,t_1]$ which shows (*) is satisfied. □

Let $q_*(s,p)\colon TM_p \to TM_{q(s,p)}$ be the isomorphism induced by $p \to q(s,p)$, for small $|s|$.

<u>Lemma 2</u>. Let $h(s) = \langle Z(p), q_*^{-1}(s,p)X(q(s,p))\rangle$. Then $H(s)$ and $h(s)$ have the same sign for s near zero.

<u>Proof</u>. Suppose $H(s) = \gamma \neq 0$. Then $X(q(s,p)) = \sum_2^n c_i Y^i(q(s,p)) + \gamma V(q(s,p))$. Now TM_r^{n-1} is spanned by $Y^2(r),\ldots,Y^n(r)$ thus $q_*^{-1}X(q(s,p)) = \sum_2^n c_i' Y^i(p) + \gamma q_*^{-1}V(q(s,p))$. Then $h(s) = \gamma\langle Z(p), q_*^{-1}V(q(s,p))\rangle$. Let $f\colon [0,1] \to \mathbb{R}^{n-1}$ with $f(0) = 0$, $f(1) = s$ and let $\mu(\sigma) = \langle Z(p), q_*^{-1}(f(\sigma),p)V(q(f(\sigma),p))\rangle$ for $0 \le \sigma \le 1$. Then $\mu(0) = 1$; μ is continuous and if there exists $\sigma_1 \in [0,1]$ such that $\mu(\sigma_1) = 0$ this would mean $V(q(f(\sigma_1),p)) \in TM^{n-1}_{q(f(\sigma_1),p)}$. But this contradicts $\langle Z(x),V(x)\rangle \equiv 1$ near p hence $\mu(\sigma)$ does not change sign showing sign $h(s) =$ sign γ when $H(s) \neq 0$. The argument is reversible. Also $H(s) = 0 \Leftrightarrow X(q(s,p)) \in TM^{n-1}_{q(s,p)} \Leftrightarrow q_*^{-1}(s,p)X(q(s,p)) = \sum_{i=2}^n \alpha_i Y^i(p) \Leftrightarrow h(s) = 0$. □

Again, let $\nu = (\nu_2,\ldots,\nu_n)$ with ν_i a non-negative integer, $|\nu| = \sum \nu_i$ and $\nu! = \nu_2!\ldots\nu_n!$. Then, by repeated use of the Campbell-Hausdorff formula,

$$q_*^{-1}(s,p)X(q(s,p)) = \sum_{|\nu|=0}^{\infty} (1/\nu!)(-s_2)^{\nu_2}\ldots(-s_n)^{\nu_n}(\mathrm{ad}^{\nu_n}Y^n,(\ldots(\mathrm{ad}^{\nu_2}Y^2,X)\ldots)(p)$$

Let

$$a(\nu) = \langle Z(p),(\mathrm{ad}^{\nu_n}Y^n,\ldots(\mathrm{ad}^{\nu_2}Y^2,X)\ldots)(p)\rangle$$

$$\varphi_r(s) = \sum_{|\nu|=r} (1/\nu!)(-s_2)^{\nu_2}\ldots(-s_n)^{\nu_n}\, a(\nu)$$

Since $X(p) = 0$ we see $a(0) = 0$ and

<u>Lemma 3</u>. $h(s) = \sum_{r=1}^{\infty} \varphi_r(s)$.

For odd r, $\varphi_r(s) = -\varphi_r(-s)$, so if the first non-vanishing r - form $\varphi_r(s)$ occurs with r odd, then h (hence H) changes sign in every nbd. of $0 \in \mathbb{R}^{n-1}$. In particular, if $r^* = 1$ we see $\langle Z(p),[Y^i,X](p)\rangle \neq 0$ for some $i = 2,\ldots,n$ which, we shall next show, is the "linear test". Let

$$\mathscr{S}^1 = \{(\mathrm{ad}^j X, Y^i)\colon\ j \ge 0,\ i = 2,\ldots,n\}$$

$$\mathcal{R} = \{Y^2,\ldots,Y^n,[X,Y^2],\ldots,[X,Y^n]\}.$$

Lemma 4. With $X(p) = 0$ and $Y^2(p),\dots,Y^n(p)$ linearly independent rank $\mathcal{S}^1(p) =$ rank $\mathcal{R}(p)$.

Proof. Clearly rank $\mathcal{S}^1(p) \geq$ rank $\mathcal{R}(p) \geq n-1$. It suffices to show rank $\mathcal{R}(p) = n-1 \Rightarrow$ rank $\mathcal{S}^1(p) = n-1$. Now rank $\mathcal{R}(p) = n-1 \Rightarrow [X,Y^i](p) \in$ span $\{Y^2(p),\dots,Y^n(p)\}$ for each $i = 2,\dots,n$. Using this, local coordinates, X_x to denote the Jacobian matrix of partial derivatives in these coordinates and $X(p) = 0$, we have:

$$[X,[X,Y^i]](p) = X_x(p)[X,Y^i](p) - [X,Y^i]_x(p)X(p) =$$

$$X_x(p)\sum_{j=2}^{n} c_j Y^j(p) = \sum_{j=2}^{n} c_j(X_x(p)Y^j(p) - Y^j_x(p)X(p))$$

$$= \sum_{j=2}^{n} c_j[X,Y^j](p) \in \text{span}\,\{Y^2(p),\dots,Y^n(p)\}.$$ Inductively

$(\mathrm{ad}^j X, Y^i)(p) \in$ span $\{Y^2(p),\dots,Y^n(p)\}$ for all $j \geq 0$,

$i = 2,\dots,n \Rightarrow$ rank $\mathcal{S}^1(p) = n-1$. □

THEOREM 2. A necessary and sufficient condition that int $\mathcal{A}(t,p,\mathcal{D}) \neq \emptyset$ $\forall t > 0$ is that there exist an integer $r \geq 1$ such that $\varphi_r(s) \not\equiv 0$. In this case let r^* be the smallest $r \geq 1$ such that $\varphi_{r^*}(s) \not\equiv 0$. A necessary and sufficient condition that $p \in$ int $\mathcal{A}(t,p,\mathcal{D})$ $\forall t > 0$ is that $\varphi_{r^*}(s)$ not be definite.

Proof. If $\exists$ an integer $r \geq 1$ such that $\varphi_r(s) \not\equiv 0$ then $\langle Z(p),(\mathrm{ad}^{\nu_n}Y^n,(\dots,(\mathrm{ad}^{\nu_2}Y^2,X)\dots)(p)\rangle \neq 0$ for some $\nu_2,\dots,\nu_n$ with $\sum \nu_i = r$. This shows $(\mathrm{ad}^{\nu_n}Y^n,(\dots,(\mathrm{ad}^{\nu_2}Y^2,X)\dots)(p)$ is linearly independent of $Y^2(p),\dots,Y^n(p)$ hence dim $L(\mathcal{S}^1)(p) = n$ and int $\mathcal{A}(t,p,\mathcal{D}) \neq \emptyset$ $\forall t > 0$ by prop. 2. Conversely, suppose $\varphi_r(s) \equiv 0$ for all $r \geq 1$. When $r = 1$ this shows $[X,Y^i](p)$ is linearly dependent on $Y^2(p),\dots,Y^n(p)$ for $i = 2,\dots,n$. By lemma 4, the same is then true for $(\mathrm{ad}^j X,Y^i)(p)$, $j \geq 0$, $i = 2,\dots,n$. From this, and $\varphi_r(s) \equiv 0$ for $r \geq 1$, we conclude dim $L(\mathcal{S}^1)(p) = n-1$ and int $\mathcal{A}(t,p,\mathcal{D}) = \emptyset$ for $t > 0$ by prop. 2.

The remainder of the theorem is an immediate consequence of lemmas 3, 2 and 1. □

§2. Examples.

For notational ease all vectors will be written as row vectors.

Example 2.1 (Elementary example.) Let $M = \mathbb{R}^2$, $k \geq 1$ an integer, $X(x) = (x_1, x_1^k)$, $Y^2(x) \equiv Y(x) = (1,0)$ and $p = 0$. (Clearly $p \in \partial\mathcal{A}(t,p,\mathcal{D})$ $\forall t > 0$ if k is even.) Computing, $[Y,X](x) = (-1,-kx_1^{k-1})$; $(\mathrm{ad}^j Y,X)(x) = (0,(-k!/(k-j)!)x_1^{k-j})$ if $j \geq 2$. Choose $V = (0,1)$; note that for application of theorem 2 (or Th. 3) we need only compute $Z(p)$ rather than $Z(x)$ for x in a nbd. of p. Here $Z(p) = (0,1)$; for $k > 1$; $\langle Z(p),[Y,X](p)\rangle = 0$ while from lemma 4,

$\langle Z(p),(ad^jX,Y)(p)\rangle = 0 \quad \forall j \geq 0$. Thus the linear test fails when $k > 1$, which we now assume.

Next, $\langle Z(p),(ad^jX,Y)(p)\rangle = 0$ if $0 \leq j \leq k-1$ while $\langle Z(p),(ad^kY,X)(p)\rangle = -k!$ showing $\varphi_r(s) = 0$ if $1 \leq r \leq k-1$ while $\varphi_k(s) = (-s)^k$. Thus $r^* = k$ and $\varphi_{r^*}(s)$ is definite ($p \in \partial \mathcal{A}(t,p,\mathcal{D})$) if k is even; $\varphi_{r^*}(s)$ changes sign ($p \in \operatorname{int} \mathcal{A}(t,p,\mathcal{D})$) if k is odd.

Example 2.2 Let $M = R^3$, $X(x) = (0,x_2,x_3^2)$, $Y^2(x) = (x_2^2,1,0)$, $Y^3(x) = (0,0,1)$, and $p = 0$. Then $X(p) = 0$; $Y^2(p)$, $Y^3(p)$ are linearly independent; $[Y^2,Y^3] = 0$ so Y^2 , Y^3 are involutive so the hypotheses of §1 are satisfied.

Computing, $[Y^2,X](x) = (2x_2^2,-1,0)$, $[Y^3,X](x) = (0,0,2x_3)$ hence $[Y^2,X](p)$ and $[Y^3,X](p)$ are linearly dependent on $Y^2(p)$, $Y^3(p)$ showing, by lemma 4 , that the linear test yields rank $\mathcal{J}^1(p) = 2$ and hence fails.

Choose $V = (1,0,0)$ giving $Z(p) = (1,0,0)$. We next compute $(ad^2Y^2,X)(x) = (-6x_2,0,0) \Rightarrow a(2,0) = 0$; $(adY^3,(adY^2X))(x) = 0 \Rightarrow a(1,1) = 0$; $(ad^2Y^3,X)(x) = (0,0,-2) \Rightarrow a(0,2) = 0$ hence $\varphi_2(s) = 0$. Next, $(ad^3Y^2,X) = (6,0,0) \Rightarrow a(3,0) = 6$; $(adY^3,(ad^2Y^2,X))(x) = 0 \Rightarrow a(2,1) = 0$; $(ad^2Y^3,(adY^2,X))(x) = 0 \Rightarrow a(1,2) = 0$; $(ad^3Y^3,X) = 0 \Rightarrow a(0,3) = 0$. Thus $\varphi_3(s) = (-s_2)^3$; $r^* = 3$ and $\varphi_{r^*}(s)$ is not definite hence $p \in \operatorname{int} \mathcal{A}(t,p,\mathcal{D}) \ \forall t > 0$. This means, for example, that the system $-\mathcal{D}$ can be controlled from a nbd. of the origin, to the origin, in finite time (i.e. null controllability.)

References

1. Hermes, H. and LaSalle, J. P.; Functional Analysis and Time Optimal Control, Academic Press, N.Y. (1969).
2. Hermes, H.; On Local and Global Controllability, SIAM J. Control, 12, (1974) 252-261.
3. Sussmann, H. and Jurdjevic, V.; Conrollability of Nonlinear Systems, J. Diff. Eqs., 12, (1972) 95-116.
4. Hermes, H.; Local Controllability and Sufficient Conditions in Singular Problems, (to appear) J.D.E.
5. Hermes, H.; Local Controllability and Sufficient Conditions in Singular Problems II.
6. Nagano, T.; Linear Differential Systems with Singularities and an Application of Transitive Lie Algebras, J. Math. Soc. Japan, 18, (1966) 398-404.

SEMIGROUP REPRESENTATIONS, BILINEAR APPROXIMATION OF INPUT-OUTPUT MAPS, AND GENERALIZED INPUTS*

Héctor J. Sussmann
Rutgers University
New Brunswick, N.J. 08903 USA

§1. INTRODUCTION

Our aim in this paper is to show how a number of interesting system-theoretic problems fit very naturally within the framework of representation theory. We will start from the rather trivial and well-known fact that a continuous-time bilinear system can be regarded as a finite-dimensional, linear representation of a semigroup, namely, the semigroup of all input functions (the semigroup operation being concatenation).

We will then proceed to ask the most obvious questions, and it will turn out that the answers lead to quite interesting ideas. Specifically, the first natural mathematical question is: which representations of the input semigroup arise from bilinear systems? This will force us to look for a reasonable topology which renders continuous all the representations arising from bilinear systems. The topology of _weak_ convergence turns out to satisfy this requirement, at least for the case when the controls are bounded by a fixed constant. Moreover, it turns out that every representation which is continuous in the weak topology necessarily arises from a bilinear system. This may seem somewhat surprising, especially if we observe that every reasonable function $A(u)$ which is defined on the control space and has $n \times n$ matrices for its values, gives rise to a representation of the input semigroup via the equation

$$\dot{x} = A(u)x, \quad x \in \mathbb{R}^n.$$

Therefore, the requirement that this representation be continuous with respect to weak convergence actually forces $A(u)$ to be of the form

$$A(u) = A + \sum_{i=1}^{m} u^i B_i,$$

*Work partially supported by N.S.F. Grant No. GP-37488.

if $u = (u^1,\ldots,u^m)$!!

A second important dividend to be obtained from our approach is that the set of all inputs bounded by a constant M and having a time duration also bounded by some constant is actually compact in the weak topology. This enables us to follow step by step the proof of the Peter-Weyl Theorem, and to prove that every input-output map F can be approximated arbitrarily close by maps which arise from bilinear systems, provided only that F satisfies the obvious necessary conditions, namely, continuity and causality (cf. §4 for the details; a similar result has been proved by M. Fliess using a different approach).

Actually, our proof of the approximation theorem referred to above really shows something much more important than the theorem itself, namely, that _the realization problem for input-output maps is basically equivalent to the main problem of group representation theory_, namely, that of realizing an arbitrary function on a group as a matrix entry of a representation. The only difference arises from the fact that

(i) here we must deal with a semigroup rather than a group, and

(ii) this semigroup does not seem to be embedded in anything reasonably close to a Lie group.

Actually, (ii) is the most serious problem. Locally compact groups have a Haar measure, and this certainly plays a role in the representation theory. Our input semigroup does not seem to have anything like Haar measure, however. When the sizes and times of the controls are bounded, we get compactness anyhow, and this makes it possible to successfully pursue the Peter-Weyl approach. When the bounds are removed, it seems clear that the realizability problem will become much more difficult, and require an excursion into infinite dimensional representation theory.

In the second half of our paper (§5 and §6) we pursue to their ultimate consequences some problems which, when stated, seem to be of a purely technical nature. We ask how to topologize the input semigroup when fixed bounds on the controls are not assumed. We then ask whether this topology has "nice" properties, such as completeness It turns out that it does not, and we then ask what the "completion" looks like. The answer to this question gives us, among other things, white noise (at least for the scalar case, but we expect to be able to prove our results for vector white noise in the near future). Actually, we construct a class of objects, called "generalized inputs", which can be used to drive ordinary differential equations, and which include the "sample paths of white noise". This class provides, in our opinion, conclusive evidence to the effect that the "purely technical" questions which we were asking actually had substance. In fact, we believe in the general principle that all "mathematically natural" questions should be asked, even those which the applied mathematician might view as belonging to the "esoteric fringe of pure mathematics". We hope that this paper will be regarded as providing some evidence in favor of the principle.

§2. BILINEAR SYSTEMS AND REPRESENTATIONS: THE BOUNDED CONTROL CASE

In what follows, all semigroups are assumed to have an identity element, which will be denoted by 1. All vector spaces are vector spaces over the field $\mathbb{R}$ of real numbers. If V is a vector space, we shall use $End(V)$ to denote the set of all linear endomorphisms of V. Then $End(V)$ is clearly a semigroup, with composition of maps as the semigroup operation. A _linear representation_ of a semigroup S is a pair (V,π), where V (the _space_ of the representation) is a vector space, and $\pi: S \longrightarrow End(V)$ is a semigroup homomorphism (i.e., $\pi(s_1s_2) = \pi(s_1)\pi(s_2)$ for s_1, s_2 in S, and $\pi(1) = 1$).

A _topological semigroup_ is a semigroup S equipped with a topology relative to which the semigroup operation is a continuous map from $S \times S$ into S.

A _continuous representation_ of a topological semigroup S is a representation (V,π) such that $\pi: S \longrightarrow End(V)$ is continuous. This requires that $End(V)$ be equipped with a topology. For general spaces V there are various ways to do so, but for _finite dimensional_ V the space $End(V)$ has a canonical topology, which clearly makes $End(V)$ into a topological semigroup. Therefore, the concept of a _continuous finite-dimensional linear representation_ (henceforth abbreviated as "c.f-d.l. representation") is unambiguously defined.

We are interested in identifying the class of bilinear systems with the set of c.f-d.l. representations of a particular semigroup. Now, a bilinear system

$$\dot{x} = (A + \sum_{i=1}^{m} u^i B_i)x, \qquad x \in \mathbb{R}^n \tag{1}$$

is completely specified by giving (a) the integer n, (b) the $n \times n$ matrices $A, B_1, \ldots, B_m$, and (c) the class of admissible controls. Here we will consider the situation when the class of controls consists of _all_ the _bounded measurable_ functions with values in a _closed_, convex subset U of $\mathbb{R}^m$. Moreover, we can assume U to have a nonempty interior in $\mathbb{R}^m$ (for otherwise we could replace (1) by a similar system with a smaller m).

Formally, let $U \subseteq \mathbb{R}^m$ be closed, convex, with nonempty interior. A U-valued input is a pair $\bar{u} = (T_u, u)$ where $T_u \geq 0$ is the duration of the input, and u is a bounded, measurable, U-valued function defined on the closed interval $[0, T_u]$. Two inputs (T_u, u), (T_v, v) are equivalent if $T_u = T_v$ and $u(t) = v(t)$ for almost all $t \in [0, T_u]$. The set of equivalence classes of U-valued inputs is the input semigroup $S(U)$. The semigroup operation is concatenation: if $\bar{u} = (T_u, u)$ and $\bar{v} = (T_v, v)$ are two inputs, then $\bar{u} * \bar{v}$, the concatenation of $\bar{u}$ and $\bar{v}$, is the input $(T_u + T_v, u * v)$ where

$$(u*v)(t) = v(t) \quad \text{for} \quad 0 \leq t < T_v$$

$$(u*v)(t) = u(t - T_v) \quad \text{for} \quad T_v \leq t \leq T_u + T_v.$$

It is clear that this operation is associative, and it is also compatible with equivalence of inputs. Hence we have a well defined associative operation on $S(U)$. Moreover, all the inputs (T_u, u) for which $T_u = 0$ form an equivalence class, which is the identity element of $S(U)$, and will be denoted by 1.

We now topologize $S(U)$. We make the additional assumption that U is bounded. We declare a sequence $\bar{u}_k = (T_{u_k}, u_k)$ of inputs to be convergent to an input (T_u, u) if

(i) $\lim_{k \to \infty} T_{u_k} = T_u$, and

(ii) $u_k \longrightarrow u$ weakly as $k \longrightarrow \infty$.

The meaning of (i) is clear. As for (ii), we remark that there exist several concepts of weak convergence of a sequence of functions, but that they all coincide when the sequence is uniformly bounded. (Indeed, if the $v_k: [0,T] \longrightarrow \mathbb{R}^m$ are uniformly bounded, then the following are equivalent:

(I) $\int_a^b v_k(t)dt \longrightarrow \int_a^b v(t)dt$ as $k \longrightarrow \infty$, for all a, b such that $0 \leq a \leq b \leq T$;

(II) $\int_0^T \psi(t)v_k(t)dt \longrightarrow \int_0^T \psi(t)v(t)dt$ as $k \longrightarrow \infty$ for every continuous $\psi: [0,T] \longrightarrow \mathbb{R}$, and

(III) same as II with "continuous ψ" replaced by "$\psi \in L^1$").

In our case, the assumption that U is a bounded set guarantees that (ii) is unambiguous. (The fact that the u_k are defined on different intervals is easily taken care of in view of (i)).

Actually, it is easy to define a metric of $S(U)$ which gives rise to the concept of convergence that has just been described. For instance, we can define the distance between two inputs $\bar{u} = (T_u,u)$, $\bar{v} = (T_v,v)$ by

$$d(\bar{u},\bar{v}) = |T_u - T_v| + \sup\{\left|\int_a^b (u(t)-v(t))dt\right| = 0 \leq a \leq b \leq T\} \tag{2}$$

where $T = \min(T_u,T_v)$.

It is clear that input concatenation is continuous relative to the topology defined above. Moreover, it follows from standard facts of Functional Analysis that, for each $T>0$, the set $S_T(U)$ of those inputs (T_u,u) such that $T_u \leq T$, is compact. Since, for each T, the union of the $S_{T'}(U)$, $T' < T$, is open, we conclude that $S(U)$ is a locally compact topological semigroup.

We now want to prove that a bilinear system with inputs in $S(U)$ (U bounded), is exactly the same as a c.f-d.l. representation of $S(U)$. To make this precise, we first observe that a bilinear system $\mathcal{B}$ given by (1) defines a c.f-d.l. representation $(V_{\mathcal{B}},\pi_{\mathcal{B}})$ of $S(U)$ as follows:

(a) we take $V_{\mathcal{B}}$ to be $\mathbb{R}^n$ and

(b) if $\bar{u} = (T_u,u) \in S(U)$, we define $\pi_{\mathcal{B}}(\bar{u})$ to be the value at $t = T_u$ of the solution $X(t)$ of the matrix differential equation

$$\dot{X} = (A + \sum_{i=1}^{m} u^i B_i)X \tag{3}$$

with initial condition $X(0) = 1$.

It is clear that $\pi_{\mathcal{B}}$ is a semigroup homomorphism from $S(U)$ into the set of $n\times n$ real matrices, which is naturally identified with End $(\mathbb{R}^n)$. Finally, the continuity of $\pi_{\mathcal{B}}$ follows from the well known fact (cf. Kučera [1] or Sussmann [3], Lemma 2) that, if the functions

u_k^i are uniformly bounded and if, for each i, $u_k^i \longrightarrow u^i$ weakly, then the corresponding solutions X_k of (3) converge to X, uniformly on compact intervals.

Now let Bin_U denote the class of all bilinear systems with inputs U, and let $\mathcal{R}_U$ denote the class of all c.f-d.l. representations of $S(U)$ whose state space is $\mathbb{R}^n$ for some n. We have just shown that $\mathcal{B} \longrightarrow (V_{\mathcal{B}}, \pi_{\mathcal{B}})$ is a map from Bin_U to $\mathcal{R}_U$. We are now ready to state the precise sense in which Bin_U and $\mathcal{R}_U$ are "the same".

Theorem 1. Let $U \subseteq \mathbb{R}^m$ be compact, convex, with nonempty interior. Then the map $\mathcal{B} \longrightarrow (V_{\mathcal{B}}, \pi_{\mathcal{B}})$ is a bijection between Bin_U and $\mathcal{R}_U$.

Proof: Let $u_0 \in U$. If $\mathcal{B}$ is a bilinear system given by (1), let $\sigma(u_0) = A + \sum_{i=1}^{m} u_0^i B_i$. Let ${}_t\bar{u}_0$ denote the constant control with duration t and value u_0. Then $\pi_{\mathcal{B}}({}_t\bar{u}_0) = e^{t\sigma(u_0)}$. Therefore $\sigma(u_0) = \frac{d}{dt}\pi_{\mathcal{B}}({}_t\bar{u}_0)\Big|_{t=0}$, which shows that the map σ is completely determined by $\pi_{\mathcal{B}}$. Since U has a nonempty interior, the matrices $A, B_1, \ldots, B_m$ are completely determined by σ. This shows that $\mathcal{B}$ can be recovered from $\pi_{\mathcal{B}}$, and therefore the map $\mathcal{B} \longrightarrow (V_{\mathcal{B}}, \pi_{\mathcal{B}})$ is one-to-one.

To prove that is is onto, we must show that every c.f-d.l. representation $(\mathbb{R}^n, \pi)$ arises from a bilinear system. For $u_0 \in U$, let ${}_t\bar{u}_0$ be defined as above. Then ${}_t\bar{u}_0$ depends continuously on t, and ${}_t\bar{u}_0 * {}_s\bar{u}_0 = {}_{t+s}\bar{u}_0$. Therefore $t \longrightarrow \pi({}_t\bar{u}_0)$ is a continuous homomorphism from the semigroup $[0,\infty)$ into $\text{End}(\mathbb{R}^n)$. Since $\pi({}_0\bar{u}_0) = 1$, it follows by continuity that $\pi({}_t\bar{u}_0)$ is invertible for sufficiently small t. Hence, for each $t>0$, $\pi({}_{t/N}\bar{u}_0)$ is invertible if N is large enough, and therefore $\pi({}_t\bar{u}_0) = [\pi({}_{t/N}\bar{u}_0)]^N$ is also invertible. It follows that the map $t \longrightarrow \pi({}_t\bar{u}_0)$ extends to a continuous homomorphism $\alpha\colon \mathbb{R} \longrightarrow GL(n,\mathbb{R})$. Therefore α is given by $\alpha(t) = e^{Mt}$ for some matrix M.

We have shown that for every $u_0 \in U$ there exists a matrix, which we will denote by $\sigma(u_0)$, with the property that

$$\pi({}_t\bar{u}_0) = e^{\sigma(u_0)t} \quad \text{for } 0 \leq t < \infty.$$

We now prove that the map $\sigma\colon U \longrightarrow \text{End}(\mathbb{R}^n)$ is affine. Recall that, if V_1, V_2 are vector space, and W is a subset of V_1, a map $\mu\colon W \longrightarrow V_2$ is said to be <u>affine</u> if there is a linear map $\mu_0\colon V_1 \longrightarrow V_2$

and a vector $v_2 \in V_2$ such that $\mu(w) = \mu_0(w)+v_2$ for all $w \in W$. If W is convex, then μ is affine iff

$$\mu(rw_1 + (1-r)w_2) = r\mu(w_1) + (1-r)\mu(w_2)$$

for all $w_1, w_2 \in W$ and all r such that $0 \leq r \leq 1$.

Now let u_0, u_1 belong to U, and let $0 \leq r \leq 1$. Choose an arbitrary $t>0$, and define a sequence $\{\bar{v}_k\}$ of inputs by

$$\bar{v}_k = \left[{}_{\frac{(1-r)t}{k}}\bar{u}_1 * {}_{\frac{rt}{k}}\bar{u}_0 \right]^k .$$

Then all the $\bar{v}_k$ have duration t. Let $u_2 = ru_0 + (1-r)u_1$. Then the inputs $\bar{v}_k$ converge in $S(U)$ to ${}_t\bar{u}_2$. Therefore

$$e^{t\sigma(u_2)} = \lim_{k\to\infty} \left[e^{\frac{1}{k}(1-r)t\sigma(u_1)} \; e^{\frac{1}{k}rt\sigma(u_0)} \right]^k .$$

By Trotter's product formula, the limit is $e^{t[(1-r)\sigma(u_1)+r\sigma(u_0)]}$. Since $t>0$ is arbitrary, we conclude that

$$\sigma((1-r)u_1 + ru_0) = (1-r)\sigma(u_1) + r\sigma(u_0).$$

Therefore σ is an affine map, and this implies that there are matrices $A, B_1, \ldots, B_m$ such that

$$\sigma(u_0) = A + \sum_{i=1}^{m} u_0^i B_i$$

for all $u_0 \in U$. Now let $\mathcal{B}$ be the bilinear system defined by A and the B_i. It follows from our construction that $\pi(\bar{u}) = \pi_{\mathcal{B}}(\bar{u})$ for all constant controls. Hence the equality holds for all piecewise constant controls, because both π and $\pi_{\mathcal{B}}$ are homomorphisms. Since the piecewise constant controls are dense in $S(U)$, and both π and $\pi_{\mathcal{B}}$ are continuous, it follows that $\pi = \pi_{\mathcal{B}}$, completing our proof.

§3. REPRESENTATIONS AND REALIZATIONS

Let S be a topological semigroup. If (V_i, π_i) $(i = 1,2)$ are linear representations of S, we can form the direct sum $(V_1 \oplus V_2, \pi_1 \oplus \pi_2)$ and the tensor product $(V_1 \otimes V_2, \pi_1 \otimes \pi_2)$. If both the (V_i, π_i) are continuous finite-dimensional, then the same is true of their direct sum and of their tensor product. If (V, π) is a c.f-d.l. representation of S, and if $v \in V$, $\lambda \in V^*$ (the dual space of V), then we shall use $f_{\pi,v,\lambda}$ to denote the real-valued function on S defined by $f_{\pi,v,\lambda}(s) = \langle\lambda,\pi(s)v\rangle$.

If two representations (V_i,π_i), $i = 1,2$ are given, as well as vectors $v_i \in V_i$ and linear functionals $\lambda_i \in V_i^*$, then the identifications $(V_1 \oplus V_2)^* \simeq V_1^* \oplus V_2^*$ and $(V_1 \otimes V_2)^* \simeq V_1^* \otimes V_2^*$ enable us to regard $\lambda_1 \oplus \lambda_2$ and $\lambda_1 \otimes \lambda_2$ as linear functionals on $V_1 \oplus V_2$, $V_1 \otimes V_2$ respectively. The identities

$$(3) \qquad f_{\pi_1\oplus\pi_2,\ v_1\oplus v_2,\ \lambda_1\oplus\lambda_2} = f_{\pi_1,v_1,\lambda_1} + f_{\pi_2,v_2,\lambda_2}$$

$$(4) \qquad f_{\pi_1\otimes\pi_2,\ v_1\otimes v_2,\ \lambda_1\otimes\lambda_2} = f_{\pi_1,v_1,\lambda_1} \cdot f_{\pi_2,v_2,\lambda_2}$$

are easily verified.

A function which is of the form $f_{\pi,v,\lambda}$ for some c.f-d.l. representation (V,π) and some choice of $v \in V$, $\lambda \in V^*$ is said to be a _matrix entry of a_ c.f-d.l. _representation_ of S. In view of formulas (3) and (4), the set $R(S)$ of all such functions is a ring, called the _representation ring_ of S. Actually, $R(S)$ _is an_ $\mathbb{R}$-_algebra, and it contains all the constant functions_.

Now let $U \subseteq \mathbb{R}^m$ be compact, convex, with a nonempty interior. If $\mathcal{B}$ is a bilinear system given by (1), and if $x_{in} \in \mathbb{R}^n$, $\lambda \in (\mathbb{R}^n)^*$, then the triple $(\mathcal{B},x_{in},\lambda)$ is called an _initialized, observed bilinear system_. The vectors x_{in},λ are called the _initial state_ and the _observation functional_ of the system. If $\bar{u} = (T_u,u)$ is an input, let $x^{\bar{u}}(t)$, $0\leq t\leq T_u$, denote the solution of the differential equation (1) with initial condition $x^{\bar{u}}(0) = x_{in}$. Then the function $\Phi_{\bar{u}}: [0,T_u] \longrightarrow \mathbb{R}$ defined by $\Phi_{\bar{u}}(t) = \langle\lambda,x^{\bar{u}}(t)\rangle$ is called the _output_ that corresponds to the input $\bar{u}$. The map Φ which to each input $\bar{u}$ assigns the function $\Phi_{\bar{u}}$ is called the _input-output map_ associated with the initialized, observed system $(\mathcal{B},x_{in},\lambda)$. Actually Φ is completely determined by a real valued function ψ defined on the semigroup $S(U)$. Indeed, let us define $\psi(\bar{u}) = \Phi_{\bar{u}}(T_u)$. Then ψ is called the _input-output function_ associated with $(\mathcal{B},x_{in},\lambda)$. To see that Φ is completely determined by ψ, define the _cutoff map_ c by letting, for $\bar{u} = (T_u,u)$ and $0\leq t\leq T_u$, $c(t,\bar{u}) = (t,u_t)$, where u_t is the restriction of u to the interval $[0,t]$. We then have

$$(5) \qquad \Phi_{\bar{u}}(t) = \psi(c(t,\bar{u})) \quad \text{for } 0\leq t\leq T_u.$$

We want to determine which maps Φ can be realized as input-output maps of some $(\mathcal{B},x_{in},\lambda)$. In view (5), this is equivalent to answering a similar question about input-output functions ψ.

<u>Definition</u>. (a) A function $\psi\colon S(U)\longrightarrow \mathbb{R}$ is <u>bilinearly realizable</u> if it is the input-output function associated with some initialized, observed bilinear system $(\mathcal{B},x_{in},\lambda)$.

(b) ψ is <u>approximately bilinearly realizable</u> if for every $\varepsilon>0$, $T>0$, there exists a bilinearly realizable ψ' such that $|\psi(\bar{u}) - \psi'(\bar{u})| < \varepsilon$ for all inputs $\bar{u}$ whose duration is $\leq T$.

<u>Theorem 2</u>. Let $U\subseteq\mathbb{R}^m$ be compact, convex, with a nonempty interior. Then (a) the set of bilinearly realizable functions on $S(U)$ is the representation ring $R(S(U))$.

(b) the approximately bilinearly realizable functions on $S(U)$ are exactly the continuous real-valued functions on $S(U)$.

<u>Proof</u>: If $(\mathcal{B},x_{in},\lambda)$ is an initialized, observed bilinear system, with $\mathcal{B}$ given by (1), and if $(\mathbb{R}^n,\pi_{\mathcal{B}})$ is the associated linear representation of $S(U)$, then the definition of the input-output function ψ can be rephrased as

$$\psi(\bar{u}) = \langle\lambda,\pi_{\mathcal{B}}(\bar{u})x_{in}\rangle.$$

Comparing this formula with the definition of matrix entries of a representation, and using Theorem 1, we see that (a) is immediate.

To prove (b), we first observe that the approximately bilinearly realizable functions are the elements of the closure $\overline{R(S(U))}$ of the representation ring in the space $F(S(U),\mathbb{R})$ of all real-valued functions on $S(U)$, provided that we give $F(S(U),\mathbb{R})$ the topology of uniform convergence on compact sets. Since every $\psi\in R(S(U))$ is continuous, and every limit of continuous functions which converge uniformly on compact sets is continuous, we conclude that every approximately bilinearly realizable function is continuous. To prove the converse, we use the fact that $R(S(U))$ is an algebra which contains the constant functions. Therefore, by the Stone-Weierstrass theorem, it suffices to prove that $R(S(U))$ separates points, i.e., that given any two different inputs $\bar{u}$, $\bar{u}'$, there exists an initialized, observed bilinear system $(\mathcal{B},x_{in},\lambda)$ such that the corresponding input-output function satisfies $\psi(\bar{u})\neq\psi(\bar{u}')$.

Let $\bar{u} = (T_u, u)$, $\bar{u}' = (T_{u'}, u')$. If $T_u \neq T_{u'}$, we can take $\mathcal{B}$ to be the system

$$\dot{x} = x \qquad x \in \mathbb{R}$$

with $x_{in} = 1$, $\lambda(x) = x$. Then $\psi(\bar{u}) = e^{T_u} \neq e^{T_{u'}} = \psi(\bar{u}')$.

If $T_u = T_{u'} = T$, then there must exist a real number r such that

$$\int_0^T e^{rt} u(t)dt \neq \int_0^T e^{rt} u'(t)dt$$

(because the linear combinations of exponentials e^{rt} are dense in the space of continuous real valued functions on $[0,T]$). Now consider the bilinear system

$$\dot{x}^1 = rx^1$$

$$\dot{x}^2 = ux^1$$

with $x_{in} = (1,0)$ and $\lambda(x^1,x^2) = x^2$. Then $\psi(\bar{u}) = \int_0^T e^{rt}u(t)dt$, and similarly for $\psi(\bar{u}')$. Therefore $\psi(u) \neq \psi(u')$, and our proof is complete.

§4. THE APPROXIMATION THEOREM

We will now translate Theorem 2 into the more familiar language of input-output maps. Let us choose a fixed time $T>0$ and a fixed bound $M>0$ for our inputs. If an initialized, observed bilinear system is given, we can associate with it the input-output map Φ which to every input u defined on $[0,T]$ assigns the output $\Phi_u: [0,T] \longrightarrow \mathbb{R}$. Therefore, Φ is a map from the set of inputs defined on $[0,T]$ to the set of continuous functions $[0,T] \longrightarrow \mathbb{R}$. Moreover, Φ clearly satisfies:

Continuity: If a sequence $\{u_k\}$ of inputs converges weakly to an input u, then the functions Φ_{u_k} converge uniformly to Φ_u.

Causality: If two inputs u,v coincide on $[0,T']$ for some $T' \leq T$, then the functions Φ_u and Φ_v also coincide on $[0,T']$.

The following is essentially due to M. Fliess 4 (cf. Appendix).

<u>Theorem 3</u>. Suppose F is an arbitrary function which to every input $u = (u^1,\ldots,u^m)$ defined on $[0,T]$ and bounded by M (i.e., $\sup\{|u^i(t)|: 0\leq t\leq T,\ i = 1,\ldots,m\} \leq M$) assigns a curve $F_u: [0,T]\longrightarrow\mathbb{R}$. Suppose moreover that F satisfies the continuity and causality conditions. Then for every $\varepsilon>0$ there is a bilinear system whose corresponding input-output map Φ is ε-close to F, in the sense that

$$|F_u(t) - \Phi_u(t)| < \varepsilon$$

for all t such that $0\leq t\leq T$ and all $u: [0,T]\longrightarrow\mathbb{R}^m$ which are measurable and bounded by M.

<u>Proof</u>: Let U denote the cube $|u^i| \leq M$, $i = 1,\ldots,m$. If $\bar{u} \in S(U)$ is of the form (T_u,u) with $T_u \leq T$, let v_u be some measurable U-valued function on $[0,T]$ which extends u. Then define $f(\bar{u}) = F_{v_u}(T_u)$. Because of the causality condition, $f(\bar{u})$ does not depend on the choice of the extension v_u. If $\bar{u} = (T_u,u)$ with $T_u > T$, let v_u be the restriction of u to $[0,T]$, and put $f(\bar{u}) = F_{v_u}(T)$.

In this way we have defined f for all $\bar{u} \in S(U)$, and it is easy to see that f is continuous. By Theorem 2, there is an initialized observed bilinear system $\mathcal{B}$ whose input-output function ψ satisfies

$$(*)\qquad |\psi(\bar{u}) - f(\bar{u})| < \varepsilon$$

for all $\bar{u} \in S(U)$ with duration $\leq T$. Now let $t\longmapsto u(t)$ be a U-valued input defined on $[0,T]$, and let Φ denote the input-output map arising from $\mathcal{B}$. If $0\leq t\leq T$, let ${}_t\bar{u} = (t,{}_tu)$, where ${}_tu$ is the restriction of u to $[0,t]$. Then u is an extension of ${}_tu$ to $[0,T]$, and therefore

$$F_u(t) = f({}_t\bar{u}).$$

Similarly $\Phi_u(t) = \psi({}_t\bar{u})$. It then follows from (*) that

$$|F_u(t) - \Phi_u(t)| < \varepsilon.$$

Since u and t were arbitrary, Φ is ε-close to F, and our proof is complete.

§5. THE TOPOLOGY OF S(U) WHEN U IS UNBOUNDED.

If $U \subseteq \mathbb{R}^m$ is closed, convex, but unbounded, it does not seem completely obvious how S(U) will be topologized. Weak convergence will not do, as shown below. But there is a property which is equivalent to weak convergence when U is bounded, but characterizes a different topology when U is unbounded. It is this topology that will be chosen. The property referred to above is that, for bounded U, the weak topology is <u>the weakest topology on</u> S(U) <u>relative to which all the representations</u> π_B <u>which arise from bilinear systems are continuous</u>.

It is the preceding characterization which makes T a good topology. (We certainly want all the π_B continuous, and we also want T as weak as possible, for the weaker the topology, the easier it is for sets to be compact.) So we carry over this characterization to the unbounded case, with a minor modification. Instead of just looking at bilinear systems, we look at arbitrary systems L of the form

$$(**) \qquad \dot{g}(t) = (X_0 + \sum_{i=1}^{m} u^i(t)X_i)(g(t)),$$

evolving on a Lie group. Here X_i, $i = 1,\dots,k$ are right-invariant vector fields on G. Given any such system, and any input $\bar{u} \in S(U)$, let $\pi_L(\bar{u})$ denote the value at T_u of the solution of the differential equation with initial condition $g(0) = 1$. We give S(U) <u>the weakest topology which makes all the maps</u> π_L <u>continuous</u>. Therefore a net $\{\bar{u}_\alpha\}$ converges in S(U) iff $\pi_L(\bar{u}_\alpha)$ converges for all L. In particular (since the π_L are homomorphisms) S(U) <u>is a topological semigroup</u>. Moreover, it is easily seen that, when U is bounded, this topology is precisely the one that was considered before.

<u>Addendum</u>. We now explain why the weak topology is not suitable for S(U) when U is unbounded. Choose a fixed time, say T=1, and take, for simplicity, $U = \mathbb{R}$. Then the set of $\mathbb{R}$-valued inputs with duration 1 is simply the space $L^\infty[0,1]$ of all bounded measurable real-valued functions on [0,1]. Let us use T to denote the weak topology of $L^\infty[0,1]$ as the dual of $L^1[0,1]$, and show that it is too weak for our purposes (i.e., it has too many convergent nets or, equivalently, it does not have sufficiently many open sets). T is not metrizable, even though its restriction to bounded sets is.

To show that T is too weak, we must construct a map F from $L^{\infty}[0,1]$ to some space, which is not continuous even though we want it to be. Now, the maps that we want to make continuous are the finite-dimensional representations that arise from bilinear systems. Hence, we must find a bilinear system

$$\dot{\vec{x}} = (A + u(t)B)\vec{x}, \qquad \vec{x} \in \mathbb{R}^n$$

and a vector $\vec{x}^0 \in \mathbb{R}^n$ such that, if $\vec{x}_u(t)$ denotes the solution of our differential equation with initial condition $\vec{x}_u(0) = \vec{x}^o$, then $\vec{x}_u(1)$ does not depend continuously on u.

We take $n=3$, and let the system be

$$\dot{x} = uy; \qquad \dot{y} = y+uz; \qquad \dot{z} = z$$

with initial condition

$$x(0) = y(0) = 0, \qquad z(0) = 1.$$

It then follows easily that

$$x_u(t) = \int_0^t e^s u(s)\left[\int_0^s u(\tau)d\tau\right]ds.$$

We show that $x_u(1)$ does not depend continuously on u, if the u's are topologized by T. Now, a fundamental system of neighborhoods of 0 in T is given by the sets

$$V(f_1,\ldots,f_k;\varepsilon) = \{u: \left|\int_0^1 u(t)f_i(t)dt\right| < \varepsilon \quad \text{for} \quad i = 1,\ldots,k\},$$

where $f_1,\ldots,f_k$ is an arbitrary finite sequence of functions in $L^1[0,1]$ and ε is any real number >0.

If $x_u(1)$ depended continuously on u, it would follow that there are $f_1,\ldots,f_k,\varepsilon$ such that $|x_u(1)| \leq 1$ for all $u \in V(f_1,\ldots,f_k,\varepsilon)$. Now let W denote the set of all $u \in L^{\infty}[0,1]$ such that $\int_0^1 u(t)dt = 0$ and that $\int_0^1 u(t)f_i(t)dt = 0$ for $i = 1,\ldots,k$. Then $W \subseteq V(f_1,\ldots,f_k;\varepsilon)$, so that $|x_u(1)| \leq 1$ for all $u \in W$. On the other hand, if $u \in W$, an easy calculation shows that

$$x_u(1) = -\frac{1}{2}\int_0^1 e^s\left[\int_0^s u(t)d\tau\right]^2 ds$$

where we have used the fact that $\int_0^1 u(s)ds = 0$.

Therefore, $x_u(1) = 0$ only if $u \equiv 0$. Hence, if $u \neq 0$, $u \in W$, we must have $x_u(1) \neq 0$. If $r \in \mathbb{R}$ is arbitrary, it is clear that $x_{ru}(1) = r^2 x_u(1)$. So, if $x_u(1) \neq 0$, it follows that $|x_{ru}(1)| > 1$ provided that r is large enough. But this is a contradiction, because $ru \in W$.

We have shown that, if $u \in W$, then necessarily $u \equiv 0$. But this is impossible because W is, by construction, a subspace of finite codimension of $L^\infty[0,1]$. This completes the proof that $x_u(1)$ does not depend continuously on u.

§6. "COMPLETION" OF $S(\mathbb{R}^m)$: GENERALIZED INPUTS

From now on we will consider the semigroup $S(\mathbb{R}^m)$ of all $\mathbb{R}^m$-valued bounded measurable inputs. We have already topologized $S(\mathbb{R}^m)$, and the time has come to ask why. Recall that, of all "good" topologies ("good" means "that makes the π_B continuous") we have chosen the weakest one, hoping that we will get as much compactness as possible. For instance, we would want all closed bounded subsets of $S(\mathbb{R}^m)$ to be compact. Now one sees easily that this is not true, and this suggests that we seek to "complete" $S(\mathbb{R}^m)$. If we succeed in doing so, we will have obtained a new class of "generalized functions".

The reader should notice the analogy with the construction of Schwartz's distributions. For the latter, one regards functions as doing a certain job, namely, that of being integrated against test functions. One then topologizes the set of functions by the weakest topology which is "good" for that job, and completes. We will do the same thing here, except for the fact that the task which we want our functions to perform is a different one, namely, that of serving as inputs for bilinear systems. There is one difference, however. The function space which one wants to complete in the Schwartz theory is a topological vector space, so that the topology gives rise to a uniform structure and it is clear what is meant by "completion". In our case, we are dealing with a topological semigroup, and it is not totally obvious (at least to us) what the meaning of "completion" should be. So we will use a somewhat more indirect procedure: we will embed $S(\mathbb{R}^m)$ in a complete topological group Γ_{m+1}, and then "complete" $S(\mathbb{R}^m)$ by taking its closure in Γ_{m+1}.

We begin by defining, for each positive integer k, the group Γ_k. For this purpose, let C_k be the category whose objects are all pairs

$\underset{\sim}{G} = (G; X_1,\ldots,X_k)$, where G is a connected Lie group and $X_1,\ldots,X_k$ are elements of the Lie algebra $L(G)$ which generate $L(G)$ as a Lie algebra. The morphisms in C_k from $\underset{\sim}{G} = (G; X_1,\ldots,X_k)$ to $\underset{\sim}{G}' = (G'; X_1',\ldots,X_k')$ are all the Lie group homomorphisms $\Phi: G \longrightarrow G'$ whose differential takes X_i to X_i' for $i = 1,\ldots,k$. Clearly, if $\underset{\sim}{G}$ and $\underset{\sim}{G}'$ are objects of C_k, and if a morphism $\Phi: \underset{\sim}{G} \longrightarrow \underset{\sim}{G}'$ exists, then Φ is unique. When such Φ exists, we denote it by $\Phi(\underset{\sim}{G},\underset{\sim}{G}')$, and we say that $\underset{\sim}{G}$ covers $\underset{\sim}{G}'$ (notation: $\underset{\sim}{G} \succ \underset{\sim}{G}'$). Clearly C_k is a small category (i.e., we can regard C_k to be a set). Moreover, the relation $\succ$ makes C_k a directed set, i.e., if $\underset{\sim}{G}$, $\underset{\sim}{G}'$ are in C_k, there is a $\underset{\sim}{G}''$ which covers both.

We then define Γ_k to be the projective limit of the $\underset{\sim}{G}$ in C_k. Precisely, an element of Γ_k is a family $g = (g_{\underset{\sim}{G}}, \underset{\sim}{G} \in C_k)$ which for every $\underset{\sim}{G}$ in C_k picks up an element of its Lie group G. Moreover, g is required to satisfy the condition that, whenever $\underset{\sim}{G} \succ \underset{\sim}{G}'$, then

$$\Phi(\underset{\sim}{G}, \underset{\sim}{G}')(g_{\underset{\sim}{G}}) = g_{\underset{\sim}{G}'}. \tag{+}$$

Examples of elements of Γ_k can be easily constructed. For instance, let $Y_1,\ldots,Y_k$ be indeterminates. Let P be any formal expression

$$P(Y_1,\ldots,Y_k) = e^{t_1 Y_{i(1)}} e^{t_2 Y_{i(2)}} \ldots e^{t_r Y_{i(r)}}$$

where r is any positive integer, the t_j are real numbers and the $i(j)$ are integers between 1 and k. If $\underset{\sim}{G} = (G; X_1,\ldots,X_k)$ is an object of Γ_k, then $P(X_1,\ldots,X_k)$ (i.e., the result of replacing the Y_i by the X_i) makes perfect sense as an element of $\underset{\sim}{G}$. In this way P defines a family $(g_{\underset{\sim}{G}}: G \in C_k)$, for which (+) clearly holds.

If $\underset{\sim}{G} \in C_k$, there is an obvious surjective homomorphism $P_G: \Gamma_k \longrightarrow G$, obtained by assigning to each family $(g_{\underset{\sim}{G}'}: \underset{\sim}{G}' \in C_k)$ the element $g_{\underset{\sim}{G}}$ of G. Clearly, $P_{\underset{\sim}{G}}$ is a group homomorphism, and (+) implies that $\Phi(\underset{\sim}{G}, \underset{\sim}{G}')P_{\underset{\sim}{G}} = P_{\underset{\sim}{G}'}$.

We give Γ_k the weakest topology which makes all the $P_{\underset{\sim}{G}}$ continuous. Since all the G are complete, it follows easily from our construction that Γ_k is complete. We can define a subset A of Γ_k to be bounded if $P_{\underset{\sim}{G}}(A)$ is relatively compact in G for all $\underset{\sim}{G}$ in C_k. (This is equivalent to: A is bounded iff for every neighborhood V of the identity in Γ_k there is an integer $n>0$ such that $A \subseteq V^n$.) Then Γ_k is a Montel group, i.e., every closed bounded subset of Γ_k is compact (by Tychonoff's Theorem).

We now embed $S(\mathbb{R}^m)$ in Γ_{m+1} as follows. If $\underset{\sim}{G} = (G, X_0, \ldots, X_m)$ is an element of C_{m+1}, and if $\bar{u} = (T_u, u) \in S(\mathbb{R}^m)$, we consider the differential equation (**) with initial condition $g(0) = 1$. Then we put $\sigma_G(\bar{u}) = g(T_u)$. It is clear that the family $(\sigma_G(\bar{u}): \underset{\sim}{G} \in C_{m+1})$ is an element of Γ_{m+1}, which we denote by $i(\bar{u})$. Clearly, $i: S(\mathbb{R}^m) \longrightarrow \Gamma_{m+1}$ is a one-to-one semigroup homomorphism.

If we compare the construction of i with the definitions of the topologies of $S(\mathbb{R}^m)$ and Γ_{m+1}, it follows easily that i enables us to identify $S(\mathbb{R}^m)$ with a topological subspace of Γ_{m+1}. So, from now on, we make this identification, and write $\bar{u}$ instead of $i(\bar{u})$.

We are now ready to give our

<u>Definition</u>. The closure $\hat{S}(\mathbb{R}^m)$ of $S(\mathbb{R}^m)$ in Γ_{m+1} is the space of <u>generalized</u> $\mathbb{R}^m$-<u>valued inputs</u>. We now give some examples.

First, take $m=1$, and define a sequence $u_1, u_2, \ldots$ of $\mathbb{R}$-valued inputs by letting u_n have duration $\frac{1}{n}$ and the constant value n on $[0, \frac{1}{n}]$. If a system

$$L: \dot{g} = (X_0 + uX_1)(g)$$

is given on a Lie group G, it is clear that

$$\pi_L(u_n) = \exp(X_1 + \tfrac{1}{n}X_0).$$

Therefore $\pi_L(u_n)$ converges for all L, as $n \longrightarrow \infty$. Hence the sequence u_n converges to a generalized input $\tilde{u}$. This input is basically the delta function.

However, generalized inputs are objects of a very different kind from distributions. We illustrate this with an example in the case $m=2$. Consider the functions defined on $[0,1]$ by

$$u_n(t) = \begin{cases} \sqrt{n} & \text{for } \frac{k}{n} \le t \le \frac{k}{n} + \frac{1}{4n} \\ -\sqrt{n} & \text{for } \frac{k}{n} + \frac{1}{2n} \le t \le \frac{k}{n} + \frac{3}{4n} \\ 0 & \text{otherwise} \end{cases}$$

$$v_n(t) = \begin{cases} \sqrt{n} & \text{for } \frac{k}{n} + \frac{1}{4n} \le t \le \frac{k}{n} + \frac{1}{2n} \\ -\sqrt{n} & \text{for } \frac{k}{n} + \frac{3}{4n} \le t \le \frac{k+1}{n} \\ 0 & \text{otherwise} \end{cases}$$

If a system L is given as above, and if we put $w_n = (u_n, v_n)$ $T_{w_n} = 1$, we find that, as $n \longrightarrow \infty$, $\pi_L(w_n)$ converges to

$$\exp(X_0 + \tfrac{1}{4}[X_2, X_1]).$$

This shows that w_n converges in $\hat{S}(\mathbb{R}^2)$ to a generalized input w. Clearly, w is not the zero input. Yet w_n does converge to zero in the sense of distributions because, if ψ is any C^1 function (actually $C^{1/2+\varepsilon}$, $\varepsilon>0$, is enough) then

$$\int_0^1 \psi(t)u_n(t)dt \quad \text{and} \quad \int_0^1 \psi(t)v_n(t)dt$$

converge to 0. The general picture suggested by this example seems to be as follows: $\hat{S}(\mathbb{R}^m)$ projects into (but not onto) the space of distributions, but a distribution which "is a generalized input" (i.e., is in the image of this projection) actually "splits" into many inputs, which are all different from each other.

We close by quoting without proof a recent result of ours which strongly indicates that this line of thought is worth pursuing. Consider the Haar functions $h_{k,n}$ defined, for all integers $n \geq 0$, and all k such that $1 \leq k \leq 2^n$, by

$$h_{k,n}(t) = \begin{cases} 2^{n/2} & \text{for } (k-1)2^{-n} \leq t \leq (k-\frac{1}{2})2^{-n} \\ -2^{n/2} & \text{for } (k-\frac{1}{2})2^{-n} \leq t \leq k2^{-n} \\ 0 & \text{otherwise} \end{cases}$$

Then the $h_{k,n}$, together with the function 1, form an orthonormal basis of $L^2[0,1]$. Any formal sum

$$H = a_0 \cdot 1 + \sum_{k,n} a_{k,n} h_{k,n}$$

with the coefficients a_0, $a_{k,n}$ in $\mathbb{R}^m$, will be called a formal Haar series. Let us call such a series convergent if, for each T such that $0 \leq T \leq 1$, the inputs $(T, {}_TH_N)$ converge in $\hat{S}(\mathbb{R}^m)$ as $N \longrightarrow \infty$. Here ${}_TH_N$ denotes the restriction to $[0,T]$ of the partial sum

$$H_N(t) = a_0 + \sum_{n\leq N} \sum_{k=1}^{2^n} a_{k,n}\, h_{k,n}(t).$$

It is not hard to see that, if two series converge in this sense to the same limit, then they are the same series. Therefore we have a bijection $\mu: \mathcal{H}_c \longrightarrow \hat{S}(\mathbb{R}^m)_H$ where $\mathcal{H}_c$ is the set of convergent Haar series and $\hat{S}(\mathbb{R}^m)_H$ is the set of generalized inputs which have a Haar series expansion.

Theorem 4. If $m=1$, let H be a Haar series whose coefficients satisfy

$$|a_{k,n}| \leq C2^{\alpha n}$$

for some fixed constants C, α, such that $\alpha < \frac{1}{6}$. Then H converges.

Now let $\mathcal{H}$ denote the set of all Haar series. We can put a measure on $\mathcal{H}$ by letting a_0 and the $a_{k,n}$ be independent normalized Gaussian random variables. Then it is well-known that, for almost all Haar series H, we have the bound

$$|a_{k,n}| \leq Cn^{1/2}$$

(cf. for instance the Lévy-Cieselski construction of the Brownian motion, as presented in McKean [2]). Therefore we can transfer our probability measure to $\hat{S}(\mathbb{R}^m)$, via μ. We conclude that $\hat{S}(\mathbb{R}^m)$ contains the "sample paths of the white noise process". These sample paths now appear as objects in their own right, which make perfect sense independently of any probabilistic considerations, and which can be used to drive a system of differential equations. The latter fact is the significant one. Ways are well known that enable us to regard the individual sample paths of white noise as "objects". For instance, they are the derivatives in the distribution sense of the sample paths of Brownian motion. However, such an approach does not lead very far, because it does not allow us to interpret an equation such as (1) as being meaningful for each sample path. Our approach does so (at least for $m=1$), and it shows that the correct way to do it requires something quite different from distributions.

§7. CONCLUSION

Viewing bilinear systems as semigroup representations is certainly natural. The results of this paper show that it is possible to gain significant insights by further pursuing this. What has been done here is very little, compared with what lies ahead. To improve upon the Approximation Theorem, it is clear that we must study infinite dimensional bilinear systems. We believe that the approach based on representation theory will be the best suited for this study. This must be so, by analogy with what is well known from group representation theory. The approach which is based on defining a bilinear system on an infinite dimensional space via the operators $A, B_1, \ldots, B_m$ will certainly require a careful study of which technical conditions are reasonable. We will want to allow A and the B_i to be unbounded, and it will not suffice to require that each operator be the infinitesimal generator of a semigroup. The natural requirement will be to ask that A and the B_i, jointly, should define a continuous representation of $S(\mathbb{R}^m)$.

As for generalized inputs, we have presented a small sample of what can be done. A much more detailed paper is now in preparation, but a lot of terrain must still be explored. For instance, it should be possible to extend Theorem 4 to arbitrary m, perhaps with the requirement of sharper bounds for the coefficients. This seems not to be too hard, and we expect to derive such an extension soon.

Finally, the appearance in our context of objects which project onto distributions, but in a many-to-one way, strongly suggests relating this work with the one known class of objects where distributions "unfold" in a similar fashion, namely, nonstandard analysis. Such work is currently in progress.

REFERENCES

[1] J. Kučera, Solution in large of control problem $\dot{x} = (A(1-u)+Bu)x$, Czech. Math. J., 16 (1966), pp. 600-622.

[2] H.P. McKean, Stochastic Integrals, Academic Press, New York, 1969.

[3] H.J. Sussmann, The bang-bang problem for certain control systems in $GL(n,\mathbb{R})$, SIAM J. Control, 10 (1972), pp. 470-476.

[4] M. Fliess, Séries de Volterra et Séries formelles non commutatives, C.R. Acad. Sci. Paris, t. 280 (1975), 965-967.

APPENDIX

(a) We now describe how our Theorem 3 is related to the work of M. Fliess. Fliess uses the topology of uniform convergence for the input semigroup, and proves that, on every compact subset of this semigroup, every continuous causal input-output map can be approximated arbitrarily close by bilinear ones.
Now, if a set is compact relative to the uniform topology, it is also compact relative to the weak topology, and both topologies coincide on it. Therefore Fliess' result follows from Theorem 3.

(b) Both our method and the one used by Fliess can be made to yield improved versions of the Approximation Theorem. The class of all bilinear systems can be replaced by any subclass B, provided only that B is closed under direct sums and tensor products, and that the corresponding ring of input-output functions separates points. In particular, it is possible to take B to be the class of bilinear systems whose matrices generate a nilpotent associative algebra, as Fliess does.

(c) If K is a finite subset of the input semigroup, the Approximation Theorem implies that the set V_o of those real functions on K that are bilinear input-output maps is dense in the set V of all functions from K to $\mathbb{R}$. Since V is a finite-dimensional vector space, and is a linear subspace, we can conclude that $V = V_o$. Therefore, given any finite set of data consisting of pairs $(\bar{u}_i, y_i)$, <u>where the u_i are inputs and the y_i are output values at the corresponding terminal times there is a bilinear system which fits these data exactly</u>.

ALGEBRAIC STRUCTURE OF INFINITE DIMENSIONAL LINEAR SYSTEMS IN HILBERT SPACE

John S. Baras*
Electrical Engineering Department
University of Maryland
College Park, Maryland 20742/USA

ABSTRACT

A module structure for a class of infinite dimensional linear systems in Hilbert space is described. The theory of invariant subspaces and of the Banach algebra H^∞ are the fundamental mathematical results used. Several results are derived which demonstrate that for this class of distributed systems, a detailed structure theory can be developed which resembles the corresponding theory for lumped systems.

1. Introduction

Recently a number of studies have been devoted in the investigation of linear distributed parameter systems in Hilbert space utilizing the methods and techniques of invariant subspace theory and of the theory of Hardy spaces. Detailed expositions of the results obtained to date can be found in Baras [1]-[5], Dewilde [6]-[9], Fuhrmann [10]-[14], Helton [15,16]. The purpose of this paper is to indicate how these results are related to the module theory of linear systems. Although the modules used are not finitely generated, the detailed knowledge about the structural properties of the Banach algebra H^∞, makes possible an analysis which is quite satisfactory from many points of view. More details about this particular subject can be found in Baras and Dewilde [5].

The systems we analyze have state space models of the form

$$\left.\begin{aligned} \frac{dx(t)}{dt} &= Ax(t) + Bu(t) \\ y(t) &= Cx(t) \end{aligned}\right\} \qquad (\Sigma)$$

where $x(t) \in \mathcal{X}$, a Hilbert space, A is a possibly unbounded operator on $\mathcal{X}$ which generates a strongly continuous semigroup of contractions [17, p. 140], $u \in L_p^2(\mathbb{R})$

This work was partially supported by a General Research Board award from the University of Maryland.

(the Hilbert space of square integrable $\mathbb{R}^p$ - valued functions) and $y \in L^2_m(\mathbb{R})$. $\mathbb{R}_+$ is the positive reals, $\mathbb{R}_-$ the negative reals. We further assume that the zero state input-output map of (Σ) is L^2 stable [18]. Certainly this is not the most general class of distributed systems. It is true, however, that many inportant classes of systems have such representations and in addition this class holds great promise for the development of a complete structure and decomposition theory. The operators B and C in (Σ) are bounded.

For notation we refer to [1], [5]. We give here only the absolutely necessary material. $\mathfrak{F}$ denotes the Fourier transform $f(t) \to \int_{-\infty}^{\infty} e^{-i\omega t} f(t)dt$. $L^2_k(\mathbb{I})$ is the space of square integrable complex vector valued functions on the imaginary axis. H^2_k is the subspace of $L^2_k(\mathbb{I})$, of functions which have analytic extensions in the right half plane. K^2_k is orthogonal complement of H^2_k in $L^2_k(\mathbb{I})$. H^∞ is the space of bounded analytic functions in the right half plane. H^∞_{mxn} is the space of bounded analytic matrix valued functions in the right half plane. K^∞ and K^∞_{mxn} are similarly defined for the left half plane. These are the Hardy spaces of analytic functions, and for details we refer to Hoffman [19]. An invariant subspace of H^2_k is a subspace invariant under multiplication by any element of H^∞. The transfer function of (Σ) is $\hat{G}(s) = C(Is-A)^{-1}B$ for $\mathrm{Re}\, s > 0$. $\{A, B, C\}$ is a realization of G whenever this equation holds or $G(t) = Ce^{At}B$, where G is the weighting pattern of (Σ). For definitions of controllability, exact controllability etc., we refer to [2], [11].

2. Module structures in infinite dimensional systems

In this section we describe a module structure for infinite dimensional linear systems that have an external description (input-output map) which is summarized by a continuous linear map f_Σ between the input space $L^2_p(\mathbb{R})$ and the output space $L^2_m(\mathbb{R})$. In addition the systems under study are time invariant and causal. That is f_Σ commutes with translations

$$f_\Sigma \sigma_\tau = \sigma_\tau f_\Sigma \tag{1}$$

whereby σ_τ denotes left translation by τ units; and $f_\Sigma(L^2_p(\mathbb{R}_+) \subseteq L^2_m(\mathbb{R}_+)$. The exposition is modelled on Kalman's original development of the module structure of linear finite dimensional systems [20, part IV, pp. 236-288]. The difference of course being that we analyze here a class of <u>infinite</u> dimensional linear systems. These systems are the well known <u>L^2 stable</u> linear time invariant systems of stability theory (see [18, p. 112]). More details about the material covered here can be found in [5], [9]. Other module theoretic approaches to infinite dimensional linear systems may be found in Kalman-Hautus [30] and Kamen [31].

An equivalent description of such systems is given by the restricted map, defined via [20, p. 243]:

$$\left.\begin{aligned} & f_{R\Sigma}: \Delta = L_p^2(\mathbb{R}_-) \to L_m^2(\mathbb{R}_+) = \Gamma \\ & f_{R\Sigma}(\delta)(t) = f_\Sigma(\delta)(t) \quad ; \quad t \geq 0 \\ & \qquad\qquad = 0 \qquad\qquad ; \quad t < 0 \end{aligned}\right\} \tag{2}$$

If we let P_+ (respectively P_-) denote the orthogonal projection from $L_k^2(\mathbb{R})$ onto $L_k^2(\mathbb{R}_+)$ (resp. onto $L_k^2(\mathbb{R}_-)$) it is easy to see that

$$P_+ \sigma_t f_{R\Sigma} = f_{R\Sigma} \sigma_t \tag{3}$$

It is also clear from causality and time invariance that f_Σ can be recaptured from $f_{R\Sigma}$.

Utilizing Fourier transforms and the Bochner-Chandrashekharan theorem [21], the input-output map in the frequency domain is described as follows:

$$\left.\begin{aligned} & \hat{f}_\Sigma \triangleq \mathfrak{F} f_\Sigma \mathfrak{F}^{-1} \\ & \hat{f}_\Sigma : L_p^2(\mathbb{I}) \to L_m^2(\mathbb{I}) \\ & \hat{y} = \hat{f}_\Sigma(\hat{u}) = \hat{G}\hat{u} \end{aligned}\right\} \tag{4}$$

where $\hat{y} = \mathfrak{F}y$, $\hat{u} = \mathfrak{F}u$ and $\hat{G} \in H^\infty_{mxp}$. $\hat{G}$ is of course the transfer function matrix associated with f_Σ. Similarly the restricted map in the frequency domain is described below:

$$\left.\begin{aligned} & \hat{f}_{R\Sigma} \triangleq \mathfrak{F} f_{R\Sigma} \mathfrak{F}^{-1} \\ & \hat{f}_{R\Sigma} : K_p^2 = \mathfrak{F}(\Delta) = \hat{\Delta} \to \hat{\Gamma} = \mathfrak{F}(\Gamma) = H_m^2 \\ & \hat{f}_{R\Sigma}\hat{\delta} = \mathcal{P}_+(\hat{G}\hat{\delta}) \end{aligned}\right\} \tag{5}$$

where $\mathcal{P}_+$ (resp. $\mathcal{P}_-$) denote the orthonogonal projection of $L_k^2(\mathbb{I})$ onto H_k^2 (onto K_k^2 respectively).

Clearly $\hat{\Delta}$ admits a module [22] structure with the ring being K^∞. The scalar multiplication is:

$$h.\hat{\delta} = \begin{bmatrix} h.\hat{\delta}_1 \\ h.\hat{\delta}_2 \\ \vdots \\ h.\hat{\delta}_p \end{bmatrix} \quad ; \quad h \in K^\infty, \; \hat{\delta} \in \hat{\Delta} = K_p^2 \tag{6}$$

Similarly $\hat{\Gamma}$ admits a K^∞ module structure where the scalar multiplication is defined as follows. For $h \in K^\infty$ and $\hat{\gamma} \in \hat{\Gamma}$ we multiply each component $\hat{\gamma}_i$ of $\hat{\gamma}$ with h (the resulting function being in $L^2(\Pi)$) and then project on H^2. We denote this by

$$h \,\square\, \hat{\gamma} = \mathcal{P}_+ \begin{bmatrix} h\hat{\gamma}_1 \\ h\hat{\gamma}_2 \\ \vdots \\ h\hat{\gamma}_m \end{bmatrix} \quad ; \quad h \in K^\infty, \; \hat{\gamma} \in \hat{\Gamma} = H_m^2 \tag{7}$$

Let us denote by e_t the function $e_t(i\omega) = e^{i\omega t}$ which is an element of K^∞, and by M_{e_t} the operator 'multiplication by e_t' on $L_j^2(\Pi)$, $j = p, m$. Then property (3) of $f_{R\Sigma}$ gives the following commutative diagram:

$$\begin{array}{ccc} \hat{\Delta} & \xrightarrow{\hat{f}_{R\Sigma}} & \hat{\Gamma} \\ M_{e_t} \downarrow & & \downarrow \mathcal{P}_+ M_{e_t} \\ \hat{\Delta} & \xrightarrow{\hat{f}_{R\Sigma}} & \hat{\Gamma} \end{array} \tag{8}$$

Let now $\chi(i\omega) = \frac{1+i\omega}{i\omega - 1}$. Clearly $\chi \in K^\infty$ and it follows [17, p. 142] that the above diagram is commutative if and only if the following diagram is commutative:

$$\begin{array}{ccc} \hat{\Delta} & \xrightarrow{\hat{f}_{R\Sigma}} & \hat{\Gamma} \\ \chi \cdot \downarrow & & \downarrow \chi \,\square \\ \hat{\Delta} & \xrightarrow{\hat{f}_{R\Sigma}} & \hat{\Gamma} \end{array} \tag{9}$$

Noting the fact that K^∞ is bounded pointwise a. e. closure of polynomials in $\frac{i\omega+1}{i\omega-1}$ [19, p. 107], we conclude that (9) remains commutative if we replace χ by any K^∞ function. We thus have the following:

<u>Theorem 1:</u> The restricted input-output map $\hat{f}_{R\Sigma}$ of a linear multivariable distributed system in this class (i. e., L^2 stable systems) is a K^∞-homomorphism (in the frequency domain and with scalar multiplications defined as in (6) and (7) above).

Concatenation of two inputs in Δ, with compact supports, is (in our setting) clearly given by (with the obvious definition for $t>0$): $\delta \circ \nu = \sigma_t \delta + \nu$, and in $\hat{\Delta}$: $(\widehat{\delta \circ \nu}) = M_{e_t} \hat{\delta} + \hat{\nu}$. It thus follows from diagram (8) that the set of Nerode equivalence classes [20, p. 250] of $f_{R\Sigma}$ is isomorphic with the quotient module $\hat{\Delta}/\mathrm{Ker}\, \hat{f}_{R\Sigma}$. So we have

Corollary 1: The "natural" state set $\hat{\Delta}/\mathrm{Ker}\, \hat{f}_{R\Sigma}$ of a linear multivariable continuous time distributed system in this class, admits the structure of a K^{∞}-module.

For our purposes, and in particular in order to make the analysis simpler, it is important to consider the Hankel operator

$$\left.\begin{aligned} & H_\Sigma = f_{R\Sigma} \sigma_- : L_p^2(\mathbb{R}_+) \to L_m^2(\mathbb{R}_+) \\ \text{or} \quad & \\ & \hat{H}_\Sigma = \hat{f}_{R\Sigma} J = \mathcal{P}_+ M_{\hat{G}} J : H_p^2 \to H_m^2 \end{aligned}\right\} \tag{10}$$

Here $\sigma_- u(t) = u(-t)$ and $J\hat{u}(i\omega) = \hat{u}(-i\omega)$ and $M_{\hat{G}}$ is multiplication by $\hat{G}$. Clearly all the input-output information is contained in H_Σ or $\hat{H}_\Sigma$. From diagram (8) we have the following commutative diagram for $\hat{H}_\Sigma$:

$$\begin{array}{ccc} H_p^2 & \xrightarrow{\quad \hat{H}_\Sigma \quad} & H_m^2 \\ M_{e_{-t}} \downarrow & & \downarrow \mathcal{P}_+ M_{e_t} \\ H_p^2 & \xrightarrow{\quad \hat{H}_\Sigma \quad} & H_m^2 \end{array} \tag{11}$$

H_p^2 has a natural H^{∞}-module structure via the scalar multiplication

$$q \cdot \hat{u} = \begin{bmatrix} q\hat{u}_1 \\ \vdots \\ q\hat{u}_p \end{bmatrix} ; \quad q \in H^{\infty}, \ \hat{u} \in H_p^2 \tag{12}$$

We define a new H^{∞}-module structure in H_m^2 via the scalar multiplication

$$q \boxtimes \hat{u} = Jq \,\square\, \hat{u}; \ q \in H^{\infty}, \ \hat{u} \in H_m^2 \tag{13}$$

whereby $\square$ is as defined in (7) (note that for $q \in H^{\infty}$, $Jq \in K^{\infty}$). Then it is easy to see from Theorem 1 that:

Collary 2: The Hankel operator in the frequency domain and with scalar multiplications as above (12), (13), becomes an H^{∞}-homomorphism.

Clearly then the kernel $\mathcal{N}_\Sigma$ of the Hankel operator $(\hat{H}_\Sigma)$ which is isomorphic to the Nerode equivalent classes, becomes an H^{∞}-submodule under the "natural" H^{∞}-module structure of H_p^2. The "natural" state set is then the quotient module

$H^2_p/\mathcal{M}_\Sigma$. This submodule plays a fundamental role in the structure theory of our class of systems.

3. H^2_k as a module over H^∞

We describe in this section H^2_k as an H^∞ - module (under the natural H^∞ - module structure), for any finite k (since H^2_k is a Hilbert space and H^∞ a Banach algebra [19], it is more correct to say that H^2_k is a topological module, but we will not use this terminology). H^∞ and H^∞_{kxk} have been extensively analyzed as Banach algebras and a great deal is known about their structure, [17] [19] [23] [24] [25]. It is this fact and the detailed results available that make possible the structural analysis of certain classes of distributed systems to a degree comparable with that of similar studies in finite dimensional systems. Trivial modifications give the corresponding facts about K^∞ and K^∞_{kxk}.

Now H^∞ is the dual of a Banach space [19, p. 137] and so we have a weak* topology. Then any w^* closed ideal A of H^∞ is of the form qH^∞ for some inner function q [19, 26], and conversely. The Beurling-Lax theorem [19, 27] asserts that any invariant subspace $\mathcal{M}$ of H^2_k has the form $\mathcal{M} = QH^2_k$ where Q (a kxk matrix valued function) is a <u>rigid</u> function [24], that is its values are partial isometries with a fixed initial space. Of particular interest to us are invariant subspaces of <u>full range</u> [26], i.e. $\mathcal{M}$ such that functions in $\mathcal{M}$ span $\mathbb{C}^k$ a.e. on the imaginary axis. In that case Q takes unitary values, i.e. is inner. Finally a subset A of H^∞_{kxk} is a w^* closed right (resp. left) ideal if and only if $A = QH^\infty_{kxk}$ (resp. $A = H^\infty_{kxk}Q$) for some rigid function Q [13] [24]. If Q_1 and Q_2 are rigid functions in H^∞_{kxk}, then Q_1 is said to <u>divide</u> Q_2 provided $Q_1H^2_k \supset Q_2H^2_k$. This is equivalent to an algebraic notion of divisibility. More important for our purposes will be the case of inner functions [26, lecture VIII].

Submodules of H^2_k (with the ring being H^∞) are now easily characterized. Thus a subset $\mathcal{M}$ of H^2_k is an H^∞ - submodule of H^2_k if and only if $\mathcal{M} = QH^2_k$, with some rigid function Q. In case $\mathcal{M}$ is of full range, the inner Q is determined uniquely modulo multiplication on the right by a constant unitary factor [27]. Notice that for Q inner detQ is a scalar inner function which is very useful in the determination of the structure of Q. Trivially for Q rigid, detQ is identically zero. For U, V inner functions in H^∞_{kxk}, in order that $UH^2_k \subset VH^2_k$ it is necessary and sufficient that $U = V.W$ for some inner function W [26, p. 60]. Two functions $F_1 \in H^\infty_{mxk}$, $F_2 \in H^\infty_{mx\ell}$ have a common left inner factor (c. ℓ. f.) if there exists an inner function $U \in H^\infty_{mxm}$ such that $F_1 = UF_3$, $F_2 = UF_4$ with $F_3 \in H^\infty_{mxk}$, $F_4 \in H^\infty_{mx\ell}$. V is their greatest common left inner factor (g.c.ℓ.f.) if $V = UW$ for any other c.ℓ.f. U. F_1, F_2 are <u>left coprime</u> if their g.c.ℓ.f. is a constant unitary matrix. Similar definitions can be given for inner factors from the right. For more details see Helson [26, p. 73-89], and [5].

As we indicated in the previous section we are interested in quotient modules $H^2_p / \mathcal{M}$ where $\mathcal{M}$ is of full range. Then $\mathcal{M} = QH^2_p$ with $Q \in H^\infty_{pxp}$ and inner. Now since $(\det Q) H^2_p \subset QH^2_p$ [26, p. 70] we have $(\det Q) v = 0$, $\forall v \in H^2_p / \mathcal{M}$. So $H^2_p / \mathcal{M}$ is a <u>torsion</u> module [22, p. 388]. Conversely suppose that $H^2_p / \mathcal{M}$ is a torsion module. Then for every $v \in H^2_p / \mathcal{M}$, $\exists$ $q_v \in H^\infty$ with $q_v v = 0$. These q_v's generate a w^* closed ideal A of H^∞ which therefore has the representation $A = q H^\infty$ with q inner and nontrivial. So $q H^2_p \subset QH^2_p$, which implies that $\mathcal{M}$ is of full range and therefore Q is inner. So we have (compare with [9]):

<u>Lemma 1:</u> $H^2_p / \mathcal{M}$ is a torsion H^∞-module if and only if $\mathcal{M}$ is of full range.

Let $\mathcal{M} = QH^2_p$ with Q inner. Then it is easy to verify that $\mathcal{M}^\perp$ (orthogonal complement in H^2_p) with scalar multiplication given by

$$q \Delta \hat{u} = P_{\mathcal{M}^\perp} (q\hat{u}) \tag{14}$$

becomes an H^∞ module isomorphic to $H^2_p / \mathcal{M}$.

<u>Theorem 2:</u> Let $\mathcal{M}$ be a submodule of H^2_p which is of full range. Then $H^2_p / \mathcal{M}$ is isomorphic to the module $\mathcal{M}^\perp$ with scalar multiplication given by (14).

4. <u>Transfer function matrices with meromorphic pseudo continuations of bounded type</u>

The transfer function matrices of the title comprise a very important class of systems, which are amenable to detailed analysis. Their importance has been recently discovered in various studies in distributed systems and networks [1, 2, 12, 13, 6, 16]. An mxp matrix valued function F is <u>meromorphic of bounded type</u> in the open left half plane if $F = \frac{G}{g}$, $G \in K^\infty_{mxp}$ and $g \in K^\infty$. $H \in H^\infty_{mxp}$ has a <u>meromorphic pseudo continuation</u> of bounded type in the open left half plane if its boundary values agree a.e. with those of a function which is meromorphic of bounded type in the open left half plane. Detailed discussions of the properties of such functions can be found in the references listed above. We summarize them in (see [4, 5, 6, 13]),

<u>Theorem 3:</u> Let $\hat{G} \in H^\infty_{mxp}$. The following areequivalent:

i) $\hat{G}$ has a meromorphic pseudo continuation of bounded type in the open left half plane.

ii) Let $\hat{H}_\Sigma$ be the Hankel operator associated with $\hat{G}$ (see (10)). Then $\mathrm{Ker}\,\hat{H}_\Sigma = (JD_\ell)^* H^2_p$ where $D_\ell \in H^\infty_{pxp}$ and inner.

iii) Let $\hat{H}_\Sigma$ as in ii). Then $(\mathrm{Range}\,\hat{H}_\Sigma)^\perp = D_r H^2_m$ with $D_r \in H^\infty_{mxm}$ and inner.

iv) $\hat{G}$ admits a left coprime factorization $\hat{G} = N^*_\ell D_\ell$, where $D_\ell \in H^\infty_{pxp}$ and inner, while $N_\ell \in H^\infty_{mxp}$.

v) $\hat{G}$ admits a right coprime factorization $\hat{G} = D_r N^*_r$, where $N_r \in H^\infty_{mxp}$ and $D_r \in H^\infty_{mxm}$ and is inner.

Moreover the factorizations in iv) and v) are unique modulo constant unitary factors from the left and right respectively.

Thus in particular for such transfer functions the natural state set $H^2_p/\mathrm{Ker}\hat{H}_\Sigma$ becomes a torsion module (Lemma 1) and isomorphic to $(\mathrm{Ker}\hat{H}_\Sigma)^\perp$ as an H^∞ - module (Theorem 2). The same remarks are of course true for the H^∞ - module $\overline{(\mathrm{Range}\,\hat{H}_\Sigma)} = (D_r H^2_m)^\perp$. One sees clearly that these factorizations are the appropriate generalizations of the polynomial matrix factorizations utilized by Rosenbrock [28]. These factorizations make possible a detailed analysis of the structure of such systems. However the full implications of Theorem 3 for continuous time distributed systems have not been exploited as yet. The main difficulties appear in the construction of state models [1] [2] [5].

It is clear from Theorem 3, that the inner function D_r (or $(JD_\ell)^*$) determines a natural candidate for state space. Whenever $\hat{G} \in H^2_{mxp}$ and has an exactly observable and controllable realization [2] [5], the following realization, which utilizes the right coprime factorization is well defined:

$$\left.\begin{aligned}
&\text{state space} = (D_r H^2_m)^\perp = \mathcal{R} \subset H^2_m \\
&e^{At} = P_{\mathcal{R}} M_{e_t} |_{\mathcal{R}} \\
&(Bu)(i\omega) = \hat{G}(i\omega)u;\ u \in \mathbb{R}^p \\
&Cx = \frac{1}{2\pi}\int_{-\infty}^{\infty} x(i\omega)\,d\omega;\ x \in \mathcal{R}
\end{aligned}\right\} \qquad (15)$$

This is the restricted translation realization. It is controllable and exactly observable and for such transfer functions, displays the common properties of all other realizations of this type [2] [5]. So we restrict our discussion to an analysis of (15). Certainly by duality similar results can be obtained utilizing the left coprime factorization. The constructions above are clearly canonical.

The inner function D_r on one hand provides the spectral analysis of the semigroup in (15) and a decomposition of the state space [17, p. 140-153 and 122-140] which is the proper generalization of the Jordan canonical form of matrix theory.

On the other hand, it determines the singularities of the transfer function $\hat{G}$. So [29] the spectrum of the infinitesimal generator of the semigroup in (15) is precisely $\mathcal{S}_{D_r}$ = { a) the points μ in the open left half plane where $D_r^*(-\mu^*)$ has non null kernel, b) the points on the imaginary axis through which D_r cannot be continued analytically to the open left half plane }. The meromorphic pseudo-continuation of $\hat{G}$ is $\frac{D_r^* N_r^*}{(\det D_r)^*}$ and its poles in the open left half plane and the points of no continuation on the imaginary axis provide the proper generalization for the concept of singularities of $\hat{G}$. The latter set is the set of singularities of $\hat{G}$ whenever $\hat{G}$ has a continuation through some point of the imaginary axis (which is the common case). This set is precisely equal to $\mathcal{S}_{D_r}$ [26] [29]. These considerations lead Baras and Brockett [1] to the concept of spectral minimality (see also [4] [5] [13]). Thus the restricted translation realization is spectrally minimal.

We close this section with a brief discussion of the decomposition mentioned above. For details we refer to [17, ch. III]. The semigroup in (15) is the prototype (and a universal model) for completely nonunitary semigroups of contractions. Its properties are best analyzed through the study of its cogenerator

$$T = (A+I)(A-I)^{-1} \tag{16}$$

where A is the generator of the semigroup. In our case T is a completely nonunitary contraction of class C_o [17, p. 122]. So let m_T be its minimal function (the generalization in our context of the concept of minimal polynomial of matrix theory). D_r and m_T are very closely related. Then we have triangular (or Jordan models) decompositions for T [17, III.7] which in turn induce the decomposition of the state space mentioned above [17, p. 149]. This decomposition and the resulting structure are in the same spirit as that for finite dimensional systems in [20, p. 270].

Finally we note that $\det D_r$ provides a measure for the size of the state space, and based on that the development of a degree theory is possible, where the above decomposition plays a central role (see [5] [14]).

5. Some suggestions for further research

It is my opinion that the ideas and methods described above hold promise of application in many problems involving distributed systems. I take this opportunity to emphasize the most important ones:

1) Approximate models. Here the decomposition theory is of fundamental importance.

2) Frequency domain methods in filtering and regulator theory. Here the factorization of positive operator valued functions plays a central role [17].

3) Suboptimal filtering and regulation. Combines 1) and 2) above.

Mostly needed is the application of the theory as developed to date to several relatively simple classes of distributed systems. This will necessarily demonstrate the power of the approach and will also generate useful examples. As the first favorable candidate I propose delay systems.

Finally some words about the restrictions imposed in the introduction. Most of them can be relaxed via standard techniques, (i.e. decomposition of semigroups of contractions to direct sums of unitary semigroups and completely nonunitary semigroups [17], use of exponential factors [1] etc.). The technical difficulties increase significantly however.

References

1. Baras, J.S. and Brockett, R.W., "H^2 functions and infinite dimensional realization theory", SIAM J. of Control, Vol. 13, No. 1, 221, 1975.

2. Baras, J.S., Brockett, R.W. and Fuhrmann, P.A., "State space models for infinite dimensional systems", IEEE Trans. on Autom. Control, Special issue on identification and time series analysis, Vol. AC-19, No. 6, 693, 1974.

3. Baras, J.S., "On canonical realizations with unbounded infinitesimal generators", Proceedings of 11th Annual Allerton Conference on Circuit and System Theory, 1, 1973.

4. Baras, J.S., "Natural models for infinite dimensional systems and their structural properties", Proc. of the 8th Annual Princeton Conf. on Information Science and Systems, 195, 1974.

5. Baras, J.S. and Dewilde, P., "Invariant subspace methods in linear multivariable distributed systems and lumped distributed network synthesis", to appear in IEEE Proc. Special Issue on Recent Trends in System Theory, Jan. 76.

6. Dewilde, P., "Roomy scattering matrix synthesis", Technical Report, Dept. of Mathematics, University of California, Berkeley, 1971.

7. Dewilde, P., "Input-output description of roomy systems", Technical Report, Kath. Univ. te Leuven, 1974.

8. Dewilde, P., "Coprime decomposition in roomy system theory", Technical Report, Kath, Univ. te Leuven, 1974.

9. Dewilde, P., "On the finite unitary embedding theorem for lossy scattering matrices", Proc. 1974 European Conf. on Circuit Theory and Design - IEE, London.

10. Fuhrmann, P.A., "On realization of linear systems and applications to some questions of stability", Math. Sys. The., Vol. 8, No. 2, 132, 1975.

11. Fuhrmann, P.A., "Exact controllability and observability and realization theory in Hilbert space", to appear J. Math. Anal. Appl.

12. Fuhrmann, P.A., "Realization theory in Hilbert space for a class of transfer functions", J. of Functional Analysis, Vol. 18, No. 4, 338, 1975.

13. Fuhrmann, P.A., "On spectral minimality of the shift realization", to appear.

14. Fuhrmann, P.A., "On series and parallel coupling of a class of discrete time infinite dimensional systems", to appear SIAM J. of Control.

15. Helton, J.W., "Discrete time systems operator models and scattering theory", J. Functional Analysis, Vol. 16, 15, 1974.

16. Douglas, R.G. and Helton, J.W., "Inner dilations of analytic matrix functions and Darlington synthesis", Acta Sci. Math. Sz. 34, 301, 1973.

17. Nagy, B.Sz. and Foias, C., Harmonic analysis of operators on Hilbert space, North Holland, 1970.

18. Desoer, C.A. and Vidyasagar, M., Feedback systems: input-output properties, Academic Press, New York, 1975.

19. Hoffman, K., Banach spaces of analytic functions, Prentice Hall, Englewood Cliffs, New Jersey, 1962.

20. Kalman, R.E., Falb, P.L. and Arbib, M.A., Topics in mathematical system theory, McGraw-Hill, New York, 1969.

21. Bochner, S. and Chandrasekharan, K., Fourier transforms, Princeton University Press, Princeton, 1949.

22. Lang, S., Algebra, Addison-Wesley, Reading, 1971.

23. Duren, P.L., Theory of H^p spaces, Academic Press, New York, 1970.

24. Sarason, D., "Generalized interpolation in H^∞", Trans. A.M.S., Vol. 127, No. 2, 179, 1967.

25. Carleson, L., "Interpolations by bounded analytic functions and the corona problem", Ann. Math., Vol. 76, No. 3, 547, 1962.

26. Helson, H., Lectures on invariant subspaces, Academic Press, New York, 1964.

27. Lax, P.D., "Translation invariant subspaces", Acta, Math. 101, 163, 1959.

28. Rosenbrock, H.H., State-space and multivariable theory, John Wiley, New York, 1970.

29. Lax, P.D. and Phillips, R.S., Scattering theory, Academic Press, New York, 1967.

30. Kalman, R.E. and Hautus, M., "Realization of continuous-time linear dynamical systems", Proc. Conf. Diff. Equ., NRL Math. Research Center, (L. Weiss editor), 151, 1972.

31. Kamen, E., "A distributional-module theoretic representation of linear continuous-time systems", Rept. SEL-71-044 (TR No. 6560-24), Stanford Electronics Lab., Stanford, Calif., 1971.

REPRESENTATION THEORY FOR LINEAR INFINITE DIMENSIONAL CONTINUOUS TIME SYSTEMS

by

A. Bensoussan
Department of Mathematics
University of Paris IX
and
IRIA-LABORIA
Paris, France

M. C. Delfour
Centre de Recherches Mathematiques
Universitè de Montreal
Montreal
Canada

S. K. Mitter
Department of Electrical Engineering
and Computer Science and
Electronic Systems Laboratory
Massachusetts Institute of Technology
Cambridge, Massachusetts, 02139, U.S.A.

0. INTRODUCTION AND SURVEY OF RESULTS

Representation theory for finite dimensional systems has been the subject of a great deal of discussion in recent years. In the case of linear systems defined over fields an account of the theory can be found in the books of BROCKETT and KALMAN-FALB-ARBIB and in the case of systems defined over rings in the book of EILENBERG, the papers of ROUCHALEAU-WYMAN-KALMAN, ROUCHALEAU-WYMAN and the thesis of JOHNSTON. For a lucid survey of results on systems defined over commutative rings, see SONTAG.

Over the past three years a body of results for linear infinite dimensional systems has been developed. Here two lines of enquiry have been pursued. On the one hand, BARAS, BROCKETT and especially FUHRMANN have developed a theory where the underlying input space is taken to be a Hilbert space (ℓ^2 or L^2) and the rich theory of restricted shift operators on ℓ^2 or L^2 is exploited to develop a theory of realization of infinite dimensional systems. On the other hand, AUBIN-BENSOUSSAN, BENSOUSSAN-DELFOUR-MITTER, KALMAN-MATSUO, KAMEN and others have tried to develop a theory of infinite dimensional which is very similar to the finite dimensional theory. For earlier work in this direction see BALAKRISHNAN and KALMAN-HAUTUS. The basic difference between these two approaches seems to be the following: If one requires that the realizations obtained be canonical (in the same sense as in the finite dimensional case, that is, reachable and completely distinguishable) and to obtain a state-space topological isomorphism theorem then it is necessary to work with a sufficiently large input function space (a space which contains sums of dirac impulses). However, it is not at all clear that the finite dimensional conceptual framework is the correct one for infinite dimensional systems. The other possibility is to fix the input space as ℓ^2 (in the discrete-time case) or L^2 and use the deep theory of Hardy spaces, invariant subspaces and canonical models. For excellent accounts of the possibilities using the latter approach see the papers by

Baras, Fuhrmann and DeWilde in this volume.

In this paper we try to show what can be done using the first approach. We present two possibilities.

Let U and Y be reflexive, separable Banach spaces which are the space of input values and output values respectively. An <u>input function</u> is a function from $(-\infty, 0)$ into U and an output function is a function from $(0, \infty)$ into Y. Suppose we are given a mapping f from the input function space to Y which is linear, continuous (in an appropriate topology) and time-invariant. Our problem of interest is to find a <u>canonical internal representation</u>, that is, a state space X (a topological vector space) and linear mappings $F:X \to X$, $G:U \to X$ and $H:X \to Y$ such that <u>the input-output map f is realized by</u>

$$\text{(A)} \qquad y(t) = Hx(t), \quad t \geq 0$$

and x(t) satisfies (in some appropriate sense) the differential equation

$$\text{(B)} \qquad \begin{cases} \dfrac{dx}{dt} = Fx + G\tilde{u}(t), & t \in (-\infty, 0) \\[2mm] \dfrac{dx}{dt} = Fx, & t \geq 0. \end{cases}$$

Furthermore, it is required that the <u>reachability</u> operator defined by (B) be surjective and that the mapping $x(0) \longmapsto y(\cdot)$ from X into the output function space be injective (<u>complete distinguishability</u>).

In the first instance we consider a very large input function space. Let Φ be the space of real infinitely differentiable functions on $(-\infty, 0]$, $\Phi = \mathcal{E}(-\infty, 0]$, then the space of input functions will be $\mathcal{L}_1(\Phi; U)$ the space of nuclear operators from $\Phi \to U$. More concretely, an input function $\tilde{u}$ will be described by

$$\tilde{u} = \sum_n \lambda_n \mu_n u_n$$

where μ_n is a bounded sequence of Φ' (which is identical to the space of distributions with compact support on $(-\infty, 0]$), u_n is a bounded sequence of U, and λ_n are real numbers such that $\sum_n |\lambda_n| < +\infty$. We will show that Φ' is a commutative algebra and $\mathcal{L}_1(\phi; U)$ is a locally convex Hausdorff space and a topological ϕ'-module. The space $\mathcal{L}_1(\Phi; U)$ is very large and contains functions in $L^1(-\infty, 0; U)$ with compact support and Dirac measures.

On $\mathcal{L}_1(\Phi; U)$ a translation operator T(t) is defined, which is a strongly continuous semi group, whose infinitesimal generator is <u>bounded</u>. We define the <u>Hankel operator</u>

$$\text{(1)} \qquad \mathcal{H}\tilde{u}(t) = f\left(T(t)\tilde{u}\right) \qquad t \geq 0$$

where $f \in \mathcal{L}\left(\mathcal{L}_1(\Phi; U); Y\right)$ and is the input-output map. $\mathcal{H}$ commutes under shifts.

The output function space will be $\Gamma = C^\infty(0, +\infty); Y)$, i.e. the space of infinitely differentiable functions from $[0, +\infty) \to Y$. It can be equipped with a structure of a Fréchet space and of a topological Φ'-module and $H \in L\left(L_1(\Phi;\ U);\Gamma\right)$ will be a homomorphism of Φ'-modules between $L_1(\Phi;\ U)$ and Γ.

A canonical internal representation can be constructed using Nerode equivalence. The state space is

(2) $$X = L_1(\Phi;\ U)/\text{Ker } H$$

and X is a locally convex Hausdorff space and a topological Φ'-module. We can define a strongly continuous semi-group on X by setting

(3) $$\Gamma(t)[\tilde{u}] = [T(t)\tilde{u}]$$

and $\Gamma(t)$ has a <u>bounded</u> infinitesimal generator F. The mappings G and H are defined by

(4) $Gu = [u\delta_o]$, where δ_o is the dirac measure concentrated at zero.

(5) $H[\tilde{u}] = f(\tilde{u})$.

(4) explains why we need to include dirac measures (with values in U) in the input function space.

Knowing $x(0) = [\tilde{u}]$, we can define the state of the system

$x(t)$ for $t \geq 0$ by

$x(t) = \Gamma(t)\ x(0)$.

In this first part we also discuss the concept of dual systems. It will then be necessary to work with weak topologies, but we will get dual algebraic as well as topological properties. However, although the Hankel operators will be duals of each other, the canonical state spaces will not be exactly duals of each other (this is a consequence of infinite dimensionality).

We then discuss the case where the input function space is taken to be $\left(H^m(-\infty,\ 0;\ U')\right)'$ the dual of the Sobolev space of order m with values in U', which is a Hilbert space if U is a Hilbert space. The state space is then a Hilbert space. In this case, we can describe the evolution of the state in terms of an operational differential equation and also prove a state-space isomorphism theorem.

1. APPROACH USING DISTRIBUTION THEORY

1.1 The Input Function Space Φ

We denote by Φ or $E(-\infty,\ 0]$ the space of infinitely differentiable functions on $(-\infty,\ 0]$. On Φ we define the countable family of semi-norms

(1.1) $$||\phi||_{m,k} = \sup_{-k \leq t \leq 0}\ \sup_{p \leq m} |\phi^{(p)}(t)| \quad , \quad \phi \in \Phi$$

which define a topology of a Fréchet space for Φ.

Let us denote by E the space of infinitely differentiable functions on $(-\infty,\ +\infty)$ provided with the countable family of semi norms

(1.2) $$||\phi||_{m,k} = \sup_{-k\leq t\leq k} \ \sup_{p\leq m} |\phi^{(p)}(t)| , \qquad \phi \in E$$

Let ρ denote the restriction of functions in E on $(-\infty, 0]$, i.e.,

(1.3) $$\rho\phi(t) = \phi(t) \qquad t \in (-\infty, 0].$$

Clearly ρ is a continuous mapping from $E \to \Phi$. Furthermore we have

Proposition 1.1: $\Phi \approx E/\ker \rho$. If $\phi \in \Phi$, there exists $\tilde{\phi}$, an extension of ϕ, $\tilde{\phi} \in E$ such that

(1.4) $$||\tilde{\phi}||_{j,k} \leq K_j \sup (1, ||\phi||_{j,k}) \qquad \forall j,k$$

where the K_j's are constants.

Let Φ' be the dual of Φ (regarded as a space of distributions). It is well known that

Φ' = set of distributions with compact support in $(-\infty, 0]$.

Moreover

Proposition 1.2: The space Φ' is isomorphic to a closed, subspace of E'. From this fact it follows that the topology on Φ' is the same as that induced by E'. One can also show

Proposition 1.3: The space Φ is reflexive.

As a topological space Φ' is the dual of a Fréchet space. Since its elements are distributions with compact support we can provide Φ' with a nice algebraic structure.

For two distributions with compact support if we define multiplication as convolution, then it can be easily shown that Φ' is a commutative algebra (with unit, the dirac distribution at the origin) and also an integral domain. Furthermore since the topology of Φ' is that induced from E', convolution is continuous.

For applications, one should note that finite sums of Dirac measures are contained in Φ'.

Let $L_1(\Phi; U)$ denote the space of nuclear operators from Φ into U (cf. TREVES, Chapter 47, for definitions and other facts). $L_1(\Phi; U)$ is a linear subspace of $L(\Phi; U)$.

On $L_1(\Phi; U)$ we can define an external multiplication as follows: for $\mu \in \Phi'$ and $\tilde{u} \in L_1(\Phi; u)$ we define

(1.5) $$\mu * \tilde{u}(\phi) = \tilde{u}(\phi * \tilde{\mu}), \text{ where } \tilde{\mu} \text{ is the distribution}$$

$$\tilde{\mu}(\phi) = \mu(\tilde{\phi}) \qquad \left(\tilde{\phi}(s) = \phi(-s)\right)$$

which has meaning since $\phi * \tilde{\mu} \in \Phi$. It can be shown that with this definition of external multiplication

Proposition 1.4: The space $L_1(\Phi; U)$ is a unitary topological Φ'-module.

Remarks: a) If $U = \mathbb{R}^n$, then $L_1(\Phi; U) = (\Phi')^n = L(\Phi; \mathbb{R}^n)$ which is a free unitary finitely generated Φ'-module. b) If Ω is a bounded open subset of $(-\infty, 0)$, then $L^1(\Omega; U) \subset L_1(\Phi; U)$ with continuous injection.

By definition

$$L_1(\Phi; U) = \left\{ \tilde{u} = \sum_n \lambda_n \mu_n u_n \,\middle|\, \mu_n \in \Phi',\ u_n \in U,\ u_n, \mu_n \text{ bounded}, \sum_n |\lambda_n| < +\infty \right\}$$

For $\phi \in \Phi$ and $\tilde{u} \in L_1(\Phi; U)$, we define

$$\tilde{\phi} * \tilde{u} = \sum_n \lambda_n (\tilde{\phi} * \mu_n) u_n, \text{ where} \tag{1.6}$$

$\tilde{\phi} * \mu_n \in \Phi'$ is defined by

$$\tilde{\phi} * \mu_n(h) = \mu_n(\phi * h).$$

Proposition 1.5: The mapping $\phi \longmapsto \tilde{\phi} * \tilde{u}$ is linear and continuous from Φ into $L_1(\Phi; U)$.

The semi group of translations

Let $\theta(-t)$, $t \geq 0$ be the semi-group of translations on Φ, defined by

$$\theta(-t)\ \phi(s) \equiv \phi(-t+s) \qquad t \geq 0, \quad s \leq 0. \tag{1.7}$$

It follows from TREVES, p. 286 that the mapping

(1.8) $t \to \theta(-t)\phi$ is infinitely differentiable from $[0, +\infty)$ into Φ, and

$$\frac{d^p}{dt^p}\ \theta(-t)\phi(s) = (-1)^p \phi^{(p)}(-t+s) \qquad s \leq 0,\ t \geq 0.$$

Therefore $\theta(-t)$ has an infinitesimal generator which is $-\frac{d}{dt}$.

Remark: The family $\theta(-t)$ is not an equicontinuous semi group on Φ. This would however be true if $\Phi = S(-\infty, 0]$ instead of $E(-\infty, 0)]$.

Since $-\frac{d}{dt} \in L(\Phi; \Phi)$ we can define its transpose $D \in L(\Phi'; \Phi')$. Let $\sigma(t)$ be the semi group of translations on Φ' defined by

$$\sigma(t)\ \mu(\phi) = \mu\Big(\theta(-t)\phi\Big) \qquad \forall\ \phi \in \Phi,\ t \geq 0. \tag{1.9}$$

We have

Proposition 1.6: The mapping $t \to \sigma(t)\mu$ is infinitely differentiable from $[0, +\infty)$

into Φ' and

$$\frac{d}{dt}\sigma(t)\mu = D\,\sigma(t)\mu = \sigma(t)\,D\mu\ , \quad \forall\, t \geq 0. \tag{1.10}$$

On $L_1(\Phi;\, U)$ we may define a semi group $T(t)$, $t \geq 0$ by setting

$$T(t)\,\tilde{u}(\phi) = \tilde{u}\Big(\theta(-t)\phi\Big) = \sum_n \lambda_n \Big(\sigma(t)\mu_n\Big)\,(\phi)\;u_n \tag{1.11}$$

hence $T(t)\tilde{u}$ has the representation

$$T(t)\tilde{u} = \sum_n \lambda_n \Big(\sigma(t)\;\mu_n\Big)\;u_n. \tag{1.12}$$

Proposition 1.7: The mapping $t \to T(t)\tilde{u}$ is infinitely differentiable from $[0, +\infty)$ into $L_1(\Phi;\, U)$.

Let us mention the useful formulas

$$\sigma(t)\mu = \delta_{-t} * \mu \tag{1.13}$$

$$T(t)\tilde{u} = \delta_{-t} * \tilde{u} \tag{1.14}$$

Remark: If $\Phi = S(-\infty, 0]$, then $\sigma(t)$ and $T(t)$ are equicontinuous semi groups of class C_o.

2. EXTERNAL REPRESENTATION OF LINEAR TIME-INVARIANT SYSTEMS

2.1 Notations and preliminary results

If $\Phi' \otimes U$ denotes the tensor product of Φ' and U, i.e., the subspace of $L(\Phi;\, U)$ of elements

$$\tilde{u} = \Sigma\, \mu_n\, u_n$$

where the sum is finite. Clearly $\Phi' \otimes U$ is dense in $L_1(\Phi;\, U)$. The topology induced on $\Phi' \otimes U$ is called the Π-topology, and provided with it, $\Phi' \otimes U$ is denoted $\Phi' \otimes_\Pi U$. An important result is the following (see TREVES, p. 438); the dual of $\Phi' \otimes_\Pi U$ is algebraically isomorphic to $B(\Phi';\, U)$, the space of bilinear continuous forms on $\Phi' \times U$. It follows from the density of $\Phi' \otimes_\Pi U$ into $L_1(\Phi;\, U)$ that

$$\Big(L_1(\Phi;\, U)\Big)' \equiv B(\Phi';\, U). \tag{2.1}$$

Denoting by

$$\Phi(-\infty, 0;\, U') = \Big\{u_*(t) \;:\; (-\infty, 0) \to U' \,\Big|\, t \longmapsto \,<u_*(t),\, u>\, \varepsilon\, \Phi \;\; \forall\, u\, \varepsilon\, U\Big\} \tag{2.2}$$

we have

Theorem 2.1: $B(\Phi' :\, U)$ and $\Phi(-\infty, 0;\, U')$ are algebraically isomorphic.

With $L_1(\Phi;\, U)$ as our choice of input function space and Y being a separable reflexive Banach space (the space of output values), we get a representation for

$f \in \mathcal{L}\left(L_1(\Phi;\ U);\ Y\right)$. This is given in

<u>Theorem 2.2</u>: If $f \in \mathcal{L}\left(L_1(\Phi;\ U);\ Y\right)$, there exists a unique family $K(t)$ of operators from $U \to Y$ which satisfy

$$K(t) \in \mathcal{L}(U:\ Y), \quad \forall\, t \geq 0, \quad \|K(t)\| \leq C_k \quad \forall\, t \in [0,\ k] \tag{2.3}$$

$$< K(-t)u,\ y_* > \in \Phi, \quad \forall\, u \in U,\ y_* \in Y', \tag{2.4}$$

such that

$$< f(\tilde{u}),\ y_* > = \sum_n \lambda_n \int_{-\infty}^{0} < K(-t)u_n,\ y_* > \mu_n(t)\ dt, \quad \forall\, y_* \in Y'. \tag{2.5}$$

<u>Remarks:</u> a) If $u(\cdot) \in L^1(\Omega;\ U)$, then

$$< f\left(u(\cdot)\right),\ y_* > = \int_{-\infty}^{0} < K(-t)u(t),\ y_* > dt \text{ and hence}$$

$$f\left(u(\cdot)\right) = \int_{-\infty}^{0} K(-t)u(t)\ dt. \tag{2.6}$$

b) If $\tilde{u} = \sum_n \delta_{t_n} u_n$ then

$$f\left(\sum_n \delta_{t_n} u_n \right) = \sum_n K(-t_n)\ u_n. \tag{2.7}$$

<u>Definition 2.1</u>: The operator $K(t)$ is called the <u>kernel</u> or <u>impulse response</u> of f.

2.2 <u>Output functions space</u>

Let Y be a reflexive separable Banach space, which will be called the space of <u>output values</u>, and will play a part parallel to U. We shall denote by

$$\Gamma = C^{\infty}[0,\ +\infty;\ Y)$$

the space of infinitely differentiable functions from $[0, +\infty)$ into Y. It is a Fréchet space for the topology defined by the family of semi norms

$$\|y(\cdot)\|_{mk} = \sup_{0 \leq t \leq k}\ \sup_{p \leq m}\ \|y^{(p)}(t)\|_Y. \tag{2.8}$$

We shall equip Γ with a structure of Φ'-module.

We first have

<u>Proposition 2.1</u>: <u>If</u> $y(\cdot) \in \Gamma$ <u>and</u> $\mu \in \Phi'$, there exists a unique $z(\cdot) \in \Gamma$ <u>such that if</u> $y_* \in Y'$ <u>and if</u> $\tilde{\phi}(t)$ <u>denotes an element of</u> E <u>which is an extension of</u> $\phi(t) = < y_*,\ y(t) >$, <u>then</u>

$$< y_*,\ z(t) > = \mu * \tilde{\phi}(t), \quad \forall\, t \geq 0. \tag{2.9}$$

Furthermore, for fixed u, the mapping

$$y(\cdot) \to z(\cdot) \qquad \text{from } \Gamma \to \Gamma$$

is continuous.

We shall set

$$z(\cdot) = \mu \cdot y(\cdot) \tag{2.10}$$

and call it the external product of $\mu \in \Phi'$ and $y(\cdot) \in \Gamma$. We have

Proposition 2.2: With the external multiplication (2.10), Γ becomes a unitary topological Φ'-module.

2.3 Structure of the external representation

The external representation of a linear invariant system is defined by a mapping $f \in L\left(L_1(\Phi;\ U) : Y\right)$. The mapping f can be extended to a mapping

$\mathcal{H}: L_1(\Phi;\ U) \to \Gamma$, by setting

$$(\mathcal{H}\,\tilde{u})(t) = f\left(T(t)\tilde{u}\right), \qquad \forall\, t \geq 0. \tag{2.11}$$

The operator $\mathcal{H}$ will be called the Hankel operator of the system.

Proposition 2.3:

$$\mathcal{H} \in L\left(L_1(\Phi;\ U);\ \Gamma\right). \tag{2.12}$$

Proposition 2.4: The operator $\mathcal{H}$ is an homomorphism of Φ'-modules between $L_1(\Phi;\ U)$ and Γ.

3. INTERNAL REPRESENTATION OF THE SYSTEM

An internal representation of the system corresponding to the Hankel operator $\mathcal{H}$ is a triple (F, G, H), where $F: X \to X$, $G: U \to X$ and $H: X \to Y$ are linear operators, with G and H being continuous, such that F is the infinitesimal generator of a semi-group $\Gamma(t)$ on X which satisfies

$$K(t) = H\Gamma(t)G, \qquad \forall\, t \geq 0$$

(recall that K(t) is the kernel (impulse response) of f). In this section we construct a canonical internal representation.

3.1 The state space

Let

$$X = L_1(\Phi;\ U)/\mathrm{Ker}\,\mathcal{H} \tag{3.1}$$

which will be called the state space. Since $\mathcal{H}$ is an homomorphism of Φ'-modules, and since $\mathcal{H}$ is continuous, Ker $\mathcal{H}$ is a topological submodule of $L_1(\Phi;\ U)$. Therefore X admits the structure of a Φ'-module with a multiplication defined by

$$\mu \cdot [\tilde{u}] = [\mu * \tilde{u}] \tag{3.2}$$

where $[\tilde{u}]$ denotes an element of X, for which $\tilde{u}$ is a representative. Moreover X can be given the structure of a locally convex Hausdorff space, the topology of which is defined by the following basis of continuous semi norms

$$\omega_\alpha([\tilde{u}]) = \inf_{\tilde{u}_1 \in [\tilde{u}]} \Pi_\alpha(\tilde{u}_1) \tag{3.3}$$

where Π_α ranges over the basis of continuous semi norms of $L_1(\Phi; U)$.

Hence X is a unitary topological Φ'-module.

3.2 Internal structure

From the definition of X, it follows that there exists a canonical factorisation of H, as follows

$$L_1(\Phi; U) \xrightarrow{H} \Gamma, \quad L_1(\Phi; U) \xrightarrow{b} X \xrightarrow{a} \Gamma \tag{3.4}$$

where b is surjective, and a is injective. More precisely we have

$$b\,\tilde{u} = [\tilde{u}] \tag{3.5}$$

$$a\,[\tilde{u}] = H\,\tilde{u} \tag{3.6}$$

It is easy to show that a and b are continuous and homomorphisms of Φ'-modules.

We then define mappings $F:X \to X$, $G:U \to X$, $H:X \to Y$, by setting

$$F[\tilde{u}] = b(\Delta\,\tilde{u}), \text{ where } \Delta \text{ is the infinitesimal generator of } T(t) \tag{3.7}$$

$$G\,u = b\,(u\delta_0) \tag{3.8}$$

$$H[\tilde{u}] = a[\tilde{u}](0) = f(\tilde{u}). \tag{3.9}$$

From properties of a and b, it follows that F, G, H are linear and continuous. Furthermore F is the infinitesimal generator of the semi group $\Gamma(t)$ on X defined by

$$\Gamma(t)[\tilde{u}] = b\,T(t)\,\tilde{u} \tag{3.10}$$

From properties of $T(t)$ it follows that

$$t \to \Gamma(t)[\tilde{u}]$$

is infinitely differentiable from $[0, +\infty)$ into X.

We shall introduce some notation. Let $x(t)$ be a mapping from $[0, +\infty)$ into X, which is infinitely differentiable, and let $\mu \in \Phi'$. If there exists an element

$\xi \in X$ such that

(3.11) $$< \xi, x_* > = \int_{-\infty}^{0} < x(-t), x_* > \mu(t)\, dt, \qquad \forall x_* \in X'$$

we will write (note that ξ is unique)

(3.12) $$\xi = \int_{-\infty}^{0} x(-t)\, \mu(t)\, dt.$$

We have

Proposition 3.1:

(3.13) $$K(t) = H\, \Gamma(t)\, G \qquad \forall t \geq 0$$

(3.14) $$b\, \tilde{u} = \sum_n \lambda_n \int_{-\infty}^{0} \Gamma(-t)\, G\, u_n\, \mu_n(t)\, dt.$$

3.3 Evolution of the state

We shall call $b\,\tilde{u}$ the state reached at time 0 and denote it $x(0)$. Hence from (3.14)

(3.15) $$x(0) = \sum_n \lambda_n \int_{-\infty}^{0} \Gamma(-s)\, G\, u_n\, \mu_n(s)\, ds.$$

We shall now define what we mean by the state of the system at any time $t \gtrless 0$. First, for $t \geq 0$ we write, by definition

(3.16) $$x(t) = \sum_n \lambda_n \int_{-\infty}^{0} \Gamma(t-s)\, G\, u_n\, \mu_n(s)\, ds = b\left(T(t)\tilde{u}\right), \qquad t \geq 0.$$

For $t < 0$, we cannot define $x(t)$ by formula (3.16), since $T(t)$ has no meaning for $t < 0$. We shall instead define $x(t)$ in the sense of distributions by giving a meaning to

$$\int_{-\infty}^{0} x(t)\, \phi(t)\, dt$$

for any $\phi \in \Phi$.

For $\phi \in \Phi$, we set if $\tilde{u} \in L_1(\Phi; U)$

(3.17) $$x_\phi = b\, (\check{\phi} * \tilde{u})$$

where $\check{\phi} * u$ has been already defined in section 1. From this, it follows that for fixed $\tilde{u}$, the mapping

$$\phi \to x_\phi \in L(\Phi; X).$$

We have

Proposition 3.2:

(3.18) $$\frac{dx}{dt} = Fx(t) \qquad \forall t \geq 0$$

$$x_{-\phi'} = Fx_{\phi} + G\,\tilde{u}(\phi) \qquad \forall\, \phi \in \Phi. \tag{3.19}$$

Setting

$$y(t) = (H\,\tilde{u})(t) \tag{3.20}$$

we can summarize the following relationships already obtained

$$y(t) = Hx(t) \qquad t \geq 0 \tag{3.21}$$

$$\frac{dx}{dt} = Fx(t) \qquad t \geq 0 \tag{3.22}$$

$$x(0) = \int_{-\infty}^{0} \Gamma(-s)\; G\, u_n\, \mu_n(s)\; ds \tag{3.23}$$

$$\int_{-\infty}^{0} \frac{dx}{dt}\; \phi(t)\; dt = F \int_{-\infty}^{0} x(t)\; \phi(t)\; dt + G\,\tilde{u}(\phi) \qquad \forall\, \phi \in \Phi. \tag{3.24}$$

(3.25) The sets $L_1(\Phi, U)$ (input functions), Γ (output functions) and X (state space) are unitary topological Φ'-modules.

The space of input functions and the state space are locally convex Hausdorff spaces. The space of output functions is a Frechet space, F, G, H are linear and continuous; F is the infinitesimal generator of the semi group $\Gamma(t)$.

(3.26) The mapping $\tilde{u} \to x(0) : L_1(\Phi;\, U) \to X$ is continuous, surjective, and is an homomorphism of Φ'-modules.

(3.27) The mapping $x(0) \to y(\cdot) : X \to \Gamma$ is continuous, injective and is an homomorphism of Φ'-modules.

Using standard terminology we will call the set (U, Y, X, F, G, H) a <u>canonical internal representation</u>. To complete this paragraph, we will point out that, using the module structure of the internal representation we have

$$H\,\tilde{u} = \sum_n \lambda_n\, \mu_n \cdot H(u_n\, \delta_0) \tag{3.28}$$

$$x(0) = \sum_n \lambda_n\, \mu_n \cdot b(u_n\, \delta_0). \tag{3.29}$$

4. DUALITY

4.1 Bilinear form associated with the external representation

In a previous section we have introduced the space $\Phi(-\infty, 0;\, U')$, which is isomorphic to the strong dual of $L_1(\Phi;\, U)$. We provide it with the topology of the strong dual. We will similarly consider $\Phi(-\infty, 0;\, Y)$. We now introduce the space

$$\Phi(Y) = C^{\infty}\big((-\infty, 0];\, Y\big)$$

provided with the topology of a Frechet space defined by the basis of continuous semi norms

$$(4.1)\qquad ||y(\cdot)||_{jk} = \sup_{-k\leq t\leq 0} \sup_{p\leq j} ||y^{(p)}(t)||.$$

We have

<u>Proposition 4.1</u>: <u>The injection of</u> $\Phi(Y)$ <u>into</u> $\Phi(-\infty, 0; Y)$ <u>is continuous</u>.

Let us now introduce the operator

$$(4.2)\qquad (\check{H}\,\tilde{u})(t) = (H\,\tilde{u})(-t) \qquad t \leq 0.$$

Clearly we have

$$(4.3)\qquad \check{H} \in L\big(L_1(\Phi;\ U);\ \Phi(Y)\big).$$

We next define on the pair $L_1(\Phi;\ U) \times L_1(\phi;\ Y')$ the bilinear form

$$(4.4)\qquad K(\tilde{u}, \tilde{y}_*) = [\check{H}\,\tilde{u}(\cdot), \tilde{y}_*]$$

From proposition 4.1, it follows that K is separately continuous. The bilinear form K can be explicitly written as follows

$$(4.5)\qquad K(\tilde{u}, \tilde{y}_*) = \sum_n \sum_m \alpha_m \lambda_n \int_{-\infty}^{0} \nu_m(t)\,dt\ \Big[\int_{-\infty}^{0} (< K(-t-s)u_n, y_m^* >)\ u_n(s)\,ds\Big]$$

$$= \sum_n \sum_m \alpha_m \lambda_n\ \nu_m * \mu_n\ (< K(-\cdot)\ u_n, y_m^* >)$$

where

$$\tilde{u} = \sum_n \lambda_n\ u_n\ \mu_n$$

$$\tilde{y}_* = \sum_m \alpha_n\ y_m^*\ \nu_m.$$

We shall use the following notation

$$(4.6)\qquad \sigma\big(L_1(\Phi;\ U);\ \phi(-\infty, 0;\ U')\big) = \text{space } L_1(\Phi;\ U)$$

provided with the weak topology.

We have

$$(4.7)\qquad \check{H} \in L\Big(\sigma\big(L_1(\Phi;\ U);\ \Phi(-\infty, 0:\ U')\big);\ \sigma\big(\Phi(-\infty, 0;\ Y):\ L_1(\Phi;\ Y')\big)\Big)$$

and

(4.8) K is separately continuous on $L_1(\Phi;\ U)$ and $L_1(\Phi;\ Y')$ for the weak topologies

$$\sigma\big(L_1(\Phi;\ U);\ \Phi(-\infty, 0, U')\big) \text{ and } \sigma\big(L_1(\Phi;\ Y');\ \Phi(-\infty, 0;\ Y)\big).$$

We denote by $S(t)$ the semi group of translations on $L_1(\Phi;\ Y')$. It readily follows from (4.5) that

$$K\left(T(t)\tilde{u},\ \tilde{y}_*\right) = K\left(\tilde{u},\ S(t),\ \tilde{y}_*\right). \qquad (4.9)$$

Furthermore the mappings

$$t \to T(t)\ \tilde{u}$$

and

$$t \to S(t)\ \tilde{y}_*$$

are <u>infinitely differentiable</u> on $\sigma\left(L_1(\Phi;\ U);\ \Phi(-\infty,\ 0;\ U;)\right)$ and $\sigma\left(L_1(\Phi;\ Y');\ \Phi(-\infty,\ 0;\ Y)\right)$ (since they already are for the strong topology). The corresponding infinitesimal generators will be denoted by Δ and Ξ.

4.2 The dual system

We consider the transpose

$$\check{H}^* \in L\left(\sigma\left(L_1(\Phi;\ Y'),\ \Phi(-\infty,\ 0;\ Y)\right);\ \sigma\left(\Phi(-\infty,\ 0;\ U');\ L_1(\Phi;\ U)\right)\right)$$

and define a linear mapping from $L_1(\Phi;\ Y') \to U'$ by setting

$$g(\tilde{y}_*) = \check{H}\ \tilde{y}_*(0). \qquad (4.10)$$

Then

$$g \in L\left(\sigma\left(L_1(\Phi;\ Y');\ \Phi(-\infty,\ 0;\ Y)\right);\ \sigma(U';\ U)\right). \qquad (4.11)$$

Let us notice that we do not have necessarily

$$g \in L\left(L_1(\Phi;\ Y');\ U'\right).$$

<u>Definition 4.1</u>: <u>The triple</u> $(Y',\ U',\ g)$ <u>defines the external representation of a system, called the dual system of</u> $(U,\ Y,\ f)$.

<u>Proposition 4.2</u>: <u>The Hankel operator</u> G <u>of the dual system satisfies</u>

$$(G\ \tilde{y}_*)(t) = (\check{H}^*\tilde{y}_*)(-t), \quad t \geq 0,\ \forall\ \tilde{y}_* \in L_1(\Phi;\ Y'). \qquad (4.12)$$

4.3 Canonical realizations of the dual systems

Let us set

$$X = L_1(\Phi;\ U)/\mathrm{Ker}\ \check{H} \qquad (4.13)$$

$$Z = L_1(\phi;\ Y')/\mathrm{Ker}\ \check{H}^* \qquad (4.14)$$

provided with the quotient topology (when $L_1(\Phi;\ U)$ and $L_1(\Phi;\ Y')$ are provided with the weak topologies

$$\sigma\Big(L_1(\phi;\ U);\ \phi(-\infty,\ 0;\ U')\Big);\ \sigma\Big(L_1(\Phi;\ Y');\ \Phi(-\infty,\ 0;\ Y)\Big).$$

Thus X and Z are <u>locally convex Hausdorff T.V.S.</u>

Let us set

$$M = \mathrm{Ker}\ \check{H} \subset L_1(\Phi;\ U) \tag{4.15}$$

$$N = \mathrm{Ker}\ \check{H} \subset L_1(\Phi;\ Y')$$

and

$$M^{\perp} = \{u_*(\cdot)\ |\ [u_*(\cdot),\ \tilde{u}] = 0 \qquad \forall\ \tilde{u} \in M\} \tag{4.16}$$

$$N^{\perp} = \{y(\cdot)\ |\ [y(\cdot),\ \tilde{y}_*] = 0, \qquad \forall\ \tilde{y}_* \in N\}.$$

It is known (cf. BOURBAKI) that there exists an algebraic isomorphism between the dual of X (respectively Z) and $M^{\perp}$ (respectively $N^{\perp}$). Furthermore the topologies on X and Z coincide with the topologies $\sigma(X,\ M^{\perp})$ and $\sigma(Z,\ N^{\perp})$. One can define a duality pairing between X and Z by setting

$$\phi([\tilde{u}],\ [\tilde{y}_*]) = K\ (\tilde{u},\ \tilde{y}_*) \tag{4.17}$$

and mappings

$$\alpha \in L\Big(\sigma(X,\ M^{\perp});\ \sigma(N^{\perp},\ Z)\Big) \tag{4.18}$$

$$\beta \in L\Big(\sigma(Z,\ N^{\perp});\ \sigma(M^{\perp},\ X)\Big)$$

such that

$$\Big[\alpha[\tilde{u}],\ [\tilde{y}_*]\Big] = \phi([\tilde{u}],\ [\tilde{y}_*]) \tag{4.19}$$

$$\Big[\beta[\tilde{y}_*],\ [\tilde{u}]\Big] = \phi([\tilde{u}],\ [\tilde{y}_*]).$$

the mappings α and β are injective. Hence

(4.20) X is algebraically isomorphic to a subspace of $N^{\perp}$ (dual of Z) = Im α, dense in $\sigma(N^{\perp},\ Z)$

Z is algebraically isomorphic to a subspace of $M^{\perp}$ (dual of X) = Im β, dense in $\sigma(M^{\perp},\ Z)$.

Considering now the canonical factorisations

$$\begin{array}{ccc} L_1(\Phi;\ U) & \xrightarrow{\ \check{H}\ } & \Phi(-\infty,\ 0;\ Y) \\ {\scriptstyle b}\downarrow & & \uparrow{\scriptstyle a} \\ X & \xrightarrow{\ \alpha\ } & Z' = N^{\perp} \end{array} \tag{4.21}$$

$$\begin{array}{ccc} L_1(\Phi;\, Y') & \xrightarrow{\tilde{H}^*} & \Phi(-\infty,\, 0;\, U') \\ {\scriptstyle b_*}\searrow & & \nearrow{\scriptstyle a_*} \\ Z & \xrightarrow{\beta} & X' = M^{\perp} \end{array}$$

We have from (4.19) and (4.21)

(4.22) $a_* = b^* \beta$

$a^* = \beta\, b_*$, (upper star denotes transpose with respect to the first diagram)

Let us then define

(4.23) $F[\tilde{u}] = [\Delta\, \tilde{u}] = b(\Delta\, \tilde{u})$

$G\, u = [u\, \delta_0]$

$H[\tilde{u}] = f(\tilde{u})$

(4.24) $A[\tilde{y}_*] = [\Xi\, \tilde{y}_*] = b_*(\Xi\, \tilde{y}_*)$

$B\, y_* = [y_*\, \delta_0]$

$C\, \tilde{y}_* = g\, (\tilde{y}_*)$

The preceding mappings have the following properties

(4.25) $F \in L\big(\sigma(X,\, M^{\perp});\ \sigma(X,\, M^{\perp})\big)$

$G \in L\big(\sigma(U,\, U');\ \sigma(X,\, M^{\perp})\big)$

$H \in L\big(\sigma(X,\, M^{\perp});\ \sigma(Y,\, Y')\big)$

(4.26) $A \in L\big(\sigma(Z,\, N^{\perp});\ \sigma(Z,\, N^{\perp})\big)$

$B \in L\big(\sigma(Y,\, Y);\ \sigma(Z,\, N^{\perp})\big)$

$C \in L\big(\sigma(Z,\, N^{\perp});\ \sigma(U',\, U)\big)$

(4.27) $\beta\, A = F^*\, \beta$

$C = G^*\, \beta$

$\beta\, B = H^*$.

The evolution of the dual systems is described by the equations

(4.28) $\frac{dx}{dt} = Fx(t) \qquad t \geq 0$

$\frac{dz}{dt} = Az(t) \qquad t \geq 0$

$$(4.29) \qquad x_{-\phi'} = Fx_{\phi} + G\tilde{u}(\phi) \qquad \forall \phi \in \Phi$$

$$z_{-\phi'} = Az_{\phi} + B\tilde{y}_*(\phi) \qquad \forall \phi \in \Phi$$

where the derivatives in (4.28) must be taken in $\sigma(X, M^{\perp})$ and $\sigma(Z, N^{\perp})$ and in (4.29), x_{ϕ} and z_{ϕ} are vector distributions with values in $\sigma(X, M^{\perp})$ and $\sigma(Z, N^{\perp})$.

4.3 Algebraic Properties of Dual Systems

The space $L_1(\Phi; U)$ can be provided with an external multiplication as we have seen before. It can be shown that

$$\tilde{u} \longmapsto \mu*u \in L\Big(\sigma\big(L_1(\Phi; U), \Phi(-\infty, 0; U')\big); \sigma\big(L_1(\Phi; U), \Phi(-\infty, 0; U')\big)\Big)$$

and therefore $L_1(\Phi; U)$ remains a unitary Φ'-module with respect to the weak topology.

Our output function space will be taken to be $\Phi(-\infty, 0; Y)$ and $\Phi(-\infty, 0; U')$ equipped with the weak topologies. It can be shown that these spaces can be provided with the structure of a unitary topological Φ'-module.

If $\check{H}$ and $\check{H}^*$ are the Hankel operators of the dual systems then it can be shown that $\check{H}: L_1(\Phi: U) \longrightarrow \Phi(-\infty, 0; Y)$ and $\check{H}^*: L_1(\Phi; Y') \longrightarrow \Phi(-\infty, 0; U')$ are Φ'-module homomorphisms.

5. APPROACH USING SOBOLEV SPACES

Introduction

We now show that working with a different input function space we can obtain a canonical internal representation where the evolution of the state can be described by an operational differential equation involving unbounded operators on a Hilbert space. The algebraic structure is however completely lost.

5.1 Notation and Assumptions

For $m \geq 1$, integer, let us set

$$(5.1) \qquad \Phi = H^m(-\infty, 0) = \{\phi \in L^2(-\infty, 0) \Big| \frac{d^j\phi}{dt^j} \in L^2(-\infty, 0), \quad j = 1, 2, \ldots m\}$$

In (5.1) the $\frac{d^j\phi}{dt^j}$ are taken in the sense of distributions. The space $H^m(-\infty, 0)$ is a Hilbert space for the norm

$$(5.2) \qquad \left(||\phi|| = |\phi|^2_{L^2(-\infty, 0)} + \sum_{j=1}^{m} \left|\frac{d^j\phi}{dt^j}\right|^2_{L^2(-\infty, 0)} \right)^{\frac{1}{2}}$$

and is called the Sobolev space of order m.

Let U and Y be separable Hilbert spaces. We denote by $\Phi(U')$ the space $H^m(-\infty, 0; U')$, that is

(5.3) $$\Phi(U') = \{u_*:(-\infty, 0) \to U' \mid u_*, \frac{d^j u_*}{dt^j} \in L^2(-\infty, 0; U'), 1 \leq j \leq m\}$$

The space $\Phi(U')$ is a Hilbert space for the norm

(5.4) $$||u_*|| = \left(|u_*|^2_{L^2(-\infty, 0:U')} + \sum_{j=1}^{m} \left|\frac{d^j u_*}{dt^j}\right|^2_{L^2(-\infty, 0;U')} \right)^{\frac{1}{2}}$$

The dual space $\Phi(U)'$ ' will be taken to be the input function space. It can be shown

Proposition 5.1:

(5.5) $$\Phi(U') = L_2(\Phi'; U')$$

(5.6) $$\left(\Phi(U')\right)' = L_2(\Phi; U)$$

where $L_2(\Phi; U)$ $\left(\text{resp. } L_2(\Phi'; U')\right)$ denotes the space of Hilbert Schmidt Operators from $\Phi \to U$ (resp. $\Phi' \to U'$).

Comparing with the distribution theory approach previously presented, we see that the space $L_1(\Phi; U)$ has been extended to $L_2(\Phi; U)$ with the major advantage that it is now a Hilbert space.

In view of the above proposition every $\tilde{u} \in \left(\Phi(U')\right)'$ has a representation of the form

(5.7) $$\tilde{u} = \sum_n \alpha_n \mu_n u_n, \text{ where}$$

$\sum_n \alpha_n^2 < +\infty$, μ_n is an orthonormal basis of Φ' and u_n an orthonormal basis of U'. For our purposes a different representation is more useful. With the aid of the mapping

$$z \longmapsto \left(z, \frac{dz}{dt}, \ldots, \frac{d^m z}{dt^m}\right),$$

the space $H^m(-\infty, 0; U')$ can be identified with a closed subspace of $L^2(-\infty, 0; U')^m$. By the Hahn Banach theorem, every element $\tilde{u} \in \left(H^m(-\infty, 0; U')\right)'$ can be represented (in a non-unique way) by

(5.8) $$< \tilde{u}, z > = \int_{-\infty}^{0} < u_o(\sigma), z(\sigma)> d\sigma + \sum_{i=1}^{m} \int_{-\infty}^{0} < u_i(\sigma), \frac{d^i z}{d^i}(\sigma) > d\sigma$$

where $u_o, u_1, \ldots, u_m$ belongs to $L^2(-\infty, 0; U')$.

On $\left(\Phi(U')\right)'$, we define a semi-group of translations by transposition, by setting

(5.9) $$< T(t)u, z > = < \tilde{u}, \left(T(t)\right)'z >, \qquad t \geq 0$$

where

(5.10) $$T(t)'z(s) = z(-t+s), \qquad s \leq 0.$$

It can be verified that $T(t)'$ is an equi-continuous semi-group of Class C° on $\Phi(U')$ and since $\Phi(U')$ is a Hilbert space, $T(t)$ is also an equi-continuous semi-group of class C° (cf. YOSIDA, p. 233 and 212).

5.2 External Representation of a Linear Time Invariant System

Let $C(0,\infty;\ Y)$ denote the space of continuous and bounded mappings from $[0,\ \infty) \to Y$ which is a Banach space for the sup norm.

A linear time invariant system is given by

$$f \in \mathcal{L}\Big(\big(\Phi(U')\big)';\ Y\Big). \tag{5.11}$$

We define the Hankel operator by

$$(\mathcal{H}\tilde{u})(t) = f\big(T(t)\tilde{u}\big), \qquad t \geq 0 \tag{5.12}$$

and $\mathcal{H} \in \mathcal{L}\Big(\big(\Phi(U')\big)';\ C(0,\ \infty;\ Y)\Big)$.

Let

$$f^*: Y' \to \Phi(U') \tag{5.13}$$

denote the transpose of f.

Let us set

$$K^*(-t)y_* = (f^* y^*)(t) \qquad t \leq 0. \tag{5.14}$$

Now $f^* y^* \in H^m(-\infty,\ 0;\ U')$ and

$$\|K^*(\cdot)y_*\|_{H^m(-\infty,\ 0;\ U')} \leq c\|y_*\|_{Y'} \tag{5.15}$$

Since $u\delta_{-t}$ (δ_{-t} is the Dirac measure concentrated at $-t$) belongs to $\big(H^m(-\infty,\ 0;U')\big)'$, we can show that

$$K(t) \in \mathcal{L}(U;\ Y) \text{ and}$$

$$\|K(t)\| \leq c \tag{5.16}$$

$$< K(-t)u,\ y_* > \in H^m(-\infty,\ 0) \qquad \forall\, u,\ y_*. \tag{5.17}$$

$K(t)$ is called the <u>kernel</u> or the <u>impulse response</u> of the system. Using the representation for the input functions, we obtain a representation for f as

$$< f(\tilde{u}),\ y_* > = \int_{-\infty}^{0} < u_o(t),\ K^*(-t)y_* > dt + \sum_{i=1}^{m} \int_{-\infty}^{0} < u_i(t),\ \frac{d^i}{dt^i} K^*(-t)y_* > dt. \tag{5.18}$$

5.3 Canonical Internal representation

We introduce the state space

(5.19) $X = \left(\Phi(U')\right)'/\mathrm{Ker}\, \mathcal{H}$

which is a Hilbert space for the quotient topology. Let us next define as usual

(5.20) $Gu = [u\delta_0]$

(5.21) $H[\tilde{u}] = f(\tilde{u})$

(5.22) $\Gamma(t)[\tilde{u}] = [T(t)\tilde{u}]$

then $G \in L(U; X)$, $H \in L(X; Y)$, and $\Gamma(t)$ is an equicontinuous semi group of class C_0 on X. We then get

(5.23) $K(+t) = H\,\Gamma(t)\,G \qquad t \geq 0.$

We will set as usual

(5.24) $b\tilde{u} = [\tilde{u}]$

(5.25) $a[\tilde{u}] = \mathcal{H}\tilde{u}$

and $b \in L\left(\left(\Phi(U')\right)'; X\right)$, $a \in L\left(X; C[0, +\infty; Y)\right)$. b and a are called the reachability and observability operators respectively.

We can prove the following theorem:

Theorem 5.1: Under the assumptions of the previous sections, the canonical internal representation of the linear time-invariant system given by f as in (5.11) is given by (U, Y, X, F, G, H), where U, Y, X are Hilbert spaces, $G \in L(U; X)$, $H \in L(X, Y)$, F is an unbounded operator on X which is the infinitesimal generator of an equicontinuous semi-group $\Gamma(t)$ of class C° on X. The Hankel Operator $\mathcal{H}$ admits the factorization

$\mathcal{H} = a \circ b$, where

$a \in L\left(X; C(0, \infty; Y)\right)$ is injective and $b \in L\left(\left(\Phi(U')\right)': X\right)$ is surjective.

The evolution of the system is described by

(5.26) $\dfrac{dx}{dt} = Fx(t) \qquad \forall\, t \geq 0,$ if $x(0) = b\tilde{u} \in D(F)$; for $t < 0$,

(5.27) $\displaystyle\int_{-\infty}^{0} \frac{dx}{dt}\,\phi\, dt = F\int_{-\infty}^{0} x\,\phi\, dt + G\tilde{u}(\phi) \qquad \forall\, \phi \in H^m(-\infty, 0)$

in the sense of vector distributions with values in X, and $\displaystyle\int_{-\infty}^{0} x\,\phi\, dt \in D(F)$, $\displaystyle\int_{-\infty}^{0} \frac{dx}{dt}\,\phi\, dt \in X$, when $x_\phi(\tilde{u})$ is given by

$$(5.28) \qquad x_\phi(\tilde{u}) = \int_{-\infty}^{0} ds\, \Gamma(-s)G\left[\int_{-\infty}^{0} u_0(t)\phi(t-s)dt + \sum_{i=1}^{m}\int_{-\infty}^{0} u_i(t)\phi^{(i)}(t-s)dt\right],$$

$$\phi \in \mathcal{D}(-\infty, 0);$$

$$(5.29) \qquad (\mathcal{H}\,\tilde{u})(t) = y(t) = Hx(t), \qquad t \geq 0.$$

5.4 The State Space Isomorphism Theorem

In this case we can prove a state space isomorphism theorem. The reason for being able to do this is that the reachability operator being onto is right invertible and since we are in a Hilbert space a right inverse (non-unique) is also continuous. This allows us to construct the required topological isomorphism.

Theorem 5.2: Let $(\Gamma(t), G, H)$ and $(\Gamma_1(t), G_1, H_1)$ be two reachable and observable realizations realizing the same kernel $K(t)$. Then there exists a boundedly invertible transformation $\chi: X \to X_1$ (X and X_1 being the corresponding state spaces) such that the following diagram is commutative

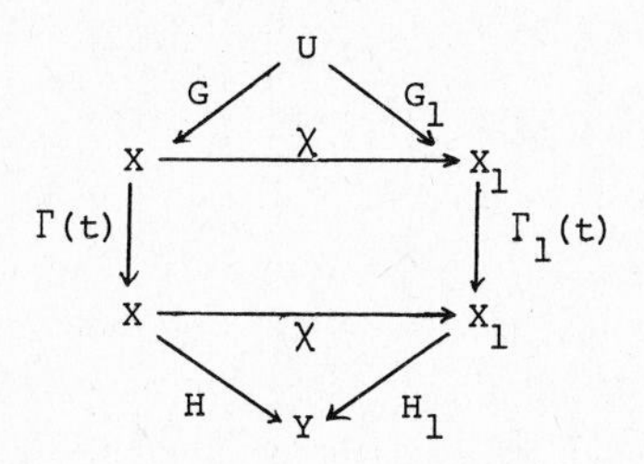

5.5 Inputs in $L^2(-\infty, 0; U')$

It is known that

$$L^2(-\infty, 0; U') \subset \left(H^m(-\infty, 0; U')\right)'$$

the canonical injection being continuous and having dense image. Therefore setting

$$(5.30) \qquad \tilde{X} = \{b(\tilde{u}) \mid \tilde{u} \in L^2(-\infty, 0; U')\},$$

$\tilde{X}$ is a dense subspace of X.

We can then write

$$(5.31) \qquad b(\tilde{u}) = \int_{-\infty}^{0} \Gamma(-t)G\tilde{u}(t)dt = x(0).$$

Then we can write (5.27) in a stronger sense. Indeed, we have

$$(5.32) \qquad \int_{-\infty}^{0} \frac{dx}{dt}\phi\, dt = F\int_{-\infty}^{0} x\,\phi\, dt + \int_{-\infty}^{0} G\tilde{u}(t)\,\phi(t)\, dt \qquad \forall \phi \in \Phi.$$

But

$$\int_{-\infty}^{0} x\,\phi\,dt = \int_{-\infty}^{0} \phi(t)\left[\int_{-\infty}^{t} \Gamma(t-s)G\tilde{u}(s)\,ds\right]$$

and hence

$$x(t) = \int_{-\infty}^{t} \Gamma(t-s)G\tilde{u}(s)\,ds \qquad t \leq 0. \tag{5.33}$$

It also follows from (5.32), $\forall\, x_* \in D(F^*)$

$$\frac{d}{dt} < x_*, x(t) > \; = \; < F^*x^*, x(t) > + < x^*, G\tilde{u}(t) > \quad \text{a.e.t.} \tag{5.34}$$

and if $G\tilde{u}(t) \in D(F)$ a.e., then $x(t) \in D(F)$, $\forall\, t \leq 0$ and

$$\frac{dx}{dt} = Fx + G\tilde{u} \quad \text{a.e.} \quad t < 0. \tag{5.35}$$

Equations (5.34) and (5.35) are operational differential equations involving unbounded operators which have been extensively studied in the literature on partial differential equations.

Moreover in this situation we capture the fact that the reachable space of most infinite dimensional systems will usually be only dense in the state space.

References

AUBIN, J.P, and A . BENSOUSSAN, to appear.

BALAKRISHNAN, A.V., Foundations of the State Space Theory of Continuous Systems I, Journal of Computer and System Sciences, 1967.

BARAS, J.S., R.W. BROCKETT, and P.A. FUHRMANN, State-Space Models for Infinite-Dimensional Systems, IEEE Transactions on Automatic Control, Vol. AC-19, No. 6, December 1974.

BENSOUSSAN, A., M.C. DELFOUR, and S.K. MITTER, Representation and Control of Infinite Dimensional Systems, monograph to appear.

BOURBAKI, N., Espaces Vectoriels Topologiques, Chapitre IV, Hermann, Paris, 1964.

BROCKETT, R.W., Finite Dimensional Linear Systems, Wiley, New York, 1970.

EILENBERG, S., Automata, Languages, and Machines, Vol. I, Academic Press, 1974.

JOHNSTON, R., Linear Systems Over Various Rings, Ph.D. dissertation, M.I.T., 1973.

KALMAN, R.E., P.L. FALB, and M.A. ARBIB, Topics in Mathematical System Theory, McGraw Hill, New York, 1969.

KALMAN, R.E. and M.L.J. HAUTUS, Realization of Continuous-time Linear Dynamical Systems: Rigorous Theory in the Style of Schwartz, in Ordinary Differential Equations, ed. L. Weiss, Academic Press, New York, 1972.

KALMAN, R.E. and T. MATSUO, to appear.

KAMEN, E.W., On an Algebraic Theory of Systems Defined by Convolution Operators, Math. Systems Theory, Vol. 9, 1975, pp. 57-74.

ROUCHALEAU, Y. and B.F. WYMAN, Linear dynamical systems over integral domains, J. Comp. Syst. Sci., 9: 129-142, 1975.

ROUCHALEAU, Y., B.F. WYMAN, and R.E. KALMAN, Algebraic structure of linear dynamical systems. III. Realization theory over a commutative ring, Proc. Nat. Acad. Sci. (USA), 69: 3404-3406, 1972.

SONTAG, E.D., Linear Systems over Commutative Rings: A Survey, to appear in Ricerche di Automatica.

TREVES, F., Topological Vector Spaces, Distributions and Kernels, Academic Press, New York, 1967.

ACKNOWLEDGEMENT:

This research was supported by the Air Force Office of Scientific Research under Grant AFOSR 72-2273 and the National Science Foundation under Grant NSF GK41647.

L^2 SYSTEMS THEORY : SOME APPLICATIONS

P. DEWILDE
Div. of Applied Mathematics
Kath. Univ. te Leuven
Louvain - Belgium

1. ABSTRACT

Some applications of L^2 systems theory are presented in this paper. We will show how the technique of coprime factorization on which the L^2 systems theory rests can be used directly to solve problems in other areas e.g. the computation of inner embeddings (which amounts to what is known as a Oono-Yasuura synthesis in network terms) and the computation of spectral factors. Two main theorems are proved, each one showing that a coprime factorization in a specific setting provides for an embedding or a spectral factorization.

2. INTRODUCTION

L^2 systems theory has now a fairly large body of results, at least in the case where the state module can be annihilated (called the "roomy" case). The theory was developed by a number of authors either from an input-output point of view [1], [2], [3] or from an intrinsic point of view [4], [5], [6], [7]. In this paper, we will need results derived mainly from the input-output point of view. First, the main theorems used are summarized.

Let T be the time set (here we will take $\mathbb{T} = \mathbb{R}$), and let $L^2_{\mathbb{C}^k}$ (T) be the Hilbert space of $\mathbb{C}^k$- valued functions on $\mathbb{T}$. Moreover let $\mathcal{S} : L^2_{\mathbb{C}^m} \to L^2_{\mathbb{C}^n}$ be an input-output map which is bounded, linear and time-invariant. Then $\mathcal{S}$ is represented by a transfer function S(p) which belongs to $H^\infty_{n\times m}$, the Hardy space of nxm matrices, whose entries are L^∞ on the imaginary axis and can be extended analytically and with uniform bound to the open right half complex plane (ORP).

Using, further, the Hardy spaces H^2_k as Fourier transforms of $L^2_{\mathbb{C}^k}$ (o,∞), we can define

$$(2.1) \qquad \mathfrak{M}_1 = \left\{F : F \in H^2_m , S(-j\omega) F \in H^2_n\right\}$$

$$(2.3) \qquad \mathfrak{M}_2 = \left\{F : F \in H^2_n , \tilde{S}(j\omega) F \in H^2_m\right\}$$

$\mathfrak{M}_1$ ($\mathfrak{M}_2$) is an "invariant subspace" of H^2_m (H^2_n) in the sense that $\phi\mathfrak{M}_1 \subset \mathfrak{M}_1$ for all $\phi \in H^\infty$. Physically, these subspaces represent the "nullspaces" of the system, members of which can be thought of as Fourier transforms of functions in the input space Ω_o = $L^2_{C^m}$ $(-\infty,o)$ which generate the zero state at time t = o. By the Beurling-Lax theorem, we have that $\mathfrak{M}_1 = U_1 H^2_k$ for some $k \leqslant m$, and some $U_1 \in H^\infty_{mxk}$ which is isometric ($\tilde{U}_1 U_1 = 1_k$).

If k = m, then $\mathfrak{M}_1$ is said to have full range, and $\mathfrak{M}_1 = U_1 H^2_k$ where U_1 is unitary. In that case, $\mathfrak{M}_2 \subset H^2_n$ also has full range and U_2 is also unitary. Systems with this property are called roomy.

Next, it is clear that $S(-j\omega)\, U_1(j\omega) = \Delta_1(j\omega)$ is analytic and we have, for roomy S, that

$$(2.4) \qquad S(-j\omega) = \Delta_1(j\omega)\, U_1^{-1}(j\omega)$$

and

$$(2.5) \qquad \tilde{S}(j\omega) = \Delta_2(j\omega)\, U_2^{-1}(j\omega) .$$

In these factorizations, Δ_1 and U_1 are right coprime, meaning that there are matrices M_1 and N_1 in appropriate H^∞ spares, so that $M_1\Delta_1 + N_1 U_1$ is outer [2] (we use "outer" in the sense of H^2_n theory : see [8], [9].

Likewise Δ_2 and U_2 are left coprime. Coprimeness in the sense used here means that there is no cancelling inner (i.e. H^∞ and unitary) factor common to the two terms, either at the right or the left. U_1 and U_2 both carry complete state information.

Their determinants are inner functions which can have a Blashke part and a singular part and whose complexity can be measured accordingly. In fact, in comparing two systems $\mathcal{S}^{(1)}$ and $\mathcal{S}^{(2)}$ with $U_1^{(1)}$ and $U_1^{(2)}$, it is perfectly reasonable and a generalisation of the finite dimensional case, to say that $\mathcal{S}^{(1)}$ has higher (generalized) degree than $\mathcal{S}^{(2)}$ if det $U_1^{(1)}$ is an H^∞ multiple of det $U_2^{(1)}$ [2]. We have the important property [2], [10] that det $U_1(-j\omega)$ = det $U_2(j\omega)$, so that the degree measured one way or the other is equivalent.

In the sequel we will not only need this type of factorization, but also another type which we now preceed to introduce. Let

$$J = \begin{bmatrix} 1_n & 0_n \\ 0_n & -1_n \end{bmatrix} \tag{2.6}$$

For a 2n x 2n matrix Σ, partitioned as follows :

$$\Sigma = \begin{bmatrix} \Sigma_{11} & \Sigma_{12} \\ \Sigma_{21} & \Sigma_{22} \end{bmatrix} \tag{2.7}$$

and with Σ_{21} not identically singular, we can define the Julia transform

$$\Theta = \begin{bmatrix} \Sigma_{21}^{-1} & -\Sigma_{21}^{-1}\Sigma_{22} \\ \Sigma_{11}\Sigma_{21}^{-1} & \Sigma_{12} - \Sigma_{11}\Sigma_{21}^{-1}\Sigma_{22} \end{bmatrix} \tag{2.8}$$

Σ is contractive in the ORP (and hence belongs to H^∞) if and only if Θ is J-expansive, meaning that

$$\tilde{\Theta} J \Theta - J \geqslant 0, \tag{2.9}$$

in the ORP. Also, Σ is unitary, if and only if Θ is J-unitary. Given a 2n x n matrix $\Theta(p)$ defined in the ORP, we will say that Θ is J-passive if it satisfies

$$\tilde{\Theta} J \Theta - 1_n \geqslant 0, \quad \text{in ORP.} \tag{2.10}$$

Likewise, it will be called J-antipassive if it is defined in the OLP and satisfies (2.10) there. Suppose a 2nxk matrix A(p) is given, then we will say that A(p) is J-factorizable if there exists a passive and J-unitary matrix Θ such that

$$A(p) = \Theta \cdot A_1 \qquad (2.10)$$

with A_1 J-antipassive and such that for any factorization $A(p) = \Theta' \cdot A_1'$ with Θ' passive, J-unitary and A_1' J-antipassive we have that Θ' is a right multiple of Θ in the multiplicative group of J-unitary passive matrices. It is clear that (2.10) is the analog of our preceeding coprime factorization situation, but this time with J-unitary matrices.

3. MAIN RESULTS

Suppose a contractive transfer matrix S(p) is given. Then S(p) always has a spectral cofactor Σ_{21} such that (* is adjoint)

$$S^*S = 1_n - \Sigma_{21}^* \Sigma_{21} \qquad (3.1)$$

a.e. on the imaginary axis [9]. Σ_{21} can always be chosen outer. In that case, we have :

Theorem 1. [2], [3]

When S is roomy, then the nullspace $\mathfrak{M}_1^{(1)}$ of the outer spectral factor Σ_{21} of S contains the nullspace $\mathfrak{M}_1$ of S.

Corollary 1

Suppose $\mathfrak{M}_1^{(1)} = U_1^{(1)} H_n^2$ and $\mathfrak{M}_1 = U_1 H_n^2$. Then U_1 is a right multiple of $U_1^{(1)}$ in the multiplicative group of nxn inner matrices. Also it follows that det U_1, is a multiple of det $U_1^{(1)}$ and hence, the generalized degree of Σ_{21} is lower than the generalized degree of S.

The main question to be adressed is whether we can produce an algorithm to compute Σ_{21} out of S. Such an algorithm can be deduced from the following theorem :

Theorem 2. [2], [3].

Given an nxn roomy scattering matrix S, then the following matrix

$$\begin{bmatrix} (1_n - S^* S)^{-1} \\ \\ S(1_n - S^* S)^{-1} \end{bmatrix} \tag{3.2}$$

is J-factorizable and its J-cofactor Θ is the Julia transform of the minimal inner embedding Σ of S.

A third theorem can be shown which is the counterpart of theorem 2 in a simpler setting. Let

$$\sigma = \begin{bmatrix} S \\ \\ \Sigma_{21} \end{bmatrix}$$

where Σ_{21} is an outer cofactor for S. Then we have :

Theorem 3. [2], [3].

Suppose U_1 is an inner function representing the nullspace $\mathfrak{M}_1$ of S. Then coprime factorizations of σ are given by :

$$\sigma(-j\omega) = \Delta_1 U_1^{-1} = \Sigma(-j\omega) \begin{bmatrix} 1_n \\ \\ 0_n \end{bmatrix}$$

where U_1 is the right inner factor and the minimal embedding $\Sigma(-j\omega)$ the left one.

Corollary : Clearly, the generalized degree of S (which is taken as det U_1) is equal to the generalized degree of Σ, since det U_1^{-1}= det $\Sigma(-j\omega)$.

4. APPLICATIONS

It follows by theorem 2 that both an inner embedding and a spectral factorization can be obtained through a coprime factorization. More specifically, suppose a positive and parahermitian function R(p) is given such that $R(p) = \frac{1}{2}(Z(p) + Z_*(p))$ (where Z_* is the parahermitian conjugate of Z), and such that Z(p) is analytic in ORP. Then, because of Theorem 2, we have that the matrix

$$(4.1) \qquad A = \frac{1}{2}\begin{bmatrix} (Z+1)\ R^{-1} \\ (Z-1)\ R^{-1} \end{bmatrix}$$

is J-factorizable, with

$$(4.2) \qquad A = \Theta \begin{bmatrix} H_*^{-1} \\ 0_n \end{bmatrix} = \begin{bmatrix} \Theta_{11} & \Theta_{12} \\ \Theta_{21} & \Theta_{22} \end{bmatrix} \begin{bmatrix} H_*^{-1} \\ 0_n \end{bmatrix}$$

where Θ is the minimal J-unitary left cofactor, and $H^{-1} = \Theta_{11} - \Theta_{21}$. The facts described above show that a coprime factorization of an appropriate matrix produces at the same time a spectral factorization.

We will now show that the above leeds to a complete algorithm for the case that R(p) is rational. The algorithm to be detailed uses what has been called "J-unitary elementary factors". These are J-unitary rational J-unitary and passive matrices of the first (Smith-Mc Millan) degree. They have one of the following forms :

$$(1) \qquad 1_{2n} + \frac{2\alpha_o\, u\, \tilde{u}\, J}{p - p_o} \qquad \text{(type 1)}$$

where $p_o \in C$, $\alpha_o = \text{Re}\ p_o < 0$

u is a 2n-vector with $\tilde{u}\, J\, u = 1$

(2) $1_{2n} - p\, r\, u\, \tilde{u}\, J$ or $1_{2n} - r.\ \dfrac{1 - p \cdot j\omega_o}{p - j\omega_o}\, u\, \tilde{u}\, J$ (type 2)

where $r \in R$, $r > 0$

$j\omega_o$ is a point on the imaginary axis

u is 2n-vector with $\tilde{u}\, J\, u = 0$.

(3) $1_{2n} + \dfrac{2\alpha_o\, u\, \tilde{u}\, J}{p - p_o}$, $\alpha_o = \mathrm{Re}\, p_o > 0$ (type 3)

u is a 2n-vector with $\tilde{u}\, J\, u = -1$

It is easily seen that elementary factors of type (1) and (3) have a simple pole at p_o and a single zero at $p = -p_o^*$, while the first factor in (2) has a pole and a zero at $p = \infty$ and the second a pole and zero at $p = j\omega_o$.

The algorithm then goes as follows :

(1) Separate the ORP and OLP poles in R(p) by computing Z(p) with poles in OLP and such that $R(p) = \frac{1}{2}\,(Z_*(p) + Z(p))$ where $Z_*(p) = \tilde{Z}(-p^*)$. This can be done e.g. through partial fraction expansion.

(2) Determine the matrix

$$\theta(p) = \begin{bmatrix} (Z + 1)\, R^{-1} \\ (Z - 1)\, R^{-1} \end{bmatrix} \tag{4.3}$$

and order its poles in the OLP and on the imaginary axis in an arbitrary fashion $(\gamma_1, \gamma_2, \ldots \gamma_\delta)$. Imaginary axis poles are always double, and listed accordingly.

(3) Extract J-unitary passive elementary factors one by one, one for each pole in the OLP and one for each double imaginary axis pole, as follows :

3a. If γ_k is in OLP, a factor can be extracted of type 1. γ_k is to be used as a pole (i.e. p_o is γ_k) and u has to be determined according to the rules for minimal factorization [11], [12].

3b. If γ_k is on the imaginary axis, a factor can be extracted of type 2. γ_k is to be used as a pole (i.e. $j\omega_o$ is γ_k)

and u has to be determined according to the rules for minimal factorization [11], [12].

(4) After repeating (3) δ times we have obtained :

$$\theta = \Theta_1\Theta_2 \cdots \Theta_\delta\, \theta_1$$

where Θ_k is an elementary factor and θ_1 has the form :

$$\theta_1 = \begin{bmatrix} \theta_{11} \\ \\ \theta_{21} \end{bmatrix} \qquad (4.4)$$

with $\theta_{11} = C_{11}\, H_*^{-1}$; $\theta_{21} = C_{21}\, H_*^{-1}$, and C_{11}, C_{21} constant matrices such that

$$\tilde{C}_{11}C_{11} - \tilde{C}_{21}C_{21} = 1_n \qquad (4.5)$$

The matrix $[\, C_{11}^T \; C_{21}^T \,]^T$ can be augmented to a 2n x 2n J-unitary matrix Θ_o and we have :

$$\phi = \Theta_1 \cdots \Theta_\delta \Theta_o \cdot \begin{bmatrix} H_*^{-1} \\ \\ O_n \end{bmatrix} \qquad (4.6)$$

which is the desired J-coprime factorization, with

$$\Theta = \Theta_1 \cdots \Theta_\delta \Theta_o \,.$$

Also H_* and $H = \theta_{11} - \theta_{21}$ have been obtained in the process.

The preceeding algorithm can be extended to a more general case, namely when R(p) can be written as a rational function of p, $e^{\lambda_1 p}$, $e^{\lambda_2 p}$, ... $e^{\lambda_k p}$. The main theorem shows that a coprime factorization does indeed exist, but the procedure may break down because the factorizability is dependent on the correct choice of parameters in an elementary factor. At this point the extra condition to guarantee the termination of the procedure does not seem to be known.

An application in another direction is the Darlington synthesis of a network. In fact, Theorem 2 solves this point directly and an algorithm similar as the above can be set up, this time on the matrix 3.2.

Finally, Theorem 3 provides for what is known in network theory as Oono-Yasuura synthesis. Again, an algorithm as previously described can be set up, but this time unitary elementary factors have to be used instead of J-unitary.

5. CONCLUSIONS

It has been indicated in the course of this paper that the technique of coprime factorization, based on L^2-systems theory, is capable of handling - also algorithmically - several important problems in systems and network theory : spectral factorization, inner embedding, Oono-Yasuura synthesis and Darlington synthesis. Proofs of the theorems have not been given but have been referenced. It is worth mentioning how the techniques developed here are related to whas has been called "partial realizations", which have proved very useful in a variety of situations, e.g. systems realizations and stability theory [13], [14].

To conclude, a few open problems are worth to be mentioned. First the question arises whether coprime factorizations can be obtained in a more general context than L^2 systems theory. This is a matter of adequate choice for a topology and algebraic setup. No satisfactory coprime theory except for the finite case (rational case) or the L^2 theory has emerged yet. Second, in case of non-rational transfer functions, the coprime factorization remains very much an abstract tool and a generalization of the algorithm described in this paper is to be sought.

6. BIBLIOGRAPHY

1. Dewilde, P., Roomy scattering Matrix Synthesis, Techn. Rept. Dept. of Math., Univ. of California, Berkeley, 1971.

2. Dewilde P., Input - Output description of Roomy Systems, SIAM Journal on Control (to appear).

3. Baras, J. and Dewilde P., "Invariant Subspace Methods in Linear Multivariable Distributed Systems and Lumped Distributed Network Synthesis". IEEE Proceedings, Jan. 1976.

4. Baras, J.S., R.W. Brockett and P.A. Fuhrmann, "State Space Models for Infinite Dimensional Systems", IEEE Trans. on Automatic Control, Special Issue on Identification and Time Series Analysis, Vol. AC-19, No. 6, pp. 693-700, Dec. 1974.

5. Fuhrmann P.A., "Realisation Theory in Hilbert Space for a Class of Transfer Functions". J. Funct. Anal., Vol. 18, 1975, pp. 338-349.

6. Baras, J.S. and R.W. Brocket, "H^2 Functions and Infinite Dimensional Realization Theory", SIAM J. on Control, Vol. 13, No. 1, Jan. 1975, pp. 221-241.

7. Helton, J.W., "Discrete Time Systems Operator Models and Scattering Theory", J. Funct. Anal., Vol. 16, 1974, pp. 15-38.

8. Hoffman, K., Banach Spaces of Analytic Functions, Prentice Hall, Englewood Cliffs, N.J. 1962.

9. Helson, H., Lectures on Invariant Subspaces, Academic Press, New York 1964.

10. Fuhrmann, P., "Factorization Theorems for a Class of Bounded Measurable Operator Valued Functions" (to appear).

11. Dewilde, P., Cascade Scattering Matrix Synthesis, Techn. Rept., Information Systems Lab., Stanford Univ., 1970.

12. Dewilde, P., "On the factorization of a Nonsingular Rational Matrix", IEEE Trans. on Circuits and Systems, Aug. 1975, pp. 637-645.

13. Rosenbrock, "State-space and Multivariable Theory", John Wiley, 1970.

14. Callier, F.M. and C.A. Desoer, "L^p Stability ($1 \quad p \quad \infty$) of Multivariable Non-Linear Time Varying Systems that are Open Loop Unistable". Int. J. of Control, 1974, No. 1, pp. 65-72.

ALGEBRAIC IDEAS IN INFINITE DIMENSIONAL SYSTEM THEORY

Paul A. Fuhrmann
Ben Gurion University of the Negev
Beer Sheva, Israel

1. Introduction

Research in the area of infinite dimensional system theory has a relatively short history. This is not surprising inasmuch as system theory as such has started to flourish only in the last two decades. Now the passage from finite dimensional linear theory to the infinite dimensional case obviously necessitated the replacement of algebraic machinery by an analytic one. Since structure theory is so intimately related with system theory then almost without exception functional analysis and especially the theory of operators and semigroups in Banach and Hilbert spaces, as well as the theory of distributions, became the source for methods and ideas to be used in a theory of infinite dimensional systems. Here there is an immediate difficulty. Structure theory for the general bounded operator in a Hilbert space, not to say Banach spaces, is an area in which not much is known. Thus research has been concentrated in special classes of operators, compact, selfadjoint, normal, spectral etc. In the case of normal operators the whole story is known given by the spectral theorem and this can be applied to system theoretic problems [6]. There is another class of operators which have been studied extensively in recent years and for which a structure theory has been developed which has so much resemblance to the finite dimensional case that it is natural to develop system theory in that context. We refer here to the class of C_0 contractions studied in detail by Sz.-Nagy and Foias [39] consisting of those contraction operators T for which $\phi(T) = 0$ for some nonzero bounded analytic function in the unit disc. In particular we will focus on the subclass of C_0 of restricted shift operators. Thus a realization theory can be developed via the theory of Hardy spaces, invariant subspaces and canonical models. The object of this paper is to point out the relation and great similarity between that theory and the theory of finite dimensional linear systems.

The invariant subspace approach to linear systems has been originated by several researchers. Some of the first results in this direction go back to Balakrishnan's papers [1,2]. More recently we want to point out the work of Helton [22,23], Baras and Brockett [3], DeWilde [7] and the author [13-18]. Most of this work is based on relatively recent advances in operator theory of which the starting points are several. They can be traced to Livsic's introduction of the concept of characteristic function as a tool in the study of nonselfadjoint operators [30,31], Beurling's characterization the invariant subspaces of the shift operator [4], Rota's short and fundamental paper concerning the shift as a canonical model [37], the approach to scattering theory used by Lax and Phillips [29], Helson's work on invariant subspaces

[21] and the deep and extensive study of contractions, unitary dilations and canonical models of Sz.-Nagy and Foias which is summarized in [39]. Some early results of the author on spectral analysis [11,12] turned out to be applicative to system theoretic questions.

Based on our experience with the development of infinite dimensional theory more insight has been gained into the finite dimensional algebraic approach. This led to an exposition [19] of algebraic system theory based on the module theoretic approach suggested by Kalman [35] but in a way in which the state space representations and the coprime factorizations of proper rational matrix functions are synthesized in a natural way. A similar approach has been used by Eckberg [10]. It is to be hoped that the interplay between system theory, algebra and operator theory which proved so fruitful in the past might give us even better results in the future. In particular the question of pole assignment through feedback in the infinite dimensional case is almost unstudied and there seem to be only scattered results in the literature.

2. Ideal Theory

Our starting point for the subsequent development is the structure of ideals in certain rings.

In fact the notion of ideals appears in linear system theory naturally even at an elementary level. The Kalman reactability and controllability criterias for a discrete time system $\{A,B\}$ can be given as

$$\mathrm{rank}[B,AB,\dots,A^{n-1}B]=n$$

and

$$\mathrm{rank}[B,AB,\dots,A^{n-1}B]=\mathrm{rank}[B,AB,\dots,A^{n-1}B,A^n]$$

or alternately as

$$\bigcap_{i=0}^{n-1} \mathrm{Ker}\, B^*A^{*i} = \{0\}$$

and

$$\mathrm{Ker}\, A^{*n} \supset \bigcap_{i=0}^{n-1} \mathrm{Ker}\, B^*A^{*i}$$

But the conditions are nothing else but the expression of the fact that I (or A^n) are in the right ideal generated by $B,AB,\dots,A^{n-1}B$.

Given a field F we let $F[\lambda]$ denote the ring of polynomials over the field F. It is well known that $F[\lambda]$ is a principal ideal ring and thus, together with the structure theory of finitely generated torsion modules over principal ideal domains, gives a most elegant approach to the structure theory of linear operators in finite dimensional vector spaces. If we look for the right extension of polynomial rings in analysis then we naturally focus on classes of analytic functions. Of course the price we pay is that we restrict ourselves to the complex field C and its subfield of real numbers. This makes sense as locally every analytic function is uniformaly approximable by polynomials. But even more is true. If we have an algebra of complex functions on a connected, locally compact Hansdorff space X which contains constants and separates points, and for which every nonconstant function is an open map and each ideal $M_x=\{f|f(x)=0\}$ is principal then X can be given a conformal structure such that each function in the algebra is analytic on X. This beautiful result is due to I. Kra [27]. Thus the notions of analyticity is closely associated with ideals being principal.

However, if we are going to develop a meaningful theory it is necessary to restrict ourselves in some ways. The class of analytic functions studied in greatest detail seems to be the Hardy spaces H^p and more generally the Nevanlinna class N of meromorphic functions of bounded type [24,9]. Thus N is H^∞/H^∞ and for functions in N, and in particular functions in H^p $1, \leq p \leq \infty$, all have factorizations in terms of inner and outer functions and for the factors we have further either integral representations in terms of boundary values or in the form of infinite Blaschke products. This will be the analytic setup in which we will develop our system theory. We want however to mention in passing that recently a beautiful extension of the Nevanlinna theory has been developed by B. Korenblum [26] and it is possible that therein we can find tools for a further extension of the infinite dimensional theory; that would not be easy though.

Going back to H^p spaces we notice that except for the case $p=\infty$ we do not have a ring, or algebra, structure with respect to multiplication. Thus in H^∞ we can develop an ideal theory whereas in the other H^p spaces we have to talk of (right) invariant subspaces that is subspaces invariant under multiplication by the identity function χ, $\chi(z) = z$, and hence invariant under multiplication by all H^∞ functions [14]. Since the product of H^p functions by H^∞ functions stays in H^p it is clear that in algebraic terms, H^p is a module over H^∞ and invariant subspaces are nothing but submodules of H^p. Since we are interested in developing an infinite dimensional theory in a Hilbert space context we will pick out the case $p=2$ for study. Also when we talk about invariant subspaces we mean norm closed invariant subspaces except in the case $p=\infty$ where the interesting subspaces, i.e. ideals, are the w^*-closed ones. These arise naturally and we have a representation theory for them.

To sum up, in the algebraic context given a field we consider $F^n[\lambda]$ the free module with n generators of vector polynomials and the noncommutative ring $F^{n\times n}[\lambda]$ of polynomial matrices, or isomorphically matrix polynomials. Of course $F^n[\lambda]$ is a module over $F[\lambda]$ and we will need a characterization of its submodules. Similarly we want to characterize the one sided ideals in $F^{n\times n}[\lambda]$. The necessary results are summed up in the following theorem [19].

Theorem 2.1. a. A subset M of $F^n[\lambda]$ is a submodule of $F^n[\lambda]$ if and only if $M = DF^n[\lambda]$ for some polynomial matrix D in $F^{n\times n}[\lambda]$.

b. A subset J of $F^{n\times n}[\lambda]$ is a right (left) ideal in $F^{n\times n}[\lambda]$ if and only if $J = DF^{n\times n}[\lambda]$ $(J = F^{n\times n}[\lambda]D)$ for some polynomial matrix D in $F^{n\times n}[\lambda]$.

Of course there is no uniqueness for the polynomial matrix D appearing in Theorem 2.1. In the sequel we will consider mostly full submodules $F^n[\lambda]$ by which we mean that $\det D$ is not the zero polynomial. If M is a full submodule then the polynomial matrix D is unique up to multiplication on the right by a unit in $F^{n\times n}[\lambda]$.

Next we pass to the analytic case. We will have to assume familiarity with the Hardy spaces and especially the vector valued ones. The best references for details are still [21,24]. By H^2 we denote the scalar H^2 space considered as a space of analytic functions on the unit disc or alternately identified with the space of its boundary value functions on the unit circle, that is with the closed subspace of L^2 of the unit circle consisting of all functions whose negative indexed Fourier coefficients vanish. By $H^2(C^n)$ we denote the C^n valued H^2 space. Similarly $H^\infty(B(C^n,C^n))$ is the Hardy space of all bounded analytic functions whose values are linear operators in C^n. Of course $H^\infty(B(C^n,C^n))$ can be identified with the ring of $n\times n$ matrices with H^∞ entries.

Again $H^2(C^n)$ is a module over H^∞ and its invariant subspaces are closed submodules. The representation theory for invariant subspaces originated by Beurling [4] and carried forward by Lax [8] and Halmos [20] yields the following.

Theorem 2.2. A subspace M of $H^2(C^n)$ is right invariant if and only if $M = QH^2(C^n)$ where Q is a rigid function.

Now a rigid function is an $H^\infty(B(C^n,C^n))$ function bounded in norm by 1 and for which its boundary values on the unit circle, existing a.e. by virtue of Fatou's theorem, are a.e. partial isometries with a fixed initial space. The most important subclass of rigid functions is the one of inner functions in which the partial isometries are actually unitary. An invariant subspace M of $H^2(C^n)$ corresponding to an inner function is an invariant subspace of full range [21] that is a.e. the values of functions in M span C^n. For the case of Hardy spaces of finite multiplicity, the case of interest to us, a rigid function Q is inner if and only if $\det Q$ does not vanish identically, and hence is necessarily a scalar inner function.

To obtain the ideal counterpart of Theorem 2.1 we use an adaptation of an argument of Helson [21, p. 26] together with some results of Sarason [38] to obtain the following [17].

Theorem 2.3. A subset M of $H^\infty(B(C^n,C^n))$ is a w^*-closed left (right) ideal of full range if and only if it has the form $H^\infty(B(C^n,C^n))Q$ $(QH^\infty(B(C^n,C^n))$ for some inner function Q. A subset M is a w^*-closed two sided ideal of full range if and only if it has the form $qH^\infty(B(C^n,C^n))$ for some scalar inner function q.

Some remarks are in order. The notion of full range is essentially the one used by Helson. For the notions of duality needed here to introduce the w^*-topology we refer to Sarason's paper. In Theorem 2.2 and 2.3 the inner functions appearing are unique up to a multiplicative constant unitary factor on the right and on the left in the case of left ideals in $H^\infty(B(C^n,C^n))$.

3. Quotient Modules, Left Invariant Subspaces and Cannonical Models

As has been mentioned previously the structure theory for linear transformation in finite dimensional vector spaces, i.e. the Jordan canonical form, and the structure theory of finitely generated torsion modules are very closely connected. This will be exploited in this section.

Let $M = DF^n[\lambda]$ be a submodule of $F^n[\lambda]$. It is natural to study with the submodule $DF^n[\lambda]$ also the quotient module $F^n[\lambda]/DF^n[\lambda]$ consisting of the set of equivalence classes modulo the submodule $DF^n[\lambda]$ with the naturally induced algebraic operations. The quotient module $F^n[\lambda]/DF^n[\lambda]$ is also finitely generated over $F[\lambda]$ and it is a torsion module if and only if the submodule $DF^n[\lambda]$ is full [19]. Moreover the quotient module $F^n[\lambda]/DF^n[\lambda]$ is a finite dimensional vector space over F if and only if it is a torsion module over $F[\lambda]$.

Now given any finite dimensional linear transformation A, let $\phi_1,\ldots,\phi_n$ be its invariant factors. Construct a polynomial matrix D by letting

$$D = \begin{pmatrix} \phi_1 & & 0 \\ & \ddots & \\ 0 & & \phi_n \end{pmatrix} \tag{3.1}$$

then the operator A' defined in $F^n[\lambda]/DF^n[\lambda]$ by

$$A'[f] = [\chi f] \tag{3.2}$$

where $[f]$ denotes the equivalence class of an element in $F^n[\lambda]$ and χ is the identity polynomial, is an operator similar to A. Thus quotient modules of $F^n[\lambda]$ and operators of the form (3.2) are natural as canonical models. Now in general it is easier to work with representatives rather than equivalence classes, so we would like to have a way of singling out a unique representative from each equivalence class in $F^n[\lambda]/DF^n[\lambda]$. To achieve this we use the fact that any rational function has a unique decomposition into the sum of a proper rational function and a polynomial. Thus there exists a projection Π mapping a rational function into its proper rational component. Next we define π_D as the map in $F^n[\lambda]$ given by

$$\pi_D f = D\Pi(D^{-1}f) \ . \tag{3.3}$$

Clearly π_D is a projection whose kernel is $DF^n[\lambda]$ and its range K_D = Range π_D is an $F[\lambda]$ module isomorphic to $F^n[\lambda]/DF^n[\lambda]$. However whereas in the scalar case there is a unique K_D in the multidimensional case there are many different representatives, one for each D corresponding to the same submodule, with all such representatives being isomorphic.

In K_D we single out the operator $S(D)$ given by

$$S(D)f = \pi_D(\chi f) \tag{3.4}$$

These operators serve as models, up to similarity for finite dimensional operators.

A similar situation occurs in the Hilbert space case. Here the situation is however simplified inasmuch as we have an orthogonal direct sum decomposition of the space. Let M be any closed subspace of a Hilbert space H, then the quotient space H/M with the usual quotient norm is unitarily equivalent to $M^{\perp}$ the orthogonal complement of M. Thus the existence of orthogonal projections guarantees the existence of a unique representative for H/M. In the special case of right invariant subspaces of $H^2(C^n)$ their orthogonal complements are left invariant. For an inner function Q we let $H(Q)$ denote the left invariant subspace $\{QH^2(C^n)\}^{\perp}$. In $H(Q)$ we define a bounded operator $S(Q)$ by

$$S(Q)f = P_{H(Q)}(\chi f) \quad \text{for} \quad f \in H(Q) \tag{3.5}$$

Now the notion of a minimal polynomial and in general that of the invariant factors can be generalized. We note that the operator $S(Q)$ is of class C_0 as $q = \det Q$ is an H^{∞} function that annihilates $S(Q)$, i.e. $q(S(Q)) = 0$. This follows easily from the inclusion $QH^2(C^n) \supset qH^2(C^n)$ which is a direct consequence of Cramer's rule [21]. Now the set of all nonzero H^{∞} functions which annihilate a given contraction is a w^*-ideal in H^{∞} and hence of the form mH^{∞} for some inner function m. This m is called the minimal inner function of the given operator. Given a set of inner functions $m_1, \ldots, m_n$ for which m_{i+1} divides m_i we can consider the matrix inner function Q defined by

$$Q = \begin{pmatrix} m_1 & & 0 \\ & \ddots & \\ 0 & & m_n \end{pmatrix} \tag{3.6}$$

then clearly $H(Q) = H(m_1) \oplus \ldots \oplus H(m_n)$ and

$$S(Q) = S(m_1) \oplus \ldots \oplus S(m_n) \tag{3.7}$$

which is the representation of $S(Q)$ as the direct sum of cyclic operators. Operators of the form (3.7) are called by Sz.-Nagy and Foias Jordan operators. Suppose however that we start with a general inner function Q, a natural question is whether there exists a Jordan operator $S(Q_J)$ for which $S(Q)$ and $S(Q_J)$ are similar. It turns out that there is too much to expect [40] but we can get close to the finite dimensional theory. It has been shown both by Nordgren [34] and by Sz.-Nagy and Foias [39] that with each inner function Q we can associate a set of invariant factors from which the diagonal inner function Q_J can be produced as in (3.6). However the natural notion of equivalence has to be replaced by the weaker notion of quasiequivalence [34]. Obviously Q_J is to Q what the Smith canonical form is to $\lambda I-A$. In fact for an inner function Q, $\det Q$ plays very much the role of the characteristic polynomial and we will return to that later. Now for less than equivalence we get less than similarity. It has been shown by Nordgren and Moore [33] that the quasiequivalence of two inner functions Q and Q_1 is equivalent to the quasisimilarity of $S(Q)$ and $S(Q_1)$. Quasisimilarity of two operators T and T_1 is a relaxation of the notion of similarity. We mean by it the existence of operators X and Y which are one-to-one and have dense range and satisfy $XT = T_1X$ and $TY = YT_1$. In particular every operator of the form $S(Q)$ is quasisimilar to its Jordan model. The question of when is $S(Q)$ actually similar to its Jordan model is difficult. It has been solved only in the case where $S(Q)$ is cyclic [39,41].

One final remark should be made about the discrepancy between Rota's result in the use of restricted left shifts as canonical models whereas we choose in (3.5) an operator which is a restriction, or rather a compression, of the right shift to a left invariant subspace of $H^2(C^n)$. It turns out that when dealing with inner functions the distinction between restricted left and right shifts is not significant inasmuch as $S(Q)$ is unitarily equivalent to $S(\tilde{Q})^*$ which is the restriction of the left shift to $H(\tilde{Q})$ where $\tilde{Q}(z) = Q(\bar{z})^*$ [12].

4. Coprime Factorizations and Realization Theory

In Rosenbrock's book [36] it is shown that every proper rational matrix function W has factorizations of the form

$$W = D^{-1}N = N_1D_1^{-1} \tag{4.1}$$

where D and N are left coprime and D_1 and N_1 are right coprime. The factors involved are unique modulo a left unit factor in the case of D and N and a right unit factor in the case of D_1 and N_1. For coprimeness and related notions we refer to [36,32]. Of course there is a third representation for W of the form

$$W = \Omega/\psi \tag{4.2}$$

where Ω is a polynomial matrix and ψ a monic polynomial. There is a unique ψ of least degree which does the job in (4.2) and it is obtained as the least common multiple of all denominators q_{ij} of the entries w_{ij} of W where $w_{ij} = p_{ij}/q_{ij}$ is a coprime representation of w_{ij}. While there are computational approaches to such factorizations the most direct route appears to be through ideal theory.

In fact let us consider for an $m\times n$ proper rational matrix function W the following three sets:

$$J = \{\phi \in F[\lambda] \mid \phi W \in F^{m\times n}[\lambda]\} \tag{4.3}$$

$$J_L = \{P \in F^{m\times n}[\lambda] \mid PW \in F^{m\times n}[\lambda]\} \tag{4.4}$$

and

$$J_R = \{Q \in F^{n\times n}[\lambda] \mid WQ \in F^{m\times n}[\lambda] \tag{4.5}$$

then J is an ideal in $F[\lambda]$, J_L a full left ideal in $F^{m\times m}[\lambda]$ and J_R a full right ideal in $F^{n\times n}[\lambda]$. Using Theorem 2.1 about the representation of ideals the factorizations (4.1) and (4.2) are easily obtained. The uniqueness of the factors, modulo unit factors, follows from the uniqueness part of Theorem 2.1.

Essentially the same approach can be carried through in the operator valued case, however, the class of (transfer) functions has to be restricted. Thus we consider the class of strictly noncyclic functions in $H^\infty(B(C^n,C^n))$ [15,16]. Strict noncyclicity of a function A is most closely associated with the range of the Hankel operator H_A induced by it. A function is strictly noncyclic if the orthogonal complement of Range H_A is a right invariant subspace of full range, and thus has an inner function associated with it. Again this restriction seems natural (in retrospect) because of the following considerations.

One of the fundamental results of realization theory in the finite dimensional case links realizability by a finite dimensional linear time invariant system with the rationality of the transfer function and the finite rank of the Hankel matrix induced by the given weighting pattern [5,25]. Now the rank of a matrix is nothing but the dimension of its range space and hence the range of a Hankel operator should be one of the fundamental objects of inquiry in system theory. This has been strangely overlooked by most researchers in the field.

How do we measure "smallness" of range of Hankel operator. We concentrate on the class of strictly noncyclic functions described above. It certainly includes all rational matrix functions. From [15,16] which generalizes the work of Douglas, Shapiro and Shields [8] it follows that a function is strictly noncyclic if and only if its boundary values on the unit circle are also the boundary values of a meromorphic function of bounded type in the domain D_e exterior to the closed unit disc. From work of Tumarkin [42,8] it follows that these functions are H^2 limits of rational functions with a restriction on the poles involved. Thus we have roughly something close to rational associated with an induced Hankel operator of small range.

For a strictly noncyclic function A we can define the three sets:

$$I = \{\phi \in H^\infty \mid \bar{\phi}A \text{ extends analytically to } D_e\} \tag{4.6}$$

$$I_L = \{P \in H^\infty(B(C^m,C^m)) \mid P^*A \text{ extends analytically to } D_e\} \tag{4.7}$$

and

$$I_R = \{R \in H^\infty(B(C^n,C^n)) \mid AR^* \text{ extends analytically to } D_e\}.$$

I is a w^*-closed ideal in H^∞ whereas I_L and I_R are right and left ideals in $H^\infty(B(C^n,C^n))$ and $H^\infty(B(C^n,C^n))$ respectively. Using Theorem 2.2 and Theorem 2.3 it follows that a strictly noncyclic function A is factorable on the unit circle in the following ways

$$A = \bar{\chi}qQ^* \tag{4.8}$$

$$A = \bar{\chi}qC^* \tag{4.9}$$

and

$$A = \bar{\chi}C_1^*P_1 \tag{4.10}$$

where q is a scalar inner function P and P_1 are inner functions in $H^\infty(B(C^m,C^m))$ and $H^\infty(B(C^n,C^n))$ respectively and C and C_1 are in $H^\infty(B(C^n,C^m))$. Moreover we can assume that P and C are right coprime which is to say that they do not have a common nontrivial inner factor. Similarly P_1 and C_1 may be taken to be left coprime. We denote the coprimeness conditions by $(P,C)_R = I$ and $(P_1,C_1)_L = I$.

We wish to add two remarks. In the analytic context there is also a strong coprimeness notion whereby we say that A and B are strongly right coprime, denoted by $[A,B]_R = I$, if there exists a $\delta > 0$ for which

$$\inf\{||A(z)\xi|| + ||B(z)\xi|| \mid ||\xi|| = 1\} \geq \delta$$

for all z in the open disc. It is a simple consequence of a generalized Corona theorem [11] that strong right coprimeness implies right coprimeness. The notion of left coprimeness is defined analogously using adjoints.

The second thing to note is that the use of ideals has enabled us to get a global factorization of the transfer function. The inner functions q,P and P_1 carry all the information about the singularities of the function. Stated roughly the ideals I, I_R and I_L measure how many zeroes are needed to cancel the singularities of A. Here zeroes is used in a generalized sense, since boundary behaviour, the singular part of an inner function, has to be taken into account. Now the same methods could be used to isolate local singularities and thus relate the fine structure of the singularities to certain factorizations of the corresponding inner function. This in turn yields information about the fine structure of the corresponding restricted shift operators. This approach has been taken in [17] and ties up nicely with factorizations of inner functions obtained by Helson [21] through a different route. This local factorization can be viewed as an approximation to the multiplicative integral representation of Potapov [35,21].

With the factorizations of transfer functions obtained above the problem of canonical realization is reduced almost to a triviality.

Consider a proper rational $m\times n$ matrix function W and its coprime factorization $W = D^{-1}N$. For the state space we take K_D defined in the previous section. Define $B:F^n \to K_D$ by

$$(4.11) \qquad (B\xi)(\lambda) = N(\lambda)\xi \quad \text{for} \quad \xi \in F^n$$

Now any proper rational function Ω has a formal expansion $\Omega(\lambda) = \sum_{i=0}^{\infty} \Omega_i \lambda^{-i-1}$. Let us define $C:K_D \to F^m$ by

$$(4.12) \qquad Cf = (D^{-1}f)_1 .$$

That the system $\{S(D),B,C\}$ is a realization is easy to check. That it is actually canonical is proved in [19]. The situation at this point is highly asymmetrical as we used only the left coprime factorization and it is natural to expect that the right coprime factorization leads to another realization. This turns out to be true and in the following way.

From the equality $D^{-1}N = N_1D_1^{-1}$ it follows that

$$(4.13) \qquad ND_1 = DN_1 .$$

The coprimeness assumptions imply that the map $X:K_{D_1} \to K_D$ given by

$$(4.14) \qquad Xf = \pi_D(Nf)$$

is invertible and moreover

$$(4.15) \qquad XS(D_1) = S(D)X$$

is satisfied. If we define now maps B_1 and C_1 in such a way as to make the following diagram commutative then $\{S(D_1),B,C_1\}$ is also a canonical realization of W. Again all details are omitted, referring the reader to [19]. We wish to remark that the first realization is essentially the standard observable realization whereas the second is the standard controllable realization [5]. The derivation of the standard controllable realization from the standard observable realization was done by an ad hoc method. This has been necessitated either by the lack of a satisfactory duality theory in this context or by the author's unawareness of its existence.

The infinite dimensional analogue of these results is even simpler in its derivation. Let $A \in H^\infty(B(C^n,C^m))$ be a strictly noncyclic function with the coprime factorizations (4.9) and (4.10).

Consider the following realization. For state space we take $H(P) = \{PH^2(C^n)\}^\perp$. Let $S(P)^*$ be the left shift restricted to the left invariant subspace $H(P)$. Define operators $B:C^n \to H(P)$ and $C:H(P) \to C^m$ by

$$(4.17) \qquad (B\xi)(z) = A(z)\xi$$

and

(4.18) $$Cf = f(0)$$

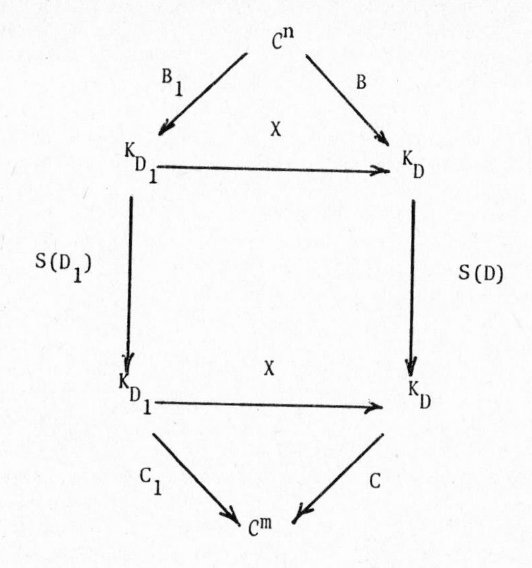

then $\{S(P)^*,B,C\}$ is a canonical realization of the transfer function A, in the sense that $A(z) = C(I-zS(P)^*)^{-1}B$ for $z, |z| < 1$. We note that the state space H(P) is actually the range closure of the Hankel operator induced by A. The same construction applied to $\tilde{A}$, where $\tilde{A}(z) = A(\bar{z})^*$, yields a realization of $\tilde{A}$ in $H(\tilde{P}_1)$. Using duality considerations we get a realization in $H(P_1)$ using the restricted left shift $S(P_1)^*$ as generator. The two realizations thus obtained turn out to be closely related and in fact to S(P) and $S(P_1)$ corresponds the same Jordan model, i.e. they are quasisimilar. Conditions for similarity involve strong coprimeness conditions and we refer to [14] for details.

The parallel development of the two theories can be pushed further as for example in the study of controllability and observability of canonical systems connected in series and parallel. The reader should consult [18] and the further references therein.

5. On the Relation Between Finite and Infinite Dimensional Realization Theory

The similarity between the finite and infinite dimensional theories as developed here is striking. This immediately brings up the question of how to relate the two. We will proceed to show how Ho's algorithm [25] can be derived from the shift realizations. For simplicity we restrict ourselves to the scalar case. So let

(5.1) $$(a_0,a_1,\dots)$$

be a single input single output discrete time wieghting pattern, that is an infinite sequence of complex numbers, and let $H = (h_{ij})$ be the infinite Hankel matrix induced by it, that is $h_{ij} = a_{i+j}$, $0 \leq i_{,j} < \infty$. We denote by H_m the principal minors i.e. the matrices (h_{ij}) $0 \leq i,j \leq m-1$. Now we assume that rank $H = n < \infty$ and, without loss of generality, that the function a defined by $a(z) = \sum_{j=0}^{\infty} a_j z^j$ is in H^∞. Let σH be the shifted Hankel matrix that is $(\sigma H)_{ij} = a_{i+j+1}$, and let $(\sigma H)_n$ be the principal $n\times n$ minor of σH. By Ho's algorithm the system $\{A,b,c\}$ with $A = (\sigma H)_n H_n^{-1}$, $b = \begin{pmatrix} a_0 \\ \vdots \\ a_{n-1} \end{pmatrix}$ and $C = (1,0,\ldots 0)$ gives a canonical realization of (5.1).

Now since $H = H_a$, the Hankel operator induced by a, has finite rank a is a rational function, and hence strictly noncyclic [8]. Let

$$\text{(5.2)} \qquad a = \bar{\chi}\, q\, \bar{h}$$

be its coprime factorization where q is inner and Range $H_a = H(q)$. Now we know that the shift realization $\{S(q)^*,B,C\}$ is a canonical realization. Here $B\alpha = \alpha a$ and $Cf = f(0)$ for all $f \in H(q)$. Consider now $\{\text{Ker } H_a\}^\perp = H(\tilde{q})$. Clearly $H|\{\text{Ker } H_a\}^\perp = H|H(\tilde{q})$ in one-to-one and onto $H(q) = \text{Range } H_a$. Let $W:H(q) \to H(\tilde{q})$ be its inverse. Thus $HW = I_{H(q)}$ and $WH = I_{H(\tilde{q})}$. From the functional equation $S^*H = HS$ characterizing Hankel operators, S being the right shift, it follows that

$$S^* H_a W|H(q) = H_a SW|H(q)$$

or

$$S^*|H(q) = S(q)^* = (H_a S)W|H(q) .$$

Thus the generator in the shift realization is recoverable directly in terms of H_a and $H_a S$.

Now the disadvantage of the shift realization is that the subspace $H(q)$ may be inconvenient to work with, especially in case numerical computation may be required. So let M be any other left invariant subspace of H^2 for which $\dim M = \dim H(q)$. Also let P_M be the orthogonal projection of H^2 on M. Then it follows that $P_M H(q) = M$ and hence $P_M|H(q)$ is invertible. Let $Q_M : M \to H(Q)$ denote the inverse of P_M. Thus $Q_M P_M|H(q) = I_{H(q)}$ and $P_M Q_M = I_M$. By reference to the basic functional equation satisfied by Hankel operators it follows that

$$P_M S^* H_a|M = P_M H_a S|\, M = P_M S^*\, Q_M P_M H_a|\, M = (P_M S(q)^* Q_M)(P_M H_a|M) .$$

Since $(P_M H_a|M) : M \to M$ is invertible it follows that

$$P_M S(q)^* Q_M = [P_M(H_a S)|M]\; [P_M H_a|M]^{-1} = F$$

Thus F is a linear operator in M which is similar to $S(q)^*$, the generator in the shift realization. Define now operators G and H such that $G = P_M B$ and $H = CQ_M$ then by similarity $\{F,G,H\}$ is also a canonical realization of our weighting pattern (5.1).

Now we specialize the choice of our left invariant subspace M by taking $M = H(\chi^n)$, thus M is the subspace of H^2 of all polynomials of degree $\leq n-1$. Obviously there is an isomorphism between $H(\chi^n)$ and $\mathbb{C}^n$ given by $\sum_{i=0}^{n-1} \xi_i z^i \to (\xi_0,\ldots,\xi_{n-1})$. Let us compute F, G and H.

Clearly in terms of this isomorphism we get $P_{H(\chi^n)} H_a|H(\chi^n) = H_n$ and $P_{H(\chi^n)}(H_aS)|H(\lambda^n) = (\sigma H)_n$ and hence $F = (\sigma H)_n H_n^{-1}$. Now $G\alpha = P_{H(\chi^n)}\alpha a = \alpha(a_0,\ldots,a_{n-1})$ and $H \sum_{i=0}^{n-1} \xi_i\chi^i = C\,Q_{H(\chi^n)} \sum \xi_i\chi^i = \xi_0$. So C is actually represented by the vector $(1,0,\ldots,0)$. It should be observed that F has necessarily a matrix representation of the form

$$F = \begin{pmatrix} 0 & 1 & 0 \\ & \cdot & 1 \\ -\alpha_o & \cdot & -\alpha_{n-1} \end{pmatrix}$$

where $\lambda^n + \alpha_{n-1}\lambda^{n-1} + \ldots + \alpha_0$ is the characteristic, and minimal, polynomial of F. Thus we are back at the standard observable realization.

References

1. Balakrishnan, A.V., Introduction to Optimization Theory in a Hilbert Space, Springer Verlag, Berlin, 1971.

2. Balakrishnan, A.V., System theory and stochastic optimization, NATO Adv. Study Inst. Network and Signal Theory.

3. Baras, J.S. and Brockett, R.W., "H^2 functions and infinite dimensional realization theory," SIAM J. Contr., 13, 221-241, 1975.

4. Beurling, A., "On two problems concerning linear transformations in Hilbert space," Acta Math., 81, 239-255, 1949.

5. Brockett, R.W., Finite Dimensional Linear Systems, Wiley, New York, 1970.

6. Brockett, R.W. and Fuhrmann, P.A., "Normal symmetric dynamical systems," to appear, SIAM J. Contr.

7. DeWilde, P., "Input output description of roomy systems," to appear, SIAM J. Contr.

8. Douglas, R.G., Shapiro, H.S., and Shields, A.L., "Cyclic vectors and invariant subspaces for the backward shift," Ann. Inst. Fourier, Grenoble 20, 1, 37-76, 1971.

9. Duren, H^p Spaces, Academic Press, New York, 1970.

10. Eckberg, A.E., "A characterization of linear systems via polynomial matrices and module theory," M.I.T. report ESL-R-528, 1974.

11. Fuhrmann, P.A., "On the corona theorem and its applications to spectral problems in Hilbert space," Trans. Amer. Math. Soc., 132, 55-66, 1968.

12. Fuhrmann, P.A., "A functional calculus in Hilbert space based on operator valued analytic functions," Israel J. Math., 6, 267-278, 1968.

13. Fuhrmann, P.A., "On realization of linear systems and applications to some questions of stability," Math. Systems Th., 8, 132-141, 1975.

14. Fuhrmann, P.A., "Realization theory in Hilbert space for a class of transfer functions," J. Funct. Anal., 18, 338-349, 1975.

15. Fuhrmann, P.A., "On Hankel operator ranges, meromorphic pseudocontinuations and factorization of operator valued functions," to appear, J. London Math. Soc.

16. Fuhrmann, P.A., "Factorization theorems for a class of bounded measurable operator valued functions," to appear.

17. Fuhrmann, P.A., "On the spectral minimality of the shift realization," to appear.

18. Fuhrmann, P.A., "On series and parallel coupling of a class of discrete time infinite dimensional systems," to appear, SIAM J. Contr.

19. Fuhrmann, P.A.,"Algebraic system theory; an analyst's point of view," to appear, J. Franklin Inst.

20. Halmos, P.R., "Shifts on Hilbert spaces," J. Reine Angew. Math. 208, 102-112, 1961.

21. Helson, H., Lectures on Invariant Subspaces, Academic Press, New York, 1964.

22. Helton, J.W., "Discrete time systems, operator models and scattering theory," J. Funct. Anal., 16, 15-38, 1974.

23. Helton, J.W., "A spectral factorization approach to the distributed stable regulator problem; the algebraic Ricatti equation," to appear.

24. Hoffman, K., Banach Spaces of Analytic Functions, Prentice Hall, Englewood Cliffs, N.J., 1962.

25. Kalman, R.E., Falb, P.L., and Arbib, M.A., Topics in Mathematical System Theory, McGraw Hill, New York, 1968.

26. Korenblum, B., "An extension of the Nevanlinna theory," to appear, Acta Math.

27. Kra, I., "On the ring of holomorphic functions on an open Riemann surface," Amer. Math. Soc., 132, 231-244, 1968.

28. Lax, P.D., "Translation invariant subspaces," Acta Math., 101, 163-178, 1959.

29. Lax, P.D., and Phillips, R.S., Scattering Theory, Academic Press, New York, 1967.

30. Livsic, M.S., "On a class of linear operators in a Hilbert space," Amer. Math. Soc. Transl., (2) 13, 61-83, 1960.

31. Livsic, M.S., "On the spectral resolution of linear non-selfadjoint operators," Amer. Math. Soc. Transl., (2) 5, 67-114, 1957.

32. MacDuffee, C.C., The Theory of Matrices, Chelsea, New York, 1946.

33. Moore III, B., and Nordgren, E.A., "On quasiequivalence and quasisimilarity," Acta Sci. Math., 34, 311-316, 1973.

34. Nordgren, E.A., "On quasiequivalence of matrices over H^∞," Acta Sci. Math., 34, 301-310, 1973.

35. Potapov, V.P., "The multiplicative structure of J-contractive matrix functions," Amer. Math. Soc. Transl., (2) 15, 131-243, 1960.

36. Rosenbrock, H.H., State Space and Multivariable Theory, Nelson-Wiley, New York, 1970.

37. Rota, G.C., "On models for linear operators," Comm. Pure Appl. Math., 13, 469-472, 1960.

38. Sarason, D., "Generalized interpolation in H^∞," Trans. Amer. Math. Soc., 127, 179-203, 1967.

39. Sz.-Nagy, B. and Foias, C., Harmonic Analysis of Operators on Hilbert Space, North Holland, Amsterdam, 1970.

40. Sz.-Nagy, B. and Foias, C., "Operateurs sans multiplicite," Acta Sci. Math. 30, 1-18, 1969.

41. Sz.-Nagy, B. and Foias, C., "On the structure of intertwining operators," Acta Sci. Math., 35, 225-254, 1973.

42. Tumarkin, G.T.s., "Description of a class of functions approximable by rational functions with fixed poles," Izv. Akad. Nauk Armyanskoi: SSR I, 89-105, 1966.

FINITENESS IN INFINITE-DIMENSIONAL SYSTEMS APPLIED TO REGULATION*

Edward W. Kamen
School of Electrical Engineering
Georgia Institute of Technology
Atlanta, Georgia 30332/U.S.A.

ABSTRACT

After describing existing mathematical representations of linear infinite-dimensional systems, we consider the problem of regulating a system's output so that it tracks a given reference. Using a general operational framework possessing certain finiteness properties, we pursue a particular aspect of this problem by first considering the concept of bounded-time controllability. New results are then given on the construction of time-invariant input/output regulators that drive the output response (resulting from nonzero initial states) to the zero function in finite time. An example is given to illustrate that such regulators can be constructed by using ideal time delays in the feedback loop.

1. Representation of Infinite-Dimensional Systems

A good deal of work has been devoted to the development of state-space representations for linear infinite-dimensional systems. There are two basic types of mathematical models. In one setup, the system dynamics are given by an equation of evolution in an infinite-dimensional linear topological space such as a Banach or Hilbert space (e.g., see the papers by Balakrishnan [1], Aubin-Bensoussan [2], and Baras-Brockett-Fuhrmann [3]). In most of these models the state is the solution of some first-order ordinary (or operational) differential equation in a locally convex space. We shall refer to these models as abstract representations since the state is usually not defined in terms of physical attributes of the given system.

*This work was supported by the U. S. Army Research Office, Durham, N.C., under Grant DA-ARO-D-31-124-73-G171.

The other basic type of mathematical representation is referred to as a hereditary model. These representations are usually given by operational differential equations (such as delay differential equations) in n-dimensional Euclidean space (see the papers by Delfour [4], Delfour-Mitter [5], and Manitius [6]). Here the derivative of the state at time t (which is an n-vector) depends in general on past values of the state and input. This concept of state is not the classical one, as the actual state consists of a function segment with values in n-dimensional space. In most cases the actual state at time t has a physical meaning such as the voltage appearing along various delay lines at time t.

Many classes of infinite systems can be represented by either hereditary or abstract models. In particular, it is often possible to construct an abstract model from a hereditary model (see [4]). On the other hand the construction of hereditary representations from abstract representations appears to be a completely open problem. The capability of going from one form to the other and vice versa is very important because some problems are much easier to study in one setting as opposed to the other. For example, the abstract setting is in general better suited for the study of optimal control with targets in function spaces (see Delfour [4] and Manitius [7]). However since the "instantaneous" state is an n-vector in the hereditary model, we can consider state-variable feedback without first having to approximate the given system by "lumped" components. We could then realize input/output regulators by interconnecting state controllers and observers.

Recently, work has been carried out on the algebraic aspects of hereditary and abstract models: Kamen [8] has shown that various classes of time-invariant hereditary systems can be characterized by finite free modules defined over Noetherian rings. The finiteness properties of this algebraic framework yield many new methods for studying hereditary systems. For instance, see [8] for results on the problem of realization and Sontag [9] for results on the construction of controllers and observers. A generalization of this algebraic setting has been carried out by Sontag [10]. Current work involves the extension of these algebraic techniques to time-varying hereditary systems (see [15]).

In the time-invariant case, various structural and dynamical properties of abstract representations have been studied via modules defined over a convolution ring of generalized functions: This approach was first formulated by Kalman (see the paper by Kalman-Hautus [11]) for the class of systems having an infinitely differentiable impulse response. It was then generalized by Kamen [12] to include a very large class of time-invariant infinite-dimensional systems. In [12] the systems are defined in a general abstract operational form, which is also utilized (with some changes) in this paper. As will be seen, finiteness properties of this framework with respect to the convolution module structure allow us to solve a particular aspect of the regulator problem.

2. The Problem of Regulation

Let Σ denote an m-input k-output causal linear system with infinite-dimensional state space X (elements of X are the actual states). We assume that Σ is given by some mathematical model which we shall not specify until later. A central problem in control is the design of an input/output regulator, denoted by Σ_{fb}, that will force Σ' s output to track a given reference signal in "steady-state operation." We assume the following closed-loop configuration:

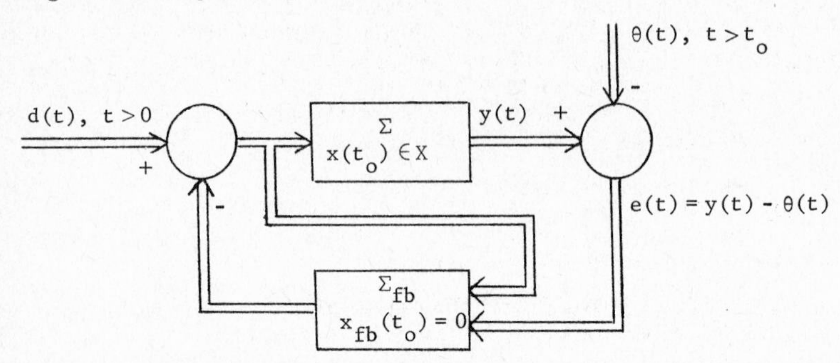

Here d(t) is a disturbance function, $x(t_o)$ (resp. $x_{fb}(t_o)$) is the initial state of the system (resp. regulator), and $e(t) = y(t) - \theta(t)$ is the error between the desired output $\theta(t)$ and the actual output y(t). Often the reference $\theta(t)$ is a k-vector of constant functions. For instance, in an aircraft control system the components of $\theta(t)$ could be the desired speed and elevation. As is the case in most applications, we assume that $x(t_o)$ is unknown and that $x_{fb}(t_o)$ is zero.

Given a set D of possible disturbance functions and a set Θ of reference signals, in the standard infinite-time tracking problem, we want e(t) to approach zero at some specified rate (usually exponential) for all $x(t_o) \in X$, all $\theta(t) \in \Theta$, and all $d \in D$. There may be additional criteria based on the minimization of a cost functional; however we will not pursue this here. We will suppose that the elements of D are of finite duration in time, so that the effect of a disturbance is reflected in the state at the time the disturbance terminates. Finally, we shall limit our attention to the case $\theta(t) = 0$ (then $e(t) = y(t)$), so our problem reduces to the construction of regulators such that $y(t) \to 0$ at some rate for all $x(t_o) \in X$. An important question here is how fast of an exponential rate of convergence can be obtained. In particular, we have the following:

<u>Question</u>: Given Σ, does there exist a causal linear regulator such that for all $t > t_o + t_d$, where $t_d > 0$, we have $y(t) \leq \left(e^{-a(t-t_o-t_d)}\right) y(t_o + t_d)$, all $x \in X$, where a can be taken arbitrarily large.

The answer to this question is yes (with t_d arbitrarily small) if Σ is finite-dimensional, time-invariant, reachable and observable. This result follows directly from the well-known pole-placement theorem (Kalman [13]). In the infinite-dimensional time-invariant case, a general answer to the above question does not exist at present. Part of the difficulty in the infinite case is a consequence of systems containing time delays, finite-time integral operators, etc., which can produce a "period of transience t_d" before the error begins to approach zero.

In approaching the above question in the time-invariant infinite case, it would be reasonable to ask if X is controllable in bounded time t_d. That is, can every state be driven to the zero state within the fixed time period $t_d < \infty$? Unlike the finite case, for an infinite system it is not true in general that if every element of X is reachable from zero, then every element is controllable to zero (or controllable in bounded time). For a general class of time-invariant systems, we study this situation in Section 4 and then apply our results in Section 5 to the construction of time-invariant regulators in the limiting case $a \to \infty$ (i.e., $y(t) = 0$, all $t > t_d > 0$, all $x \in X$, where we have set $t_o = 0$).

3. System Description

In this section we define the class of systems which we shall study. The definitions and constructions given below are a slight modification of those given in [12]. The reader is referred to [12] for many details which are not included here.

Although the emphasis is on infinite-dimensional systems, our theory will be primarily algebraic in nature. Questions involving topology are not considered in order that full attention can be given to the algebraic aspects of system behavior.

Let R denote the field of real numbers and let $\underline{\mathcal{D}}$ denote the linear space of R-valued infinitely differentiable functions defined on R with support bounded on the right. With the Schwartz topology [14] on $\underline{\mathcal{D}}$, let V denote the dual of $\underline{\mathcal{D}}$ with the strong topology (i.e., V is the space of all linear continuous functionals on $\underline{\mathcal{D}}$). The R-linear space V is the space of R-valued distributions on R with support bounded on the left. With the operations of addition (+) and convolution (*), V is also a commutative ring with no divisors of zero (i.e., $u * v = 0$ implies that $u = 0$ or $v = 0$). We say that V is an integral domain. Given $u, v \in V$, convolution is defined by $\langle u * v, \varphi \rangle = \langle u, \langle v, \varphi(t + \tau) \rangle \rangle$ all $\varphi \in \underline{\mathcal{D}}$. Let V^n denote the n-fold direct sum of V and let $V^{k \times m}$ denote the set of $k \times m$ matrices over V. Then we have the following fundamental concept.

<u>Definition 1</u>: A time-invariant linear causal input/output (i/o) operator is a map $f: V^m \to V^k$ given by $f(v) = W * v$, all $v \in V^m$, where $W = (w_{ij}) \in V^{k \times m}$ is the impulse response matrix with support w_{ij} contained in $[0, \infty)$, all i, j.

The i/o operator $f: V^m \to V^k$ characterizes the input/output behavior of some m-input k-output linear time-invariant system (yet to be defined). It is very important to note that in addition to the R-linear structure on f, there is also a V-module structure. More precisely, V^m and V^k are finite free modules over the ring V with the usual componentwise operations, and with respect to this structure f is a V-module homomorphism (i.e., f is additive and for every $\alpha \in V$, $f(\alpha * v) = \alpha * f(v)$, all $v \in V^m$). Our approach is based on the convolution module structure of f.

We first need to define a restricted form of f in the following manner. Let Ω denote the subring of V given by $\Omega = \{v \in V: \text{supp } v \subseteq (-\infty, 0]\}$. Note that the elements of Ω have compact support since all elements in V have support bounded on the left. Let Γ denote the linear space $\Gamma = \left\{v_{|(0,\infty)} : v \in V\right\}$ where $_{|(0,\infty)}$ denotes restriction to the interval $(0,\infty)$ in the sense of distributions.

With the componentwise operations, Ω^m is a free m-dimensional Ω-module. We can also place an Ω-module structure on Γ^k by defining multiplication by

$$\Omega \times \Gamma^k \to \Gamma^k : (\alpha, \gamma) \mapsto \alpha\gamma = (\alpha * \bar{\gamma})_{|(0,\infty)} \tag{1}$$

where $\bar{\gamma}$ is any extension of γ to V^k (see [12] for a proof of this result).

Now define the injection $I: \Omega^m \to V^m : \omega \mapsto \omega$ and the "projection" $P: V^k \to \Gamma^k : v \mapsto v_{|(0,\infty)}$. Then given an i/o operator f, let $f^*: \Omega^m \to \Gamma^k$ denote the composition $f^* = PfI$. The map f^* characterizes the input/output behavior relative to the time reference $t = 0$. Since f is a V-module homomorphism and Ω is a subring of V, f is also an Ω-module homomorphism. It is also clear that I and P are Ω-module homomorphisms, and thus $f^* = PfI$ is an Ω-module homomorphism.

Even though f is causal, it is not necessarily true that knowledge of f is equivalent to knowledge of f^* (see [12]). The two maps do carry the same information when f is strictly causal; that is, $f: v \mapsto W * v$, $W = (w_{ij})$, is causal and for each i,j, there exists a neighborhood U_{ij} of $\{0\}$ such that $w_{ij}|_{U_{ij}}$ is regular (generated by a locally integrable function).

We need the following definition and then we can give the "internal" definition of the class of systems we shall consider here. Let ℓ denote the map

$$\ell: V \to R: v \mapsto \ell(v) = \begin{cases} \inf_t \{t \in \text{supp } v\}, & v \neq 0 \\ 0, & v = 0 \end{cases} \tag{2}$$

Note that $\ell(\alpha) \leq 0$ for all $\alpha \in \Omega$. We can extend ℓ to $V^{n \times m}$ by defining $\ell: V^{n\times m} \to R: S = (s_{ij}) \mapsto \ell(S) = \min_{i,j} \ell(s_{ij})$.

Definition 2: An m-input k-output strictly causal linear time-invariant system Σ (in operational form) is a septuple $\Sigma = (\Omega^m, X, \Gamma^k, \mu, \eta, \phi(t), \psi)$ where X is the R-linear space of states, $\mu: \Omega^m \to X$ and $\eta: X \to \Gamma^k$ are R-linear maps with $\eta\mu = PfI$ for some strictly causal f, $\{\phi(t): X \to X, t \geq 0\}$ is an algebraic semigroup of R-linear maps, and ψ is a map defined by

$$\psi:\Omega^m \times X \times R^- \to X:(\omega,x,a) \mapsto \phi(-a-\ell(\omega))x + \mu(\omega)$$

where $R^- = \{a \in R: a < 0\}$.

In this definition, $\mu(\omega)$ is the state at time $t = 0$ due to input $\omega \in \Omega^m$, $\eta(x)$ is the output response on $(0,\infty)$ due to state x at time $t = 0$, $\phi(-t)x$ is the state at $t = 0$ due to state x at time $t < 0$, and ψ is the state transition operator.

In many cases X admits an Ω-module structure with respect to which μ and η are Ω-module homomorphisms. For example, this is true if X is defined via an Ω-module factorization of the Ω-module homomorphism f^*. Here we shall consider the more general situation in which only the reachable subsystem of Σ admits an Ω-module structure. More precisely, let X_r denote the linear subspace of X consisting of all reachable states, i.e., $X_r = \left\{x \in X: x = \mu(\omega_x),\ \text{some } \omega_x \in \Omega^m\right\}$. Now define $\mu_r:\Omega^m \to X_r:\omega \mapsto \mu(\omega)$ and $\eta_r:X_r \to \Gamma^k:x \mapsto \eta(x)$. Then the reachable subsystem of Σ is the system given by $\Sigma_r = (\Omega^m, X_r, \Gamma^k, \mu_r, \eta_r, \phi_r(t), \psi_r)$. We shall assume that X_r admits an Ω-module structure such that with respect to this structure the following diagram is a commutative diagram of Ω-module homomorphisms

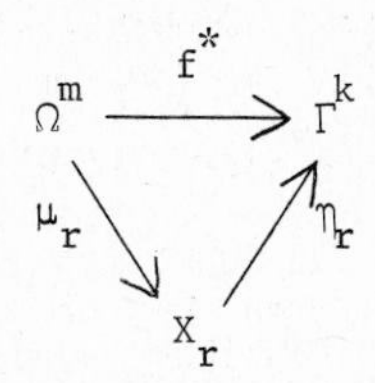

By time-invariance it follows that for every $\tau < 0$ and $x \in X_r$, $\phi_r(-\tau)x = \delta_\tau \cdot x$ where δ_τ is the Dirac distribution at the point $\{\tau\}$ and $\cdot$ denotes multiplication in the Ω-module structure on X_r. From here on we denote the reachable subsystem (with the Ω-module structure) by $\Sigma_r = (X_r, \mu_r, \eta_r)$.

4. Controllability

Let (X_r, μ_r, η_r) be the reachable subsystem of a system Σ with Ω-module structure. Let O_r denote the submodule of Γ^k given by $O_r = f^*(\Omega^m)$ where $f^* = \eta_r \mu_r$. The elements of O_r are the output functions on $(0,\infty)$ that are reachable from input functions in Ω^m. We then have the following basic

<u>Definition 3</u>: X_r (resp. O_r) is controllable (to zero) if for each $x \in X_r$ (resp. $\gamma \in O_r$) there exists $\tau < 0$ and $\omega \in \Omega^m$ with $\ell(\omega) > \tau$ such that $\delta_\tau \cdot x + \mu(\omega) = 0$ (resp. $\delta_\tau \gamma + f^*(\omega) = 0$).

In words, X_r is controllable if every element of X_r can be driven to the zero state at time $t = 0$ by some control function in Ω^m. Controllability of O_r has the following system-theoretic interpretation: Given $\gamma \in O_r$, suppose that there exist $\tau < 0$, $\omega \in \Omega^m$ with $\ell(\omega) > \tau$ such that $\delta_\tau \gamma + f^*(\omega) = 0$. By definition of $f^* = PfI$, we have that $\gamma_{|(-\tau,\infty)} + f(\delta_{-\tau} * \omega)_{|(-\tau,\infty)} = 0$. Therefore viewing γ as an output response due to some input in Ω^m, there exists a control $\delta_{-\tau} * \omega$ with support contained in $(0,-\tau]$ such that the total output response for $t > -\tau$ is zero. Here we have taken the time reference to be $t = 0$; however by time invariance we can consider controls setup with respect to any desired reference.

Unless X_r (resp. O_r) is finite-dimensional as a linear space over R, it is not necessarily true that X_r (O_r) is controllable. General conditions for controllability, given in terms of the R-linear structure, appear to be difficult to formulate because of the infinite dimensionality of X_r and O_r. In contrast, the problem is tractable with respect to the Ω-module structure since both X_r and O_r are finitely-generated Ω-modules. These modules are finite because they are homomorphic images of the finite free module Ω^m (recall that $X_r = \mu(\Omega^m)$ and $O_r = f^*(\Omega^m)$). If $\{e_1, e_2, \ldots, e_m\}$, where $e_i = (0\ 0 \ldots \delta_0\ 0 \ldots 0)^{TR}$, TR = transpose, denotes the standard basis of Ω^m, then $X_r = \sum_{i=1}^{m} \Omega\mu(e_i)$ and $O_r = \sum_{i=1}^{m} \Omega f^*(e_i)$.

A necessary and sufficient condition for controllability can be constructed in terms of the annihilating ideals of X_r and O_r, given by Ann $X_r = \{\alpha \in \Omega : \alpha \cdot x = 0, \text{ all } x \in X_r\}$ and Ann $O_r = \{\alpha \in \Omega : \alpha\gamma = 0, \text{ all } \gamma \in O_r\}$. Both Ann X_r and Ann O_r are ideals of the ring Ω. As a consequence of the finiteness of X_r and O_r, Ann $X_r = \bigcap_i$ Ann $(\mu(e_i))$ and Ann $O_r = \bigcap_i$ Ann$(f^*(e_i))$.

<u>Theorem 1</u>: X_r (resp. O_r) is controllable if and only if Ann X_r (Ann O_r) contains an element of the form $\delta_\tau + \sigma$ with $\ell(\sigma) > \tau$.

A proof of this result is given in [12, Theorem 5.1].

In many control problems (e.g., the regulator problem), we would like the elements of X_r and O_r to be reachable and controllable in bounded time. That is, every element of X_r and O_r can be reached within some fixed time period and can be driven to zero within a fixed period. A sufficient condition (which as we shall see is also necessary for bounded-time reachability and controllability is given in the following

Theorem 2: If Ann X_r (Ann O_r) contains an element β which has an inverse β^{-1} in the overring V, then X_r (O_r) is reachable and controllable in bounded time $-\ell(\beta) + a$ where a is arbitrarily small.

In [12] this result is proven by giving a procedure for constructing controls from β.

By using the fact that every element in Ω of the form $\delta_\tau + \sigma$, $\ell(\sigma) > \tau$, has an inverse in V, from Theorems 1 and 2 we get the following very surprising result.

Theorem 3: X_r (O_r) is reachable and controllable in bounded time if and only if X_r (O_r) is controllable.

In order to use the above criteria for determining controllability, we need to compute Ann X_r and Ann O_r. As we now show, information on these annihilating ideals can be obtained from a certain ideal associated with the impulse response matrix.

Let $W = (w_{ij})$ denote the impulse response matrix of the given system, and for each i,j define $A_{ij} = \{\alpha \in \Omega : \alpha * w_{ij} \in \Omega\}$. Each A_{ij} is an ideal of the ring Ω. Using the relationships $O_r = \eta_r(X_r)$ and $f^*(\omega) = (W * \omega)_{|(0,\infty)}$, we obtain the following useful results.

Proposition 1: Ann $X_r \subseteq$ Ann $O_r = \bigcap_{i,j} A_{ij}$

Corollary 1: Ann X_r = Ann O_r if and only if η_r is injective on each submodule $\Omega\mu(e_i) \neq \{0\}$.

Corollary 2: X_r controllable implies that O_r is controllable.

By Theorems 2-3 and Proposition 1, we see that X_r (resp. O_r) is controllable only if (resp. if and only if) there exists an element in $\bigcap_{i,j} A_{ij}$ which has an inverse in V. For some classes of systems it is always true that $\bigcap_{i,j} A_{ij}$ contains invertible elements. For instance, this is the case for the class of systems whose input/output behavior is characterized by delay differential equations. The proof of this result, given in [12], is based on the fact that any delay differential operator viewed as a distribution in V has an inverse in V.

Given that X_r and O_r are controllable and reachable in bounded time, an important question is what is the minimal time period in which all elements can be controlled and reached. If X_r (O_r) is finite dimensional as a linear space over R, it is easy to show that all elements in X_r (O_r) can be reached and controlled in an arbitrarily small time interval. By Theorem 2, in general we have that the minimal time required for control is bounded above by $\inf\{-\ell(\beta) : \beta \in \text{Ann } X_r \text{ (or Ann } O_r),\ \beta^{-1} \in V\}$. Additional results on the minimal time are given in [12].

5. Regulator Construction

Given a system $\Sigma = (X,\mu,\eta)$, consider the feedback configuration shown below. Here W_{fb} is an impulse response matrix and K is a matrix of gains whose values are determined by Σ. The gain matrix K is needed to insure the existence of regulators that are strictly causal (see Section 3). As in Section 2, we assume that the initial state of the regulator is zero. The objective is to construct a strictly causal linear time-invariant regulator such that when $d(t) = 0$, $t > 0$,

$$y(t) = 0, \text{ all } t > t_d, \text{ some } t_d > 0, \text{ all } x \in X_r . \tag{3}$$

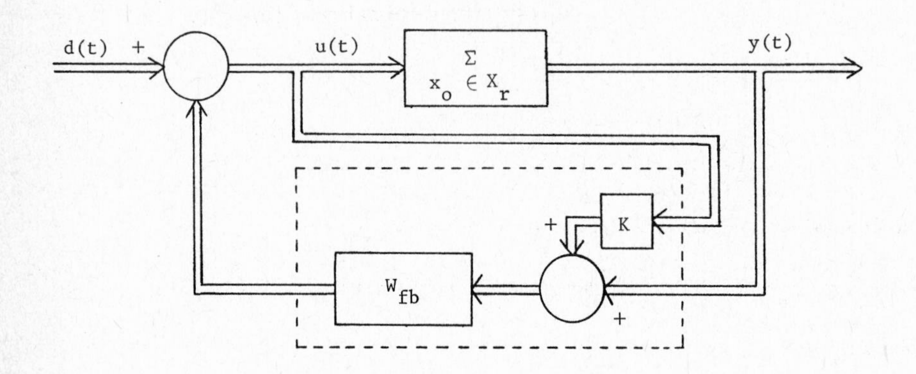

A crucial point here is that W_{fb} must be strictly causal so that there are no direct paths or differentiations from inputs to outputs. A direct feed (from inputs to outputs) in W_{fb} is not allowable since the path through the gain matrix K would result in a feedback loop containing only gains.

By definition of controllability of the submodule O_r of reachable output functions, it is clear that for there to exist a W_{fb} such that (3) holds, it is necessary that O_r be controllable in bounded time t_d. With the addition of an invertibility condition, we get a sufficient condition as follows.

Given Σ with impulse response matrix W and given a gain matrix $K = (k_{ij}\ \delta_o)$, define the set

$$E(K,W) = \{(\tau,\sigma,D) \in R^- \times \Omega \times \Omega^{k\times m} : (\delta_\tau + \sigma) * (K+W) = D,\ \ell(\sigma) > \tau\} .$$

If $(\tau,\sigma,D) \in E(K,W)$, then $(\delta_\tau + \sigma) * W \in \Omega^{k\times m}$ since $(\delta_\tau + \sigma) * K \in \Omega^{k\times m}$. Then by Proposition 1 the element $\delta_\tau + \sigma$ belongs to Ann O_r, and thus O_r is controllable in bounded time $-\tau + a$, where $a > 0$ is arbitrarily small. Conversely, if O_r is controllable, by Theorem 1 there exists $\delta_\tau + \sigma \in \text{Ann } O_r$ with $\ell(\sigma) > \tau$. Then by Proposition 1, $(\delta_\tau + \sigma) * W \in \Omega^{k\times m}$ which implies that $(\delta_\tau + \sigma) * (K+W) = D$ for some $D \in \Omega^{k\times m}$. Hence $(\tau,\sigma,D) \in E(K,W)$. Therefore the set E(K,W) is not empty if and only if O_r is controllable. We then have the following new result.

<u>Theorem 4</u>: Suppose that O_r is controllable so that E(K,W) is not empty. If there exists an element $(\tau,\sigma,D) \in E(K,W)$ such that D has left inverse D^{-1} and $\sigma * D^{-1}$ is strictly causal, then with $W_{fb} = -\sigma * D^{-1}$ (3) is satisfied with $t_d = -\tau$. Further, the resulting closed-loop impulse response matrix W_{cl} consists of elements having compact support and is equal to $\delta_{-\tau} * [D - (\delta_\tau + \sigma)K]$.

<u>Proof</u>: Suppose that (τ,σ,D) satisfies the hypothesis of the theorem. In the following we omit the convolution symbol *.
Now we have that

$$\begin{aligned}(\delta_\tau + \sigma)W_{fb}(K+W) &= W_{fb}[(\delta_\tau + \sigma)(K+W)] \\ &= W_{fb}D \text{ by definition of } E(K,W) \\ &= -\sigma D^{-1}D = -\sigma I \quad .\end{aligned}$$

Thus

$$W_{fb}(K+W) = \left(\frac{-\sigma}{\delta_\tau + \sigma}\right)I \tag{4}$$

From the above feedback configuration (with $x_o = 0$)

$$y = Wd - WW_{fb}(K+W)u$$

Hence from (4),

$$\begin{aligned} y &= Wd - W\left(\frac{-\sigma}{\delta_\tau + \sigma}\right)u \\ \Rightarrow y &= Wd + \left(\frac{\sigma}{\delta_\tau + \sigma}\right)y \text{ since } y = Wu \\ \Rightarrow y &= \left(\frac{\delta_\tau + \sigma}{\delta_\tau}\right)Wd \end{aligned}$$

Then since $(\delta_\tau + \sigma)\ (W+K) = D$,

$$(\delta_\tau + \sigma)W = D - (\delta_\tau + \sigma)K$$

Therefore, $y = \delta_{-\tau}[D - (\delta_\tau + \sigma)K]d$

Thus the closed-loop impulse response matrix W_{cl} is equal to $\delta_{-\tau}[D - (\delta_\tau + \sigma)K]$. Given $x \in X_r$, let $y \in V^k$ denote the resulting response of the closed-loop system (supp $y \subseteq [0,\infty)$). Again from the above feedback configuration,

$$y_{|(0,\infty)} = \eta(x) - [WW_{fb}\ (K+W)u]_{|(0,\infty)} \tag{5}$$

Now since $\delta_\tau + \sigma \in \Omega$, with respect to the Ω-module structure on Γ^k; we can operate by $\delta_\tau + \sigma$ on both sides of (5). Using (4) and the fact that $(\delta_\tau + \sigma)\eta(x) = 0$, we get

$$[(\delta_\tau + \sigma)y]_{|(0,\infty)} = -[W(-\sigma)u]_{|(0,\infty)} = (\sigma y)_{|(0,\infty)}$$

Hence

$$(\delta_\tau y)_{|(0,\infty)} = 0 \text{ which implies that } y_{|(-\tau,\infty)} = 0. \blacksquare$$

In any application of the result in Theorem 4, we must be able to construct a suitable realization of $W_{fb} = -\sigma * D^{-1}$. In many instances W_{fb} can be realized by using integrators and ideal delay lines. We shall show that this is the case for the class of finite-dimensional systems.

First we need to enlarge the set $E(K,W)$. In the definition of $E(K,W)$, in general we cannot allow $\ell(\sigma)$ to be equal to τ since σ may contain Dirac distributions at the point $\{\tau\}$. However, if σ is regular on a neighborhood of $\{\tau\}$ and $(\delta_\tau + \sigma)^{-1} \in V$, the case $\ell(\sigma) = \tau$ is allowable, so we can enlarge $E(K,W)$ to include this case. In the remainder of this paper we shall work with the enlarged set, also denoted by $E(K,W)$.

<u>Theorem 5</u>: If Σ is finite-dimensional, there exist a gain matrix K and an element $(\tau,\sigma,D) \in E(K,W)$ that satisfies the hypothesis of Theorem 4 and such that $W_{fb} = -\sigma * D^{-1}$ can be realized by an interconnection of integrators, ideal delay lines, scalors, and summers. Further, τ can be taken as small as desired.

<u>Proof</u>: We sketch a constructive proof for the single-input single-output $(m = k = 1)$ case only. Our approach is based on the operational calculus of Kamen [8]. Let p denote $\delta_o^{(1)}$, the first derivative of the Dirac distribution δ_o. For any $v \in V$, $p * v = v^{(1)}$ = first derivative of v in the sense of distributions. Let $R[p]$ denote the ring of all differential polynomials with coefficients in R. Since $p^n = \delta_o^{(n)}$, $R[p]$ can be viewed as a subring of V. Now since V is an integral domain, we can construct its quotient field $Q = \{\frac{u}{v}: u,v \in V, v \neq 0\}$. The quotient field of $R[p]$ is $R(p)$, the field of rational functions in p, which is a subfield of Q. Since Σ is finite dimensional, its impulse response $w \in V$ can be expressed as an element $\frac{\alpha(p)}{\beta(p)}$ of $R(p)$ where $\beta(p)$ is monic and $\deg \alpha(p) < \deg \beta(p)$ since Σ is strictly causal. Now pick some $K > 0$. Then $w + K = \frac{\alpha(p) + K\beta(p)}{\beta(p)}$ has the general form

$$w + K = \frac{Kp^n + c_{n-1}p^{n-1} + \dots + c_1 p + c_o}{p^n + a_{n-1}p^{n-1} + \dots + a_1 p + a_o} \tag{6}$$

Let $H(t)$ denote the Heaviside (step) function and fix $\tau < 0$. It can be shown that there exists a polynomial $p(t)$ in time t such that defining $\psi(t) = p(t)[H(t-\tau) - H(t)] \in \Omega$, we have that $\psi^{(n)}(t) = \delta_\tau + \lambda$, $\ell(\lambda) = \tau$, and λ is piecewise continuous.

From (6), we have

$$w + K = (w + K) * \frac{\psi}{\psi} = \frac{K\psi^{(n)} + c_{n-1}\psi^{(n-1)} + \dots + c_1\psi^{(1)} + c_o\psi}{\delta_\tau + \lambda + a_{n-1}\psi^{(n-1)} + \dots + a_1\psi^{(1)} + a_o\psi}.$$

By definition of ψ, we can take $\sigma = \lambda + a_{n-1}\psi^{(n-1)} + \dots + a_1\psi^{(1)} + a_o\psi$. Then let

$$w_{fb} = \frac{-(\lambda + a_{n-1}\psi^{(n-1)} + \dots + a_1\psi^{(1)} + a_o\psi)}{K\psi^{(n)} + c_{n-1}\psi^{(n-1)} + \dots + c_1\psi^{(1)} + c_o\psi}$$

By differentiating, we can express w_{fb} as a ratio of Dirac distributions and their derivatives at the points $\{\tau\}$ and $\{0\}$. Since λ and $\psi^{(n-1)}$ are piecewise continuous, it follows that in this expression for w_{fb} the order of the numerator is less than the order of the denominator, which implies that w_{fb} is strictly causal (we must also consider the supports of the elements comprising w_{fb}). Finally, using the results on realization given in [8] it is possible to realize w_{fb} via integrators and ideal delay lines having time delay equal to $-\tau$. ■

Theorem 5 can be extended to include a large class of systems given by delay differential equations. A complete version of this theory will appear in a separate paper. The following example serves to illustrate the constructions presented in the proof of Theorem 5.

Example: Consider a single-input single-output finite-dimensional system whose input/output differential equation is

$$\frac{d^2y(t)}{dt^2} = u(t)$$

In operational form, we have $p^2 * y = u$ where $p = \delta_o^{(1)}$ as before. The operational form of the impulse response w is equal to $\frac{1}{p^2}$ where $1 = p^o$. Taking $K = 1$, we have $w + 1 = \frac{p^2+1}{p^2}$.

Fix $\tau < 0$, set $p(t) = \frac{t^2}{\tau}(\frac{t}{\tau} - 1)$, and let $\psi(t) = p(t)[H(t-\tau) - H(t)]$.

Then $\psi^{(2)}(t) = \delta_\tau + \left(\frac{6t}{\tau^2} - \frac{2}{\tau}\right)[H(t-\tau) - H(t)]$.

Hence $(w+1) * \frac{\psi}{\psi} = \dfrac{\psi^{(2)} + \psi}{\delta_\tau + \left(\frac{6t}{\tau^2} - \frac{2}{\tau}\right)[H(t-\tau) - H(t)]}$

We then have $w_{fb} = \dfrac{-\left(\frac{6t}{\tau^2} - \frac{2}{\tau}\right)[H(t-\tau) - H(t)]}{\psi^{(2)} + \psi}$

This gives

$$w_{fb} = w_{fb} * \frac{\delta_{-\tau}p^4}{\delta_{-\tau}p^4} = \frac{-ap^3 - bp^2}{p^4 + ap^3 + cp^2 + ap + b}$$

where $a = \frac{2}{\tau}(2\delta_o + \delta_{-\tau})$

$$b = \frac{6}{\tau^2}(\delta_o - \delta_{-\tau})$$

$$c = \delta_o + b\ .$$

In this case w_{fb} can be realized using four integrators and six delay lines each having a time delay of $-\tau > 0$ seconds. The closed-loop impulse function w_{cl} is

$$w_{cl} = \delta_{-\tau}[D - \delta_\tau - \sigma] = \delta_{-\tau}[\psi^{(2)} + \psi - \delta_\tau - \sigma]$$

In this case, $\psi^{(2)} = \delta_\tau + \sigma$, so $w_{cl} = \delta_{-\tau}\psi$.

References

1. Balakrishnan, A. V., "Foundation of the State Space Theory of Continuous Systems," J. Comp. & Sys. Sci., Vol. 1, No. 1, pp. 91-116, March 1967.

2. Aubin, J. and Bensoussan, A., "Models of Representation of Linear Invariant Systems in Continuous-Time," Report #1286, Math. Research Center, University of Wisconsin, Sept. 1972.

3. Baras, J., Brockett, R., and Fuhrmann, P., "State-Space Models for Infinite-Dimensional Systems," IEEE Trans. Automatic Control, Vol. AC-19, No. 6, pp. 693-700, Dec. 1974.

4. Delfour, M., "State Theory of Linear Hereditary Differential Systems," Report CRM-395, Centre de recherches mathématiques, Université de Montréal, Aug. 1974.

5. Delfour, M. and Mitter, S., "Hereditary Differential Systems with Constant Delays. I. General Case," J. Diff. Equ., Vol. 12, pp. 213-235, 1972.

6. Manitius, A., "Mathematical Models of Hereditary Systems," Report CRM-462, Centre de recherches mathématiques, Université de Montréal, Nov. 1974.

7. Manitius, A., "Optimal Control of Hereditary Systems," Report CRM-472, Centre de recherches mathématiques, Université de Montréal, Dec. 1974.

8. Kamen, E. W., "On an Algebraic Theory of Systems Defined by Convolution Operators," J. Math. Systems Theory, Vol. 9, No. 1, pp. 57-74, 1975.

9. Sontag, E. D., "Linear Systems over Commutative Rings: A Survey," to appear in Ricerche di Automatica.

10. Sontag, E. D., "On Linear Systems and Noncommutative Rings," to appear in J. Math. Systems Theory.

11. Kalman, R. and Hautus, M., "Realization of Continuous-Time Linear Dynamical Systems," Proc. Conf. Diff. Equ., NRL Math. Research Center, June 1971.

12. Kamen, E. W., "Module Structure of Infinite-Dimensional Systems with Applications to Controllability," to appear in SIAM J. Control.

13. Kalman, R., "Lectures on Controllability and Observability," CIME Summer Course, 1968, Cremonese, Roma.

14. Schwartz, L., Theorie des distributions, Hermann, Paris, 1966.

15. Kamen, E. W., "Representation and Realization of Operational Differential Equations with Time-Varying Coefficients," to appear in the Special Issue on Realization Theory, Journal of the Franklin Institute.

Exponential stabilization of Functional Differential Equations

Luciano Pandolfi
Mathematical Institute
University of Florence

1. Introduction

Let us consider the autonomous functional differential equation

$$\dot{x} = \int_{-h}^{0} dN(z)\, x\,(t+z) \qquad t \geq 0 \tag{1}$$

where: $z \to N(z)$ is an $n \times n$ matrix of bounded variation on the interval $[-h,0]$, $t \to x(t)$ is a continuous vector function for $t \geq -h$.
A solution of eq. (1) is a function $t \to x(t)$ which is continuous for $t \geq -h$ (and absolutely continuous for $t > 0$) and satisfies the equation. Let $t \to x(t)$ be a solution of eq. (1). Then $t \to x_t(\cdot)$ is the function from R_+ (the set of non negative real numbers) in $C([-h,0]; R^n) =_{def} C$, defined by $x_t(\cdot) = x(t+z) \quad z \in [-h,0]$.
A well known theorem ([1]) asserts that for any function $\varphi(\cdot) \in C$, there exists only one solution $t \to x(t,\varphi)$ of eq. (1) such that $x_0(\cdot;\varphi) = \varphi(\cdot)$. The function $x(t) = 0$ is a solution of eq. (1) and we say that it is exponentially stable if it is possible to find two positive numbers k, γ such that

$$\|x_t(\cdot;\varphi)\| \leq k\, e^{-\gamma t} \|\varphi\|$$

for every function $\varphi(\cdot) \in C$. The symbol $\|\cdot\|$ indicates the supremum norm in C.
Suppose now that the null solution of eq. (1) is not exponentially

stable. If an $n \times m$ matrix B is given, we are going to find necessary and sufficient conditions for the existence of a continuous linear operator (feedback)

$$\varphi(\cdot) \to \int_{-h}^{0} dL(z)\,\varphi(z)$$

$C \to R^m$ ($L(z)$ $m \times n$ matrix of bounded variation), such that the null solution of the equation

$$\dot{x} = \int_{-h}^{0} dN(z)\,x(t+z) + B \int_{-h}^{0} dL(z)\,x(t+z) \qquad t \geq 0 \tag{2}$$

is exponentially stable. In this case we say that eq. (1) is <u>stabilizable</u>.

2. Preliminaries.

From [1] we recall some properties of eq. (1). Define

$$\Delta(l) = l\, I_n - \int_{-h}^{0} dN(z)\, e^{lz}$$

(I_n $n \times n$ identity matrix, l complex number).

The roots of the equation

$$\det \Delta(l) = 0$$

are the eigenvalues of eq. (1). Let σ be the set of all eigenvalues of (1). The set σ may be finite or not. However $\sigma \cap \{l:\ \mathrm{Re}\ l \geq a\}$ is a finite set for any real number a.

Then the null solution of eq. (1) is exponentially stable when $l \in \sigma$ implies $\mathrm{Re}\ l < 0$.

So we have to study when it is possible to shift the eigenvalues of eq. (1) into the half-plane $\{l : \mathrm{Re}\ l < 0\}$.

Now we recall (from [1]) that eq. (1) is equivalent to the following dif ferential equation in the Banach space C

$$\frac{d\, x_t(\cdot)}{dt} = \frac{d\, x_t(\cdot)}{dz} =_{def} \mathcal{A}\, x_t(\cdot) \tag{3}$$

with

$$D(\mathcal{A}) = \{\varphi(\cdot) \in C : \dot{\varphi}(\cdot) \in C , \quad \dot{\varphi}(0) = \int_{-h}^{0} dN(z)\,\varphi(z) \ .$$

$\overline{D(\mathcal{A})} = C$. $\mathcal{A}$ is the infinitesimal generator of the semigroup of operators defined by

$$T(t)\varphi(\cdot) = x_t(\cdot;\varphi) \qquad\qquad \varphi(\cdot) \in C \ .$$

$\sigma(\mathcal{A})$ (the spectrum of the operator $\mathcal{A}$) is only a point spectrum and $\sigma(\mathcal{A}) = \sigma$.

If $l_0 \in \sigma$,

$$\mathcal{N} =_{def} \bigcup_{1}^{\infty}{}_k \ker(l_0 I - \mathcal{A})^k$$

is a subspace of C of finite dimension K . (I is the identity operator in C).

So there are K linearly independent functions $\varphi_1(\cdot),\ldots,\varphi_K(\cdot)$ such that $\mathcal{N} = \operatorname{span}\{\varphi_1(\cdot),\ldots,\varphi_K(\cdot)\}$.

Let $\Phi(\cdot)$ be the $n \times K$ matrix which has the $\varphi_i(\cdot)$ as columns.

Then $\varphi(\cdot) \in \mathcal{N}$ if there is a K-vector c such that $\varphi(\cdot) = \Phi(\cdot)c$.

It is known that there exists a subspace $\mathcal{M} \subset C$ such that $C = \mathcal{N} \oplus \mathcal{M}$.

$\mathcal{M}$ is a subspace of C of finite codimension. Then it is the kernel of a linear functional which has a finite dimensional range.

In particular, there exists a $K \times n$ matrix $\Omega(\cdot)$ of continuous function such that

$$\mathcal{M} = \{\varphi(\cdot) \in C , \qquad (\Omega(\cdot), \varphi(\cdot)) = 0\},$$

with

$$(\Omega(\cdot), \varphi(\cdot)) =_{def} \Omega(0)\varphi(0) - \int_{-h}^{0}\int_{0}^{z} \Omega(s-z)\,dN(z)\varphi(s)\,ds \tag{4}$$

In [1] it is proved that the rows of $\Omega(z)$ are functions of the form

$$\sum_{1}^{K}{}_{i}\; \alpha_i \frac{(-z)^{K-i}}{(K-i)!} e^{-l_o z} \qquad 0 \leq z \leq h$$

for some vectors α_i .

Obviously, if $\varphi(\cdot) \in \mathcal{N}$, $(\Omega(\cdot), \varphi(\cdot)) \neq 0$. Hence, it is not restrictive to assume that $(\Omega(\cdot), \varphi(\cdot)) = I_K$ (I_K $K \times K$ identity matrix) . Then, if $\varphi(z) \in \mathcal{N}$, $\varphi(z) = \Phi(z)\,(\Omega(\cdot), \varphi(\cdot))$. The function

$$\varphi(z) \rightarrow \Phi(z)\,(\Omega(\cdot), \varphi(\cdot))$$

is said the projection of C onto $\mathcal{N}$ along $\mathcal{M}$.

3. Spectrum assignment problem.

Now we start to study when it is possible to stabilize eq. (1). As we have just noted, we have to shift all the eigenvalues of eq. (1) on the half-plane Re $l < 0$.

In order to simplify the exposition, we assume that only the eigenvalue $l_o \in \sigma$ verifies Re $l_o \geq 0$, referring to ([2]) for the detailed treatment of the problem.

For any matrix $L(z)$ of bounded variation on $[-h,0]$ and any initial condizion $\varphi(\cdot) \in C$, eq. (2) has the unique solution $t \rightarrow x(t;\varphi,L)$. The function $x_t(\cdot;\varphi,L)$ may be projected on $\mathcal{N}$. Let $c(t;\varphi,L) = (\Omega(\cdot), x_t(\cdot;\varphi,L))$. It is simple to prove that $c(t;\varphi,L)$ is the solution of the equation

$$\dot{c} = \mathcal{B}\,c + \Omega(0)\,B \int_{-h}^{0} dL(z)\,x(t+z) \tag{5}$$

with initial condition $c(0) = (\Omega(\cdot), \varphi(\cdot))$.

Here $\mathcal{B}$ is a $K \times K$ matrix defined by $\mathcal{A}\Phi(z) = \Phi(z)\mathcal{B}$.

Then $\mathcal{B}$ has the unique eigenvalue l_o.

Now we define $L(z)$ by

$$\int_{-h}^{0} dL(z)\, \varphi(z) = F\, (\Omega(\cdot), \varphi(\cdot)) \qquad \varphi(\cdot) \in C \tag{6}$$

F $m \times K$ matrix. Then $\int_{-h}^{0} dL(z)\, x(t+z) = F\, C(t)$.

So eq. (5) becames

$$\dot{c} = (\mathcal{B} + \Omega(0)\, BF)\, c(t) \tag{7}$$

Suppose that there exists a matrix F for which $(\mathcal{B} + \Omega(0)\, BF)$ has only one eigenvalue $l' \notin \sigma$, $\mathrm{Re}\, l' < 0$. Then the null solution of eq. (2), with $L(z)$ defined by (6) with such a matrix F, is exponentially stable.

In fact let $x_F(t, \varphi)$ be solution of eq. (2).

Then $T_F(t)$ is the semigroup defined by $T_F(t)\varphi(\cdot) = x_{F_t}(\cdot\,;\varphi)$, $\mathcal{A}_F$ its infinitesimal generator.

$$\mathcal{A}_F\big|_{\mathcal{M}} = \mathcal{A}\big|_{\mathcal{M}}, \quad \text{since} \quad (\Omega(\cdot), \varphi(\cdot)) = 0 \quad \text{if} \quad \varphi(\cdot) \in \mathcal{M}.$$

Then, if σ_F denotes the spectrum of eq. (2), $\sigma_F \supset \sigma - \{l_o\}$. We prove that $l' \in \sigma_F$. Let $c_F(t) = e^{l't}v$ be a solution of eq. (7). Some calculation proves that $\Phi(z)\, v\, e^{l't}$ is the projection on $\mathcal{P}$ of the function

$$x_t(z) = \left[\Phi(z)v + e^{l'z}\, v'\right] e^{l't}$$

where $v' = \Delta(l')^{-1}\left[I_n - \Phi(0)\,\Omega(0)\right] B\, F\, v$ (Note that $\det \Delta(l') \neq 0$ since $l' \notin \sigma$), which is a solution of (2).

Then $e^{l't} \in \sigma(T_F(t))$. But $T_F(t)$ is a semigroups of compact operators (if $t>h$) with only point spectrum. So $l' \in \sigma(\mathcal{A}_F)$ ([3]).

Now it remains only to prove that σ_F has no more eigenvalue then those in $\sigma - \{l_o\}$ and l'. But this is obvious since $C = \mathcal{M} \oplus \mathcal{N}$. If eq. (2) has a solution $e^{\bar{l}t} v$, $\bar{l} \notin \sigma - \{l_o\}$, $e^{\bar{l}(t+z)} v$ has a non null projection on $\mathcal{N}$, and so $\bar{l} = l'$, since the matrix $(\mathcal{B} + \Omega(0) BF)$ has the unique eigenvalue l'.

So the null solution of eq. (2), with $L(z)$ just defined, is exponentially stable.

Hence we can assert that:

Lemma 1. The stabilization problem formulated in section 1 (with respect to the unique unstable eigenvalue l_o) has a solution if

$$\operatorname{rank} [\mathcal{B} - l_o I_K, \Omega(0) B] = K \tag{8}$$

since condition (8) is sufficient (as well necessary) for the existence of a matrix F such that the matrix $(\mathcal{B} + \Omega(0) BF)$ has only one eigenvalue, with negative real part ([4]).

The condition given by lemma 1 is not easy to handle. However observe that it is satisfied if and only if for every K-vector v such that

$$v^* \mathcal{B} = l_o v^*, \quad \text{we have} \quad v^* \Omega(0) B \neq 0 .$$

It is simple to prove that if $v^* \mathcal{B} = l_o v^*$, then $v^* \Omega(z) = v^* \Omega(0) \exp\{-l_o z\}$, and conversely ([2]).

So condition (8) is equivalent to say: $v^* \Omega(0) B \neq 0$ when $v^* \Omega(z) = v^* \Omega(0) \exp\{-l_o z\}$.

The matrix $\Omega(z)$ has rows of the form $z^j \alpha_j \exp\{l_o z\}$ $0 \leqslant i \leqslant k-1$, where the α_j's are row vectors.

So, if $v^* \Omega(z) = v^* \Omega(0) \exp\{-l_o z\}$ only the entries of v which corresponds to the vectors α_o may be different from zero.

It is an easy computation now to prove that $v^* \Omega(0) B = 0$ implies $\alpha_o B = 0$ for any vector α_o. But (from [1]) it is known that the vectors α_o are all the vectors in $[R\Delta(l_o)]^{\perp}$ ($R\Delta(l_o)$ is the range of the matrix $\Delta(l_o)$).

So if $\operatorname{rank}[\mathcal{B} - l_o I_K, \Omega(0)B] < K$, also $\operatorname{rank}[\Delta(l_o), B] < n$.

This proves the sufficient part of the following

<u>Theorem 1</u>. The instable eigenvalue l_o of eq. (1) may be shifted on the left of the immaginary axis if and only if

$$\operatorname{rank}[\Delta(l_o), B] = n .$$

The necessary part of the theorem is easyly proved, since if $v^* \Delta(l_o) = 0$ implies $v^* B = 0$, then also

$$v^* \Delta(l_o) + v^* B \int_{-h}^{0} dL(z)\, e^{l_o z} = 0$$

for any matrix $L(z)$. So l_o is an eigenvalue also of eq. (2), for any choice of the feedback.

In general, when eq. (1) has many instable eigenvalues (but in a finite number, obviously) repeated application of theorem 1 proves

<u>Theorem 2</u>. Eq. (1) is (exponentially) stabilizable if and only if

$$\operatorname{rank}[\Delta(l), B] = n \quad \text{when} \quad \operatorname{Re} l \geqslant 0 .$$

4. Concluding remarks.

A particular case of eq. (1) is the differential-difference equation

$$\dot{x} = A_0 x(t) + A_1 x(t-h) \qquad t \geq 0 \tag{9}$$

(A_0, A_1 are $n \times n$ matrices).

In this case, from (4), we have that the stabilized equation has the form

$$\dot{x} = (A_0+BF)\, x(t) + A_1\, x(t-h) + BF \int_{-h}^{0} \Omega(h+z)\, A_1\, x(t+z)\, dz \qquad t \geqslant 0 \tag{10}$$

and, with the feedback (6) the stabilized equation is no longer a differential-difference one, but contains a distributed lag.

However,

$$\int_{-h}^{0} \Omega(h+z)\, A_1\, x(t+z)\, dz \simeq \sum_{1}^{N}{}_i\, H_i\, x(t-h_i) \qquad 0 < h_1 < \ldots < h_i < h_{i+1} < \ldots < h$$

H_i $i = 1,\ldots,N$ $n \times n$ matrices.

Now

$$\dot{x} = (A_0+BF)\, x(t) + A_1\, x(t-h) + BF \sum_{1}^{N}{}_i\, H_i\, x(t-h_i) =$$

$$= (A_0+BF)\, x(t) + A_1\, x(t-h) + BF \sum_{1}^{N}{}_i\, H_i\, x(t-h_i) + BF \Big\{ \int_{-h}^{0} \Omega(h+z)\, A_1\, x(t+z) dz -$$

$$- \sum_{1}^{N}{}_i\, H_i\, x(t-h_i) \Big\} \tag{11}$$

Equation (11) may be seen as a perturbation of equation (10). Since the null solution of equation (10) is exponentially stable, the null solution of equation (11) is exponentially stable also, at least if N is sufficiently large ([5]).

In Section 3 we have shown how to change one eigenvalue of eq. (1), and so a finite number of eigenvalues, by a repeated application of the same method.

Exactly the same way may be used if one has to change a finite number of eigenvalues of the neutral functional differential equation

$$\dot{x} = \int_{-h}^{0} dM(z)\, \dot{x}(t+z) + \int_{-h}^{0} dN(z)\, x(t+z) \qquad (12)$$

where the matrix $M(z)$ is an $n\times n$ matrix of bounded variation on $[-h,0]$, with suitable properties (for exemple without jumps for $z=0$). However eq. (12) may have infinitely many eigenvalues in a vertical strip (but there exists a constant a such that any eigenvalue of eq. (12) satisfies $\operatorname{Re} l < a$ [6]). Hence, in order to stabilize the null solution of eq. (12) we have to apply theorem 1 infinitely many times. Suitable hypothesis on the matrices $M(z)$ and $N(z)$ may be assumed in order to justify this procedure. These hypothesis are sufficiently general to include the case of the equation

$$\dot{x} = \sum_{1}^{N}{}_{i}\, \bar{A}_i\, \dot{x}(t-h_i) + \sum_{0}^{N}{}_{i}\, A_i\, x(t-h_i) \qquad (13)$$

(A_i, $\bar{A}_i$ $n\times n$ constant matrices, $h_o = 0$ $h_{i-1} < h_i$, h_N bounded)

and we have the following

<u>Theorem 3</u>. Given an $n\times m$ matrix B, there exists a continuous operator $L : C^1([-h_N,0];R^n) \to R^m$ (feedback) such that the null solution of the equation

$$\dot{x} = \sum_{1}^{N}{}_{i}\, \bar{A}_i\, \dot{x}(t-h_i) + \sum_{0}^{N}{}_{i}\, A_i\, x(t-h_i) + BL(x_t(\cdot))$$

is exponentially stable if and only if there exists a positive number d such that

$$\operatorname{rank}\,[\Delta(l),B] = n \qquad \text{when} \qquad \operatorname{Re} l \geqslant -d\ .$$

For the proof of this theorem we refer to [7].

We remember only that for eq. (13)

$$\Delta(l) = l I_n - \sum_1^N {}_i \, l \bar{A}_i e^{-lh_i} - \sum_0^N {}_i A_i e^{-lh_i} .$$

R E F E R E N C E S

1. J. Hale, Functional Differential Equations.
Springer Verlag - Applied Mathematical Sciences n. 3, New York, 1971.
2. L. Pandolfi, On feedback Stabilization of Functional Differential Equations.
to appear in: Boll. Un. Mat. It. suppl. 12 (4) (1975).
3. E. Hille - R.S. Phillips, Functional Analysis and semigroups.
Am. Math. Soc. Coll. Pub. Vol. 31 (1957).
4. M.L.J. Hautus, Stabilization, Controllability and Observability of Linear Autonomous Systems.
Indag. Math. 32 (1970 448-455.
5. Krasovski, Stability of motion.
Stanford University Press, Stanford, 1963.
6. J.K. Hale - K.R. Mayer, A Class of Functional Equations of Neutral Type.
Memoirs of the Am. Math. Soc. n. 76, 1967.
7. L. Pandolfi, Stabilization of Neutral Functional Differential Equations.
To appear.

CODES AS ELEMENTS IN A GROUP ALGEBRA

J. M. Goethals
MBLE Research Laboratory
Brussels, Belgium.

1. Introduction.

Let F be a finite field with q elements, where q is a prime power, and let $V:=F^n$ be the vector space of all n-tuples over F. A q-ary code of length n is a subset C of V containing at least two elements. The vector space V has, with respect to vector addition, the structure of an elementary Abelian group. We shall describe the group algebra $\mathbb{C}(V)$ of this group over the complex field $\mathbb{C}$ as follows. To each element $u \in V$ we associate a basis element E(u) of C(V). The product of the basis elements E(u), E(v) is defined by

$$E(u)\, E(v) = E(u+v) \quad , \tag{1}$$

so that multiplication by a given E(u) induces a permutation on the q^n basis elements. It is easily seen that these permutations correspond to the regular representation of the additive group of V. Thus, $\mathbb{C}(V)$ is the q^n-dimensional linear associative and commutative algebra over $\mathbb{C}$ afforded by the regular representation of this group. Every subset $C \subseteq V$ is represented by the element

$$C := \sum_{u \in C} E(u) \tag{2}$$

in $\mathbb{C}(V)$, for which we use the same symbol. In this way, a code C can be viewed as an element of $\mathbb{C}(V)$. The aim of this paper is to show the usefulness of this representation in some classical problems of coding theory. In section 2, we shall study the MacWilliams identities in connection with the Fourier transform. In section 3, we shall given an account of Delsarte's linear programming method, and finally, in section 4, we shall show how Lloyd's theorem for perfect and unformly packed codes can be obtained. For more details on these subjects, we refer the reader to the list of references at the end of this paper.

2. The Fourier transform and the MacWilliams identities.

Let $\chi: F \to \mathbb{C}$ be any nonprincipal character of the additive group of the field F. Then, we shall define a symmetric bilinear mapping from V×V into $\mathbb{C}$ by

$$\langle u,v\rangle := \chi((u,v)),\ \forall\, u,v \in V \quad , \tag{3}$$

and, for every $u \in V$, a linear mapping χ_u from $\mathbb{C}(V)$ into $\mathbb{C}$ by :

$$\text{for } A = \sum_{v \in V} a(v)\ E(v) \in C(V) \quad , \tag{4}$$

$$\chi_u(A) := \sum_{v \in V} a(v) \quad \langle u,v \rangle \ . \tag{5}$$

Then, for every $A,B \in \mathbb{C}(V)$, we have

$$\chi_u(A)\ \chi_u(B) = \chi_u(AB) \quad . \tag{6}$$

For every element $A \in \mathbb{C}(V)$, its Fourier transform, denoted by $\hat{A}$ is defined by

$$\hat{A} := \sum_{u \in V} \chi_u(A)\ E(u) \quad .$$

Note that the map $\phi : A \to \phi(A) := q^{-n/2}\hat{A}$ is an involution in $\mathbb{C}(V)$, i.e. it satisfies $\phi(\phi(A))=A$.

Of basic importance in coding theory are the concepts of weigths and distances. The weight $w(u)$ of an element $u \in V$ is defined to be the number of its nonzero components, and for any $u,v \in V$, their distance $d(u,v)$ is defined to be the number of components in which they differ, that is, we have

$$d(u,v) := w(u-v) \quad .$$

The weight-enumerator $W_C(z)$ of a code C is the formal polynomial

$$W_C(z) = \sum_{u \in C} z^{w(u)} \quad . \tag{8}$$

Thus, the coefficient of z^i in its expansion is the number of elements of weight i in C. More generally, for an arbitrary element A like (4) in $\mathbb{C}(V)$, let us define its weight-enumerator $W_A(z)$ to be the formal polynomial

$$W_A(z) = \sum_{v \in V} a(v)\ z^{w(v)} \quad . \tag{9}$$

Note that, for a code C like (2), the two definitions (8) and (9) coïncide, and that, for the element A, the coefficient of z^i in $W_A(z)$ is the complex number A_i defined by

$$A_i = \sum_{w(v)=i} a(v) \quad . \tag{10}$$

Theorem 1. The weight-enumerators of an element A and of its Fourier transform $\hat{A}$ in $\mathbb{C}(V)$ are related by (the MacWilliams identity)

$$W_{\hat{A}}(z) = (1+(q-1)z)^n \, W_A\left(\frac{1-z}{1+(q-1)z}\right) .$$

For a proof of this theorem, we refer to [5].

Let C be an (n,k) linear code, i.e. a k-dimensional subspace of V. Then the Fourier transform $\hat{C}$ of C is related to the orthogonal complement $C^{\perp}$ of C in V by

$$\hat{C} = q^k C^{\perp} = q^k \sum_{u \in C^{\perp}} E(u) .$$

The orthogonal complement $C^{\perp}$, called the dual code of C, is itself a linear (n,n-k) code, and by theorem 1 we have

$$W_{C^{\perp}}(z) = q^{-k}(1+(q-1)z)^n \, W_C\left(\frac{1-z}{1+(q-1)z}\right) ,$$

which is the original form of the MacWilliams identity.

Let us now investigate the implications of theorem 1 for an arbitrary code C. To that end, let us define C^* to be the element

$$C^* := \sum_{u \in C} E(-u) ,$$

and let $A = CC^*$ in C(V). Then, the coefficient a(v) in the expression (4) for A is equal to the number of ordered pairs u, u' of elements of C such that u-u'=v holds. Thus, the weight-enumerator of A is identical to the distance-enumerator of C defined by

$$D_C(z) := \sum_{u \in C} \sum_{u' \in C} z^{d(u,u')} . \tag{11}$$

On the other hand, for the weight-enumerator of the Fourier transform $\hat{A}$ of A,

$$W_{\hat{A}}(z) = \sum_{j=0}^{n} B_j z^j , \quad \text{say} ,$$

we have, by definition,

$$B_j = \sum_{w(u)=j} \chi_u(A) .$$

Note that, for all $u \in V$, we have $\chi_u(C^*) = \chi^*(C)$, whence,

$$\chi_u(A) = \chi_u(C)\, \chi_u(C^*) = |(\chi_u(C)|^2 .$$

It follows that for j=0,1,2,...,n, B_j is a nonnegative real number. Note that B_0 is equal to $|C|$, the number of code elements. Now, by theorem 1, we have

$$\sum_{j=0}^{n} B_j z^j = \sum_{i=0}^{n} A_i (1-z)^i (1+(q-1)z)^{n-i} , \tag{12}$$

where

$$B_j = \sum_{w(u)=j} |\chi_u(C)|^2 , \tag{13}$$

and A_i, as noted above, is equal to the number of ordered pairs of code elements at distance i from each other.

3. Delsarte's linear programming method.

As noted above, given the distance-enumerator (11) of any code C, $D_C(z)=\sum_i A_i z^i$, say, the coefficients B_j appearing in the MacWilliams transform (12) of $D_C(z)$ have to be nonnegative real numbers, cf. (13). This observation led Delsarte [1] to the formulation of the following linear programming problem.

Let the length n of a code C over F be given, and let us require that, for any two distinct code elements u,u', $d(u,u')\geq d$ holds. Then, in the distance-enumerator of C, we have for the first few coefficients A_i,

$$A_0 = |C| , \text{ and } A_1 = A_2 = \dots = A_{d-1} = 0 .$$

Let $P_j(i)$ denote the coefficient of z^j in the expansion of $(1-z)^i(1+(q-1)z)^{n-i}$. Note that, for i=0, we have $P_j(0)=\binom{n}{j}(q-1)^j$. Then, with the above assumptions, we deduce from (12) that the following inequalities hold :

$$B_j=|C| \binom{n}{j} (q-1)^j + \sum_{i=d}^{n} A_i P_j(i) \geq 0 , \tag{14}$$

for j=1,2,...,n, and that

$$B_0 = |C| + \sum_{i=d}^{n} A_i = |C|^2 . \tag{15}$$

If we want to maximize the size $|C|$ of a code with given length n and minimum distance d, the above inequalities impose some restrictions on its distance-enumerator. Let us consider the numbers

$$x_i := |C|^{-1} A_i , \; i=d, d+1, \dots, n ,$$

as variables subject to the following linear constraints :

$$x_i \geq 0 , \quad i=d, d+1, \dots, n,$$

$$\sum_{i=d}^{n} x_i P_j(i) \geq 0 , \quad j=1,2,\dots, n .$$

Then, the maximum value of the sum

$$1 + x_d + x_{d+1} + \dots + x_n$$

for any set of values of the x_i's satisfying the above inequalities is an upper bound on the size $|C|$ of the code, cf. (14), (15). The problem of finding this maximum value can be solved by the methods of linear programming, thus providing an upper bound on the size of a code with given length and minimum distance. For more details, we refer to [1] .

4. Lloyd's theorem for perfect and uniformly packed codes.

Let X_k be the element of $\mathbb{C}(V)$ representing the set of all elements u of weight k in V, that is, $X_k = \sum_{w(u)=k} E(u)$. These elements X_k, thus defined for $k=0,1,2,\dots,n$, generate an (n+1)-dimensional subalgebra of $\mathbb{C}(V)$, and, for any $u \in V$ of weight i, we have

$$\chi_u(X_j) = P_j(i) \quad , \quad j=0,1,2,\dots,n, \tag{16}$$

where $P_j(x)$, a polynomial of degree j in x, is formally defined by

$$(1-z)^x \, (1+(q-1)z)^{n-x} = \sum_j P_j(x) \, z^j \quad . \tag{17}$$

For a proof of these properties, we refer to van Lint [6].
Let C be a code represented as in (2) in $\mathbb{C}(V)$, and let us consider the element

$$X_k C = \sum_{u \in V} a_k(u) \, E(u) \quad . \tag{18}$$

Then $a_k(u)$ is equal to the number of code elements v at distance k from u, as it follows from the definition of X_k and of the product in $\mathbb{C}(V)$. If the code has a minimum distance $d \geq 2t+1$, then there is at most one code element at distance less than or equal to t from every $u \in V$. Let S_t be the element $\sum_{k=0}^{t} X_k$, and let us consider the product

$$S_t C = \sum_{u \in V} b_t(u) \, E(u) \quad . \tag{19}$$

Then, from the above, we conclude that

$$\forall u \in V, \quad b_t(u) = \sum_{k=0}^{t} a_t(u) = 0 \text{ or } 1.$$

A code with minimum distance $d \geq 2t+1$ is said to be <u>t-error-correcting</u>. A t-error-correcting code is said to be <u>perfect</u> if $\forall u \in V$, $b_t(u)=1$ holds, and <u>quasi-perfect</u> if $\forall u \in V$, $b_{t+1}(u)=1$ holds, with $b_t(u)$ defined as above, cf. (19). A quasi-perfect t-error-correcting code is said to be <u>uniformly packed</u> with parameters λ, μ , if the following two conditions are satisfied :

(i) every $u \in V$ which is at distance t+1 or more from every code element is at distance t+1 from exactly μ code elements :

(ii) every $u \in V$ which is at distance t from some code element is at distance t+1 from exactly λ code elements.

These conditions can be stated as follows :

(i) $\forall u \in V, [b_t(u)=0] \Rightarrow [a_{t+1}(u) = \mu]$,

(ii) $\forall u \in V, [a_t(u)=1] \Rightarrow [a_{t+1}(u) = \lambda]$,

From the above discussion, it appears that we have the following theorem :

Theorem 2. A t-error-correcting code C is (i) perfect iff $S_tC=S_n$ holds,
(ii) uniformly packed quasi-perfect iff

$$(\mu S_t - \lambda X_t + X_{t+1})C=S_n \text{ holds.}$$

Note that $S_n = \sum_{k=0}^{n} X_k = \sum_{u \in V} E(u)$; then, from (16) and (17), we deduce

$$\chi_u(S_n) = \begin{cases} q^n & \text{if } w(u)=0 \text{ (i.e. } u=0) , \\ 0 & \text{otherwise (i.e. if } u\neq 0) . \end{cases}$$

Theorem 3. (Lloyd's theorem).

(i) If there exists a perfect t-error-correcting code of length n over F, then the polynomial

$$L_t(x) := \sum_{k=0}^{t} P_k(x) \tag{20}$$

of degree t has t integral zeros in the interval $[1,n]$.

(ii) If there exists a quasi-perfect t-error-correcting code of length n over F, which is uniformly packed with parameter λ,μ, then the polynomial

$$P_{t+1}(x) - \lambda P_t(x) + \mu L_t(x) \tag{21}$$

of degree t+1 has t+1 integral zeros in the interval $[1,n]$.

We shall briefly indicate how this theorem can be obtained from theorem 2. Let C be a perfect t-error-correcting code. Then, by theorem 2, $S_tC=S_n$ holds, whence, by (6), we have

$$\forall u \in V, \quad \chi_u(S_t)\, \chi_u(C) = \chi_u(S_n) \quad .$$

Hence, if for some $u\neq 0$ we have $\chi_u(C)\neq 0$, we must have

$$\chi_u(S_t) = \sum_{k=0}^{t} \chi_u(X_k) = 0 \quad .$$

Now, by use of (16), we deduce

$$\chi_u(S_t) = \sum_{k=0}^{t} P_k(w(u)) \quad .$$

Therefore, for any u such that $\chi_u(C)\neq 0$, the weight of u, $w(u) \in [1,n]$, has to be a zero of the polynomial $L_t(x)$. It remains to be shown that there exists at least t distinct values of w(u) for which an u exists having the property that $\chi_u(C)\neq 0$. This follows from the linear independence of X_0C, X_1C, ..., X_tC. For more details, we refer to van Lint [6] . the second part of the theorem is proved similarly. It is perhaps worth mentioning that the zeros of the above polynomials (20), resp. (21), are the integers $j \geq 1$ such that the numbers B_j defined by (13) are non-zero for the corresponding perfect or uniformly packed code.

References.

[1] P. Delsarte, Bounds for unrestricted codes by linear programming, Philips Res. Repts. 27(1972), 272-289.

[2] P. Delsarte, Four fundamental parameters of a code and their combinatorial significance, Inf. Control 23(1973), 407-438.

[3] J. M. Goethals and S. L. Snover, Nearly perfect binary codes, Discrete Math. 3(1972), 65-88.

[4] J. M. Goethals and H. C. A. van Tilborg, Uniformly packed codes, Philips Res. Repts. 30(1975), 9-36.

[5] F. J. MacWilliams, N. J. A. Sloane and J. M. Goethals, The MacWilliams identities for nonlinear codes, Bell System tech. J. 51(1972), 803-819.

[6] J. H. van Lint, "Coding Theory", Lecture Notes in Mathematics N° 201, Springer-Verlag, Berlin (1971).

[7] J. H. van Lint, Recent results on perfect codes and related topics, in "Combinatorics, Part I ", M. Hall and J. H. van Lint (eds.), Mathematical Centre Tracts N°55, Mathematisch Centrum, Amsterdam (1974), pp. 158-178.

AN INTRODUCTION TO ALGEBRAIC CODING THEORY

Giuseppe Longo
Istituto di Elettrotecnica ed Elettronica
University of Trieste
34100 TRIESTE, ITALY

1. Introduction

The following is a short introduction to some aspects of that part of coding and decoding theory which uses algebraic tools, and is therefore called algebraic coding theory. If the channel over which information is to be transmitted from source to user is noisy, and if it is not possible to modify the channel itself, then coding and decoding can help the user receive a better (i.e. less corrupted) reproduction of the channel input. Our basic assumption will be that information consists of a (potentially infinite) sequence of symbols (or digits) belonging to a finite alphabet (often binary). The input alphabet of a channel need not coincide with the output alphabet, but this is often the case. The effect of the noise is then of transforming, with some positive probability, every input symbol into a different symbol, and the mathematical description of the channel consists precisely of the set of all conditional probabilities of an output given an input. If the channel is memoryless, i.e. if past history has no influence on the current noise effect, the channel description is simplified enormously.

2. Codes

If $|A|$ denotes the size of the finite (input and output) alphabet A, then at most $|A|^n$ different messages can be communicated if one uses n-length sequences (or vectors) from A. To transmit information at a high rate (bits/channel symbol), it is desiderable that the number M of different messages transmissible with n-vectors is as close as possible to $|A|^n$, but if each of the $|A|^n$ n-sequences corresponds to a message, then every channel error (change in at least one of the n components) causes a decoding error, since the received n-tuple is a message, and is therefore accepted, but it differs from the message sent. On the other hand one of the recipes of Information Theory is that to achieve good results one should make n go to infinity. As n increases the probability that any n-tuple will contain at least one channel error increases too, and this causes decoding errors with very high probability. In order to avoid this unpleasant effect, one has to give up a portion of the transmission rate, i.e. one has to make M less than $|A|^n$ so that the M n-tuples to be associated with the M messages have to be chosen in some way among the $|A|^n$ n-tuples available. These M meaningful n-vectors are called <u>codewords</u> and they constitute a <u>code</u>. The codewords should be chosen in such a way as to maximize the minimum (Hamming) <u>distance</u> d between any two of them, this distance being the number of places where the two codewords differ. This is a consequence of

<u>Theorem 2.1</u>. If d is the minimum distance of a code(i.e. between any two distinct codewords), then no set of $[(d-1)/2]$ or less channel

errors can cause a decoding error, no matter what the transmitted codeword is.

Actually at least $[(d-1)/2] + 1$ errors are needed to make the received vector closer to a codeword different from the one transmitted. We then say that a code with minimum distance d is a $[(d-1)/2]$ -error correcting code. A code possessing M codewords each of length n and with minimum distance d will be denoted as a $[M, n, d]$ -code. The number $R = \frac{1}{n} \log_{|A|} M$ is called the rate of the code; always $R \leq 1$, and $R = 1$ iff $M = |A|^n$.

The coding theorist looks for $[M, n, d]$ -codes having large M (i.e. large rate), large d and small n. These are conflicting aims, and coding theory is largely an attempt to find a reasonable (and constructive) tradeoff between these aims.

If the number of errors e is close to $[(d-1)/2]$, one might feel that the decision implied by the correction procedure is too risky, although possible, and be inclined not to take any decision. The simple aknowledgement that an error has occurred is called error detection, and since detection can be performed as long as the received n-tuple is not a codeword, a minimum distance d allows detecting up to d-1 errors. An $[M, n, d]$ -code can be used to detect and/or correct errors. For example all received n-tuples lying at distance e or less from some codeword are decoded into that codeword, but if no codeword exixts at such a distance from the received n-tuple, the latter is not decoded. Therefore all n-tuples which contain at least e+1 but not more than d-(e+1) channel errors are recognized as possessing errors. Therefore if one chooses to correct e or less errors, and $e \leq d/2 - 1$, then a number t of errors can still be detected, with $e + 1 \leq t \leq d - e - 1$.

With respect to this mixed strategy the following theorem is relevant.

Theorem 2.2 An $[M, n, d]$ -code can correct e errors and detect t errors iff $d \geq t + e + 1$.

3. Linear codes

The minimum distance has been shown to be a very important parameter of a code, but its computation, for a given code, can be extremely difficult. Consider a binary (i.e. $|A| = 2$) $[M, 100, d]$ - code ($n = 100$ is a very reasonable assumption in practice) with rate $R = \frac{1}{2}$. The number of its codewords is then $M = 2^{50}$, and finding the minimum distance can be very difficilt indeed if no restriction is put upon the structure of the code.

We shall now proceed to impose such a structure, and arrive at the notion of linear code, which is the simplest and most general algebraic code. Assume the alphabet A to be a finite field and write q instead of $|A|$ (q is some prime power), i.e. $A = GF(q)$, and consider the set F_n of all n-tuples over A to be a vettor space over GF(q), with the sum defined componentwise mod q.

Definition 3.1 A linear code of length n and dimension k over GF(q) is a linear subspace F (n, k) of F_n, of dimension k, $0 < k \leq n$.

A linear code can therefore be assigned through any of its bases, i.e. through any set of k linearly independent codewords (or generators), which are usually arranged in a k x n matrix G called the <u>generator matrix</u> of the code. Remark that a linear code of dimension k has $M = q^k$ codewords. It will be often indicated by "a linear (n,k) code".

The linear structure af these codes implies that given any two codewords $\underline{v}_1$, $\underline{v}_2$ their difference $\underline{v}_1 - \underline{v}_2$ belongs to the code, and if d (. , .) indicates the distance:

(3.1) $$d(\underline{v}_1 , \underline{v}_2) = d(\underline{v}_1 - \underline{v}_2 , \underline{0}) \triangleq \text{weight } (\underline{v}_1 - \underline{v}_2)$$

Therefore the <u>weight</u> (wt) of a vector $\underline{v}$ is the number of non-zero coordinates in it. As a consequence, the set of all the distances between pairs of codewords of a linear code coincides with the set of all the weights of its codewords.

A linear subspace F(n, k) of F_n can be also assigned by means of n - k = r linear homogeneous equations which must be obeyed by the vectors of F (n, k). Such eqaations (parity-check equations) are often given by means of the r x n matrix H of the coefficients, whose rank is r. H is called a <u>parity-check matrix</u> for F (n, k). Consequently

(3.2) $$G \cdot H^T = \underline{0}$$

for every generator matrix G of F(n, k). Eq. (3.2) shows that the parity-check matrix H is by no means unique, as it is clearly seen if one considers the linear code F (n, n-k) generated by H: every generator matrix of F (n, n-k) is a parity-check matrix for the code F (n, k) generated by G. Considered as subspaces of F_n, F(n, k) and F (n, n-k) are orthogonal spaces, and they are called <u>dual codes</u>. If F (n, k) = F(n,n-k) ($k = \frac{n}{2}$, n even), then the code is called <u>self-dual</u>.

Equation (3.2) is the characteristic equation of the code

generated by G, since it implies

(3.3) $\quad \underline{v} \cdot H^T = \underline{0}$

for all and only the vectors $\underline{v}$ in the space spanned by G. For an arbitrary n-vector $\underline{u}$, the r-vector

(3.4) $\quad \underline{u} \cdot H^T = \underline{s}\ (\underline{u})$

is called syndrome of $\underline{u}$. Therefore:

Theorem 3.1 An n-vector belongs to the (n, k) linear oode checked by H iff its syndrome is zero.

Consider now in F_n the relation $\mathcal{R}$ defined by

(3.5) $\quad \underline{u}_1 \mathcal{R}\, \underline{u}_2 \quad \text{iff} \quad \underline{u}_1 - \underline{u}_2 \in F(n, k)$

where F (n, k) is an (n, k) linear code. $\mathcal{R}$ is an equivalence relation, compatible in the additive group of F_n and its classes are the cosets of F (n, k). To each cosets corresponds a different syndrome, and based on this we have the following algorithm, due to Slepian, which corresponds to a maximum likekihood procedure (optimal, i.e. minimizing the error probability, if the messages are equiprobable):

Decoding algorithm 3.1:

Step 1: Let $\underline{u}$ be the received vector; compute its syndrome $\underline{s}(\underline{u})$ by (3.4). Go to Step 2.

Step 2: If $\underline{s}(\underline{u}) = \underline{0}$, $\underline{u}$ belongs to the code, and is considered as the transmitted vector. If $\underline{s}(\underline{u}) \neq \underline{0}$ go to Step 3.

Step 3: Find the coset corresponding to $\underline{s}(\underline{u})$, and let $\underline{v}$ be (one of) the minimum weight vector(s) in the coset (coset leader). Go to step 4.

Step 4: Add the coset leader to the received vector u. The resulting vector is in the code and is considered as the transmitted vector.

Remark that the error capabilities of a linear code depend on the weights of the coset leaders since the latter are precisely the correctable error patterns. There are $\binom{n}{t}(q-1)^t$ n-vectors of weight t, and q^{n-k} cosets; therefore if a code can correct all error patterns of weight 0, 1, ..., e, then necessarily the following upper bound (due to Hamming) on the code rate $\frac{k}{n}$ holds:

$$q^{n-k} \geq \sum_{t=0}^{e} \binom{n}{t} (q-1)^t \qquad (3.6)$$

A code is said to be _perfect_ (or close-packed) if it satisfies eq. (3.6) with equality for some e, and therefore can correct all errors of weight $\leq e$ and no error of greater weight. As an example, for q=2 and m any integer ≥ 2, consider the codes having an $(m \times 2^m - 1)$ parity-check matrix whose columns are the binary expansions of the integers from 1 to $2^m - 1$. These codes (binary Hamming codes) are $(2^m-1, 2^m-m-1)$ linear codes having minimum distance 3, and can therefore correct 1 error in $2^m - 1$ positions. They satisfy (3.6) with equality (e=1), and constitute an infinite class of perfect codes. Hamming codes can be defined for an arbitrary (i.e. nonbinary) alphabet, and again they are perfect. Another infinite class of perfect codes is provided by the trivial repetition codes, and no other class of perfect codes exists.

4. Weight enumerators

Let now A_i be the number of codewords having weight i $(0 \leq i \leq n)$ in a given code. If the code is linear with minimum weight d, then $A_o=1$, and $A_1 = A_2 = \dots = A_{d-1} = 0$, $A_d \neq 0$. The polynomial

$$(4.1) \qquad W(x, y) = \sum_{i=0}^{n} A_i \, x^{n-i} \, y^i$$

is called the weight enumerator of the code; it is a homogeneous polynomial of degree n. The weight distribution, or weight spectrum, of the code, i.e. the set $[A_o, \dots, A_n]$, gives a great deal of information about the code, e.g. its minimum weight, but for nonbinary codes it is sometimes useful to heve more details. If C is a code over GF(q) $\equiv \{0, w_1, \dots, w_{q-1}\}$, the composition of an n-vector $\underline{v} \in F_n$ is defined as $s(\underline{v}) = [s_o(\underline{v}), s_1(\underline{v}), \dots, s_{q-1}(\underline{v})]$, where $s_i(\underline{v})$ is the number of times that w_i appears in $\underline{v}$. If A(s) denotes the number of codewords $\underline{u} \in C$ having composition s, then the complete weight enumerator of C is

$$(4.2) \qquad W^*(x_o, \dots, x_{q-1}) = \sum_{s} A(s) \, x_o^{so} \, x_1^{s1} \dots x_{q-1}^{s_{q-1}}$$

the sum being over all possible composition vectors $s = [s_o, \dots, s_{q-1}]$.

One of the most important results of coding theory, due to Mrs. MacWilliams, states that the weight enumerator of the dual code $C^{\perp}$ of a linear code C is completely determined by the weight enumerator of C. More precisely, we have

Theorem 4.1 If C is a linear (n, k) code over GF(q), and $C^{\perp}$ is the dual of C, then

$$(4.3) \qquad W_{C^\perp}(x, y) = \frac{1}{q^k} W_C(x + (q-1)y, x - y).$$

In particular, for binary codes ($q = 2$), the following corollary is true

<u>Theorem 4.2</u> For a binary (n, k) code C, the weight enumerator of $C^\perp$ is given by

$$(4.4) \qquad W_{C^\perp}(x, y) = \frac{1}{2^k} W_C(x + y, x - y)$$

Both theorem 4.1 and theorem 4.2 refer to the weight enumerators. A similar result holds also for the complete weight enumerators. Let $q = p^a$, then $GF(q)$ can be obtained extending $GF(p)$ by means of an irreducible polynomial $f(x)$ of degree a; any element α of $GF(q)$ can then be expressed uniquely as

$$(4.5) \qquad \alpha = \alpha_o + \alpha_1 \beta + \ldots + \alpha_{a-1} \beta^{a-1} \qquad (\alpha_i \in GF(p))$$

where β is a root of $f(x)$. Let now $\xi = e^{2\pi i/p}$ be a complex p-th root of unity; the mapping $\chi : \alpha \rightarrow \xi^{\alpha_o}$ is a character of $GF(q)$, and is a homomorphism from the additive group of $GF(q)$ to the multiplicative group of the p-th roots of unity. We can now state

<u>Theorem 4.3</u> Let C be an (n, k) code over $GF(q)$. Then the complete weight enumerator of $C^\perp$ is given by

$$(4.6) \qquad W^*_{C^\perp}(x_o, \ldots, x_{q-1}) = \frac{1}{q^k} W^*_C\left(\sum_{j=0}^{q-1} \chi(w_o w_j) x_j, \ldots, \sum_{j=0}^{q-1} \chi(w_{q-1} w_j) x_j\right).$$

For $q = 2$ eq.(4.6) reduces obviously to eq. (4.4).

When the code C is self-dual, i.e. $C = C^\perp$, the MacWilliams theorems 4.1, 4.2, 4.3 yield identities that the weight enumerator must satisfy. From Theorem 4.2 and from the homogeneous nature of $W(x, y)$, one immediately gets in the binary case:

$$W(x, y) = W\left(\frac{x+y}{\sqrt{2}}, \frac{x-y}{\sqrt{2}}\right) \qquad (4.7)$$

From this it is possible to prove the following remarkable result (due to Gleason):

Theorem 4.4 The weight enumerator $W_C(x, y)$ of a binary self-dual code C is a polynomial in the polynomials

$$\varphi_2 = x^2 + y^2 \quad \text{and} \quad \varphi_8 = x^8 + 14\, x^4 y^4 + y^8. \qquad (4.8)$$

Incidentally, remark that φ_2 and φ_8 of (4.8) are the weight enumerators of the [2, 2, 2] binary code and of the [16,8,4] binary extended Hamming code, respectively.

Theorem 4.4, along with other results of the same type allows computing the weight distribution (and consequently the minimum weight) for a variety of codes.

We only mention that the proof of Gleason's theorem 4.4 is based upon the theory of invariants of finite groups.

5. Cyclic codes

Although the decoding algorithm 3.1, based upon the syndrome, is simple and optimal for equiprobable messages, step 3 involves the use of a dictionary of $2^{n-k} = 2^{n(1-R)}$ entries, and therefore the complexity of this step grows too fast with n.

To avoid this shortcoming, it is desirable to use codes having more structure, so that the use of dictionaries is avoided. Such are the cyclic codes. Consider the principal ideal ring R of all the polynomials over GF(q), and let as usual F_n be the n-dimensional vector space over GF(q). Assume (n, q) = 1. The polynomial $x^n - 1$ generates in R an ideal S, and the elements of the residue class ring R/S can be represented by polynomials of degree $< n$ over GF(q). Consider the mapping which associates a vector $\underline{a} = (a_o, a_1, \dots a_{n-1}) \in F_n$ to the polynomial $a_o + a_1 x + \dots + a_{n-1} x^{n-1}$. This mapping is an isomorphism between F_n and the additive group of R/S, and moreover it is also possible to "multiply two vectors" of F_n, simply by multiplying the two corresponding polynomials mod $(x^n - 1)$.

Notice that multiplying a vector by the polynomial x corresponds to shift it cyclically, i.e. $(a_o, a_1, \dots, a_{n-1})$ is transformed into $(a_{n-1}, a_o, \dots, a_{n-2})$.

After these preliminaries, it is possible to give the following definition:

Definition 5.1 A linear code C is called a cyclic code if it is a cyclic subspace of F_n, i.e. if $a(x) = a_o + a_1 x + \dots + a_{n-1} x^{n-1} \in C$ implies $x\, a(x) = a_o x + \dots + a_{n-2} x^{n-1} \in C$.

An algebraic characterization of cyclic codes is offered by the following theorem:

Theorem 5.1 C is a cyclic code iff C is an ideal of R/S.

We now state without proof a number of facts about cyclic codes.

Given n and q relatively prime it is possible to find all

cyclic codes of length n over GF(q) just factoring $x^n - 1$ into its irreducible factors, i.e.

(5.1) $$x^n - 1 = f_1(x)\ f_2(x)\ \dots\ f_t(x)$$

and taking any one of the 2^t factors of $x^n - 1$ as generator. The t cyclic codes generated by the irreducible factors $f_i(x)$ are maximal cyclic codes in the lattice of the ideals of R/S ordered by inclusion.

If g(x) is a factor of $x^n - 1$, i.e. if it is the generator polynomial of an n-length cyclic code, then $h(x) = (x^n - 1)/g(x)$ is a parity-check for the code generated by g(x) in the sense that $a(x).h(x) \equiv 0 \bmod (x^n - 1)$ for all polynomials a(x) of the code. The degree k of h(x) equals the dimension of the code.

The description of a cyclic code is more economical than that of a non-cyclic linear code, in that to identify the cyclic code, it is sufficient to assign its generator polynomial and actually a generator matrix can be constructed with k successive cyclic shifts of the generator polynomial. Similarly, a parity-check matrix can be assigned by means of the check polynomial h(x) constructing its n-k successive cyclic shifts.

Clearly the generator polynomial can be assigned either through its coefficients or through its n-k zeros in a suitable extension field of GF(q). Since the zeros of g(x) always appear in sets of conjugates, it is sufficient to give one zero for each of these sets. Every set of conjugate zeros corresponds to an irreducible factor $f_i(x)$ of $x^n - 1$ appearing in g(x).

Since a polynomial a(x) belongs to a cyclic code C generated by g(x) iff a(x) is a multiple of g(x), then a(x) belongs to the code iff the zeros of $g(x)$ are also zeros of $a(x)$. Let $\alpha_1, \alpha_2, \ldots, \alpha_r$ be the zeros of $g(x)$; then the equations

(5.2) $$a(\alpha_1) = 0, \quad a(\alpha_2) = 0, \ldots, a(\alpha_r) = 0$$

are a set of necessary and sufficient conditions for $a(x)$ to belong to C, and are precisely a particular case of eq. (3.3). More than that, the components of the syndrome of a vector $a(x)$, defined by (3.4), are now expressed by

(5.3) $$a(\alpha_1), \ a(\alpha_2), \ldots, a(\alpha_r)$$

and actually the parity-check matrix can be constructed by expressing the zeros $\alpha_1, \ldots, \alpha_r$ by means of a primitive element of GF(q).

To give an example, the following theorem characterizes the binary Hamming codes, already defined in section 3, as cyclic codes.

<u>Theorem 5.2</u> The binary cyclic code of length $n = 2^m - 1$ whose generator polynomial is the minimal polynomial of a primitive element of $GF(2^m)$ is the Hamming code of length n and dimension $n - m$.

As we already saw, Hamming codes of any length $n = 2^m - 1$ correct only one error, and therefore their error-correction capability is very scanty. Their rate $R = \frac{k}{n} = 1 - m/2^m - 1$, however, approaches 1 as n, and consequently m, goes to infinity. A somewhat complementary behaviour is exibited by the $1/n$ - rate repetition codes (n odd), which can correct up to $(n-1)/2$ errors. The efforts of many coding theorists have been addressed to fill the gap between these two extremal classes of codes.

Filling this gap amounts to find a constructive (infinite)

class of codes for which, given a positive error-correction capability, the rate does not vanish as $n \to \infty$; or, conversely, for which, given a positive rate, the error correction capability does not vanish as $n \to \infty$. Most attempts along this direction have failed, although some of the existing classes of codes are reasonably good for intermediate values of the block-length n. One such family of codes are the BCH codes, which form a particular infinite class of cyclic codes. More precisely:

<u>Definition 5.2</u> A (primitive narrow sense) BCH code of length n over GF(q), where $n = q^m - 1$, is a cyclic code whose generator polynomial is the monic polynomial of lowest degree over GF(q) having $\alpha, \alpha^2, \ldots, \alpha^{d-1}$ as roots, where α is a primitive element of $GF(q^m)$, and d is some positive integer, called the designed distance.

It can be proved that the minimum distance of such a BCH code is not smaller than the designed distance d.

Since BCH codes form an infinite class, it is appropriate to ask how they behave asymptotically. A classical result in coding theory (due to Gilbert and Varshamov) states that if we fix the rate R ($0 < R < 1$) then there exist binary codes of length n, with k information digits and minimum distance d for which: i) $k/n \geq R$ and ii) $d/n \geq H^{-1}(1-R)$, where $H^{-1}(x)$ is the inverse of the usual entropy function $H(x) = -x\log x - (1 - x) \log(1 - x)$. On the other hand only in the cases R = 0 (repetition codes) and R = 1 (Hamming codes) families of codes meeting this bound are known.

Unfortunately there exists no infinite sequence of (primitive) BCH codes over GF(q) for which both the rate k/n and d/n are bounded away from zero. Therefore long BCH codes are said to be bad.

An open question is whether such good sequences of codes can be found in the larger family of cyclic codes.

6. Other families of codes

A particular subclass of the (nonbinary) BCH codes are the Reed-Solomon (RS) codes, for which the length n is q-1. They can be used to construct certain concatenated codes, which are defined as follows: consider kK binary information symbols divided into K k-tuples, each of which is thought of as an element of $GF(2^k)$. These K elements of $GF(2^k)$ are encoded into an N-tuple $a_o\ a_1 \ldots a_{N-1}$ over $GF(2^k)$ by an outer encoder. Each a_i (which is still a binary k-tuple) is now encoded by an inner encoder into a binary n-tuple b_i. Then $b_o\ b_1 \ldots b_{N-1}$ is the codeword to be transmitted over the channel. Frequently a RS code is used as outer code. The overall code is a binary code of length nN, dimension kK and rate kK/nN.

Some classes of concatenated codes perform very well. One example is provided by Justesen's codes, which can be thought of as concatenated codes where the inner encoder uses N distinct codes. Justesen codes form an infinite class of good codes, in the sense specified above. It is true, however, that they are not as good as the good codes promised by the Gilbert-Varshamov theorem.

Another important family of codes, defined in a way that has little to do with cyclic codes, are the Goppa codes. Goppa codes are linear codes which are best defined through a polynomial g(z) with coefficients in $GF(q^m)$, q a prime power, m any integer. More precisely:

Definition 5.3 Let L be the set of all elements of $GF(q^m)$ which are not roots of g(z). Then there exists a Goppa code with symbol field GF(q) and length (L), whose codewords $\underline{w}$ satisfy the condition

$$\sum_{i \in L} \frac{w_i}{z-i} \equiv 0 \mod g(z)$$

It is interesting to remark that the class of Goppa codes include the class of primitive BCH codes, and, in fact, no other cyclic codes is a Goppa code. It is, however, possible to slightly modify these codes to find some close relationships with cyclic codes. An important feature of irreducible Goppa codes (namely, those whose polynomial g(z) is irreducible) is that most of them asymptotically meet the Gilbert - Varshamov bound.

Bibliographical Note

It was felt that the reader of this introductory survey could benefit more from a list of selected text-books and review papers than from a more conventional bibliography containing the papers where the original results first appeared. Consequently what follows is a short list of essential items suitable for the non-coding theorist seriously interested in coding theory.

A - Books

[1] Berlekamp, E.R., Algebraic Coding Theory, McGraw-Hill, New York, 1968. (Quoted as the best book available. Suited for mathematicians rather than engineers.)

[2] van Lint, J.H., Coding Theory, Springer-Verlag, Berlin, 1971. (Excellent and concise introduction for mathematicians.)

[3] Peterson, W.W., Error-Correcting Codes, MIT Press, Cambridge, Mass.

1961. (Good introduction for engineers.)

[4] Peterson, W.W. and Weldon E.J.,Jr., Error-Correcting Codes, Second Ed., MIT Press, Cambridge, Mass. 1972. (A much expanded version of [3].)

[5] MacWilliams, F.J. and Sloane, N.J.A., Combinatorial Coding Theory. In preparation. (Very promising thorough treatment.)

B - Chapters from Books

[6] Gallager, R.G., Information Theory and Reliable Communication,Wiley, New York, 1969, Ch. 6.

[7] Stiffler, J.J., Theory of Synchronous Communications, Prentice-Hall, 1971, Ch. 13.

[8] Wozencraft, J.M. and Jacobs, I.M., Principles of Communication Engineering, Wiley, New York, 1965, Ch. 6.

C - Short Surveys

[9] Berlekamp, E.R., A Survey of Algebraic Coding Theory, CISM - Springer Verlag, Udine-Vienna, 1970.

[10] Sloane, N.J.A., A Short Course on Error Correcting Codes, CISM - Springer Verlag, Udine - Vienna, 1975.

[11] Solomon, G., Algebraic Coding Theory, Ch. 6 of Communication Theory, A.V. Balakrishnan, editor,McGraw-Hill, New York, 1968.

D - Collections of Papers

[12] Mann, H.B., editor, Error Correcting Codes, Wiley, New York, 1968.

[13] Golomb, S.W., editor, Digital Communications with Space Applications, Prentice-Hall, 1964.

[14] Longo, G., editor, Coding and Complexity, CISM - Springer Verlag, in preparation.

E - Periodicals

[15] IEEE (Institute of Electrical and Electronics Engineers) Transactions on Information Theory.

[16] Information and Control.

[17] Problemy Peredachi Informatsii (Russian).

[18] Discrete Mathematics.

ALGEBRAIC STRUCTURE AND FINITE DIMENSIONAL NONLINEAR ESTIMATION*

Steven I. Marcus
Department of Electrical Engineering
University of Texas at Austin
Austin, Texas 78712, U.S.A.

Alan S. Willsky
Department of Electrical Engineering
and Computer Sciences
Massachusetts Institute of Technology
Cambridge, Mass. 02139, U.S.A.

INTRODUCTION

Optimal recursive state estimators have been derived for very general classes of nonlinear stochastic systems.[1,2] The optimal estimator requires, in general, an infinite dimensional computation to generate the conditional mean of the system state given the past observations. This computation involves either the solution of a stochastic partial differential equation for the conditional density or an infinite set of coupled ordinary stochastic differential equations for the conditional moments. However, the class of linear stochastic systems with linear observations and white Gaussian plant and observation noises has a particularly appealing structure, because the optimal state estimator consists of a finite dimensional linear system -- the Kalman-Bucy filter.[3]

In this paper we exploit the algebraic structure of certain other classes of systems, in order to prove that the optimal estimators for these systems are finite dimensional. The general class of systems is given by a linear Gauss-Markov process ξ which feeds forward into a nonlinear system with state x. Our goal is to estimate ξ and x given noisy linear observations of ξ. Specifically, consider the system

$$d\xi(t) = F(t)\xi(t)dt + G(t)dw(t) \tag{1}$$

$$dx(t) = a_0(x(t))dt + \sum_{i=1}^{N} a_i(x(t))\,\xi_i(t)dt \tag{2}$$

$$dz(t) = H(t)\xi(t)dt + R^{1/2}(t)\,dv(t) \tag{3}$$

where $\xi(t)$ is an n-vector, x(t) is a k-vector, z(t) is a p-vector, w and v are independent standard Brownian motion processes, $R > 0$, $\xi(0)$ is a Gaussian random variable independent of w and v, x(0) is independent $\xi(0)$, w, and v, and $\{a_i, i=0,\ldots,N\}$ are

*This work was supported by NSF under Grant GK-42090 and by AFOSR under Grant 72-2273.

analytic functions of x. It will be assumed that $[F(t), G(t), H(t)]$ is completely controllable and observable. Also, we define $Q(t) \triangleq G(t)G'(t)$.

The optimal estimate, with respect to a wide variety of criteria, of $x(t)$ given the observations $z^t \triangleq \{z(s), 0 \leq s \leq t\}$, is the conditional mean $\hat{x}(t|t)$ (also denoted by $E^t[x(t)]$ or $E[x(t) \mid z^t]$)[4]. Thus our objective is the computation of $\hat{\xi}(t \mid t)$ and $\hat{x}(t \mid t)$. The computation of $\hat{\xi}(t \mid t)$ can be performed by the finite dimensional (linear) Kalman-Bucy filter; moreover, the conditional density of $\xi(t)$ given z^t is Gaussian with mean $\hat{\xi}(t \mid t)$ and nonrandom covariance $P(t)$.[3,4] However, the computation of $\hat{x}(t \mid t)$ requires in general an infinite dimensional system of equations. The purpose of this paper is to show that if $x(t)$ is characterized by a certain type of Volterra series expansion, or if $x(t)$ satisfies a certain type of bilinear equation, then $\hat{x}(t \mid t)$ can be computed with a finite dimensional nonlinear estimator. We will only state the major results and present an example here; the detailed proofs may be found in Marcus.[5]

VOLTERRA SERIES AND FINITE DIMENSIONAL ESTIMATION

As shown by Brockett[6,7] and d'Alessandro, Isidori, and Ruberti[8] in the deterministic case, considerable insight can be gained by considering the Volterra series expansion of the system (2). The Volterra series expansion for the ith component of x is given by

$$x_i(t) = w_{0i}(t) + \sum_{j=1}^{\infty} \int_0^t \cdots \int_0^t \sum_{k_1,\ldots,k_j=1}^{n} w_{ji}^{(k_1,\ldots,k_j)}(t,\sigma_1,\ldots,\sigma_j) \cdot \xi_{k_1}(\sigma_1)\ldots\xi_{k_j}(\sigma_j)\,d\sigma_1\ldots d\sigma_j \tag{4}$$

where the <u>jth order kernel</u> $w_{ji}^{(k_1,\ldots,k_j)}$ is a locally bounded, piecewise continuous function. We will consider, without loss of generality,[6] only <u>triangular kernels</u> which satisfy $w_{ji}^{(k_1,\ldots,k_j)}(t,\sigma_1,\ldots,\sigma_j)=0$ if $\sigma_{\ell+m} > \sigma_m$; $\ell, m=1,2,3,\ldots$ We say that a kernel $w(t,\sigma_1,\ldots,\sigma_j)$ is <u>separable</u> if it can be expressed as a finite sum

$$w(t,\sigma_1,\ldots,\sigma_j) = \sum_{i=1}^{m} \gamma_0^i(t)\,\gamma_1^i(\sigma_1)\,\gamma_2^i(\sigma_2)\ldots\gamma_j^i(\sigma_j) \tag{5}$$

With these preliminary concepts, the major results can be stated.

<u>Theorem 1</u>: Consider the linear system described by (1), (3), and define the

scalar-valued process

$$x(t) = e^{\xi_j(t)} \eta(t) \tag{6}$$

where η is a finite Volterra series in ξ (i.e., the expansion (4) has a finite number of terms) with separable kernels. Then $\hat{\eta}(t|t)$ and $\hat{x}(t|t)$ can be computed with a finite dimensional system of nonlinear stochastic differential equations driven by the innovations $d\nu(t) \triangleq dz(t) - H(t)\hat{x}(t|t)dt$.

Theorem 2: Consider the linear system (1), (3), and define the scalar-valued processes

$$\eta(t) = \int_0^t \int_0^{\sigma_1} \cdots \int_0^{\sigma_{j-1}} \xi_{k_1}(\sigma_{m_1}) \cdots \xi_{k_i}(\sigma_{m_i}) \gamma_1(\sigma_1) \cdots \gamma_j(\sigma_j) d\sigma_1 \cdots d\sigma_j \tag{7}$$

$$x(t) = e^{\xi_\ell(t)} \eta(t) \tag{8}$$

where $\{\gamma_i\}$ are deterministic functions of time and $i>j$. Then $\hat{\eta}(t|t)$ and $\hat{x}(t|t)$ can be computed with a finite dimensional system of nonlinear stochastic differential equations driven by the innovations.

The distinction between Theorems 1 and 2 lies in the fact that $i>j$ in (7) -- i.e., there are more ξ_k's than integrals. On the other hand, each term in the finite Volterra series in (6) has $i=j$ and the σ_{m_k} are distinct. As Brockett[6] remarks, we can consider (7) as a single term in a Volterra series if the kernel is allowed to contain impulse functions. It can be shown[5] that a term (7) with $i>j$ (more integrals than ξ_k's) can be rewritten as a Volterra term with $i=j$, so Theorem I also applies in this case.

The basic technique employed in the proofs[5] of Theorem 1 and 2 is the augmentation of the state of the original system with the processes which are required in the nonlinear filtering equation[1,2,4] for $\hat{x}(t|t)$. For the classes of systems considered here, it is shown that only a finite number of additional states are required. The theorems are proved by induction on the order of the Volterra term. We now sketch the proof of the first induction step for Theorem 1; the complete proofs are presented in Marcus.[5]

Proof of Theorem 1 (Sketch): We consider one term in the finite Volterra series; since the kernels are separable, we can assume without loss of generality that this term has the form

$$\eta(t) = \int_0^t \int_0^{\sigma_1} \cdots \int_0^{\sigma_{j-1}} \xi_{k_1}(\sigma_1)\ldots\xi_{k_j}(\sigma_j)\,\gamma_1(\sigma_1)\ldots\gamma_j(\sigma_j)\,d\sigma_1\ldots d\sigma_j \tag{9}$$

The theorem is proved by induction on j, the order of the Volterra term (9).

If j=1, then

$$\eta(t) = \int_0^t \gamma_1(\sigma_1)\xi_{k_1}(\sigma_1)\,d\sigma_1 \tag{10}$$

and $\eta(t)$ is linear function of ξ. Hence, if the state ξ of (1) is augmented with η, the resulting system is also linear. Then the Kalman-Bucy filter for the system described by (1), (3), (10) generates $\hat{\xi}(t|t)$ and $\hat{\eta}(t|t)$. In order to prove that $\hat{x}(t|t)$ is "finite dimensionally computable" (FDC), we need the following lemma.[5] First we define, for $\sigma_1, \sigma_2 \leq t$, the conditional cross-covariance matrix

$$P(\sigma_1,\sigma_2,t) = E[(\xi(\sigma_1) - \hat{\xi}(\sigma_1|t))(\xi(\sigma_2) - \hat{\xi}(\sigma_2|t))' \,|\, z^t] \tag{11}$$

(where $\hat{\xi}(\sigma|t) = E[\xi(\sigma)|z^t]$.

Lemma 1: The joint conditional density $p_{\xi(\sigma_1),\xi(\sigma_2)}(\nu,\nu'|z^t)$ is Gaussian with nonrandom conditional cross-covariance $P(\sigma_1,\sigma_2,t)$--i.e., $P(\sigma_1,\sigma_2,t)$ is independent of $\{z(s), 0 \leq s \leq t\}$.

This lemma allows the off-line computation of $P(\sigma_1,\sigma_2,t)$ via the equations of Kwakernaak[15] (for $\sigma_1 \leq \sigma_2$)

$$P(\sigma_1,\sigma_2,t) = P(\sigma_1)\,\Psi'(\sigma_2,\sigma_1)$$

$$- P(\sigma_1)\left[\int_{\sigma_2}^t \Psi'(\tau,\sigma_1)\,H'(\tau)R^{-1}(\tau)\,H(\tau)\Psi(\tau,\sigma_2)\,d\tau\right]P(\sigma_2) \tag{12}$$

$$\frac{d}{dt}\Psi(t,\tau) = [F(t) - P(t)H'(t)R^{-1}(t)H(t)]\,\Psi(t,\tau);\quad \Psi(\tau,\tau) = I \tag{13}$$

where the Kalman filter error covariance matrix $P(t) \triangleq P(t, t, t)$ is computed via the Riccati equation

$$\dot{P}(t) = F(t)P(t) + P(t)F'(t) + Q(t) - P(t)H'(t)R^{-1}(t)H(t)P(t)$$

$$P(0) = P_0 \tag{14}$$

Recall[4] that the characteristic function of a Gaussian random vector y with mean m and covariance P is given by

$$M_y(u) = E[\exp(iu'y)] = \exp[iu'm - \frac{1}{2} u'Pu] \tag{15}$$

Hence, by taking partial derivatives of the characteristic function, we have

$$E^t[x(t)] = \int_0^t \gamma_1(\sigma)\, E^t[e^{\xi_j(t)}\, \xi_{k_1}(\sigma)]d\sigma$$

$$= \int_0^t \gamma_1(\sigma)[\hat{\xi}_{k_1}(\sigma | t) + P_{k_1,j}(\sigma,t,t)]e^{\hat{\xi}_j(t|t) + \frac{1}{2}P_{jj}(t)} d\sigma$$

$$= \left\{\int_0^t \gamma_1(\sigma)\, P_{k_1,j}(\sigma,t,t)d\sigma + E^t\left[\int_0^t \gamma_1(\sigma)\xi_{k_1}(\sigma)d\sigma\right]\right\}$$

$$\cdot e^{\hat{\xi}_j(t|t) + \frac{1}{2}P_{jj}(t)}$$

$$= \left\{\int_0^t \gamma_1(\sigma)\, P_{k_1,j}(\sigma,t,t)d\sigma + \hat{\eta}(t|t)\right\} e^{\hat{\xi}_j(t|t) + \frac{1}{2}P_{jj}(t)} \tag{16}$$

Since the first term in (16) is nonrandom and $\hat{\eta}(t|t)$ and $\hat{\xi}(t|t)$ can be computed with a Kalman-Bucy filter, $\hat{x}(t|t)$ is indeed FDC for the case j=1.

The general induction step in the proof of Theorem 1 relies heavily on further properties of Gaussian random processes,[16] linear smoothing,[9] and the realization of finite Volterra series with bilinear systems.[6] The proof of Theorem 2 is almost identical. The following example illustrates the basic concepts of these theorems; this example is a special case of Theorem 2. However, we will need one preliminary lemma.[5]

Lemma 2: The conditional cross-covariance satisfies

$$P(\sigma, t, t) = K(t,\sigma)P(t) \tag{17}$$

where

$$\frac{d}{dt}K'(t,\sigma) = -[F'(t) + P^{-1}(t)\, Q(t)]\; K'(t,\sigma);\; K'(\sigma,\sigma) = I \tag{18}$$

Example 1: Consider the system described by

$$\begin{bmatrix} d\xi_1(t) \\ d\xi_2(t) \end{bmatrix} = \begin{bmatrix} -\alpha & 0 \\ 0 & -\beta \end{bmatrix} \begin{bmatrix} \xi_1(t) \\ \xi_2(t) \end{bmatrix} dt + \begin{bmatrix} dw_1(t) \\ dw_2(t) \end{bmatrix} \tag{19}$$

$$dx(t) = (-\gamma x(t) + \xi_1(t)\xi_2(t))dt \tag{20}$$

$$\begin{bmatrix} dz_1(t) \\ dz_2(t) \end{bmatrix} = \begin{bmatrix} \xi_1(t) \\ \xi_2(t) \end{bmatrix} dt + \begin{bmatrix} dv_1(t) \\ dv_2(t) \end{bmatrix} \tag{21}$$

where $\alpha, \beta, \lambda > 0$, w_1, w_2, v_1, and v_2 are independent, zero mean, unit variance Wiener processes, $\xi_1(0)$ and $\xi_2(0)$ are independent Gaussian random variables which are also independent of the noise processes, and $x(0) = 0$ (see Figure 1).

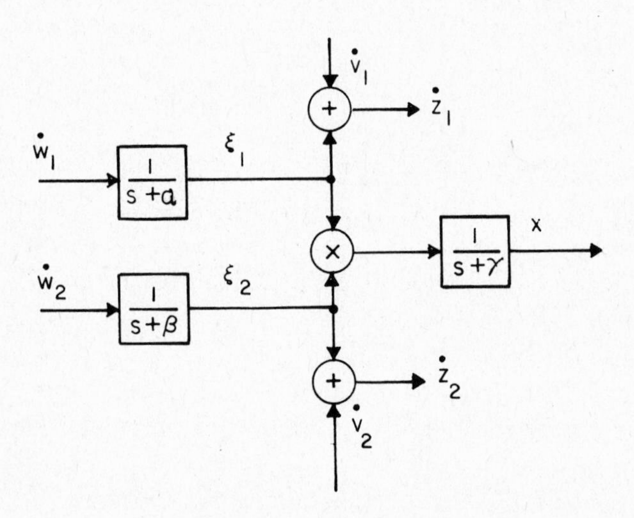

Figure 1. Block Diagram of the System of Example 1

The conditional expectation $\hat{x}(t|t)$ satisfies the nonlinear filtering equation[1,2,4]

$$d\hat{x}(t|t) = E^t[-\gamma x(t) + \xi_1(t)\,\xi_2(t)]dt$$

$$+ \{E^t[\int_0^t e^{-\gamma(t-s)}\xi_1(s)\,\xi_2(s)ds \cdot \xi'(t)]$$

$$- E^t[\int_0^t e^{-\gamma(t-s)}\xi_1(s)\,\xi_2(s)ds]\,\hat{\xi}'(t|t)\}\,d\nu(t) \qquad (22)$$

where $\xi(t) = [\xi_1(t), \xi_2(t)]'$ and the innovations process ν is given by

$$d\nu(t) = dz(t) - \hat{\xi}(t|t)dt \qquad (23)$$

Recall that the conditional covariance $P(t)$ of $\xi(t)$ given z^t satisfies the Riccati equation (14). Since $\xi_1(0)$ and $\xi_2(0)$ are independent, it is not difficult to show that $P_{12}(t) = P_{21}(t) = 0$ for all t. From (17) - (18) we can compute

$$P(\sigma,t,t) = \begin{bmatrix} P_{11}(t)\exp[\alpha(t-\sigma) - \int_\sigma^t P_{11}^{-1}(s)ds] & 0 \\ 0 & P_{22}(t)\exp[\beta(t-\sigma) - \int_\sigma^t P_{22}^{-1}(s)ds] \end{bmatrix} \qquad (24)$$

These facts and an identity from Miller[16] imply that the transpose of the gain term in (22) is

$$E^t[\int_0^t e^{-\gamma(t-s)}\xi_1(s)\,\xi_2(s)\,\xi(t)ds] - E^t[\int_0^t e^{-\gamma(t-s)}\xi_1(s)\,\xi_2(s)ds]\,\hat{\xi}(t|t)$$

$$= \int_0^t e^{-\gamma(t-s)}(E^t[\xi_1(s)\,\xi_2(s)\,\xi(t)] - E^t[\xi_1(s)\,\xi_2(s)]\,E^t[\xi(t)])ds$$

$$= E^t\left\{\int_0^t e^{-\gamma(t-s)}\begin{bmatrix} 0 & P_{11}(s,t,t) \\ P_{22}(s,t,t) & 0 \end{bmatrix}\begin{bmatrix} \xi_1(s) \\ \xi_2(s) \end{bmatrix} ds\right\} \qquad (25a)$$

$$= E^t\begin{bmatrix} \eta_1(t)P_{11}(t) \\ \eta_2(t)P_{22}(t) \end{bmatrix} \qquad (25b)$$

where

$$\begin{bmatrix} \dot{\eta}_1(t) \\ \dot{\eta}_2(t) \end{bmatrix} = \begin{bmatrix} \alpha-\gamma-P_{11}^{-1}(t) & 0 \\ 0 & \beta-\gamma-P_{22}^{-1}(t) \end{bmatrix} \begin{bmatrix} \eta_1(t) \\ \eta_2(t) \end{bmatrix} + \begin{bmatrix} 0 & 1 \\ 1 & 0 \end{bmatrix} \begin{bmatrix} \xi_1(t) \\ \xi_2(t) \end{bmatrix} \tag{26}$$

$$\eta_1(0) = \eta_2(0) = 0$$

In other words, the argument of the conditional expectation in (25a) can be realized as the output of a finite dimensional linear system with state $\eta(t) = [\eta_1(t), \eta_2(t)]'$ satisfying (26).

Thus the finite dimensional optimal estimator for the system (19) - (21) is constructed as follows (see Figure 2). First we augment the state ξ of (19) with the state η of (26). Then the Kalman-Bucy filter for the linear system (19), (26), with observations (21), computes the conditional expectations $\hat{\xi}(t|t)$ and $\hat{\eta}(t|t)$. Finally,

$$d\hat{x}(t|t) = [-\gamma\hat{x}(t|t) + \hat{\xi}_1(t|t)\hat{\xi}_2(t|t)]dt + \hat{\eta}'(t|t)P(t)d\nu(t)$$

$$\hat{x}(0|0) = 0 \tag{27}$$

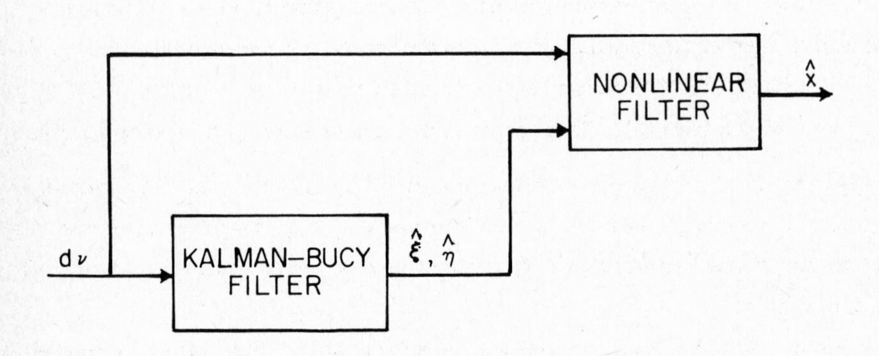

Figure 2. Block Diagram of the Optimal Filter for Example 1

The steady-state performance of this filter is studied in Marcus.[5]

FINITE DIMENSIONAL ESTIMATORS FOR BILINEAR SYSTEMS

In this section the results of the previous section are applied, with the aid of some concepts from the theory of Lie algebra,[10] to prove that the optimal estimators for certain bilinear systems are finite dimensional. Consider the system described by (1), (3), and the bilinear system[5,11,12]

$$\dot{X}(t) = (A_0 + \sum_{i=1}^{N} \xi_i(t)A_i)X(t) \; ; \qquad X(0) = I \tag{28}$$

where X is a kxk matrix. We associate with (28) the Lie algebra $L \triangleq \{A_0, A_1, \ldots, A_N\}_{LA}$, the smallest Lie algebra containing $A_0, A_1, \ldots, A_N$, and the ideal L_0 in L generated by $\{A_1, \ldots, A_N\}$.[5,12] A Lie algebra L is said to be <u>nilpotent</u> if the series of ideals L^n defined by $L^0 = L$, $L^{n+1} = [L, L^n]$ is the trivial ideal $\{0\}$ for some n; L is <u>abelian</u> if $L^1 = \{0\}$.[10]

It is easy to shown, using Brockett's results[6] on finite Volterra series, that each term in (6) can be realized by a bilinear system of the form

$$\dot{x}(t) = \xi_j(t)x(t) + \sum_{k=1}^{n} A_k(t)\xi_k(t)x(t) \tag{29}$$

where x is a k-vector and the A_j are strictly upper triangular (zero on and below the main diagonal). For such systems, the Lie algebra L_0 is nilpotent. Conversely, it can be shown[5] that if the Lie algebra L_0 corresponding to the bilinear system (28) is nilpotent, then each component of the solution to (28) can be written as a finite sum of terms of the form (6). This result and Theorem 1 yield the following theorem.

<u>Theorem 3</u>: Consider the system described by (1), (3), and (28), and assume that L_0 is a nilpotent Lie algebra. Then the conditional expectation $\hat{x}(t|t)$ can be computed with a finite dimensional system of nonlinear differential equations driven by the innovations.

We note that Theorem 3 provides a generalization of the work of Lo and Willsky[13] (in which L is abelian) and Willsky.[14]

It may appear that the classes of systems described by (6) or by (28) where L_0 is nilpotent represent a very restricted class of systems. However, the papers of Fliess[17] and Sussmann[18] in this Proceedings show that, in the deterministic case with bounded inputs, any causal and continuous input-output map can be uniformly approximated by a bilinear system of the form (28) in which $A_0, A_1, \ldots, A_N$ are all strictly upper triangular. For such a bilinear system both L_0 and L are nilpotent Lie algebras. Stochastic analogs of this result are currently being investigated. The implication of such a result would be that suboptimal estimators for a large class of nonlinear stochastic systems could be constructed using the results of this paper.

ACKNOWLEDGEMENT

The authors would like to thank Professor Roger Brockett of Harvard University for many helpful discussions and for suggesting the use of Volterra series in the present context.

REFERENCES

1. Kushner, H. J., "Dynamical Equations for Optimal Nonlinear Filtering," J. Differential Equations, 3, 179, 1967.

2. Fujisaki, M., Kallianpur, G., and Kunita, H.,"Stochastic Differential Equations for the Nonlinear Filtering Problem," Osaka J. Math., 9, 19, 1972.

3. Kalman, R. E. and Bucy, R. S., "New Results in Linear Filtering and Prediction Theory," J. Basic Engr. (Trans. ASME), 83D, 95, 1961.

4. Jazwinski, A. H., Stochastic Processes and Filtering Theory, Academic Press, New York, 1970.

5. Marcus, S. I., Estimation and Analysis of Nonlinear Stochastic Systems, Ph.D. Thesis, Dept. of Electrical Engineering, M.I.T., Cambridge, Mass., June 1975; also M.I.T. Electronic Systems Laboratory Report No. ESL-R-601, June 1975.

6. Brockett, R. W.,"Volterra Series and Geometric Control Theory," in Proc. 1975 IFAC Congress, ISA, Philadelphia, 1975.

7. Brockett, R. W.,"Nonlinear Systems and Differential Geometry," Proc. IEEE, January 1976.

8. d'Alessandro, P., Isidori, A., and Ruberti, A., "Realizations and Structure Theory of Bilinear Dynamical Systems, SIAM J. Control, 12, 517, 1974.

9. Kailath, T. and Frost, P., "An Innovations Approach to Least-Squares Estimation, Part II: Linear Smoothing in Additive White Noise," IEEE Trans. Automatic Control, AC-13, 655, 1968.

10. Sagle, A. A. and Walde, R. E., Introduction to Lie Groups and Lie Algebras, Academic Press, New York, 1973.

11. Brockett, R. W., "System Theory on Group Manifolds and Coset Spaces," SIAM J. Control, 10, 265, 1972.

12. Jurdjevic, V. and Sussman, H. J.,"Control Systems on Lie Groups," J. Differential Equations, 12, 313, 1972.

13. Lo, J. T. and Willsky, A. S., "Estimation for Rotational Processes with One Degree of Freedom I: Introduction and Continuous Time Processes, IEEE Trans. Automatic Control, AC-20, 10, 1975.

14. Willsky, A. S., "Some Estimation Problems on Lie Groups," in Geometric Methods in System Theory, Brockett, R. W. and Mayne, D. W., Eds., Reidel, The Netherlands, 1973.

15. Kwakernaak, H., "Optimal Filtering in Linear Systems with Time Delays," IEEE Trans. Automatic Control, AC-12, 169, 1967.

16. Miller, K. S., Multidimensional Gaussian Distributions, John Wiley, New York, 1965.

17. Fliess, M., "Un Outil Algebrique: Les Series Formelles Non Commutatives," in Proc. of the CNR-CISM Symposium on Algebraic System Theory, Udine, Italy, June 16-27, 1975.

18. Sussmann, H. J., "Semigroup Representations, Bilinear Approximation of Input-Output Maps, and Generalized Inputs," in Proc. of the CNR-CISM Symposium on Algebraic System Theory, Udine, Italy, June 16-27, 1975.

19. Wong, E., Stochastic Processes in Information and Dynamical Systems, McGraw-Hill, New York, 1971.

FILTERING FOR RANDOM FINITE GROUP HOMOMORPHIC SEQUENTIAL SYSTEMS

Alan S. Willsky
Massachusetts Institute of Technology
Cambridge, Massachusetts

I. INTRODUCTION

The nonlinear filtering problem has proven to be an extremely difficult one. When the filtering problem is described by a set of stochastic differential equations, the optimal nonlinear filter in general requires the solution of a stochastic partial differential equation for the conditional density or the solution of an infinite set of stochastic differential "moment" equations.[1] In the case of discrete-time partially observable finite-state Markov processes (POFSMP), the solution is conceptually simpler, as the conditional distribution can be computed sequentially via straightforward finite-dimensional difference equations.[2] However, even in this conceptually simple case, the nonlinear filtering problem can be computationally nontrivial. Specifically, we note that if we are considering an n-state POFSMP, a straightforward implementation of the conditional distribution update equations requires $0(n^2)$ multiplications. For n of reasonable size this becomes an extremely demanding computational task.

In this paper we investigate the inherent computational limitations in the nonlinear filtering problem for a special class of POFSMP's. Specifically, we consider the class of dynamical systems called finite group homomorphic sequential systems (FGHSS).[3,4] By considering the inputs to be stochastic and by including multiplicative observation noise, we obtain the class of POFSMP's of interest. By viewing probability distributions on a finite group as elements of the group algebra, we obtain a form for the nonlinear filtering equations that exposes their inherent structure. We then utilize several of the key results from the representation theory of finite groups to obtain a "dual" form for the nonlinear filtering equations.

Our use of representation theory in the study of stochastic processes on groups is very much in the spirit of the work of Grenander,[5] who utilized Fourier analysis to study the problem of the multiplication of independent random variables on a compact group. Depeyrot[6,7] and Paz[13] have also studied a "dual automaton" approach to stochastic automata (without measurements). In our work, we extend the analyses of these authors in order to study the nonlinear filtering problem, and we obtain a more complete picture of the duality between diffusion

* This work was supported by NSF under Grant GK-42090.

and measurement updates and the two formulations of the nonlinear filtering equations. In this manner we are able to uncover some of the key computational issues in the filtering problem. For the sake of brevity, we limit our present development to a description of the basic concepts involved in our treatment and to the consideration of two illustrative examples. A more thorough development containing the detailed derivations of our results is given in Willsky.[8]

II. DEFINITIONS AND BACKGROUND

Let U,X, and Y be finite groups, and let a: $X \to X$, b: $U \to X$, and c: $X \to Y$ be group homomorphisms. A <u>random finite group homomorphic sequential system</u> (RFGHSS) is a system of the form

$$x(k+1) = b[u(k)]a[x(k)], \quad k \geq 0 \tag{1}$$

$$y(k) = v(k)c[x(k)], \quad k \geq 1 \tag{2}$$

where $\{u(k)\}$ and $\{v(k)\}$ are sequences of independent random variables in U and Y, respectively, independent of each other and of the random variable $x(0)$. We let $\eta(k)$, $\xi(k)$, and $\rho(0)$ denote the probability distributions for $u(k)$, $v(k)^{-1}$, and $x(0)$, respectively. We regard these distributions as functions on the respective groups. For example, $\eta(k)_\mu$ is the probability that $u(k)=\mu$.

Let $\rho(k|\ell)$ denote the conditional probability distribution for $x(k)$ given the measurements $y(1),\ldots,y(k)$. By convention we set $\rho(0|0)=\rho(0)$. The nonlinear filtering problem consists of finding an algorithm for the sequential computation of the distributions $\rho(k+1|k)$ and $\rho(k|k)$. One could also consider the prediction problem (compute $\rho(k|\ell)$, $\ell<k$) and the smoothing problem ($\ell>k$). These problems will be studied elsewhere.[8]

Before we study the filtering problem, we must recall some of the relevant concepts from the theory of group representations. Let X be a finite group with $|X|=n$, and let C[X] denote the complex group algebra of X.[9] Here we consider an arbitrary element ρ of C[X] to be represented as a formal sum

$$\rho = \sum_{g \in X} \rho_g \cdot g \quad , \rho_g \in C \tag{3}$$

The operations of pointwise addition, scalar multiplication, and the convolution product $\rho * \eta$[9] provide C[X] with the structure of a complex algebra. We can regard X as a subset of C[X] with the obvious identification of g with $1 \circ g$. Finally, we can give C[X] the structure of a commutative algebra by endowing it with the pointwise product $(\rho\eta)_g = \rho_g \eta_g$.

Let $T^1,\ldots,T^s$ be a complete set of inequivalent irreducible matrix representations of X over C, with dim $T^i = Z_i$ (we will always take $T^1 \equiv 1$). We note the fundamental relation

$$n = \sum_{i=1}^{s} z_i^2 \qquad (4)$$

Let t^i_{jk} denote the element in the jth row and kth column of T^i, and let $\phi \in C[X]$. One can then compute the transform pair[8,9]

$$C^i(\phi) = \frac{z_i}{n} \sum_{g\in X} \phi_g [T^i(g^{-1})]' \qquad (5)$$

$$\phi_g = \sum_{i=1}^{s} \sum_{j,k=1}^{z_i} C^i_{jk}(\phi) t^i_{jk}(g) \qquad (6)$$

The C^i are called the <u>transform matrices</u> of ϕ.

The motivation for the introduction of this framework is computational. We begin to see this in the following. Suppose $\phi, \psi \in C[X]$. Then one can show[5]

$$C^i(\phi*\psi) = C^i(\phi)C^i(\psi) \qquad (7)$$

Note that the number of multiplications involved in computing $\phi*\psi$ is n^2, while the number in (7) is always less than n^2.

A second computation of importance in the sequel is the calculation of $C^i(\phi\psi)$ from $\{C^j(\phi)\}$ and $\{C^k(\psi)\}$. The general finite group case requires an involved computation and the introduction of several additional group-theoretic concepts. We defer the general discussion to a lengthier exposition on this subject[8]. We will give several examples later in the paper.

III. FILTERING FOR RFGHSS's

Consider the system in (1),(2). For simplicity in the present discussion, we consider only the case X=U=Y, a=b=c identity. The general case is studied in Willsky[8]. As described in Section II, we wish to compute the distributions $\rho(k+1|k)$ and $\rho(k|k)$. We regard all distributions as elements of $C[X]$. One can then compute[8].

<u>Theorem 1</u>: The following equations define an algorithm for the solution of the filtering problem:

<u>Diffusion Update</u>

$$\rho(k+1|k) = \eta(k)*\rho(k|k) \qquad (8)$$

<u>Measurement Update</u>

$$\lambda(k) = \xi(k)*y(k) \qquad (9)$$

$$\gamma(k|k) = \lambda(k)\rho(k|k-1) \qquad (10)$$

$$N(k|k) = \sum_{g\in X} \gamma(k|k)_g \qquad (11)$$

$$\rho(k|k) = \frac{\gamma(k|k)}{N(k|k)} \tag{12}$$

Thus the diffusion update consists of a C[X] convolution, while the measurement update involves the convolution $\xi * y$, followed by a pointwise multiplication and the normalization (12). We therefore see that the computational load for one time step consists of the n^2 multiplications for the convolution (8), the permutation induced by the convolution $\xi * y$, the n multiplications in the pointwise product in (10), and the normalization (11),(12).

Using results from the preceding section, we now consider a "dual" solution to the filtering problem. Let us associate the various probability distributions of interest with their transform matrices:

$$\begin{array}{ll} \rho(k|\ell) \leftrightarrow \{A^i(k|\ell)\} & , \eta(k) \leftrightarrow \{D^i(k)\} \\ \xi(k) \leftrightarrow \{E^i(k)\} & \lambda(k) \leftrightarrow \{F^i(k)\} \\ \gamma(k|k) \leftrightarrow \{G^i(k|k)\} & \end{array} \tag{13}$$

<u>Theorem 2</u>: The following equations define a "dual" algorithm for the solution of the filtering problem:

<u>Diffusion Update</u>

$$A^i(k+1|k) = D^i(k)A^i(k|k) \tag{14}$$

<u>Measurement Update</u>

$$F^i(k) = E^i(k)[T^i(y(k)^{-1})]' \tag{15}$$

$$G^i(k|k) = C^i[\lambda(k)\rho(k|k-1)] \tag{16}$$

$$A^i(k|k) = \frac{G^i(k|k)}{nG^1(k|k)} \tag{17}$$

We comment on the relationship between these two sets of equations. We first note that the calculations in (14)-(17) correspond exactly to equations (8)-(12). Comparing (8) and (14), we see that the convolution in (8) corresponds to the multiplications of the s transform matrices in (14), which is a simpler computation. Equations (9) and (15) indicate that a permutation corresponds to the matrix products in (15). If the T^i are <u>monomial</u> representations ($T^i(g)$ has only one non-zero element in each row and column),[9] the computational load (measured by the number of multiplications) in (15) is at most of the order of n. Equation (10) is a pointwise product, and the transform version can be calculated as indicated in the preceding section (see the examples). Finally, the normalizations of the conditional distribution is given by (12) and (17). In Theorem 2 we have called (14)-(17) a "dual" algorithm, although the motivation for this terminology is not readily apparent from the equations. In the next two sections we will go into more detail for two specific cases, and this analysis will help clarify our somewhat loose use of the term duality. Roughly, we will find that the diffusion update leads to a convolution in the group algebra domain and a "pointwise" multiplication for the transforms. On the other hand the measurement update corresponds to a pointwise multiplication in the group algebra and a "convolution" in the transform domain. In addition, our analyses for these specialized cases will point out the computational significance of the dual algorithms.

IV. THE CYCLIC GROUP Z_n

In this section we describe some analysis first reported in Willsky.[10] Consider the cyclic group Z_n, which we will interchangeably identify with the integers and with the set $\{a^k\}$, where integer addition is defined modulo n. We use the notation

$$\phi = \sum_{k=0}^{n-1} \phi_k a^k \tag{18}$$

for elements of $C[Z_n]$. All of the irreducible representations are one-dimensional, and a complete set of these, $T^0,\ldots,T^{n-1}$, is specified by the equation $(j=\sqrt{-1})$

$$T^i(a) = e^{j2\pi i/n} = \gamma^i \tag{19}$$

We now consider the various steps of the two filtering algorithms. We use the identification specified in (13) and note that in the Z_n case the expansions (5), (6) yield the finite Fourier series

$$A^\nu(k|\ell) = \frac{1}{n} \sum_{m=0}^{n-1} \rho(k|\ell)_m \gamma^{-\nu m} \tag{20}$$

$$\rho(k|\ell)_m = \sum_{\nu=0}^{n-1} A^\nu(k|\ell)\gamma^{\nu m} \tag{21}$$

In the following, all integers are to be interpreted modulo n.

Step 1: Equations (8) and (14) become

$$\rho(k+1|k)_m = \sum_{\nu=0}^{n-1} \eta(k)_\nu \rho(k|k)_{m-\nu} \tag{22}$$

$$A^\nu(k+1|k) = D^\nu(k)A^\nu(k|k) \tag{23}$$

Step 2: Equations (9) and (15) become

$$\lambda(k)_m = \xi(k)_{m-y(k)} \tag{24}$$

$$F^{\nu}(k) = \gamma^{-\nu y(k)} E^{\nu}(k) \tag{25}$$

Step 3: Equations (10) and (16) become

$$\gamma(k|k)_m = \lambda(k)_m \, \rho(k|k-1)_m \tag{26}$$

$$G^{\nu}(k|k) = \sum_{i=0}^{n-1} F^{i}(k) A^{\nu-i}(k|k-1) \tag{27}$$

Step 4: The normalizations are as in (12),(17) except in this case the denominator in (17) should be $nG^0(k|k)$.

Comparing (22),(23) with (26),(27), we see that these do represent dual operations as described at the end of Section III. In order to perform these calculations efficiently, we can utilize FFT techniques in order to reduce the number of multiplications required to calculate (22) or (27) from n^2 to $O(n\log(n))$.[10,11]

V. THE DIHEDRAL GROUP D_n

Let D_n denote the group of order 2n generated by the two elements a and b, which satisfy the relations

$$a^n = b^2 = 1, \quad aba = b \tag{28}$$

(note that $Z_n \approx \langle a \rangle$). We remark that D_n is an example of a metacyclic group, a class which is examined in detail in Willsky[8]. For $\phi \in C[D_n]$ we write

$$\phi = \sum_{m=0}^{n-1} \sum_{\ell=0}^{1} \phi_{m\ell} \, a^m b^{\ell} \tag{29}$$

If n is odd, there are two inequivalent one-dimensional representations T^0 and T^1, defined by

$$T^0(g) = 1 \quad \forall g \tag{30}$$

$$T^1(a) = 1, \; T^1(b) = -1 \tag{31}$$

If n is even, we have two additional 1-D representations:

$$T^k(a^{\ell} b^m) = (-1)^{\ell}(-1)^{km} \quad k=2,3 \tag{32}$$

The remaining irreducible representations S^i, $i=1,\ldots,\left\lfloor \frac{n-1}{2} \right\rfloor$ (here $\lfloor x \rfloor$ is the largest integer less than or equal to x), are two dimensional and are given by

$$S^r(a^k b^j) = \begin{bmatrix} \gamma^{kr} & 0 \\ 0 & \gamma^{-kr} \end{bmatrix} \begin{bmatrix} 0 & 1 \\ 1 & 0 \end{bmatrix}^j \tag{33}$$

where γ is given by (19).

Let $\phi \varepsilon C[D_n]$. We introduce a new notation for its transform

$$B^i(\phi) = \frac{1}{2n} \sum_{g \varepsilon D_n} \phi_g T^i(g^{-1}) \tag{34}$$

$$R^i(\phi) = \frac{1}{n} \sum_{g \varepsilon D_n} \phi_g [S^i(g^{-1})]' \tag{35}$$

We also define the two Z_n - transforms

$$\alpha_k = \frac{1}{n} \sum_{m=0}^{n-1} \phi_{m,0} \gamma^{-mk} \tag{36}$$

$$\beta_k = \frac{1}{n} \sum_{m=0}^{n-1} \phi_{m,1} \gamma^{-mk} \tag{37}$$

The one can show[8]

$$\begin{gathered} B^0(\phi) = \frac{1}{2}(\alpha_0+\beta_0), \quad B^1(\phi) = \frac{1}{2}(\alpha_0-\beta_0) \\ B^2(\phi) = \frac{1}{2}(\alpha_{\frac{n}{2}}+\beta_{\frac{n}{2}}), \quad B^3(\phi) = \frac{1}{2}(\alpha_{\frac{n}{2}}-\beta_{\frac{n}{2}}) \end{gathered} \tag{38}$$

$$R^i(\phi) = \begin{bmatrix} \alpha_i & \beta_i \\ \beta_{n-i} & \alpha_{n-i} \end{bmatrix} \tag{39}$$

Thus, we see that we can devise a fast algorithm for the computation of the transform of ϕ by performing the two FFT's (36),(37).

We now consider the two filtering algorithms. In this case we identify the various distributions with both their D_n transforms and their two Z_n - transforms

$$\begin{aligned} &\rho(k|\ell) \text{<->} \{A^i(k|\ell),\ U^i(k|\ell)\} \text{<->} \{\alpha_r(k|\ell),\ \beta_r(k|\ell)\} \\ &\eta(k) \text{<->} \{D^i(k),\ V^i(k)\} \text{<->} \{\delta_r(k),\ \varepsilon_r(k)\} \\ &\xi(k) \text{<->} \{E^i(k),\ W^i(k)\} \text{<->} \{\psi_r(k),\ \mu_r(k)\} \\ &\lambda(k) \text{<->} \{F^i(k),\ Y^i(k)\} \text{<->} \{\theta_r(k),\ \nu_r(k)\} \\ &\gamma(k|k) \text{<->} \{G^i(k|k),\ Z^i(k|k)\} \text{<->} \{\phi_r(k|k),\ \tau_r(k|k)\} \end{aligned} \tag{40}$$

Step 1: Equation (8) can be calculated easily and we refer the reader to

Willsky[8] for details. Equation (14) becomes

$$A^{i}(k+1|k) = D^{i}(k)A^{i}(k|k) \tag{45}$$

$$U^{i}(k+1|k) = V^{i}(k)U^{i}(k|k) \tag{46}$$

In terms of the Z_n transforms, one can show[8] that in order to calculate $\alpha(k+1|k)$, and $\beta(k+1|k)$, one must only compute the four $C[Z_n]$ pointwise products

$$\alpha(k|k)\delta(k),\ \alpha(k|k)\varepsilon(k),\ \beta^{rev}(k|k)\delta(k),\ \beta^{rev}(k|k)\varepsilon(k) \tag{47}$$

$$\beta_r^{rev} = \beta_{n-r} \tag{48}$$

Step 2: The permutation (9) of $\xi(k)$ can be evaluated in a straightforward manner[8]. As for (15), suppose

$$y(k) = a^p b^\ell, \quad 0\leq p\leq n-1, \quad 0\leq \ell\leq 1 \tag{49}$$

Then we have, for $\ell=0$

$$\Theta_r(k) = \gamma^{-pr}\psi_r(k), \quad \nu_r(k) = \gamma^{pr}\mu_r(k) \tag{50}$$

and for $\ell=1$

$$\Theta_r(k) = \gamma^{-pr}\mu_r(k), \quad \nu_r(k) = \gamma^{pr}\psi_r(k) \tag{51}$$

One can work out analogous equations in terms of E,W,F, and Y.[8]

Step 3: Equation (10) is simply the pointwise product

$$\gamma(k|k)_{m,\ell} = \lambda(k)_{m,\ell}\ \rho(k|k-1)_{m,\ell} \tag{52}$$

Equation (16) yields the cyclic convolution of the Z_n - transforms

$$\phi_r(k|k) = \sum_{m=0}^{n-1} \Theta_m(k)\alpha_{r-m}(k|k-1) \tag{53}$$

$$\tau_r(k|k) = \sum_{m=0}^{n-1} \nu_m(k)\beta_{r-m}(k|k-1) \tag{54}$$

The equations in terms of G,Z,F,Y,A, and U are given in Willsky.[8]

Step 4: The normalizations are as in (12),(17) except in this case the denominator of (17) becomes

$$2nG^0(k|k) = n[\phi_0(k|k) + \tau_0(k|k)] \tag{55}$$

From this example, we find further substantiation for the use of the term "duality" for the two filtering algorithms. In terms of computational complexity,

(8) requires $4n^2$ multiplications, while (47) requires 4n (we can reduce this further if we use fast matrix multiplication techniques[8,12]). Thus the use of the fast D_n- transform algorithm described earlier in this section can again reduce the required computational requirements for the calculation of a $C[D_n]$ convolution from $O(n^2)$ to $O(n\log(n))$. Examining (52)-(54), we find that Step 3 requires 2n multiplications for (52) and $2n^2$ for (53),(54). Comparing to Step 1, we see that (neglecting Steps 2 and 4), the diffusion update requires twice as much computation as the measurement update.

VI. CONCLUSIONS

In this paper we have studied a class of estimation problems on finite groups. By viewing probability distributions as elements of a group algebra and by taking the Fourier transforms of such elements, we were able to uncover the underlying structure of the filtering problem. We have illustrated this structure by means of two examples, which vividly display the duality between the two proposed filtering algorithms. Also, by utilizing fast Fourier transform techniques and a generalization of the FFT to dihedral groups, we have been able to point out an efficient realization of the filtering solution. We note that Depeyrot[6,7] has considered several of these issues, but he has limited himself to elements of the group algebra which are constant on conjugacy classes of the group. His transforms are the so-called character transforms which utilize only the irreducible group characters[9] as basis functions. Thus, unless the group is abelian, the transform is not a 1-1 map (a given transform can correspond to many elements of the algebra).

Our analysis in this paper has been brief, and the detailed derivations and a number of extensions and additional topics are contained in Willsky[8].

REFERENCES

1. Jazwinski, A.H., Stochastic Processes and Filtering Theory, Academic Press, New York, 1970.

2. Astrom, K.J., "Optimal Control of Markov Processes with Incomplete State Information," J. Math. Anal. Appl., Vol. 10, p.174, 1965.

3. Brockett, R.W., and Willsky, A.S., "Finite Group Homomorphic Sequential Systems," IEEE Trans. on Automatic Control, Vol. AC-17, p.483, 1972.

4. Willsky, A.S., Dynamical Systems Defined on Groups: Structural Properties and Estimation, Ph.D. thesis, Dept. of Aeronautics and Astronautics, M.I.T., Cambridge, Mass., June 1973.

5. Grenander, U., Probabilities on Algebraic Structures, John Wiley, New York, 1963.

6. Depeyrot, M., Operand Investigation of Stochastic Systems, Ph.D. thesis, Stanford Univ., May 1968.

7. Depeyrot, M., Marmorat, J.P., and Mondelli, J., "An Automaton Theoretic Approach to the F.F.T.," Centre D'Automatique De L'Ecole Nationale Superieure Des Mines De Paris, Fountainebleau, France, April 1971.

8. Willsky, A.S., "On the Algebraic Structure of Certain Partially Observable Finite State Markov Processes," to appear.

9. Curtis, C.W. and Reiner, I., Representation Theory of Finite Groups and Associative Algebras, Interscience, New York, 1966.

10. Willsky, A.S., "A Finite Fourier Transform Approach to Estimation on Cyclic Groups," Proc. of the Fifth Symposium on Nonlinear Estimation and Its Applications, San Diego, Calif., Sept. 1974.

11. Stockham, T.G.., Jr., "High Speed Convolution and Correlation," 1966 Spring Joint Computer Conf., AFIPS Proc., Vol. 28, p.229, 1966.

12. Hopcroft, J. and Kerr, L., "On Minimizing the Number of Multiplications Necessary for Matrix Multiplication," SIAM J. on Appl. Math., Vol. 20, p.30, 1968.

13. Paz, A., Introduction to Probabilistic Automata, Academic Press, New York, 1971.

CATEGORICAL APPROACH TO GRAPHIC SYSTEMS AND GRAPH GRAMMARS

Hartmut Ehrig

Hans-Jörg Kreowski

TECHNISCHE UNIVERSITÄT BERLIN

Fachbereich 20 - Kybernetik
1 Berlin 10
Otto-Suhr-Allee 18/20

Abstract: The algebraic approach of graph grammars using homomorphisms and pushout constructions given in /Eh-Pf-Sch 73/ and /Ros 74/ is extended to graphic systems which are graphs in a suitable category $\underline{K}$ including partial graphs, multigraphs, stochastic and topological graphs. These are useful models in computer science, biology, chemistry, network theory and ecology.
Several new results concerning efficient pushout constructions, pushout complements and enlargements are given in the framework of category theory.

INTRODUCTION

The basic idea of graph grammars is to generalize Chomsky-grammars, generating string languages, to higher dimensional structures which can be represented as labeled graphs. Hence to give a grammar which allows to replace subgraphs by other subgraphs using the graph-productions of the grammar. Motivated by first approaches of J.L. Pfaltz - A. Rosenfeld (1969), H.-J. Schneider (1970), and T.-J. Pratt (1971) in /Eh-Pf-Sch 73/ an algebraic approach of graph grammars was given using homomorphisms and pushout constructions to specify embeddings and direct derivations constructively. Meanwhile this approach is generalized by B.K. Rosen in /Ros 74/ to partially labeled graphs and noninjective embeddings, in /Sch-Eh 75/ to partial and in /Eh-Ti 75/ to stochastic graphs.
Applications to flowcharts, recursive definitions and trees are given in the above papers and in separate papers there are studied applications of graph grammars to programming languages in /Pra 71/, /De-Fr-St 74/, incremental compilers in /Sch 74/ and biological organisms in /Eh-Ti 74/. A summary of applications, problems and results is given in /Eh 75/.
In this paper we will extend graph grammars to graphs in a suitable category $\underline{K}$ including ordinary graphs, multigraphs - edges are allowed to have several sources and targets - , stochastic and topological graphs. Due to /Ri 74/ these graphical systems are useful models in biology, chemistry, network theory and ecology for example. Moreover we will give in section 2 an efficient procedure for pushout constructions and in section 4 a complete enlargement analysis - construction of all "p.o. complements" D such that the equation $D \amalg_K B = G$ holds for given graphs B, K and G. This includes the injective analysis in /Eh-Pf-Sch 73/ and the natural analysis in /Ros 74/. Moreover the embedding of enlargements given in /Eh-Pf-Sch 73/ §5 are generalized to noninjective embeddings.
Our theory is based on two categorical lemmas, the splitting and the devision lemma given in section 2, and on colimit constructions for $\underline{K}$-graphs which are very similar to those of automata in pseudoclosed categories (cf. /E-K^3 74/ chapters 6 and 7).

The reader is supposed to be familiar with elementary category theory. The basic ideas of graph grammars as given in /Eh-Pf-Sch 73/ are the following (stated for unlabeled graphs for simplicity):

1. Gluing of graphs: In generalization of the concatenation of strings the gluing G of graphs D and B, with gluing points specified by an auxiliary graph K and graph morphisms $d: K \longrightarrow D$ and $p: K \longrightarrow B$, is given by the following pushout in the category of graphs:

2. Productions: A production $\pi = ('B \xleftarrow{'p} K \xrightarrow{p'} B')$ is not only a pair of graphs ('B, B') but also a specification of vertices and edges of 'B and B' given by an auxiliary graph K and graph morphisms $'p: K \longrightarrow 'B$ and $p': K \longrightarrow B'$ respectively.

3. Direct Derivations: Given a production $\pi = ('B \xleftarrow{'p} K \xrightarrow{p'} B')$ and an "enlargement" $\varepsilon = (K \xrightarrow{d} D)$, which is a graph D together with a graph morphism d, with the same K we get a direct derivation

$$(\pi, \varepsilon) : D \amalg_K {'B} \Longrightarrow D \amalg_K B'$$

defined by the following pushouts

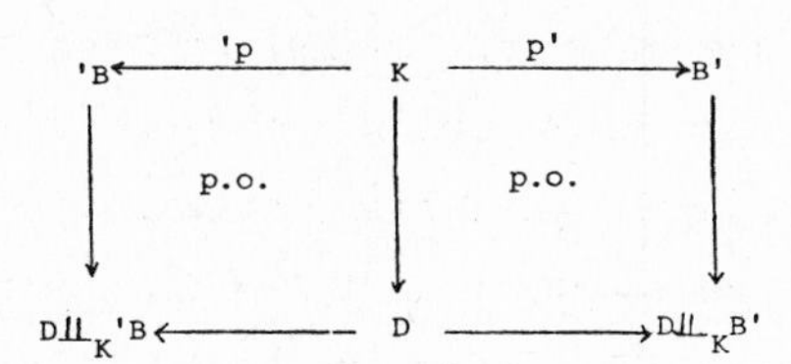

Conversely a graph H can be directly derivated from a graph G if there is a production π and an enlargement ε as above such that G and H are isomorphic to $D \amalg_K {'B}$ and $D \amalg_K B'$ respectively.

4. Derivations and Graph Language: H is called to be derivable from G, written $G \overset{*}{\Longrightarrow} H$, if $G \cong H$ or there is a sequence of direct derivations $G \Longrightarrow G_1 \Longrightarrow G_2 \Longrightarrow \dots \Longrightarrow G_n \Longrightarrow H$. $(n \geq o)$

Given a start graph S the graph language is the class, or more precisely the set of all isomorphism classes, of all graphs derivable from S.
To illustrate the gluing and derivation concept for graphs let us consider the following simple example which shows a direct derivation of graphs $G \Longrightarrow H$ where 'B in G is replaced by B' to obtain H. Graph morphisms are indicated by numbering.

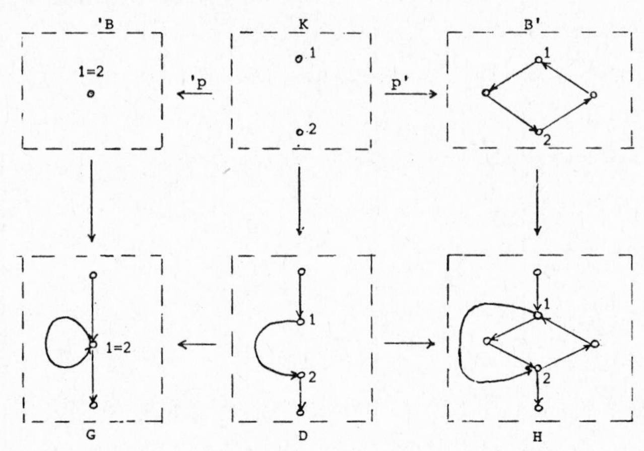

1. $\underline{K}$ - GRAPHS

In general a $\underline{K}$-graph is a diagram of the shape

$$E \overset{s}{\underset{t}{\Longrightarrow}} V$$

in an arbitrary category $\underline{K}$, i.e. E, called edges, and V, called vertices, are objects and s, source, and t, target, are morphisms in $\underline{K}$.
Since our categories $\underline{K}$ in consideration, e.g. the category of relations, don't have pushouts in general we only consider those graph morphisms such that the components belong to a coreflexive subcategory $\underline{K}'$ of $\underline{K}$ which has pushouts.

1.1 GENERAL ASSUMPTIONS AND NOTATIONS

a) Let $\underline{K}'$ be a coreflexive subcategory of $\underline{K}$ and $\mathcal{M}$ a class of monomorphisms in $\underline{K}'$ and $\underline{K}$.

b) $\underline{K}'$ has pushouts, an initial object - written $\emptyset$ -, and an $\mathcal{E}$-$\mathcal{M}$-factorization (cf. /He-St 74/, /E-K^3 74/).

c) $\underline{K}'$ has unique coproduct complements up to isomorphism, i.e. for each coproduct injection $u_1: A_1 \longrightarrow A$ there is a $u_2: A_2 \longrightarrow A$ such that

(u_1,u_2) is a coproduct of A and if for $u_2':A_2' \longrightarrow A$ (u_1,u_2') is also a coproduct of A then there is a unique isomorphism $i:A_2 \longrightarrow A_2'$ such that $u_2=u_2' \circ i$.

d) The set of morphisms Mor $\underline{K}'(A,B)$ is nonempty if B is no initial object in $\underline{K}'$, $\emptyset \longrightarrow A$ belongs to $\mathcal{M}$ and $A \longrightarrow \emptyset$ implies $A \cong \emptyset$. It also sufficies to assume that $\underline{K}'$ has a zero object.

For section 5 we need in addition:

e) The class $\mathcal{S}$ of splitting morphisms defined below is closed under composition.

Notation: The class of coproduct injections in $\underline{K}'$ is denoted by $\mathcal{J}$ and u_2 in c) is called coproduct complement of u_1.

The class $\mathcal{S}$ of splitting morphisms in $\underline{K}'$ is defined by:

$f \in \mathcal{S}$ iff $f_2 \in \mathcal{J}$ in each $\mathcal{E}$-$\mathcal{M}$-factorization $f=f_2 \circ f_1$ of f.

$|\underline{K}'|$ is the class of objects in $\underline{K}'$.

1.2 EXAMPLES

In most of our considerations $\underline{K}'$ is the category Sets of sets and assumptions 1.1 b) - 1.1 d) are wellknown for this case because $\mathcal{J} = \mathcal{M}$ is the class of all injective functions, $\mathcal{E}$ the class of all surjective functions and $\mathcal{S}$ is the class of all functions because $\mathcal{J} = \mathcal{M}$.

For $\underline{K}$ we can take the following categories

a) Sets = category of sets (leading to ordinary graphs)

b) PD = category of partial functions (partial graphs)

c) ND = category of nondeterministic functions (multigraphs)

d) Rel = category of relations (partial multigraphs)

e) Stoch= category of stochastic channels (stochastic graphs)

f) ORel = category of ordered relations (ordered multigraphs).

The exact definitions of examples a) - e) and the verification of 1.1.a) are given in /E-K[3] 74/ in chapters 1 and 6. The category ORel is defined to be the Kleisli category of the monad of monoids, i.e. objects of ORel are sets and morphisms $f:A \longrightarrow B$ in ORel are functions $f':A \longrightarrow B^*$ where B^* is the free monoid of B (cf. /ML 72/). In fact examples b) - e) are also Kleisli categories of the corresponding monads on Sets and in general we can take for $\underline{K}$ an arbitrary Kleisli category $\underline{K}$ over a category $\underline{K}'$ satisfying 1.1. The category Top of topological spaces and continuous functions for example satisfies 1.1 b)-1.1 d) but not 1.1 e). Thus our theory - except of section 5 - is also applicable to topologically structured graphs and multigraphs. Some suitable categories K for K' = Top are discussed in /E-K^3 74/ chapter 6.

1.3 DEFINITION

Let K and K' as given in 1.1 then a K-graph is a diagram G of the shape $G = E \overset{s}{\underset{t}{\rightrightarrows}} V$ in K with E, V objects in K' and a graph morphism $f:G \longrightarrow G'$ is a pair of K'-morphisms $(f_E:E \longrightarrow E',\ f_V:V \longrightarrow V')$ such that $f_V \circ s=s' \circ f_E$ and $f_V \circ t=t' \circ f_E$.

$$\begin{array}{ccc} G & \quad & E \overset{s}{\underset{t}{\rightrightarrows}} V \\ \downarrow f & & \downarrow f_E \qquad \downarrow f_V \\ G' & & E' \overset{s'}{\underset{t'}{\rightrightarrows}} V' \end{array}$$

K-graphs and graph-morphisms constitute a category K-Graph, or more precisely K-K'-Graph, called category of K-Graphs.

Given a fixed pair of label objects $\Omega = (\Omega_E, \Omega_V)$ in K an Ω-graph G, or labeled graph, is a diagram of the shape

$$\Omega_E \xleftarrow{l_E} E \overset{s}{\underset{t}{\rightrightarrows}} V \xrightarrow{l_V} \Omega_V$$

in K. An Ω-graph-morphism $f:G \longrightarrow G'$ is a graph morphism $f=(f_E,f_V)$ such that $l_E=l'_E \circ f_E$ and $l_V=l'_V \circ f_V$. Ω-graphs and Ω-graph-morphisms lead to the category K-Graph_Ω, or more precisely K-K'-Graph_Ω.

Examples for K-graphs are already given in 1.2.

1.4 LEMMA

The categories K-Graph and K-Graph_Ω have pushouts and coproducts which are constructed in K' componentwise.

Proof: In each component these constructions exist in K' by 1.1 b) and hence also in K-Graph because the inclusion functor $J:K' \longrightarrow K$ preserves colimits by 1.1 a). Moreover all the constructions are compatible with the label morphisms l_E, l_V leading to pushouts and coproducts in K-Graph_Ω. □

2. SPLITTING AND DIVISION LEMMA

In this section we are concerned with properties of pushouts in K' which are needed for pushout constructions in K-Graph and K-Graph_Ω in the following sections.

2.1 SPLITTING LEMMA

Let (1) be a commutative diagram with $d \in \mathcal{J}$ and complement $d':D' \longrightarrow D$,

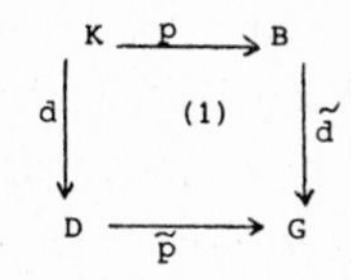

then (1) is a pushout iff $\tilde{d} \in \mathcal{J}$ with complement $\tilde{p} \circ d':D' \longrightarrow G$.

Corollary: $\mathcal{J}$ is closed under pushouts.

Proof: Let (1) be a pushout and $f_1:D' \longrightarrow X$, $f_2:B \longrightarrow X$ arbitrary K-morphisms then there is a unique $f_3:D \longrightarrow X$ such that

(2) $f_3 \circ d' = f_1$ and $f_3 \circ d = f_2 \circ p$,

because D is a coproduct, and hence a unique $f:G \longrightarrow X$ satisfying

(3) $f \circ \tilde{p} = f_3$ and $f \circ \tilde{d} = f_2$

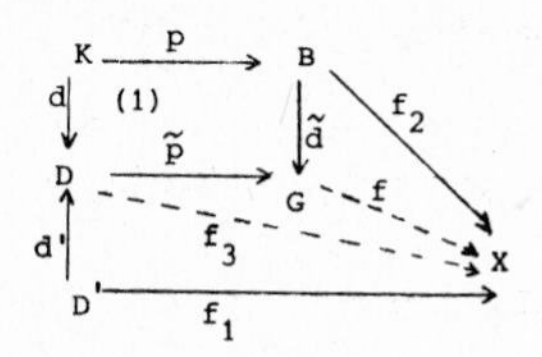

Thus we have

(4) $f \circ \tilde{p} \circ d' = f_1$ and $f \circ \tilde{d} = f_2$

and f is unique with respect to (4) which is an easy consequence of the uniqueness of f_3 in (2) and f in (3). Hence G has coproduct injections $\tilde{p} \circ d'$ and $\tilde{d}$.

Vice versa let $\tilde{d} \in \mathcal{J}$ with complement $\tilde{p} \circ d':D' \longrightarrow G$ and $f_3:D \longrightarrow X$, $f_2:B \longrightarrow X$ K'-morphisms satisfying $f_3 \circ d = f_2 \circ p$. Now let $f_1 = f_3 \circ d'$. By assumption there is a unique $f:G \longrightarrow X$ satisfying (4) and hence also unique with respect to (3) using the uniqueness of f_3 with respect to (2). □

2.2 DIVISION LEMMA

Let (1) in 2.1 be a pushout with $d \in \mathcal{S}$, i.e. we have $d = m_1 \circ e_1$ with $e_1 \in \mathcal{E}$ and $m_1 \in \mathcal{J}$. Moreover let $\tilde{d} = m_2 \circ e_2$ be an $\mathcal{E}$-$\mathcal{M}$-factorization of $\tilde{d}$.

Then there is a unique diagonal morphism $p':D' \longrightarrow G'$ such that (2) and (3) below are pushouts

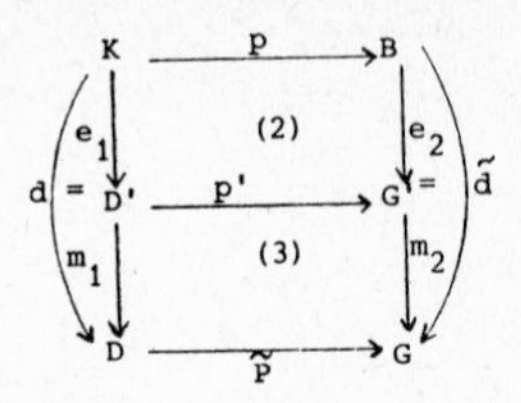

<u>Corollary:</u> $\mathcal{S}$ is closed under pushouts.

<u>Proof:</u> Since $e_1 \in \mathcal{E}$ and $m_2 \in \mathcal{M}$ there is a unique morphism $p':D' \longrightarrow G'$ such that (2) and (3) are commutative using the diagonal lemma (cf. /He-St 74/ and /E-K^3 74/). Using the fact that $e_2 \in \mathcal{E}$ is an epimorphism and (1) is a pushout it is easy to show that (3) is a pushout. In order to show that (2) is a pushout let X be an arbitrary object and $g_1:D' \longrightarrow X$, $g_2:B \longrightarrow X$ arbitrary morphisms such that

(4) $\quad g_1 \circ e_1 = g_2 \circ p.$

Now we have to consider two cases:

i) If X is an initial object we have by 1.1 d) that also D', B and K are initial objects. Hence $e_2:\emptyset \longrightarrow G'$ belongs to $\mathcal{M}$ and $\mathcal{E}$. Thus e_2 is an isomorphism showing that G' is also initial object and hence (2) a pushout.

ii) If X is noninitial let $m_1'':D'' \longrightarrow D$ be the complement of $m_1:D' \longrightarrow D$, which belongs to $\mathcal{F}$ by assumption, and $g_3:D'' \longrightarrow X$ an arbitrary morphism which exists by 1.1 d). Hence there is a unique $g_4:D \longrightarrow X$ satisfying

(5) $\quad g_4 \circ m_1 = g_1 \quad$ and $\quad g_4 \circ m_1'' = g_3$.

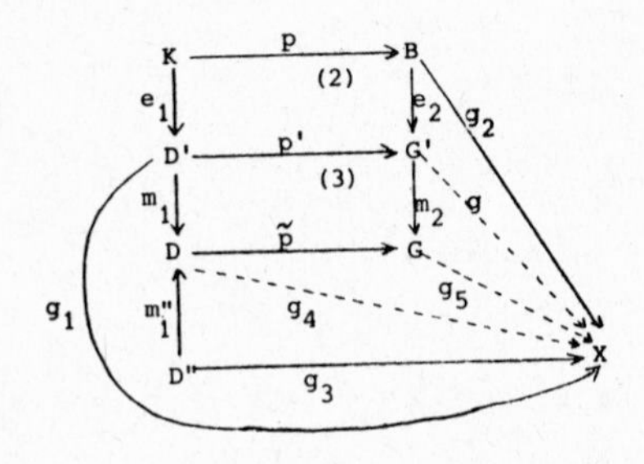

Hence we have $g_4 \circ m_1 \circ e_1 = g_1 \circ e_1 = g_2 \circ p$ and thus there is a unique $g_5 : G \longrightarrow X$ satisfying

$$(6) \quad g_5 \circ \tilde{p} = g_4 \quad \text{and} \quad g_5 \circ m_2 \circ e_2 = g_2$$

Now let $g := g_5 \circ m_2$. We have to show

$$(7) \quad g \circ p' = g_1 \quad \text{and} \quad g \circ e_2 = g_2$$

and uniqueness of g with respect to (7):

$$g \circ p' = g_5 \circ m_2 \circ p' \overset{(3)}{=} g_5 \circ \tilde{p} \circ m_1 \overset{(6)}{=} g_4 \circ m_1 \overset{(5)}{=} g_1$$

$$g \circ e_2 = g_5 \circ m_2 \circ e_2 \overset{(6)}{=} g_2$$

Now for each $g' : G' \longrightarrow X$ satisfying (7) we have to show $g' = g$. But this follows because $g' \circ e_2 = g_2 = g \circ e_2$ and $e_2 \in \mathcal{E}$ is epimorphism. Hence (2) is a pushout. Finally using the splitting lemma we have $m_2 \in \mathcal{F}$ and hence $\tilde{d} \in \mathcal{S}$ which shows the corollary. □

2.3 THEOREM (Analysis of Pushouts)

Let (1) in 2.1 be a pushout with $d, p \in \mathcal{S}$ and $d = m_1 \circ e_1$, $\breve{d} = m_2 \circ e_2$, $p = m_3 \circ e_3$, $\tilde{p} = m_4 \circ e_4$ the corresponding $\mathcal{E}$-$\mathcal{M}$-factorizations. Then in the following diagram all subdiagrams are pushouts and moreover we have for the complements $m_1'' : D'' \longrightarrow D$ of m_1 and $m_3'' : B'' \longrightarrow B$ of m_3:

a) $D = D' \amalg D''$ and $G'' = D'' \amalg X$

b) $B = B' \amalg B''$ and $G' = B'' \amalg X$

c) $G = D'' \amalg X \amalg B''$

Moreover all morphisms m_i $(i = 1, \ldots 6)$ belong to $\mathcal{F}$.

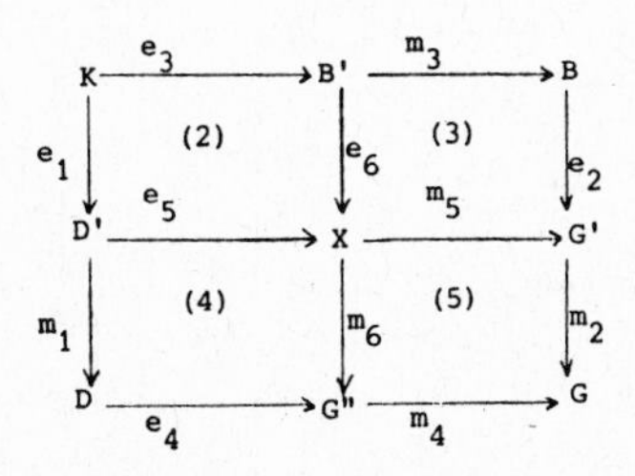

Remark: The decomposition of G in c) for K'=Sets is also given in /Ros 74/ meaning that G is the disjoint union of the non-gluing points D" in D, the gluing points X, and the non-gluing points B" in B.

Proof: The construction and pushout properties of the subdiagrams (2) - (5) follow from the division lemma 2.2 applied twice and the splitting lemma, applied to (4) and (5), leads to

a') $D = D' \amalg D''$, $G'' = D'' \amalg X$, $G = D'' \amalg G'$

and symmetrically to (3) and (5)

b') $B = B' \amalg B''$, $G' = B'' \amalg X$, $G = G'' \amalg B''$

a), b), c) are now consequences of a') and b') showing associativity of the construction of G.

2.4 COROLLARY (Construction of Pushouts)

Given morphisms $d: K \longrightarrow D$ and $p: K \longrightarrow B$ the pushout of d and p can be constructed in the following steps:

1. Construction of the $\mathcal{E}$-$\mathcal{M}$-factorizations

   $d = m_1 \circ e_1$ and $p = m_3 \circ e_3$

2. Construction of the pushout (2) of the epimorphisms e_1, e_3.

   If K' has kernel pairs, coequalizers and e_1 is a coequalizer, which is true in Sets and several other interesting examples, there is a very simple pushout construction (cf. /Ha 74/ 2.6):

   2.1 Construct the kernel pair $K_1 \overset{\pi_0}{\underset{\pi_1}{\rightrightarrows}} K$ of e_1

   2.2 Let $e_6: B' \longrightarrow X$ be the coequalizer of $e_3 \circ \pi_0$, $e_3 \circ \pi_1$

   2.3 There is a unique $e_5: D' \longrightarrow X$ such that (2) commutes because e_1 is the coequalizer of π_0, π_1 by 2.1.

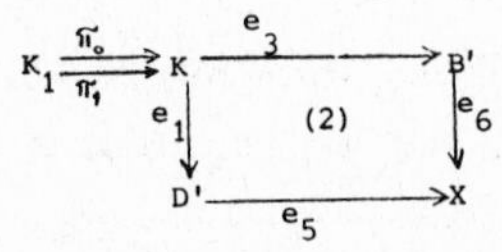

3. Construction of the complements D" of D' and B" of B'

4. Construction of the pushout object G as coproduct $G = D'' \amalg X \amalg B''$

5. Construction of $\tilde{p}: D \longrightarrow G$ induced by $u_{D''}: D'' \longrightarrow G$ and $u_X \circ e_5: D' \longrightarrow G$ and of $\tilde{d}: B \longrightarrow G$ induced by $u_{B''}: B'' \longrightarrow G$ and $u_X \circ e_6: B' \longrightarrow G$ where $u_{D''}$, u_X and $u_{B''}$ are the coproduct injections of G.

Remark: In the case K'=Sets we have for the main step 2 the following algorithmic construction: X is the quotient set $B'/_{\bar{R}}$ where $\bar{R}$ is the induced equivalence relation of

$R = \{(e_3(k), e_3(k')) \,/\, k, k' \in K, e_1(k) = e_1(k')\}$.

Since all the other steps - image of the functions d and p in step 1, complements in step 3 and disjoint unions in step 4 and 5 - are given by very simple and short procedures, the above algorithmic construction leads to a much mor efficient algorithm than the usual one where G is the quotient $D \dot{\cup} B/_{\overline{R_o}}$ with $R_o = \{(x,y)/\exists\, k \in K \;\; d(k) = x \wedge p(k) = y\}$. In both constructions we have to build the equivalence closure of a relation, but in our construction on the set B' which is much smaller than B and $D \dot{\cup} B$ of course.

3. GRAPH GRAMMARS

We begin with the definition of the labeled gluing of graphs which - performed twice with the left hand and right hand side of production - leads to a direct derivation in a graph grammar.

3.1 GENERAL ASSUMPTION

Let U:K-Graph $\longrightarrow$ K'^2 be the forgetful functor defined by UG=(E,V) and $U(f_E, f_V) = (f_E, f_V)$ which preserves pushouts and coproducts. To indicate that (E,V) belongs to G we also write (G_E, G_V). In the following we only consider those graph-morphisms $f: G \longrightarrow G'$ such that $f_E: G_E \longrightarrow G'_E$ and $f_V: G_V \longrightarrow G'_V$ are splitting, in other words $Uf: UG \longrightarrow UG'$ is splitting in K'^2 but not f in K-Graph in general. By the corollary of 2.2 those graph-morphisms are closed under pushouts but they are closed under composition iff $\mathcal{S}$ in K' is closed under composition. Thus for K'=Top, where $\mathcal{S}$ is not composition closed, we have to take care that the splitting property is not used for composite morphisms. However, the composition closure is necessary for section 5 only. All constructions and notations in 1.1 for K' we will frequently use also for K'^2 since we have the corresponding properties in each component.

3.2 CONSTRUCTION (Labeled Gluing of Graphs)

Let $p: K \longrightarrow B$ and $d: K \longrightarrow D$ be graph-morphisms with pushout G in K-Graph.

$$\begin{array}{ccc} K & \xrightarrow{p} & B \\ {\scriptstyle d}\downarrow & \text{p.o.} & \downarrow{\scriptstyle \tilde{d}} \\ D & \xrightarrow[\tilde{p}]{} & G \end{array}$$

Furthermore let $l_B: UB \longrightarrow \Omega$ be a label morphism for B in K'^2, i.e. $l_B=(l_{B,E}:B_E \longrightarrow \Omega_E, l_{B,V}:B_V \longrightarrow \Omega_V)$, and $m_1 \circ e_1=Ud$ an $\mathcal{E}$-$\mathcal{M}$-factorization in $\underline{K}'^2$ with complement $m_1'':D'' \longrightarrow UD$ of $m_1:D' \longrightarrow UD$ in $\underline{K}'^2$, and $l_{D''}:D'' \longrightarrow \Omega$ a label morphism for D".

Note that D" is no $\underline{K}$-graph in general.

In order to get a unique label morphism $l_G:UG \longrightarrow \Omega$ for the pushout graph we have to assume that e_1 satisfies the following label condition with respect to (B, l_B).

Label Condition: $l_B \circ Up \circ \pi_1 = l_B \circ Up \circ \pi_2$

where $(\pi_1, \pi_2):Kp(e_1) \Longrightarrow UK$ is the kernel pair of e_1.

Now consider the following diagram in $\underline{K}'^2$:

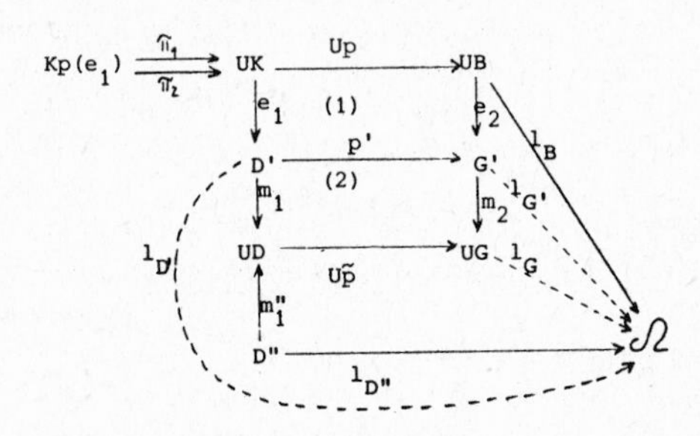

First of all (1) and (2) are pushouts by the division lemma and hence $UG=D''\amalg G'$ by the splitting lemma. Now e_1 is the coequalizer of (π_1, π_2) - if $\mathcal{E} \subseteq \underline{K}'$-coequalizers, which is true for $\underline{K}'$-$\underline{\text{Sets}}$, otherwise the label condition has to assure the existence of l_D, - such that by the label condition there is a unique $l_{D'}:D' \longrightarrow \Omega$ satisfying $l_{D'} \circ e_1 = l_B \circ Up$. Since (1) is pushout we obtain a unique $l_{G'}:G' \longrightarrow \Omega$ with $l_{G'} \circ e_2 = l_B$ and $l_{G'} \circ p'=l_{D'}$. Finally $UG=D''\amalg G'$ leads to a unique $l_G:UG \longrightarrow \Omega$ satisfying $l_G \circ U\tilde{p} \circ m_1''=l_{D''}$ and $l_G \circ m_2=l_{G'}$. Hence defining $l_D:=l_G \circ U\tilde{p}$ and because $l_G \circ U\tilde{d}=l_B$ $\tilde{p}:D \longrightarrow G$ and $\tilde{d}:B \longrightarrow G$ are Ω-graph morphisms.

Notation: $(G, l_G)=(D, l_{D''})\amalg_K(B, l_B)$ is called labeled gluing of (B, l_B) and $(D, l_{D''})$ along K.

3.3 LEMMA

Given the assumption of 3.2 there are unique label morphisms $l_K:UK \longrightarrow \Omega$ $(l_K=l_B \circ Up)$ and $l_D:UD \longrightarrow \Omega$ (induced by $l_{D'}$ and $l_{D''}$) such that (G, l_G), with l_G constructed in 3.2, is pushout object in $\underline{K}$-$\underline{\text{Graph}}_\Omega$

in the pushout

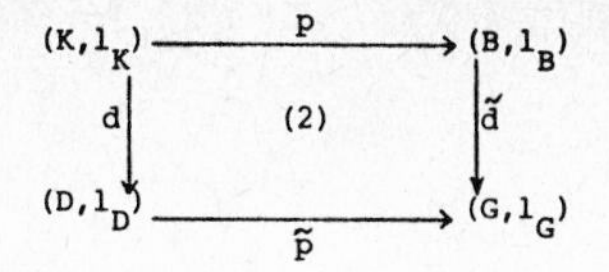

Vice versa for each pushout (2) 1_B satisfies the label condition and there is a unique $1_{D''}:D''\longrightarrow\Omega$ such that $(G, 1_G)=(D,1_{D''}) \amalg_K (B, 1_B)$.

Proof: cf. /Eh-Pf-Sch 73/ lemma 1.7 and the constructions in 3.2.

3.4 DEFINITION

Let $\Omega = (\Omega_E, \Omega_V) \in \underline{K}'^2$ be a label alphabet and $T=(T_E, T_V)$ a subobject of Ω. Then (Ω, T) is called graph alphabet, T_E and T_V are called terminal alphabet for the edges and the vertices respectively. $\underline{K}$-$\underline{\text{Graph}}_T$ can be regarded as a full subcategory of $\underline{K}$-$\underline{\text{Graph}}_\Omega$.

A graph grammar is a 4-tuple $Q=(\Omega, T, S, 1_s, P)$ where (Ω, T) is a graph alphabet, $(S, 1_s) \in |\underline{K}\text{-}\underline{\text{Graph}}_\Omega|$ a start graph with $(S, 1_s) \notin |\underline{K}\text{-}\underline{\text{Graph}}_T|$ and P a finite set of productions π of the following shape

$$\pi = ((B, 1_B) \xleftarrow{p} K \xrightarrow{p'} (B', 1_{B'}))$$

with $(B, 1_B), (B', 1_{B'}) \in |\underline{K}\text{-}\underline{\text{Graph}}_\Omega|$, $K \in |\underline{K}\text{-}\underline{\text{Graph}}|$

$(B, 1_B) \notin |\underline{K}\text{-}\underline{\text{Graph}}_T|$, $p, p' \in \underline{K}\text{-}\underline{\text{Graph}}$.

An Ω-Graph $(H, 1_H)$ is called directly derivable from $(G, 1_G)$ via $\pi \in P$, written $(G, 1_G) \overset{\pi}{\Longrightarrow} (H, 1_H)$ if there is an enlargement $\varepsilon = (D, (D, 1_{D''}))$ of π, i.e. $D \in |\underline{K}\text{-}\underline{\text{Graph}}|$, $d:K \longrightarrow D$ a graph morphism and $1_{D''}:D''\longrightarrow\Omega$ as given in 3.2 satisfying the label condition with respect to $(B, 1_B)$ and $(B', 1_{B'})$, such that there are $\underline{K}$-$\underline{\text{Graph}}_\Omega$ isomorphisms

$$(G, 1_G) \cong (D, 1_{D''}) \amalg_K (B, 1_B)$$
$$(H, 1_H) \cong (D, 1_{D''}) \amalg_K (B', 1_{B'})$$

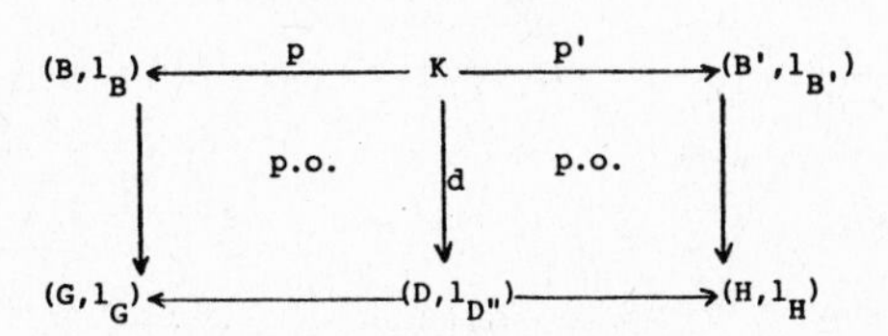

As usual $(H, 1_H)$ is called derivable from $(G, 1_G)$, written $(G, 1_G) \overset{*Q}{\Longrightarrow} (H, 1_H)$, if $(G, 1_G) \cong (H, 1_H)$ or there is a sequence of direct derivations $(G, l_G) \Longrightarrow (G_1, l_{G_1}) \Longrightarrow \ldots \Longrightarrow (G_n, l_{G_n}) \Longrightarrow (H, l_H)$ with $n \geq 0$, and the <u>graph language</u> $L(Q)$ is the class of all terminal labeled graphs (G, l_G), i.e. $(G, l_G) \in |\underline{K}\text{-}\underline{Graph}_T|$, derivable from the start graph (S, l_S).

4. ENLARGEMENT ANALYSIS

One basic problem in the theory of graph grammars is the construction and uniqueness of suitable enlargements D in order to apply a production $\pi = ((B, l_B) \xleftarrow{p} K \xrightarrow{p'} (B', l_{B'}))$ to a given graph (G, l_G), i.e. to find for given $p:K \longrightarrow B$ and $\tilde{d}:B \longrightarrow G$ a graph D and graph morphisms $d:K \longrightarrow D$, $\tilde{p}:D \longrightarrow G$ such that (1)

$$\begin{array}{ccc} K & \xrightarrow{p} & B \\ {\scriptstyle d}\downarrow & (1) & \downarrow{\scriptstyle \tilde{d}} \\ D & \xrightarrow[\tilde{p}]{} & G \end{array}$$

becomes a pushout in <u>K-Graph</u>. More general to find all possible $D, d, \tilde{p}$ such that (1) becomes a pushout. After that construction the label morphism for D is necessarily $l_D := l_G \circ Up:UD \longrightarrow \Omega$ such that (1) becomes a pushout in $\underline{K}\text{-}\underline{Graph}_\Omega$.

Clearly we have $l_{D''} = l_D \circ m_1'' : D'' \longrightarrow \Omega$ (cf. 3.2) such that we get the derived graph (H, l_H) by

$$(H, l_H) = (D, l_{D''}) \amalg_K (B', m_{B'}).$$

Since pushouts in <u>K-Graph</u> are constructed componentwise we first study the enlargement problem in the category <u>K</u>' leading to the general result in 4.5.

4.1 PROBLEM AND DEFINITION

Given $p:K \longrightarrow B$ and $\tilde{d}:B \longrightarrow G$ in <u>K</u>' with $p, \tilde{d} \in \mathcal{S}$ find up to isomorphism all tripels $(D, d, \tilde{p})$ with $d, \tilde{p} \in \mathcal{S}$, called <u>p.o.-complement</u> of $p, \tilde{d}$, such that diagram (1) above becomes a pushout in <u>K</u>' and characterize all those $p, \tilde{d}$ such that at least one p.o.-complement exists.

4.2 THEOREM (p.o.-Complements)

Given $p:K \longrightarrow B$, $\tilde{d}:B \longrightarrow G$ with $p, \tilde{d} \in \mathcal{S}$ and $\mathcal{E}$-$\mathcal{M}$-factorizations $p = K \xrightarrow{e_3} B' \xrightarrow{m_3} B$, $\tilde{d} = B \xrightarrow{e_2} G' \xrightarrow{m_2} G$ there is a p.o.-complement of $p, \tilde{d}$ iff G' is a coproduct $G' = X \amalg B''$ with injections $X \xrightarrow{m_5} G'$ and $B'' \xrightarrow{m_3''} B \xrightarrow{e_2} G'$ where

$$B' \xrightarrow{e_6} X \xrightarrow{m_5} G' = B' \xrightarrow{m_3} B \xrightarrow{e_2} G'$$

is the $\mathcal{E}$-$\mathcal{M}$-factorization of $e_2 \circ m_3$ and m_3'' the complement of $m_3 \in \mathcal{F}$. Moreover in this case, each p.o.-complement of $p, \tilde{d}$ is uniquely determined up to isomorphism by an p.o.-complement of $K \xrightarrow{e_3} B', B' \xrightarrow{e_6} X$ using $\mathcal{E}$-morphisms and there are two canonical p.o.-complements of e_3, e_6:

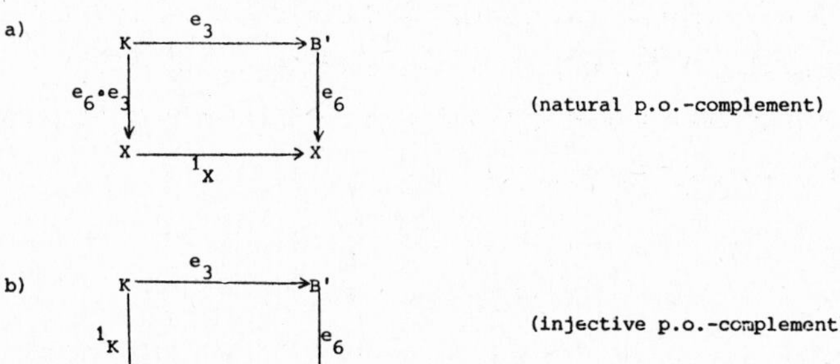

Case b) is possible only in the case $\tilde{d} \in \mathcal{F}$ which implies that e_2, e_6 are isomorphisms.

<u>Construction:</u> Given an arbitrary p.o.-complement e_1, $e_5 \in \mathcal{E}$ of e_3, e_6

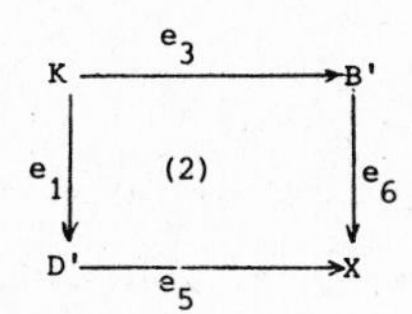

the corresponding p.o.-complement of p, $\tilde{d}$ can be obtained in the following way:

1. $m_2'' : D'' \longrightarrow G$ is the complement of $m_2 : G' \longrightarrow G$,
2. $G'' = D'' \amalg X$ with injections $m_6 : X \longrightarrow G''$ and $m_6'' : D'' \longrightarrow G''$,
3. $m_4 : G'' \longrightarrow G$ induced by $m_2'' : D'' \longrightarrow G$ and $m_2 \circ m_5 : X \longrightarrow G$,
4. $D = D' \amalg D''$ with injection $m_1 : D' \longrightarrow D$,
5. $e_4 : D \longrightarrow G''$ induced by $m_6 \circ e_5 : D' \longrightarrow G''$ and $m_6'' : D'' \longrightarrow G''$.

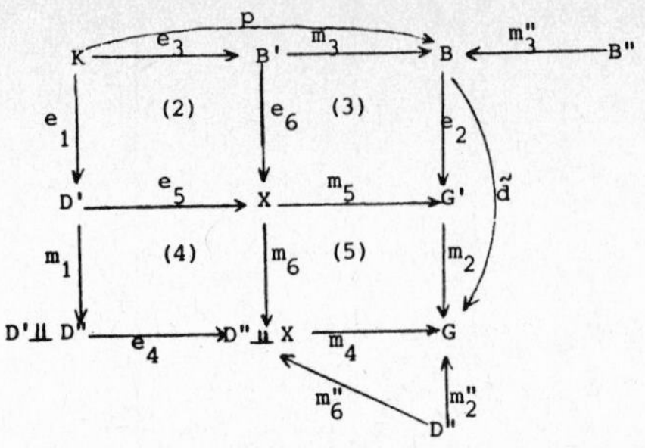

Proof: First let us assume that we have a p.o.-complement d, $\tilde{p}$ of p, $\tilde{d}$ then by theorem 2.3 (3) is a pushout and hence by the splitting lemma we have G' = X ⊔ B" with the injections given above. This shows the necessity of the condition.

Vice versa given this condition (3) is a pushout again by the splitting lemma. Moreover (2) is a pushout by construction and (5) and (4) are pushouts by the splitting lemma. Defining $d=m_1 \circ e_1$ and $p=m_4 \circ e_4$ we obtain a p.o.-complement of $\tilde{d}$,p because the composition of pushouts is a pushout.

Given an arbitrary p.o.-complement d, $\tilde{p}$ of p, $\tilde{d}$ we have by theorem 2.3 that (2) is a pushout of $\mathcal{E}$-morphisms and the construction in steps 1-5 is necessary and unique up to isomorphism by 2.3 a').

It remains to show that the canonical diagrams a) and b) are pushouts but this is easy to see checking the universal properties. □

4.3 COROLLARY

According to 4.2 a) the natural p.o.-complement of p, $\tilde{d}$ is given by $d = m_6 \circ e_1$, $\tilde{p} = m_4 \in \mathcal{J}$, using $e_5 = 1_X$, where X is the image of $\tilde{d} \circ p$ and

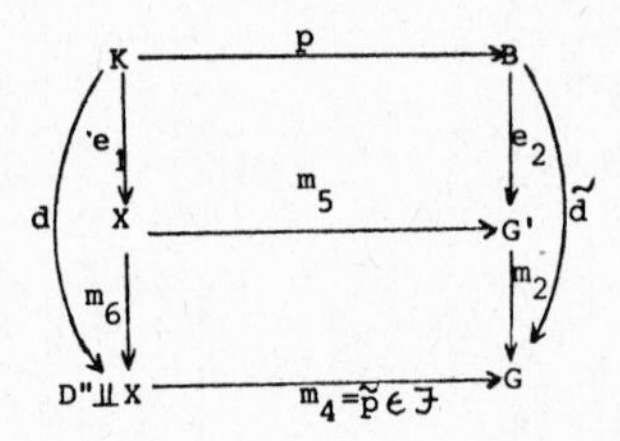

this p.o.-complement is unique up to isomorphism with respect to the property $\tilde{p} \in \mathcal{J}$ (cf. /Ros 74/).

On the other hand starting with $\tilde{d} \in \mathcal{J}$ the injective p.o.-complement of p, $\tilde{d}$ is given by $d=m_1:K \longrightarrow K ⊔ D'' \in \mathcal{J}$ using $e_1=1_K$, and $\tilde{p}:K ⊔ D'' \longrightarrow G$ induced by $\tilde{d}\circ p:K \longrightarrow G$ and $m_2'':D'' \longrightarrow G$ which is now the complement of $\tilde{d}$.

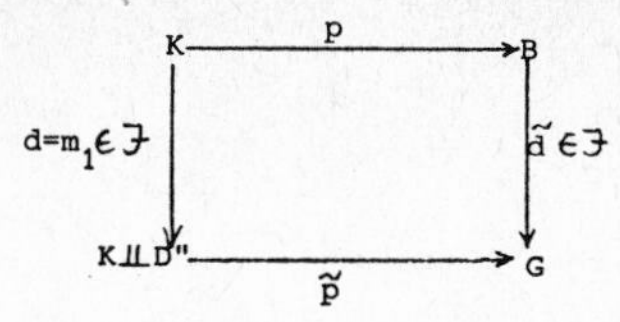

Moreover this p.o.-complement is unique up to isomorphism with respect to the property $d \in \mathcal{F}$ (cf. /Eh-Pf-Sch 73/).

Proof: It suffices to show the uniqueness properties:

a) Given a p.o.-complement d, $\tilde{p}$ with $\tilde{p} \in \mathcal{F}$
   we have by theorem 2.3 that e_4 is an isomorphism. Hence
   $m_6 \circ e_5 = e_4 \circ m_1 \in \mathcal{M}$ which implies $e_5 \in \mathcal{E} \cap \mathcal{M}$ to be an isomorphism.

b) In the second case $\tilde{d} \in \mathcal{F}$ implies immediately that e_1 is an isomorphism. □

It remains to characterize all p.o.-complements of $\mathcal{E}$-morphisms.

4.4 PROPOSITION (Construction of p.o.-complements for -morphisms)

Assume that K' has kernel pairs and coequalizers of kernel pairs and $\mathcal{E}$ is a class of regular epimorphisms.

Let $p:K \longrightarrow B$ and $\tilde{d}:B \longrightarrow G$ be $\mathcal{E}$-morphisms then each pushout complement d, $\tilde{p}$ of p, $\tilde{d}$ can be constructed as follows

1. Let $U \overset{u_o}{\underset{u_1}{\rightrightarrows}} K$, $V \overset{v_o}{\underset{v_1}{\rightrightarrows}} K$ be the kernel pairs
   of $p:K \longrightarrow B$ and $\tilde{d}\circ p:K \longrightarrow G$ respectively.

2. Let $W \overset{w_o}{\underset{w_1}{\rightrightarrows}} K$ be an arbitrary kernel pair such that
   there is $w:W \longrightarrow V$ satisfying $w_o = v_o \circ w$, $w_1 = v_1 \circ w$
   and the kernel pair (v_o, v_1) is the supremum of
   (u_o, u_1) and (w_o, w_1), written $V = U \vee W$.

3. Let $d:K \longrightarrow D$ be the coequalizer of $W \overset{w_o}{\underset{w_1}{\rightrightarrows}} K$

4. Let $\tilde{p}:D \longrightarrow G$ be the unique morphism such that (1) commutes using step 3.

Remark: In $\underline{K}'$=Sets step 2 means to construct an equivalence relation W on K such that $U \vee W := \overline{U \cup W} = V$. All such equivalence relations can be constructed by standard procedures although a specific procedure would be useful in this case.

Proof: It suffices to show that $V=U \vee W$ iff (1) is a pushout (cf. /Ha 74/ 2.7). If (1) is pushout and a kernel pair (x_o, x_1) with g_1, g_2 are given we obtain f_o = coequ (x_o, x_1) and morphisms f_1, f_2, f and hence g. Vice versa if $V=U \vee W$ morphisms f_1, f_2 and hence f_o are given leading to (x_o, x_1)= kernel pair (f_o) and hence to g_1, g_2, g and f_1 using that $\tilde{d} \circ p$ = coequ (v_o, v_1).

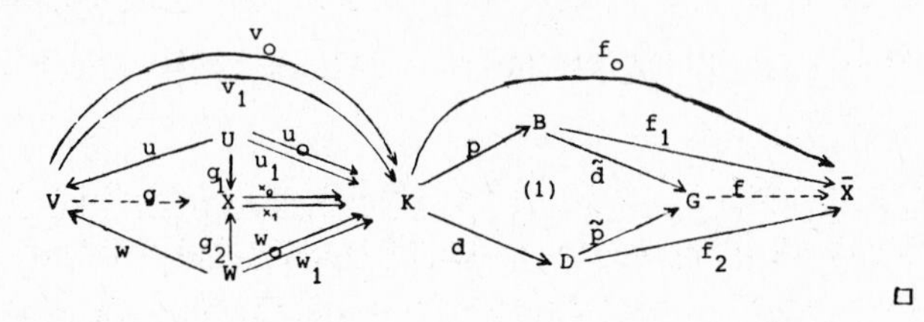

□

4.5 THEOREM (p.o.-Complements for $\underline{K}$-Graphs)

Given graph-morphisms $p:K \longrightarrow B$, $\tilde{d}:B \longrightarrow G$ there is a p.o.-complement of p, $\tilde{d}$ iff there are p.o.-complements of p_E, $\tilde{d}_E$ and p_V, $\tilde{d}_V$ in $\underline{K}'$ and the following gluing condition is satisfied:

Gluing Condition: There are $\underline{K}$-morphisms $\sigma, \tau : D''_E \longrightarrow D'_V \amalg D''_V$ such that the following diagram (1) commutes with σ, s_G and τ, t_G respectively where s_G, t_G are source and target maps of G and the other objects and

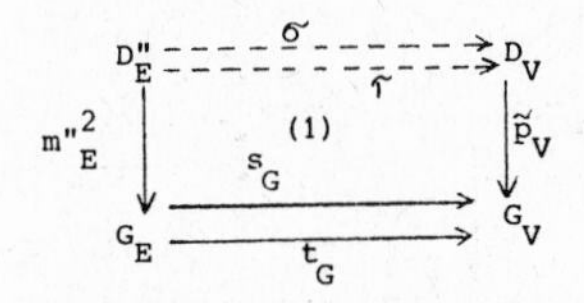

morphisms are defined in analogy to 4.3 with $D_V=D'_V \amalg D''_V$ (D'_V ="gluing points", D''_V = complement of G'_V in G_V).

Moreover all p.o.-complements of p, $\tilde{d}$ can be obtained by all those p.o.-complements in the components which have e^1_E and e^1_V satisfying the

Compatibility Condition: There are (unique) $s_{D'}$, $t_{D'}$ making diagram(2)

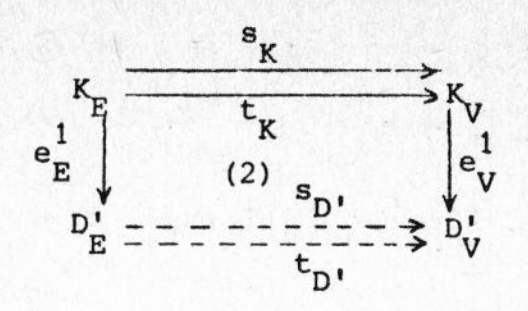

commutative for $s_{K'}$ $s_{D'}$ and $t_{K'}$ $t_{D'}$ respectively.

If we take natural p.o.-complements in both components the graph D is unique up to isomorphism. In the case of injective p.o.-complements - possible if $\tilde{d}_E, \tilde{d}_V \in \mathcal{J}$ - the compatibility condition is also satisfied but there are several choices for σ and τ in (1) such that D is not unique up to isomorphism in general.

The construction of $s_{D'}$ $t_D : D_E \longrightarrow D_V$ is given in the proof.

Note that we always consider graph morphisms with splitting components (cf. 3.1).

Proof: Suppose that we have determined p.o.-complements d_E, $\tilde{p}_E$ of p_E, $\tilde{d}_E$ and d_V, $\tilde{p}_V$ of p_V, $\tilde{d}_V$ satisfying the compatibility and the gluing condition then we have the following commutative diagram in K where $s_D : D_E \longrightarrow D_V$ is obtained as induced morphism of $m_V^1 \circ s_{D'}$ and σ.

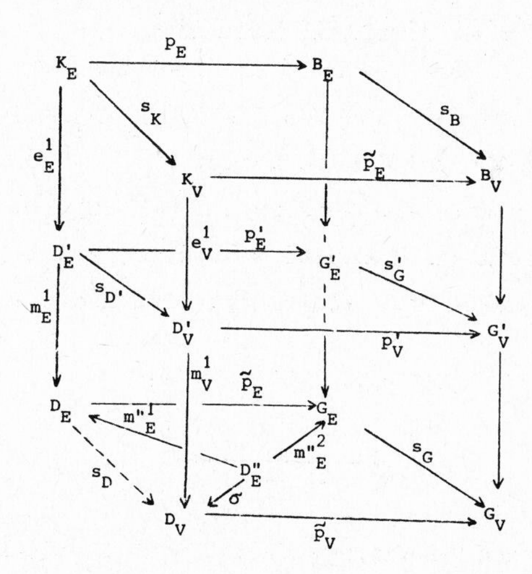

By the corresponding argument for targets we obtain $t_D : D_E \longrightarrow D_V$ such that $D = D_E \overset{s_D}{\underset{t_D}{\rightrightarrows}} D_V$ becomes a K-graph and $d = (d_E, d_V)$ with $d_E := m_E^1 \circ e_E^1$, $d_V := m_V^1 \circ e_V^1$ and $\tilde{p} = (\tilde{p}_E, \tilde{p}_V)$ become graph-morphisms

$d:K \longrightarrow D$, $\tilde{p}:D \longrightarrow G$ and hence a p.o.-complement of p, $\tilde{d}$.
In the case of natural p.o.-complements in both components we have $p'_E, p'_V \in \mathcal{M}$ (cf. 4.3) such that in this case the compatibility condition is satisfied because $s_{D'}$ is the unique diagonal morphism in

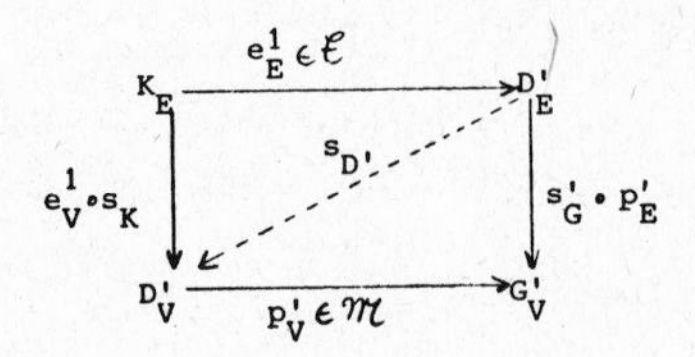

Moreover we have in this case $\tilde{p}_V \in \mathcal{I}$ (cf. 4.3) such that s_D, t_D and hence the graph D is uniquely determined up to isomorphism.
(The same is true for σ, τ in (1)).
If we have injective p.o.-complements we have $e^1_E = 1_{K_E}$, $e^1_V = 1_{K_V}$ showing that $s_{D'} = s_K$ and $t_{D'} = t_K$ satisfies the compatibility condition and for all possible s_D, $s'_D: D_E \longrightarrow D_V$ we have

$$\tilde{p}_V \circ s_D = s_G \circ \tilde{p}_E = \tilde{p}_V \circ s'_D$$

and the same for the target morphisms.
It remains to show that in the general case the gluing and the compatibility condition are necessary.
For this purpose let $d:K \longrightarrow D$, $\tilde{p}:D \longrightarrow G$ be an arbitrary p.o.-complement of p, $\tilde{d}$ and define $\sigma := s_D \circ m''^1_E$, $\tau := t_D \circ m''^1_E$ which satisfy the gluing condition. Also the compatibility condition is satisfied because $s_{D'}$ and $t_{D'}$ can be defined as suitable diagonal morphisms using $e^1_E \in \mathcal{E}$, $m^1_V \in \mathcal{M}$. □

5. EMBEDDING OF DERIVATIONS

In this section we generalize the results of /Eh-Pf-Sch 73/ §5 concerning embeddings of derivations to K-graphs and not necessarily injective enlargement graph morphisms $d:K \longrightarrow D$. Since this theory is already formulated in categorical terms except the label problems it suffices to treat these problems. In addition to assumptions 1.1 a) - 1.1 d) and 3.1 we also need 1.1 e) which implies that the graph morphisms with splitting components, which are only considered, are closed under composition. This is essential for the following constructions, but a real restriction for the choice of K'. K'=Sets satisfies this condition but K'=Top not. But we do not need 1.1 e) if we only consider injective enlargements $\bar{\mathcal{E}}$ in 5.2 - 5.4.

5.1 LEMMA

Given a graph-morphism $p:K \longrightarrow (B, 1_B)$, an injective enlargement $\mathcal{E} = (K \xrightarrow{d} (D, 1_{D''}))$ leading to the following pushout (1) in K-Graph, a further graph morphism $q:\bar{K} \longrightarrow D$ together with an injective

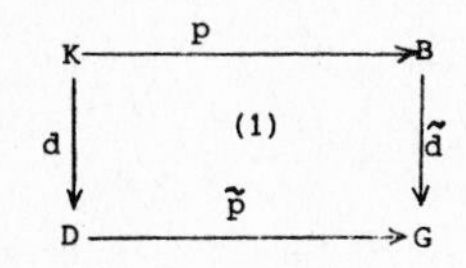

enlargement

$\bar{\mathcal{E}} = (\bar{K} \xrightarrow{\bar{d}} (\bar{D}, 1_{\bar{D}''}))$ then there is an injective enlargement

$\hat{\mathcal{E}} = (K \xrightarrow{\hat{d}} (\hat{D}, 1_{\hat{D}''}))$ such that

$(\bar{D}, 1_{\bar{D}''}) \amalg_{\bar{K}} ((D, 1_{D''}) \amalg_K (B, 1_B)) \cong (\hat{D}, 1_{\hat{D}''}) \amalg_K (B, 1_B)$.

An enlargement is called injective if the components of the graph morphism belong to the class $\mathcal{J}$ of coproduct injections.

Addendum: The K^2-morphism $1_{\hat{D}''}: D'' \longrightarrow \Omega$ is independent of 1_B because $1_{\hat{D}''}$ is the induced morphism of $1_{\bar{D}''}$ and $1_{D''}$.

Proof: The first part follows from the following $K\text{-Graph}_\Omega$ -pushouts using lemma 3.3 where the enlargement $\hat{\mathcal{E}}$ is defined by $\hat{d} = \tilde{\tilde{d}} \circ d \in \mathcal{J}$, $\hat{d}'': D'' \longrightarrow U\hat{D}$ complement of $U\hat{d}$ and $1_{\hat{D}''} = 1_{\hat{D}} \circ \hat{d}''$.

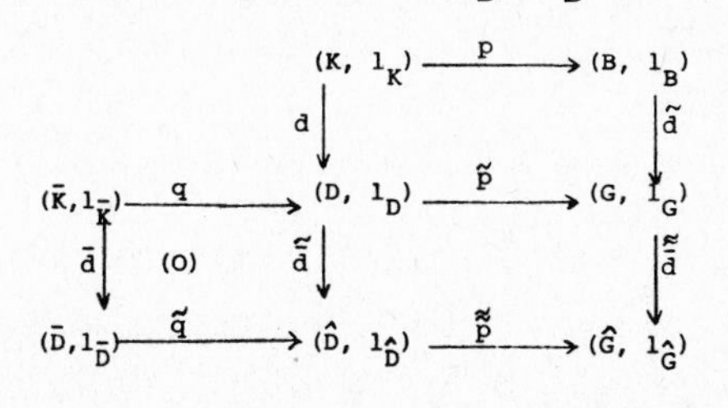

In order to verify the addendum we consider the following K^2-diagram where (1) is a pushout so that by the splitting lemma $U\hat{D} = \bar{D}'' \amalg UD$ and by construction $UD = UK \amalg D''$, $U\hat{D} = UK \amalg \hat{D}''$ which implies $U\hat{D} = \bar{D}'' \amalg UK \amalg D''$ and hence by uniqueness of complements (cf. 1.1 c) there is an isomorphism $i: \bar{D}'' \amalg D'' \xrightarrow{\sim} \hat{D}''$ such that (5) commutes .

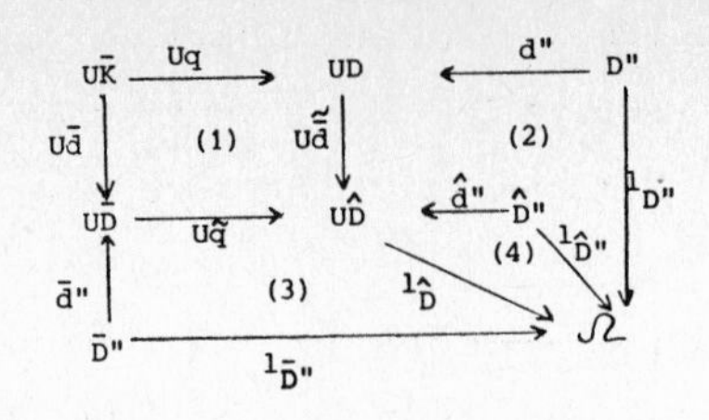

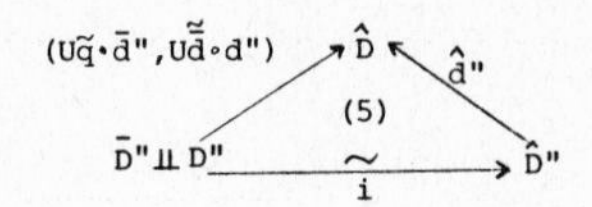

Now let i_1, i_2 be the coproduct injections $\bar{D}'' \xrightarrow{i_1} \bar{D}''\amalg D'' \xleftarrow{i_2} D''$

then $\bar{D}'' \xrightarrow{i\circ i_1} \hat{D}'' \xleftarrow{i\circ i_2} D''$ is a coproduct and it remains to show

$$l_{\hat{D}''}\circ i\circ i_1 = l_{\bar{D}''}\ ,\ l_{\hat{D}''}\circ i\circ i_2 = l_{D''}.$$

But this follows from:

$$l_{\bar{D}''} \overset{(3)}{=} l_{\hat{D}}\circ U\tilde{q}\circ\bar{d}'' \overset{(5)}{=} l_{\hat{D}}\circ\hat{d}''\circ i\circ i_1 \overset{(4)}{=} l_{\hat{D}''}\circ i\circ i_1$$

$$l_{D''} \overset{(3)}{=} l_{\hat{D}}\circ U\tilde{\bar{d}}\circ d'' \overset{(5)}{=} l_{\hat{D}}\circ\hat{d}''\circ i\circ i_2 \overset{(4)}{=} l_{\hat{D}''}\circ i\circ i_1 \quad . \square$$

5.2 LEMMA

Lemma 5.1 remains true also for non-injective enlargements $\mathcal{E}$ and $\bar{\mathcal{E}}$ which satisfy the label conditions with respect to (B, l_B) resp. (G, l_G) and hence also resp. (D, l_D).

Addendum: $l_{\hat{D}''}$ is independent of l_B because it only depends on $l_{\bar{D}''}$ and $l_{D''}$.

Remark: For this lemma we need the following stronger version of uniqueness of coproduct complements in K' which is satisfied for K'=Sets and K'=Top.

Strong Complement Condition: Let $D_1 \longrightarrow D \longleftarrow D_2$ be a coproduct in K' and $D_3 \longrightarrow D \in \mathcal{M}$ with $D_1 \cap D_3 = \emptyset$. Then there is a unique $m: D_3 \longrightarrow D_2$ such that $D_3 \longrightarrow D = D_3 \xrightarrow{m} D_2 \longrightarrow D$. Moreover we assume that K' has pullbacks, $\mathcal{E}$ and $\mathcal{M}$ are closed under pullbacks and $D_1 \cap D_2 = \emptyset$ for all coproduct components.

Clearly this condition implies the uniqueness of coproduct complements cf. 1.1 c).

Corollary: Given coproducts $D=D' \amalg D''$ and $D'=D''' \amalg \tilde{D}$ and $f,f' \in \mathcal{E}$ with (1) commutative then there exist D''', m and $f''' \in \mathcal{E}$ such that (2) commutes.

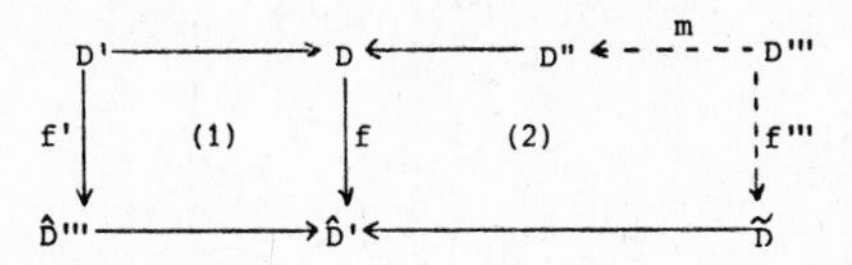

Proof of the Corollary: Let

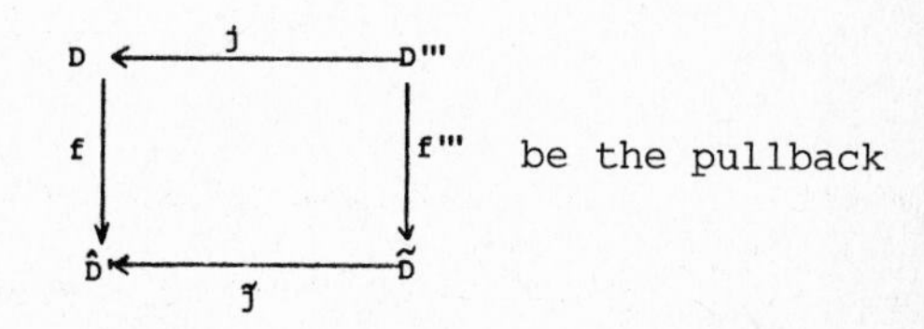

be the pullback of $f \in \mathcal{E}$ and $\tilde{j} \in \mathcal{M}$. Then we have $f''' \in \mathcal{E}$, $j \in \mathcal{M}$ and $D' \cap D'''=\emptyset$ because $\hat{D}''' \cap \tilde{D}=\emptyset$ implies a unique morphism $D' \cap D''' \longrightarrow \emptyset$ which is an isomorphism by 1.1 d).

Now the strong complement condition yields a unique $m:D''' \longrightarrow D''$ such that (2) commutes.

Proof of the Lemma: The first part is similar to lemma 5.1. In order to show the addendum we consider the following diagram with $\mathcal{E}$-$\mathcal{M}$-factorizations of Ud, $U\tilde{\bar{d}}$, $U(\tilde{\bar{d}} \circ d)$ and $U\bar{d}$ where $\hat{m} \in \mathcal{M}$ and $f' \in \mathcal{E}$ are uniquely defined diagonal morphisms and D'', $\tilde{D}$, $\hat{D}''$ complements in $\underline{K}'^2$ defined by $UD=D' \amalg D''$, $\hat{D}'=\hat{D}''' \amalg \tilde{D}$, $U\hat{D}=\hat{D}''' \amalg \hat{D}''$. Arrows $\rightarrowtail$ and $\twoheadrightarrow$ denote morphisms in $\mathcal{M}$ and $\mathcal{E}$ respectively. Since we have $U\hat{D}=\bar{D}'' \amalg \hat{D}'$ we obtain in analogy to 5.1 $\hat{D}'' \cong \bar{D}'' \amalg \tilde{D}$ and $1_{\hat{D}''}$ induced by $1_{\bar{D}''}$ and $1_{\tilde{D}}$ (but not by $1_{\bar{D}''}$ and $1_{D''}$). It remains to show that $1_{\tilde{D}}$ depends only on $1_{D''}$. Since (3) corresponds to (1) in the corollary above we

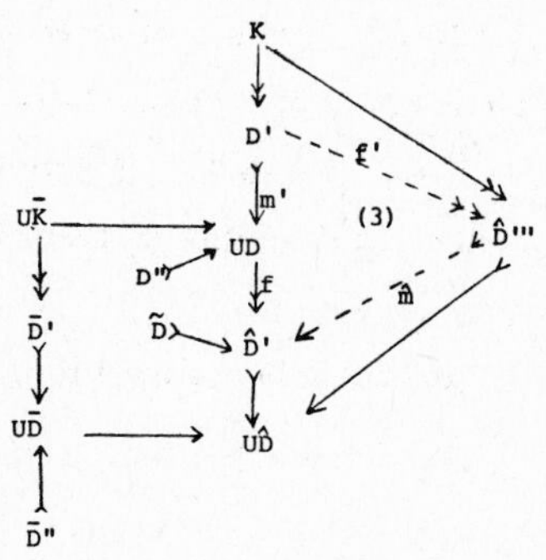

obtain commutative diagrams (4) corresponding to (2), and (6), (7) and the outer diagram by definition of $l_{D''}$, $l_{\tilde{D}}$, and $l_{\hat{D}'}$. Hence also (8)

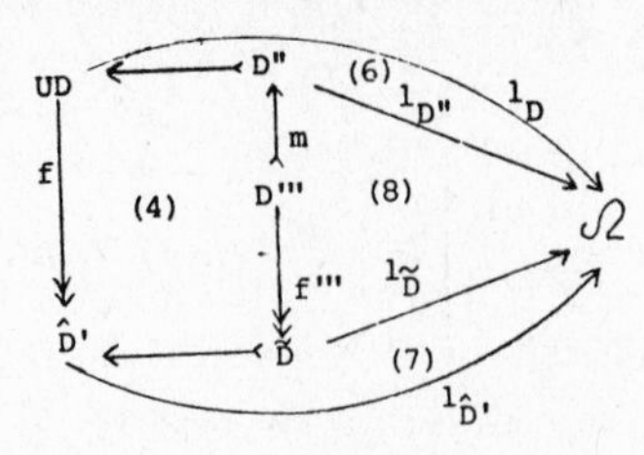

commutes with $f''' \in \mathcal{E}$ epimorphism showing that $l_{\tilde{D}}$ only depends on the label morphism $l_{D''}$. Hence $l_{\hat{D}''}$ only depends on $l_{\bar{D}''}$ and $l_{D''}$. □

For the rest of the paper we assume that $\underline{K}'$ satisfies the strong complement condition.

5.3 PROPOSITION

Let $Q = (\Omega, T, S, P)$ be a graph grammar, $(\pi, \varepsilon): (G, l_G) \Longrightarrow (H, l_H)$ a direct derivation with enlargement $\varepsilon = (K \xrightarrow{d} (D, l_{D''}))$, $\bar{\varepsilon} = (\bar{K} \xrightarrow{\bar{d}} (\bar{D}, l_{\bar{D}''}))$ an enlargement satisfying the label condition with respect to (G, l_G) and (H, l_H), and $f: \bar{K} \longrightarrow G$, $f': \bar{K} \longrightarrow H$ graph morphisms. Then there is an enlargement $\hat{\varepsilon} = (K \xrightarrow{\hat{d}} (\hat{D}, l_{\hat{D}''}))$ yielding a direct derivation

$$(\pi, \hat{\varepsilon}): (\bar{D}, l_{\bar{D}''}) \amalg_{\bar{K}} (G, l_G) \Longrightarrow (\bar{D}, l_{\bar{D}''}) \amalg_{\bar{K}} (H, l_H)$$

provided that the following coherence condition is satisfied:

There is a graph morphism $q: \bar{K} \longrightarrow D$ such that

$$f = \tilde{p} \circ q \quad \text{and} \quad f' = \tilde{p}' \circ q.$$

Proof: According to lemma 5.2 we have:

$$(\hat{G}, l_{\hat{G}}) := (\bar{D}, l_{\bar{D}''}) \amalg_{\bar{K}} (G, l_G) \cong (\hat{D}, l_{\hat{D}''}) \amalg_K (B, l_B)$$
$$(\hat{H}, l_{\hat{H}}) := (\bar{D}, l_{\bar{D}''}) \amalg_{\bar{K}} (H, l_H) \cong (\hat{D}, l_{\hat{D}''}) \amalg_K (B', l_{B'}).$$

The construction of pushout (o) in 5.1 - regarded as graph diagram only - depends on q and $\bar{d}$, hence $\hat{D}$ and by the addendum also $l_{\hat{D}''}$ are the same in both constructions showing that there is a direct derivation $(\pi, \hat{\varepsilon}): (\hat{G}, l_{\hat{G}}) \Longrightarrow (\hat{H}, l_{\hat{H}})$ with $\hat{\varepsilon} = (K \xrightarrow{\bar{d} \circ d} (\hat{D}, l_{\hat{D}''}))$. □

5.4 EMBEDDING THEOREM

Let $(G_o, l_{G_o}) \xrightarrow{(\pi_1, \varepsilon_1)} (G_1, l_{G_1}) \Longrightarrow \dots \xrightarrow{(\pi_n, \varepsilon_n)} (G_n, l_{G_n})$ be a derivation in a graph grammar $Q = (\Omega, T, S, P)$ and let

$q:\bar{K} \longrightarrow G_o$, $q':\bar{K} \longrightarrow G_n$ be graph morphisms and $\bar{\varepsilon} = (\bar{K} \xrightarrow{\bar{d}} (\bar{D}, l_{\bar{D}''}))$ an enlargement satisfying the label condition with respect to (G_o, l_{G_o}) and (G_n, l_{G_n}).

Then there is a derivation

$$(\bar{D}, l_{\bar{D}''}) \amalg_{\bar{K}} (G_o, l_{G_o}) \overset{*}{\Longrightarrow} (\bar{D}, l_{\bar{D}''}) \amalg_{\bar{K}} (G_n, l_{G_n})$$

provided q, q' satisfy the following condition:

Coherence Condition: There are morphisms $q_i:\bar{K} \longrightarrow D_i$ $(1 \leq i \leq n)$ such that

(1) $q=\tilde{p}_1 \circ q_1$

(2) $q'=\tilde{p}'_n \circ q_n$

(3) $\tilde{p}'_i \circ q_i = \tilde{p}_{i+1} \circ q_{i+1}$ $(1 \leq i \leq n)$

(4) $\bar{\varepsilon}$ satisfies the label condition with respect to (G_i, l_{G_i}) for $i=1,\ldots,n-1$

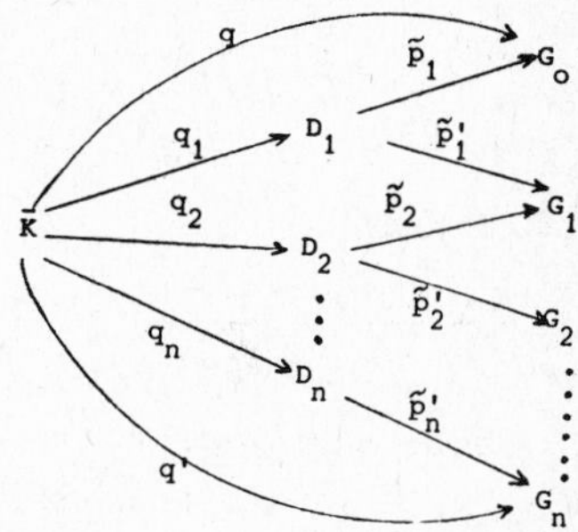

where $\tilde{p}_i$, $\tilde{p}'_i$ are defined by the following pushouts $(i=1,\ldots,n)$

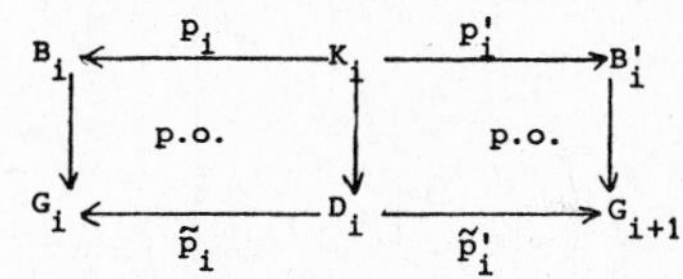

Remark: If $\bar{\varepsilon}$ is an injective embedding condition (4) is always satisfied.

Proof: According to (3), (4) and proposition 5.3 there are direct derivations, in complete notation (cf. /Eh-Pf-Sch 73/).

$$(\bar{D}, l_{\bar{D}''}) \underset{\bar{d},\tilde{p}_i \circ q_i}{\amalg}_{\bar{K}} (G_{i-1}, l_{G_{i-1}}) \Longrightarrow (\bar{D}, l_{\bar{D}''}) \underset{\bar{d},\tilde{p}'_i \circ q_i}{\amalg}_{\bar{K}} (G_i, l_{G_i})$$

for $1 \leq i \leq n$.

In the complete notation $(\bar{D}, l_{\bar{D}''}) \underset{\bar{d},\tilde{p}_i \circ q_i}{\amalg}_{\bar{K}} (G_{i-1}, l_{G_{i-1}})$ we also note the corresponding graph morphisms in the pushout below.

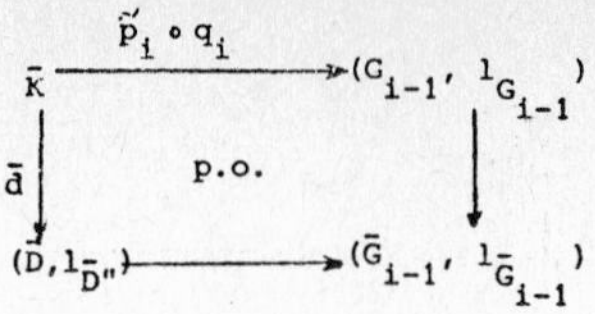

$$\begin{array}{ccc} \bar{K} & \xrightarrow{\tilde{p}_i \circ q_i} & (G_{i-1}, l_{G_{i-1}}) \\ \bar{d} \downarrow & \text{p.o.} & \downarrow \\ (\bar{D}, l_{\bar{D}''}) & \longrightarrow & (\bar{G}_{i-1}, l_{\bar{G}_{i-1}}) \end{array}$$

Taking $(\bar{G}_i, l_{\bar{G}_i}) := (\bar{D}, l_{\bar{D}''}) \underset{\bar{d}, \tilde{p}_i' \circ q_i}{\sqcup\!\sqcup}_{\bar{K}} (G_i, l_{G_i})$

we have according to (3) for $1 < i \leq n$ and define for $i = 1$

$$(\bar{G}_{i-1}, l_{\bar{G}_{i-1}}) = (\bar{D}, l_{\bar{D}''}) \underset{\bar{d}, \tilde{p}_i \circ q_i}{\sqcup\!\sqcup}_{\bar{K}} (G_{i-1}, l_{G_{i-1}}).$$

We thus have a sequence of direct derivations

$$(\bar{G}_o, l_{\bar{G}_o}) \Longrightarrow (\bar{G}_1, l_{\bar{G}_1}) \Longrightarrow \ldots \Longrightarrow (\bar{G}_n, l_{\bar{G}_n})$$

which coincides with the desired derivations by (1) and (2). □

6. FURTHER DEVELOPMENT

In order to formulate unambiguity and parsing algorithms for graph grammars it is desirable to have canonical derivation sequences like leftmost or rightmost derivations in the string cases which can be shown to be equivalent using the concept of parallel derivations. For graphs only the parallel derivation concept seems to be useful. In any case, the main step is to define independence of occurences of two left hand sides of productions p_1 and p_2 in a graph G such that for natural direct derivations $\pi_1 : G \Longrightarrow H_1$ and $\pi_2 : G \Longrightarrow H_2$ there are a unique graph $\tilde{H}$ and derivations $\pi_1 : H_2 \Longrightarrow \tilde{H}$ and $\pi_2 : H_1 \Longrightarrow \tilde{H}$. This means that for independent productions the order of sequential applications is arbitrary, or, in other words, the productions can be applied in parallel. Such an independence result, usually called Church-Rosser-Theorem, will be stated here without proof for the case $\underline{K} = \underline{K}''$ = Sets, the proof will be published in /Eh-Ros 75/.

CHURCH-ROSSER-THEOREM:

Let $(G, l_G) \in \underline{\text{Graph}}_\Omega$ and $\tilde{\pi}_i : (G, l_G) \Longrightarrow (H_i, l_{H_i})$ two natural direct derivations with productions

$$\tilde{\pi}_i = ((B_i, l_{B_i}) \xleftarrow{p_i} K_i \xrightarrow{p_i'} (B_i', l_{B_i'})) \text{ and}$$

enlargements $\mathcal{E}_i = (K_i \xrightarrow{d_i} (D_i, l_{D_i}))$ for $i = 1,2$

which are independent in the following sense:

a) There are graph morphisms $\tilde{d}_1' : B_1 \longrightarrow D_2$ and $\tilde{d}_2' : B_2 \longrightarrow D_1$ such that $\tilde{p}_2 \circ \tilde{d}_1' = \tilde{d}_1$ and $\tilde{p}_1 \circ \tilde{d}_2' = \tilde{d}_2$.

b) Let (1) pullback of $\tilde{p}_1$, $\tilde{p}_2$ leading to f_i and unique d_i' such that $f_i \circ d_i' = d_i$ $(i = 1,2)$ and (2) pullback of d_1', d_2' in the diagram below then we assume for $i = 1,2$:

$$l_{B_i} \circ U(p_i \circ k_i) = l_{B_i'} \circ U(p_i' \circ k_i) : U\tilde{K} \longrightarrow \Omega$$

$$\begin{array}{ccccc}
\tilde{K} & \xrightarrow{k_1} & K_1 & \xrightarrow{p_1} & B_1 \\
{\scriptstyle k_2}\downarrow & (2) & \downarrow{\scriptstyle d_1'} & & \downarrow{\scriptstyle \tilde{d}_1'} \\
K_2 & \xrightarrow{d_2} & \tilde{D} & \xrightarrow{f_2} & D_2 \\
{\scriptstyle p_2}\downarrow & & \downarrow{\scriptstyle f_1} & (1) & \downarrow{\scriptstyle \tilde{p}_2} \\
B_2 & \xrightarrow[\tilde{d}_2']{} & D_1 & \xrightarrow[\tilde{p}_1]{} & G
\end{array}$$

Given such independent direct derivations productions $\tilde{\pi}_1$ and $\tilde{\pi}_2$ can also be applied to (H_2, l_{H_2}) and (H_1, l_{H_1}), respectively, leading to the same graph $(\tilde{H}, l_{\tilde{H}})$, i.e. there are the following direct derivations:

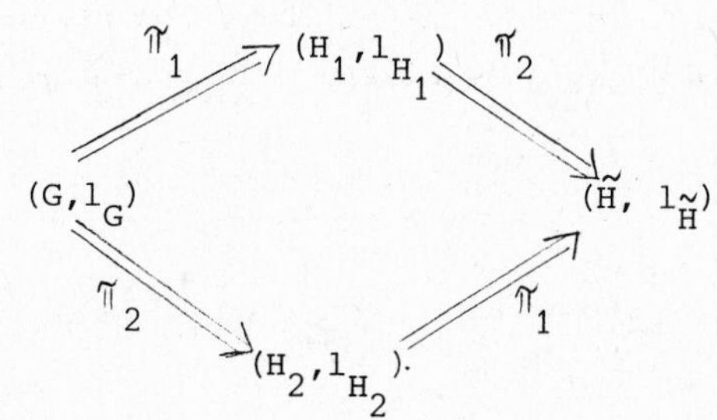

Moreover, there is a direct derivation

$$\tilde{\pi}_1 \amalg \tilde{\pi}_2 : (G, l_G) \Longrightarrow (\tilde{H}, l_{\tilde{H}})$$

using the parallel production

$$\tilde{\pi}_1 \amalg \tilde{\pi}_2 = ((B_1 \amalg B_2, (l_{B_1}, l_{B_2})) \xleftarrow{p_1 \amalg p_2} K_1 \amalg K_2 \xrightarrow{p_1' \amalg p_2'} (B_1' \amalg B_2', (l_{B_1'}, l_{B_2'})))$$

with enlargement $\mathcal{E} = ((K_1 \amalg K_2) \xrightarrow{(d_1', d_2')} (\tilde{D}, l_{\tilde{D}''}))$.

This Church-Rosser-Theorem hopefully will lead to canonical derivation sequences for graph grammars and later on to a syntax-analysis algorithm for context-free graph grammars, which gives for each graph in the graph language a canonical derivation sequence.

On the other hand, different notions of context-free graph grammars will be studied with respect to Normal-Form-Theorems, decidability problems and pumping lemmas for example.

Finally, let us note that basing on this approach of sequential graph grammars we also have started to develop a theory for parallel grammars in /Eh-Kr 75/ which generalizes Lindenmayer-systems to graphs.

REFERENCES

/De-Fr-St 74/ E.Denert,R.Franck,W.Streng: PLAN2D - Towards a 2-dimensional Programming Language, in Lecture Notes in Comp. Sci. 26, 202-213

/Eh 75/ H.Ehrig: Graph Grammars: Problems and Results in View of Computer Science Applications, Forschungsbericht 75-21 (1975), Technische Universität Berlin, Fachbereich 20

/Eh-K³ 74/ H.Ehrig,W.Kühnel,H.-J.Kreowski,K.D.Kiermeier: Universal Theory of Automata, Teubner, Stuttgart 1974

/Eh-Kr 75/ H.Ehrig,H.-J.Kreowski: Parallel Graph Grammars, in "Automata, Languages and Developement", to be published by North-Holland 1975/76

/Eh-Pf-Sch 73/ H.Ehrig,M.Pfender, H.-J.Schneider: Graph Grammars: An Algebraic Approach, Proc.IEEE Conf.on Automata and Switching Theory,Iowa City 1973, 167-180

/Eh-Ros 75/ H.Ehrig,B.K.Rosen: A Church-Rosser-Theorem for Graph Grammars, to appear

/Eh-Ti 74/ H.Ehrig,K.W.Tischer:Graph Grammars and Applications to Specialization in Biology,Proc.Conf.Biolog.Mot. Aut.Th.,Virginia 1974, extended version in Journ. Comp.Syst.Science 11 (1975)

/Eh-Ti 75/ H.Ehrig,K.W.Tischer:Derivations of Stochastic Graphs, Proc.Intern.Symp.on Uniformly Structured Automata and Logic, Tokyo 1975

/Ha 74/ H.Hansen: Kategorielle Betrachtungen zu den Sätzen von Jordan-Hölder und Schreier, Diplom-Arbeit FB 3, TU Berlin 1974

/He-St 74/ H.Herrlich, Strecker: Category Theory, Allyn and Bacon, Boston 1974

/ML 72/ S.MacLane: Categories for the Working Mathematician, Springer, New York-Heidelberg-Berlin 1972

/Pra 71/ T.W.Pratt: Pair Grammars, Graph Languages and String to Graph Translation, Journ.Comput.Syst.Science, 5(1971), 560-595

/Ri 74/ Riquet: Graphical Models in Biochemistry, Lecture in Oberwolfach (Tagung Automatentheorie und Formale Sprachen) 1974

/Ros 74/ B.Rosen: Deriving Graphs from Graphs by Applying a Production, IBM Research Report RC 5163,Dec 1974 to appear in Acta Informatica

/Sch 74/ H.-J.Schneider: Syntax-Directed Description of Incremental Compilers, Tagungsband 4. GI-Jahrestagung Berlin 1974, Springer Lecture Notes in Computer Science 26, 192-201

/Sch-Eh 75/ H.-J.Schneider,H.Ehrig: Grammars on Partial Graphs, Arbeitsberichte Inst.Mat.Masch.Dat.verarb. Univ. Erlangen, vol 8, No 1 (1975), 64-91, to appear in Acta Informatica

CORRECTNESS AND EQUIVALENCE OF DATA TYPES*

J. A. Goguen
Computer Science Department
University of California
Los Angeles, California 90024, U.S.A.

1. Introduction

In this paper, the term "data type" is intended to include both the items of data (of that type) and the permissible operations upon them (such as selectors, boolean tests for non-triviality and the like); this idea is developed (informally) in Knuth (1968), Dahl and Nygard (1966), and Liskow and Zilles (1974) among other places. The starting point for the algebraic approach is to realize that the data representations together with the operations on them constitute an algebra. But the really interesting point is that this algebra is in some sense free, that is, it is determined (up to isomorphism) by the operations, plus some equations; Zilles (1975) seems to have been the first to use this observation. Goguen, Thatcher, Wagner and Wright† (1975) gave a particularly clean formulation using their (1973) notion of many-sorted algebra and their machinery of initial algebra semantics (1974), with applications to proving correctness of both implementations and abstract types. Zilles (1975A) has developed a notion of "data algebra" along similar lines. This paper is intended to argue that many problems can be handled in a still more satisfactory way with algebraic theories, than with just algebras.

The crucial practical problem is to show that some implementation (e.g., of stack) is correct. This demands comparison with some standard. The most concrete approach is to compare code with some more abstract program, assumed correct, as in Hoare (1972). This entirely sidesteps the issue of what is a data type, and is not discussed further here. Rather, we note that proving correctness of an implementation can be regarded as proving equivalence to the abstract type itself, so that the crucial theoretical problem is equivalence of types. This becomes the center point of our discussion, in that we argue that a superior notion of equivalence is available with theories; we also discuss a still weaker notion of equivalence associated with presentations.

2. Presentations of Data Types

We quickly review basic concepts of S-sorted Σ-algebras; see ADJ (1973, 1974, or 1975) for more leisurely presentations. Let S be a (not necessarily finite) set of sorts, and let Σ be a S^*xS- indexed family of sets called an S-sorted operator

*This research was supported by the National Science Foundation, Grant No. DCR72-03633 A03.

†This set of authors is hereafter referred to as "ADJ".

domain, with $\Sigma_{w,s}$ being the set of operator symbols of arity $w \in S^*$ and sort $s \in S$ (note that $\Lambda \in S^*$ is the empty sequence, so that $\sigma \in \Sigma_{\Lambda,s}$ is a constant symbol). An (S-sorted) Σ-algebra A is then an S-indexed family of sets $\langle A_s \rangle_{s \in S}$, together with a function $\sigma_A: A^w \to A$ for each $\sigma \in \Sigma_{w,s}$, where $A^w = A_{s_1} \times \ldots \times A_{s_n}$ when $w = s_1 \ldots s_n \in S^*$ (and A^Λ is some one point set).

We now recall (from ADJ (1974)) that a Σ-term of sort s is an element of $T_{\Sigma,s}$, where $\langle T_{\Sigma,s} \rangle_{s \in S}$ is defined to be the smallest S-indexed family of subsets of $(\Sigma \cup \{\mathbf{(},\mathbf{)}\})^*$, where Σ temporarily denotes $\cup_{w,s \in S^* \times S} \Sigma_{w,s}$ and $\{\mathbf{(},\mathbf{)}\}$ is a two element set disjoint from Σ satisfying the following conditions:

(T0) $\Sigma_{\Lambda,s} \subseteq T_{\Sigma,s}$; and

(T1) if $w = s_1 \ldots s_n$, $n > 0$, $\sigma \in \Sigma_{w,s}$, and $t_i \in \Sigma_{w,s_i}$ (for $i = 1, \ldots, n$) then the sequence $\sigma\mathbf{(}t_1 \ldots t_n\mathbf{)}$ is in $T_{\Sigma,s}$.

The family $\langle T_{\Sigma,s} \rangle_{s \in S}$ is made into a Σ-algebra, T_Σ, by defining operations, here denoting σ_{T_Σ} by just σ_T for convenience:

(0) for $\sigma \in \Sigma_{\Lambda,s}$ σ_T is the element σ of $T_{\Sigma,s}$; and

(1) for $\sigma \in \Sigma_{w,s}$, $w = s_1 \ldots s_n$, $n > 0$ and $t_i \in T_{\Sigma,s_i}$ (for $i = 1, \ldots, n$), $\sigma(t_1, \ldots, t_n)$ is the sequence $\sigma\mathbf{(}t_1 \ldots t_n\mathbf{)}$ in $T_{\Sigma,s}$.

(Both conditions may be subsumed under the condition $\sigma_T(t) = \sigma\mathbf{(}t\mathbf{)}$ for $t \in T_\Sigma{}^w$, but this may appear overly compact.) T_Σ is called the initial Σ-algebra.

If Σ is an S-sorted operator domain, and $X = \langle X^{(s)} \rangle_{s \in S}$, where $X^{(s)} = \{x_n^{(s)} | n \in N\}$ (here N is the non-negative natural numbers, $\{0, 1, 2, \ldots\}$, then let $\Sigma(X)$ be the domain with $\Sigma(X)_{\Lambda,s} = \Sigma_{\Lambda,s} \cup X^{(s)}$, and $\Sigma(X)_{w,s} = \Sigma_{w,s}$ for $w \neq \Lambda$. (We may often find it convenient to omit the $^{(s)}$ on $x^{(s)}$, or even to use entirely other letters for variables.) Let $T_\Sigma(X)$ be $T_{\Sigma(X)}$ viewed as a Σ-algebra, by "forgetting" about the additional constants in X. Elements of $T_\Sigma(X)$ are Σ-terms with variables.

Now, a Σ-equation of sort $s \in S$ is a pair $\langle LHS,RHS \rangle$ in $T_\Sigma(X)_s$, and is satisfied by a Σ-algebra A iff $LHS(\alpha) = RHS(\alpha)$ for all assignments $\alpha: X \to A$ of values to variables. (See ADJ (1974) for a more detailed discussion of the meaning of $t(\alpha)$ for $t \in T_\Sigma(X)$.)

We can now give the main concept of this section.

Definition. A presentation is a triple $\langle S,\Sigma,E \rangle$, where Σ is an S-sorted operator domain, and E is a set of Σ-equations.

The most convenient way to give an abstract data type is by presentation. We give a few examples; see ADJ (1975) for more and for the motivation.

Example 1. string-of-d. Let $S = \{d,t\}$; let $\Sigma_{\Lambda,t} = \{\Lambda\}$, $\Sigma_{dt,t} = \{*\}$, and $\Sigma_{w,s} = \emptyset$ otherwise; let $E = \emptyset$. Here d,t,* are alphabet, string, and insert, respectively.□

Example 2. list-of-d. Let $S = \{d,\ell\}$; let $\Sigma_{\Lambda,\ell} = \{\Lambda\}$, $\Sigma_{d\ell,\ell} = \{*\}$, $\Sigma_{\ell\ell,\ell} = \{\cdot\}$, and $\Sigma_{w,s} = \emptyset$ otherwise; let E contain the equations

$$A \cdot (B \cdot C) = (A \cdot B) \cdot C$$

$$A \cdot \Lambda = A$$
$$\Lambda \cdot A = A$$

Here $d, \ell, *, \cdot$ are alphabet, list, insert, and concatenate, respectively. It will be convenient to let $D*\Lambda$ be abbreviated by just D. □

Example 3: bag-of-d. Let $S = \{d,b\}$; let $\Sigma_{\Lambda,b} = \{\emptyset\}$, $\Sigma_{db,b} = \{*\}$, $\Sigma_{bb,b} = \{\cup\}$, and $\Sigma_{w,s} = \emptyset$ otherwise; let E contain the equations

$$A\cup(B\cup C) = (A\cup B)\cup C$$
$$A\cup B = B\cup A$$
$$A\cup\emptyset = A$$
$$\emptyset\cup A = A$$

Here $d,b,*,\cup$ are element, bag, insert, and union, respectively; and $D*\Lambda$ may be abbreviated D. □

Example 4. bag-of-d (again). Let $S = \{d,g\}$; let $\Sigma_{\Lambda,g} = \{\text{NUL}\}$, $\Sigma_{dg,g} = \{\text{INSERT}\}$, $\Sigma_{gg,g} = \{\text{UNION}\}$, and $\Sigma_{w,s} = \emptyset$ otherwise; let E contain the equations

$$\text{UNION }(A, \text{UNION }(B,C)) = \text{UNION }(\text{UNION }(A,B),C)$$
$$\text{UNION }(A,B) = \text{UNION }(B,A)$$
$$\text{UNION }(A,\text{NUL}) = A$$
$$\text{UNION }(\text{NUL},A) = A.$$

Clearly this is just a notational variant of Example 3. □

Example 5. bag-of-d (still again). Let S, Σ be as in Example 3, but let E contain the equations

$$A\cup(B\cup C) = (A\cup C)\cup B$$
$$\emptyset\cup A = A$$

which is in some sense (to be developed later) equivalent. □

Now Examples 1 and 2 ought not be be equivalent, because they just have different operational capabilities; concatenation is impossible for strings (incidentally, these designations "string" and "list" are arbitrary and just for convenience). But "forgetting" concatenation should lead to something.

Examples 4 and 5 are equivalent in a somewhat subtle way, but Examples 3 and 4 are presentations which differ only in notation; and no information is needed about the abstract types which they define to see it. This leads to our first, and most concrete, kind of type equivalence.

Definition. An isomorphism $\langle S,\Sigma,E\rangle \to \langle S',\Sigma',E'\rangle$ of presentations is a triple $\langle f,g,h\rangle$, where $f\colon S \to S'$ is a set isomorphism, $g\colon \Sigma(X) \to \Sigma'(X)$ is a $S^* \times S$-indexed family of set isomorphisms, $g_{w,s}\colon \Sigma(X)_{w,s} \to \Sigma'(X)_{f(w),f(s)}$ (where $f(s_1 \ldots s_n) = f(s_1) \ldots f(s_n)$) and $h\colon E \to E'$ is a set isomorphism such that $h(\langle LHS,RHS\rangle) = \langle g(LHS),g(RHS)\rangle$ (where $g(t)$ for $t\in T_\Sigma{}^W$ is defined recursively by $g(\sigma(t)) = g(\sigma)(g(t))$).

(Since all variables are dummy anyway, a more general definition could be given which permits different rearrangements of variables in different equations; however a simpler way around this problem is to require that E always use distinct variables in distinct equations.) The reader may wish to formulate a notion of homomorphism of presentations; however, we have no use for it in this paper. Do note, before passing to the next section, that the presentations of Example 3 and 4 are isomorphic.

Data types on this level really aren't very abstract at all. (It might be parenthetically noted that computer science has a tendency to use the word "abstract" for what appear as very low levels of generality in category theory.)

3. Data Types as Initial Algebras

Again, we quickly run through the algebra we need. A Σ-congruence $\equiv$ on a -algebra A is a family $<\equiv_s>_{s\in S}$ of equivalence relations, $\equiv_s$ on A_s for $s\in S$, such that if $\sigma\in\Sigma_{w,s}$ where $w = s_1 \dots s_n$ if $a_i, a_i'\in A_{s_i}$ and if $a_i \equiv_{s_i} a_i'$ (for i = 1, ...,n), then

$$\sigma_A(a_1, \dots, a_n) \equiv_s \sigma_A(a_1', \dots, a_n').$$

If A is a Σ-algebra then $<A_s/\equiv_s>_{s\in S}$ is also a Σ-algebra (here $A_s/\equiv_s$ is the set of $\equiv_s$-equivalence classes of A_s), denoted $A/\equiv$. Now if $<E_s>_{s\in S}$ is a family of relations on $<A_s>_{s\in S}$, that is, $E_s \subseteq A_s \times A_s$ for $s\in S$, then $<E_s>_{s\in S}$ is contained in a least congruence on A, called the Σ-congruence generated by $<E_s>_{s\in S}$.

Recall that a Σ-homomorphism h: A → A' is a family $<h_s>_{s\in S}$ of functions h_s: $A_s \to A_s'$ such that

(H0) if $\sigma\in\Sigma_{\Lambda,s}$ then $h_s(\sigma) = \sigma_{A'}$; and

(H1) if $\sigma\in\Sigma_{w,s}$ and $a\in A^w$ then $h_s(\sigma_A(a)) = \sigma_{A'}(h^w(a))$, where $h^{s_1\dots s_n}(a_1, \dots a_n) = <h_{s_1}(a_1), \dots, h_{s_n}(a_n)>$.

Let $\underline{\underline{Alg}}_\Sigma$ denote the category of all Σ-algebras and Σ-homomorphisms; and let $\underline{\underline{Alg}}_{\Sigma,E}$ denote the category of all Σ-algebras satisfying E together with all Σ-homomorphisms among them. Finally, recall that an object I in a category $\underline{\underline{C}}$ is initial iff for every object A in $\underline{\underline{C}}$ there is a unique morphisms I → A in $\underline{\underline{C}}$.

Theorem. Let $<S,\Sigma,E>$ be a presentation. let E_s be the set of equations in E of sort s, and let $\equiv$ be the Σ-congruence generated by $<E_s>_{s\in S}$. Then $T_\Sigma/\equiv$ is initial in $\underline{\underline{Alg}}_{\Sigma,E}$; hereafter it is denoted $T_{\Sigma,E}$.

Definition. Call presentations $<S,\Sigma,E>$ and $<S',\Sigma',E'>$ algebraically isomorphic iff there is a set isomorphism f: S → S' such that $f(\Sigma) = \Sigma'$ and $T_{f(\Sigma),f(E)}$ and $T_{\Sigma',E'}$ are isomorphic Σ'-algebras. (Here the application of f to Σ and E merely changes the sorts involved.)

Then an "abstract data type" can be taken to be an "algebraic isomorphism class" of presentations; this essentially the approach taken in ADJ (1975). For it is the same thing to define an abstract data type to be (the isomorphism class of) an initial algebra, by the Theorem above. For example, the presentations of Examples 3 and 5 are algebraically isomorphic, since the sets of equations involved generate congruences. Let Σ_1 and Σ_2 denote the operator domains of Examples 1 and 2 respectively. Since $\Sigma_1 \subseteq \Sigma_2$, we can regard any Σ_2-algebra as a Σ_1-algebra, just by forgetting the extra operation involved. Doing this to the Σ_2-algebra T_{Σ_2,E_2}

of Example 2 gives a Σ_1-algebra isomorphic to that of Example 1. Thus, these data types are Σ_1-algebraically isomorphic.

The point seems to be that the notion of presentation is far too weak to be very useful in itself, and even the notion that an "abstract data type" is an equivalence class of presentations is too awkward to be of much practical value. Regarding abstract data types as initial algebras, with presentations merely convenient ways to present them, is a big improvement. See ADJ (1975) for specific algebraic methods for proving correctness based on the initial algebra view. But this approach can be improved in both theoretical and practical directions, as we try to show in the next section.

4. Data Types as Algebraic Theories

If A is an S-sorted Σ-algebra, we let $\underline{\underline{T}}_A$, the theory of A, be the S*xS*-indexed family $\underline{\underline{T}}_A(w,w')$, for $w,w' \in S^*$, of functions $A^w \to A^{w'}$ defined by the following conditions:

(Th1) if $\sigma \in \Sigma_{w,s}$ then $\sigma \in \underline{\underline{T}}_A(w,s)$;

(Th2) if $w = s_1 \dots s_n$, then $p^w \in \underline{\underline{T}}_A(w,s_i)$ for $i = 1, \dots, n$, where $p_i^w(a_1, \dots, a_n) = a_i$ (denote this p_i, or sometimes p_{s_i}, for convenience);

(Th3) if $t \in \underline{\underline{T}}_A(w,w')$ and $t' \in \underline{\underline{T}}_A(w',w'')$, then $tt' \in \underline{\underline{T}}_A(w,w'')$ (here tt' denotes the set theoretic composition of functions);

(Th4) if $w' = s_1 \dots s_n$ and $t_i \in \underline{\underline{T}}_A(w,s_i)$ for $i = 1, \dots, n$, then $[t_1, \dots, t_n] \in \underline{\underline{T}}_A(w,w')$ (here $[t_1, \dots, t_n]$ denotes the "target tupling of the functions t_i: $A_w \to A_{s_i}$ to get a function t: $A_w \to A_{w'}$ defined by $t(a) = \langle t_1(a), \dots, t_n(a)\rangle$.)

Note that for $w = s_1 \dots s_n$, $1_w = [p_1^w, p_2^w, \dots, p_n^w] \in \underline{\underline{T}}_A(w,w)$ functions as an identity function for composition, in that $1_w t = t$ and $t1_w = t$ whenever these compositions are defined. And of course, composition is associative, since it is just composition of set functions. Thus $\underline{\underline{T}}_A$ is a category, with objects $|\underline{\underline{T}}_A| = S^*$. Note the following additional properties of $\underline{\underline{T}}_A$: $[t_1, \dots, t_n]p_i = t_i$ for $i = 1, \dots, n$; and for $t \in \underline{\underline{T}}_A(w, s_1 \dots s_n)$, $[tp_1, \dots, tp_n] = t$.

Now given a presentation $\langle S,\Sigma,E\rangle$, we can form the algebra $T_{\Sigma,E}$. The next step is then

Definition. Presentations $\langle S,\Sigma,E\rangle$ and $\langle S',\Sigma',E'\rangle$ are theoretically isomorphic iff the categories $\underline{\underline{T}}_{\Sigma,E}$ and $\underline{\underline{T}}_{\Sigma',E'}$ are isomorphic; and the presentations are theoretically equivalent iff the categories $\underline{\underline{T}}_{\Sigma,E}$ and $\underline{\underline{T}}_{\Sigma',E'}$ are equivalent (as categories).

Notice that algebraic isomorphism of presentations implies theoretical isomorphism implies theoretical equivalence. Moreover, theoretical isomorphism automatically gives an isomorphism $f: S \to S'$ of sorts (from the isomorphism of $|\underline{\underline{T}}_{\Sigma,E}|$ with $|\underline{\underline{T}}_{\Sigma',E'}|$), but theoretical equivalence permits quite different sort sets.

The intuitive idea is that theoretically isomorphic presentations can do exactly the same things to essentially the same data, but it may be done with different combinations of primitive operations; theoretical equivalence permits the added twist of extra sorts which behave the same as others already given. This last becomes particularly interesting in the case of semantics of programming languages, where sometimes "extra" productions and non-terminals appear which have no meaning, but which help with parsing.

As was the case with algebraic isomorphism, we do not want to have merely a more powerful equivalence relation on presentations, but we want to have specific mathematical objects which embody the concepts; that is, we are led to *define* an abstract data type to be a many-sorted algebraic theory. This requires, first of all, an abstract notion of many-sorted theory, and leads to a lot of technicalities, which it has been our purpose here to motivate, rather than to explain in detail.

Before launching into some technicalities, let us at least evoke a few more of the advantages of this approach, It is easy to consider multiple valued operations; for example division might appear as the target tuple [QUOTIENT, REMAINDER]; or for *stack*, POP is a target two-tuple, one component the old top and the other the new stack. We can define a *sub-abstract data type* to be a theory equivalent to a subtheory of the given one. In fact, we can define morphisms of abstract data types and use this as a basis for making such constructions as *list-of-d* into *functors* of the type variable *d*. Finally, the various ways of forming compounding types (such as union, array, and struct) are clarified.

Now, following ADJ (1975A), but with more compatible notation, let $\underline{\underline{St}}_S$ be the category with: objects functions $w\colon n \to S$ for some integer n, that is $|\underline{\underline{St}}_S| = S^*$; and with morphisms $w \to w'$ set functions $n \to n'$ such that

$$\begin{array}{ccc} n & \longrightarrow & n' \\ & {\scriptstyle w}\searrow \quad \swarrow{\scriptstyle w'} & \\ & S & \end{array}$$

commutes. Call a functor *firm* iff it is bijective on objects. Then an *S-sorted algebraic theory* is a firm coproduct preserving functor $J\colon \underline{\underline{St}}_S \to \underline{\underline{T}}$. Note that coproduct in $\underline{\underline{St}}_S$ is just concatenation. A *morphism* $J \to J'$ of many-sorted *algebraic theories* is a pair $\langle f,F\rangle$, where $f\colon S \to S'$ is a set function, and $F\colon \underline{\underline{T}} \to \underline{\underline{T}}'$ is a functor, such that

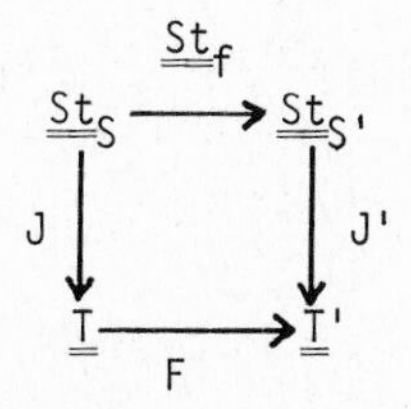

commutes (in the category of categories) where $\underline{\underline{St}}_f$ sends $w\colon n \to S$ to $wf\colon n \to S'$, and sends $m\colon w \to w'$ to $m\colon wf \to wf$ by $\underline{\underline{St}}_f(m\colon n \to n') = m\colon n \to n'$. Let $\underline{\underline{MSAT}}$ be the resulting category of many-sorted algebraic theories. This, or some similar, place is where we propose to do data type theory. Thus the following

Definition: An abstract data type is a many-sorted algebraic theory, and a morphism of abstract data types is a morphism of many-sorted algebraic theories.

In particular, a concrete data type (all the data representations together with the programs that manipulate them) constitute an algebra of the many-sorted theory which is the abstract type (such an algebra appears as a functor from the theory (or really its opposite) to the category of sets, satisfying some conditions; see ADJ (1975A)). However, to go over even the ground covered in ADJ (1975) in this framework is a major undertaking; our purpose here has been to show why it should be undertaken.

References

ADJ†(1973) "A Junction between Computer Science and Category Theory, I: Basic Concepts and Examples, Part 1," IBM T. J. Watson Research Center Report RC 4526.

ADJ (1974) "Initial Algebra Semantics," Proceedings of the 15th IEEE Symposium on Switching and Automata Theory, pp. 63-77 (a more complete version is to appear in the Journal of the ACM).

ADJ (1975) "Abstract Data Types as Initial Algebras and the Correctness of Data Representations," Proceedings of Conference on Computer Graphics, Pattern Recognition and Data Structures, pp. 89-93.

ADJ (1975A) "An Introduction to Categorics, Algebraic Theories and Algebras," IBM T. J. Watson Research Center Report.

Dahl and Nygard (1966) "SIMULA - An ALGOL-Based Simulation Language," Communications of the ACM, 9, pp. 115-133.

Hoare (1972) "Proof of Correctness of Data Representations," Acta Informatica, 1, pp. 271-281.

Knuth (1968) The Art of Computer Programming, Vol. 1, Addison-Wesley.

Liskow and Zilles (1974) "Programming with Abstract Data Types," SIGPLAN Notices, 9.

Zilles (1975) "Algebraic Specifications for Data Types," draft for Project MAC Progress Report 1973-1974.

Zilles (1975A) "Data Algebra" IBM San Jose Research Center Draft Working Paper.

†Recall that this means "Goguen, Thatcher, Wagner and Wright."

MINIMIZATION CONCEPTS OF AUTOMATA IN PSEUDOCLOSED CATEGORIES

H.-J. KREOWSKI
H. EHRIG
TECHNISCHE UNIVERSITÄT BERLIN
FACHBEREICH 20 (KYBERNETIK)
1 Berlin 10, Otto-Suhr-Allee 18/20

ABSTRACT

In this paper starting in section 1 with a review of the basic notions and constructions of automata in pseudoclosed categories as given in /Eh-Kr 73/ and /Eh-K^3 74/, in section 2 we give an improved approach to minimization of automata in closed categories based on a terminal object in the category of automata, we build up a hierarchy of minimization concepts and give a characterization of minimal automata in the pseudoclosed case in section 3, and finally in section 4 an algorithm for the scoop minimization of nondeterministic automata and some open problems are formulated.

INTRODUCTION

This paper is based on a theory of automata in pseudoclosed categories including the cases of nondeterministic, relational, stochastic and several kinds of relational topological automata which is developed in /Eh-Kr 73/, /Eh-Kr 74/ and /Eh-K^3 74/ and closely related to the non-deterministic approach of machines from M.A. Arbib and E.G. Manes in /Ar-Ma 73/, /Ar-Ma 74/ and /Ma 75/. Whereas in the case of automata without initial state the main problems of minimization and realization are already solved, the exact relations between the different minimization concepts are studied in this paper, but for the case of automata with initial state the main problem of minimization remains unsolved. As pointed out in /Eh-Kr 74/ scoop minimization seems to be a suitable approximation and therefore we give an algorithm for scoop minimization of finite nondeterministic automata in this paper.
Although automata in closed categories are well-known meanwhile (cf. /Go 71/, /Ar-Ma 73/, /Eh-K^3 74/) we give an improved approach of minimization and realization which is very smooth, only using a suitable terminal object and the factorization theorem in the category of automata. In fact this is only a suitable modification of our constructions in /Eh-K^3 74/ which is similar to the approach of S. Bainbridge in /Ba 72/ in some sense.

1. BASIC NOTATIONS AND CONSTRUCTIONS

1.1 GENERAL ASSUMPTIONS

$(\underline{K}, \otimes)$ will be a pseudoclosed category relative $(\underline{K}', \otimes)$, (i.e. $(\underline{K}, \otimes)$ is a symmetric monoidal category and $(\underline{K}', \otimes)$ is a coreflexive closed monoidal subcategory of $(\underline{K}, \otimes)$ with the same class of objects. Hence we have in particular

i) an internal hom functor $\langle K, - \rangle : \underline{K}' \longrightarrow \underline{K}'$ right adjoint to the tensor product $- \otimes K : \underline{K}' \longrightarrow \underline{K}'$ with counit ev: $\langle K, - \rangle \otimes K \longrightarrow 1_{\underline{K}'}$ for each object K in $\underline{K}'$;

ii) a power functor $P : \underline{K} \longrightarrow \underline{K}'$ right adjoint to the inclusion $J : \underline{K}' \longrightarrow \underline{K}$ with counit $v : JP \longrightarrow 1_{\underline{K}}$;

iii) the composite adjunction $- \otimes K \dashv \langle K, P \rangle : \underline{K} \longrightarrow \underline{K}'$).
Moreover we assume that $\underline{K}'$ has countable coproducts and a canonical $\mathcal{E}$ - $\mathcal{M}$ -factorization.
In order to simplify notation all associativity, symmetry, right unit and left unit morphisms of the tensor product and also the inclusion $J : \underline{K}' \longrightarrow \underline{K}$ will not be given explicitly.

1.2 EXAMPLES

The following categories are pseudoclosed relative to the closed category ($\underline{Sets}$, $\times$) of sets, where in each case the tensor product is the cartesian one:

($\underline{PD}$, $\times$) :partial functions

($\underline{ND}$, $\times$) :nondeterministic functions

($\underline{FND}$, $\times$) :finite nondeterministic functions,
i.e. f(a) is finite for each $a \in$ dom (f)

($\underline{Rel}$, $\times$) :relations

($\underline{FRel}$, $\times$) :finite relations

($\underline{Stoch}$, $\times$) :(discrete) stochastic channels.

Furthermore the categories ($\underline{Rel}$, $\times$) and ($\underline{FRel}$, $\times$) are also pseudoclosed relative ($\underline{PD}$, $\times$) and clearly each closed category is pseudoclosed.

For explicit definitions and topological examples of pseudoclosed categories cf. /Eh-K^3 74/.
Corresponding to the above examples we obtain deterministic, partial, nondeterministic, relational, stochastic and several kinds of relational topological automata as examples of automata in pseudoclosed categories due to the following definition.

1.3 DEFINITION

An automaton A=(I, O, S, d, l) in a pseudoclosed category ($\underline{K}$, $\otimes$) consists of $\underline{K}$-objects I, O, S and $\underline{K}$ - morphisms d: $S \otimes I \longrightarrow S$ and l: $S \otimes I \longrightarrow O$, an automata morphism f: $A \longrightarrow A'$ is defined to be a $\underline{K}'$-morphism f: $S \longrightarrow S'$, such that $d'(f \otimes I) = f\, d$ and $l'(f \otimes I) = l$ (I, O fixed). Both together lead to the category $\underline{K}$-$\underline{K}'$-$\underline{Aut}$ of automata in ($\underline{K}$, $\otimes$), written $\underline{K}'$-$\underline{Aut}$ in the closed case ($\underline{K}$, $\otimes$)=($\underline{K}'$, $\otimes$).

1.4 BASIC CONSTRUCTIONS

Given an automaton $A=(I, O, S, d, l)$ in $(\underline{K}, \otimes)$ we consider the following basic constructions:

1) The last output $l^+: S \otimes I^+ \longrightarrow O$ is defined by $l_1 = l$ and $l_{n+1} = l_n (d \otimes I^n)$, using the coproduct properties of the free semigroup I^+ and $S \otimes I^+$ in $\underline{K}$.

2) Using the adjunction $- \otimes I^+ \dashv \langle I^+, P - \rangle : \underline{K} \longrightarrow \underline{K}'$ the last output l^+ defines the machine morphism $M(A): S \longrightarrow \langle I^+, PO\rangle$ in $\underline{K}'$:

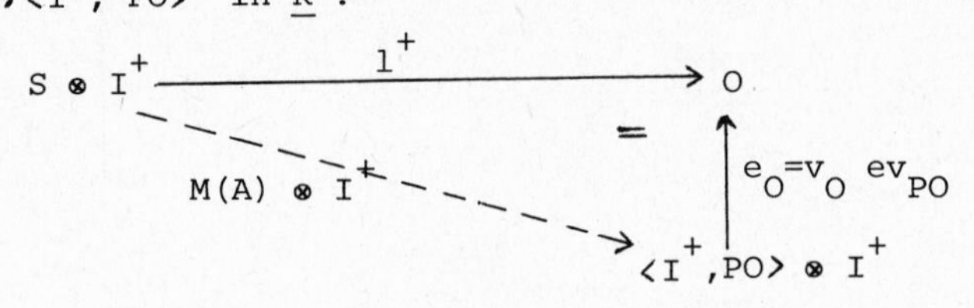

3) The $\mathcal{E}$-$\mathcal{M}$-factorization of the machine morphism $M(A)$

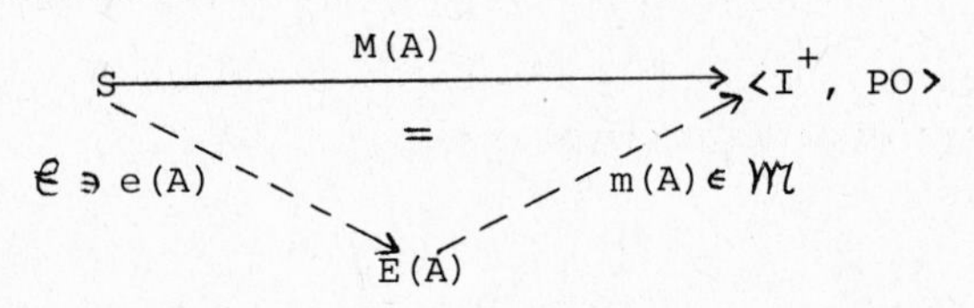

leads to the behavior $E(A)$ of A (or more precisely $m(A): E(A) \longrightarrow \langle I^+, PO\rangle$, where $m(A)$ is assumed to be a canonical representative in $\mathcal{M}$, e.g. an inclusion).

Note that the construction of l^+, $M(A)$ and $E(A)$ can be extended to suitable functors.

4) Using again the adjunction $- \otimes I^+ \dashv \langle I^+, P - \rangle$ we get the following operations on $\langle I^+, PO\rangle$ called left shift $L: \langle I^+, PO\rangle \otimes I \longrightarrow \langle I^+, PO\rangle$ and union $u: P\langle I^+, PO\rangle \longrightarrow \langle I^+, PO\rangle$ in $\underline{K}'$:

$$\langle I^+,PO\rangle \otimes I \otimes I^+ \xrightarrow{\langle I^+,PO\rangle \otimes i} \langle I^+,PO\rangle \otimes I^+ \xrightarrow{e_O} O$$

$$\langle I^+,PO\rangle \otimes I \otimes I^+ \xrightarrow{L \otimes I^+} \langle I^+,PO\rangle \otimes I^+ \xrightarrow{e_O} O \quad (=)$$

where $i: I \otimes I^+ \longrightarrow I^+$ is the restriction of the multiplication

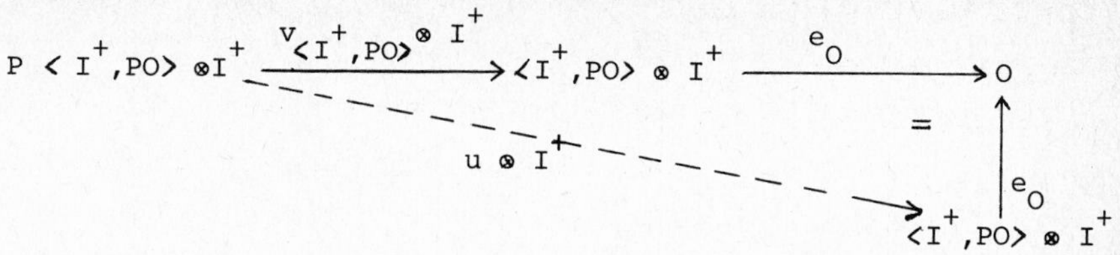

$$\begin{array}{ccccc} P\langle I^+,PO\rangle \otimes I^+ & \xrightarrow{v_{\langle I^+,PO\rangle} \otimes I^+} & \langle I^+,PO\rangle \otimes I^+ & \xrightarrow{e_O} & O \\ & \searrow^{u \otimes I^+} & & = & \uparrow e_O \\ & & & & \langle I^+,PO\rangle \otimes I^+ \end{array}$$

5) The <u>power automaton</u> $\hat{P}A$ of A is defined by

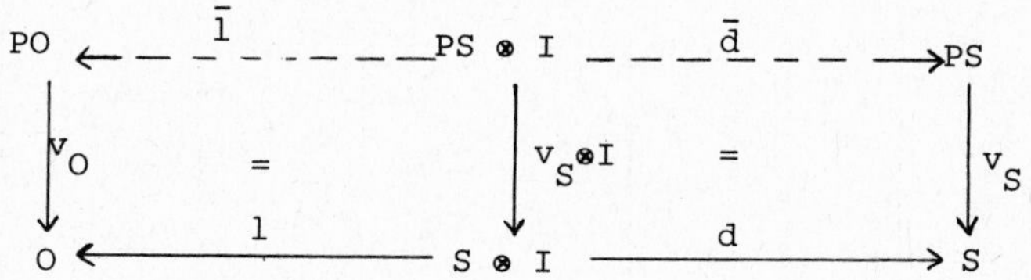

$$\begin{array}{ccccc} PO & \xleftarrow{\bar{l}} & PS \otimes I & \xrightarrow{\bar{d}} & PS \\ \downarrow v_O & = & \downarrow v_S \otimes I & = & \downarrow v_S \\ O & \xleftarrow{l} & S \otimes I & \xrightarrow{d} & S \end{array}$$

using the adjunction $J \dashv P : \underline{K} \longrightarrow \underline{K}'$, which can be lifted to an adjunction between corresponding categories of automata.
Moreover the machine morphism of $\hat{P}A$ satisfies:

$$M(\hat{P}A) = u\, PM(A),$$

and the behavior $E(\hat{P}A)$ of $\hat{P}A$ is the union closure of $E(A)$.

6) A pair of $\underline{K}'$-morphisms $(m: S' \longrightarrow S,\ n: S \longrightarrow PS')$ with m in $\mathfrak{M}$ is called <u>scoop</u> of A, if the following diagram commutes:

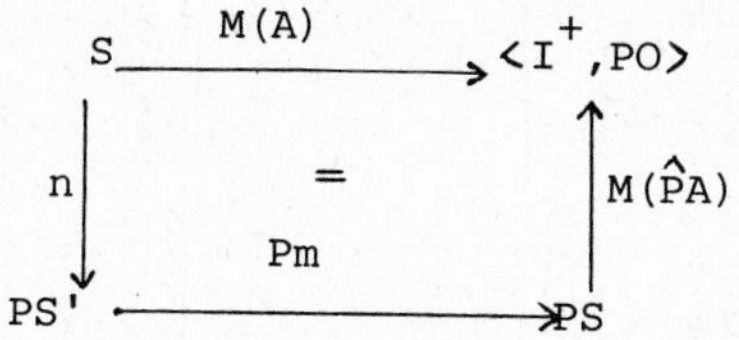

$$\begin{array}{ccc} S & \xrightarrow{M(A)} & \langle I^+,PO\rangle \\ \downarrow n & = & \uparrow M(\hat{P}A) \\ PS' & \xrightarrow{Pm} & PS \end{array}$$

Each scoop (m, n) of A defines a <u>scoop automaton</u> A(m, n) given by

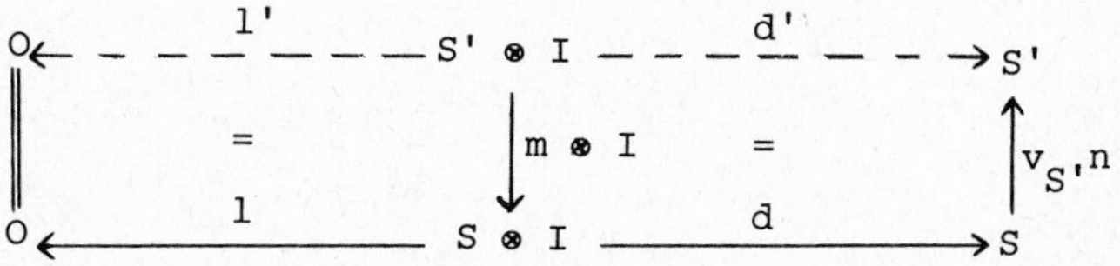

$$\begin{array}{ccccc} O & \xleftarrow{l'} & S' \otimes I & \xrightarrow{d'} & S' \\ \| & = & \downarrow m \otimes I & = & \uparrow v_{S'} n \\ O & \xleftarrow{l} & S \otimes I & \xrightarrow{d} & S \end{array}$$

the behavior of which satisfies:

$$E(A(m, n)) \subseteq E(A) \subseteq E(\hat{P}A(m, n)).$$

Finally we will state some notions, each of them expressing minimality of automata from another point of view. The relationship between these notions will be studied in the following sections.

1.5 DEFINITION

An automaton A=(I, O, S, d, l) in ($\underline{K}$, $\otimes$) is called

1) observable, if $M(A) \in \mathcal{M}$;
2) strong observable, if A and $\hat{P}A$ are observable;
3) scoop minimal, if for each scoop (m, n) of A m is an isomorphism;
4) reduced, if each reduction $r: A \longrightarrow A'$ (i.e. $r \in \mathcal{E}$) to an arbitrary automaton A' is an isomorphism;
5) minimal, if for each automaton A' with $E(A') \subseteq E(A)$ there exists a unique automata morphism $f: A' \longrightarrow A$, which is a reduction in the case $E(A') = E(A)$;
6) state minimal, if for each equivalent automaton A' (i.e. $E(A')=E(A)$) we have

   $$\text{card } (S') \geq \text{card } (S), \text{ where card}$$

   is a cardinality functor in the sense (cf. also /Eh-K^3 74/).

2. IMPROVED APPROACH TO MINIMIZATION OF AUTOMATA IN CLOSED CATEGORIES

In this section we show that there is a canonical terminal object in the category $\underline{K}'$-Aut of automata in a closed category $\underline{K}'$ which together with the well-known factorization theorem for automata morphisms immediately leads to the main minimization-realization results (cf. /Eh-K^3 74/, chapter 5) and to the equivalence of the minimization concepts stated in 1.5 (cf. /Eh-Kr 73/) for automata in closed categories which will be given in 3.1.

2.1 THEOREM (A_∞ is terminal object)

The automaton $A_\infty = (I, O, \langle I^+, O\rangle, L, e_1)$, with L and e_1 defined in 1.4.4 and 3.2 respectively, is a terminal object in the category $\underline{K}'$-Aut. Moreover for each automaton A the unique morphism $f: A \longrightarrow A_\infty$ coincides with the machine morphism M(A) of A.

Proof: The proof is an immediate consequence of the characterization of machine morphisms which will be given and proved in 3.2.1 specialized to the closed case, i.e. $P=1_{\underline{K}'}$, $D=d$, and $u=1_{\langle I^+, O\rangle}$. ▭

2.2 COROLLARY 1 (Minimization-Realization-Theory)

Given an automaton $A=(I, O, S, d, l)$ in $\underline{K}'$-$\underline{Aut}$ the machine morphism $M(A)$ of A is an automata morphism $M(A): A \longrightarrow A_\infty$ and moreover the factorization theorem for automata morphisms (cf./Eh-K[3]74/, 11.9) assures that the $\mathcal{E}$-$\mathcal{M}$-factorization of the machine morphism in 1.4.2 can be extended to an $\mathcal{E}$-$\mathcal{M}$-factorization of automata morphisms

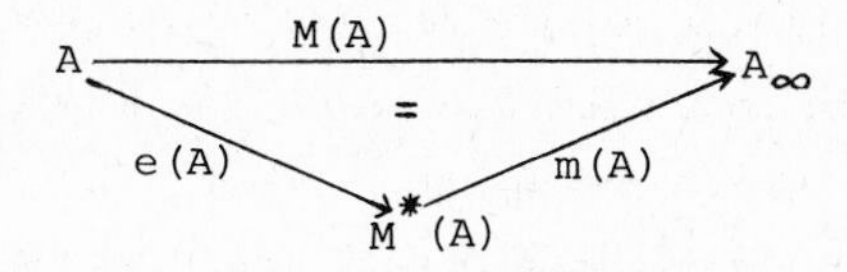

with $M^*(A)=(I, O, E(A), L^*, e_1^*)$ where $L^*: E(A) \otimes I \longrightarrow E(A)$ and $e_1^*: E(A) \otimes I \longrightarrow O$ are restrictions of L and e_1 respectively.

Now theorem 2.1 implies:

1) Minimization: $M^*(A)$ is observable and equivalent to A.

2) Reduction: $e(A): A \longrightarrow M^*(A)$ is a reduction and, provided that $\underline{K}'$ has kernel pairs, we have

$$M^+(A) \cong A / \equiv$$

where $\equiv$ is the kernel pair of $M(A)$, called Nerode-equivalence, and $A/_\equiv$ the corresponding quotient automaton.

3) Behavior characterization: A subobject B of $\langle I^+, O\rangle$, or more precisely a canonical $\mathcal{M}$-morphism $m: B \longrightarrow \langle I^+, O\rangle$, is the behavior of a suitable automaton A' iff B is closed under left shift, i.e. there is a restriction $L': B \otimes I \longrightarrow B$ of L.

4) Uniqueness: $M^*(A)$ is minimal and equivalent minimal automata are isomorphic.

Remark: 1.-4. are equivalent to the fact that the construction $M^*(A)$ for each A can be extended to a minimal realization functor M^* which is right adjoint and right inverse to the behavior functor (cf. /Eh-K[3] 74/, theorems 5.3 and 5.5).

Proof: 1) $M(M^*(A)) = m(A) = m(M^*(A))$

2) Using the factorization theorem we have $e(A) \in \underline{K}'$-$\underline{Aut}$ and $\mathcal{E}$-$\mathcal{M}$-factorizations are unique up to isomorphism.

3) $m(A) \in \underline{K}'\text{-}\underline{Aut}$ implies that $m(A): E(A) \longrightarrow \langle I^+, O\rangle$ is closed under left shift. Vice versa define $A'=(I,O,B,L',e_1')$ with e' restriction of e_1.

4) $M^*(A)$ is minimal because $E(A') \subseteq E(A)$ implies $M^+(A') \subseteq M^+(A)$ and hence an automata morphism $A' \xrightarrow{e(A')} M^+(A') \subseteq M^+(A)$ which is unique because $m(A) \in \mathfrak{M}$. Uniqueness up to isomorphism of minimal automata follows from the definition of minimality. ▭

2.3 <u>COROLLARY 2 (Equivalence of Minimization Concepts)</u>

All the different minimization concepts in 1.5 are equivalent for automata in closed categories. However, we have to assume that each $e \in \mathcal{E}$ is a retraction to show scoop minimality and, of course, finite cardinality to show the equivalence with state minimality.

<u>Proof:</u> For simplicity we make use of theorem 3.1, although some proofs are much easier in the closed case. Since $P = 1_{\underline{K}'}$ strong observable coincides with observable and it remains to show that reduced implies observable: If A is reduced the reduction $e(A):A \longrightarrow M^*(A)$ is an isomorphism and hence $M(A)=m(A) \circ e(A) \in \mathfrak{M}$. ▭

3. HIERARCHY OF MINIMIZATION CONCEPTS

In /Eh-K^3 74/ chapter 7 we have shown, that for each automaton A in a pseudoclosed category there is a reduction $u(A):A \longrightarrow R(A)$ onto a reduced automaton $R(A)$ given as cofibre product of all reductions with source A. On the other hand at least one observable automaton A', equivalent to A but not unique up to isomorphism, can be constructed by a coretraction of $e(A)$ in 1.4.3. Moreover in /Eh-K^3 74/, 10.6 and 10.8 also sufficient conditions are given for a scoop minimization of A.

But the results of 2.2 and 2.3 do not remain true for automata in pseudoclosed categories in general. Instead of equivalence in the close case we now get a proper hierarchy of all minimization notions.

3.1 <u>THEOREM (Hierarchy of Minimization Concepts)</u>

The minimization notions given in 1.5 relate each other in the following way:

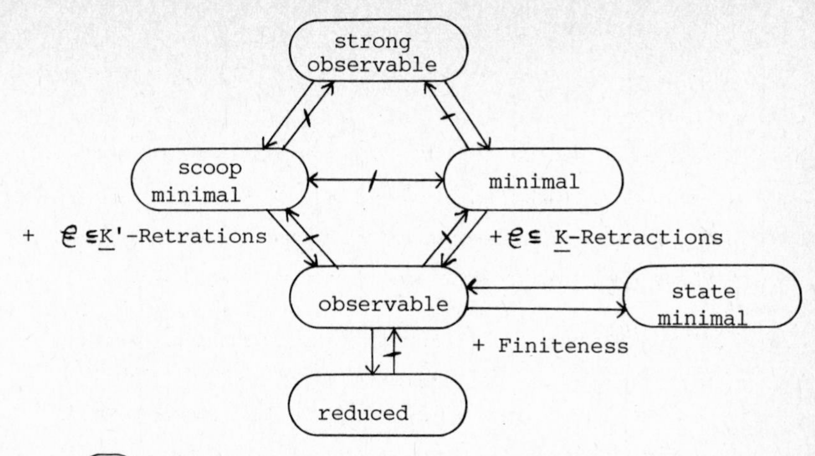

(p) ⟶ (q) means, that an arbitrary automaton in ($\underline{K}$, ⊗) with property p has also property q, and (p) ⟶̸ (q) means, that (p) ⟶ (q) is not satisfied in general.

<u>Proof:</u> First we give the counterexamples.

Consider the following nondeterministic automata

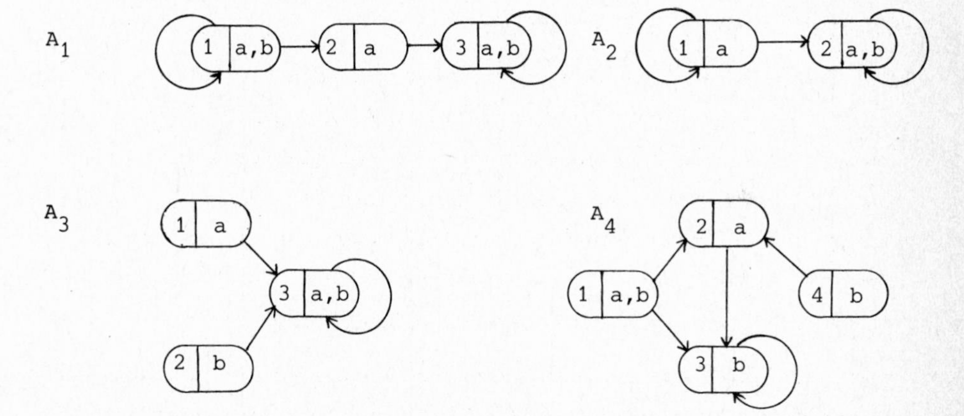

(A_1, ..., A_4 are assumed to have single input symbol x, and (m | a,b) ⟶ (n | b) means, that starting in state m with input x we have possible outputs a and b and n is a possible next state)

Automaton A_1 is not observable, because states 1 and 3 are equivalent, but it is reduced, because states 1 and 3 have different state transition with respect to equivalent states, and hence they cannot be identified by automata morphisms.

Especially there is no morphism from A_1 to A_2. Thus A_2, which is observable and also scoop minimal, is not minimal and also not strong observable, because its power automaton has two equivalent states.

Automaton A_3 is not scoop minimal, because state 3 can be scooped by states 1 and 2, but A_3 is observable, of course.

Finally automaton A_4 is minimal (cf. 3.3); however, state 1 is equivalent to state set $\{2, 4\}$ (i.e. $M(A_4)(1) = M(\hat{P}A_4)(\{2, 4\})$), and hence A_4 is not scoop minimal and hence also not strong observable.

Now we will prove the conclusions in the hierarchy. Given a strong observable automaton A it is an easy consequence of $M(\hat{P}A) \in \mathfrak{M}$, that A is also scoop minimal. Moreover minimality of A is shown in /Eh-K^3 74/,7.9.

If A is scoop minimal, then A is also observable, because the pair of $\underline{K}'$-morphisms $(c, e(A))$ with $e(A)c = 1_{E(A)}$, which exists by the additional assumption, can be interpreted as a scoop of A, such that c and hence also e(A) are isomorphisms and therefore $M(A) = m(A)e(A) \in \mathfrak{M}$.

If A is minimal we obtain an equivalent observable automaton A' by /Eh-K^3 74/, 7.4, where the additional assumption is used. Now we have a reduction $r: A' \longrightarrow A \in \mathfrak{E}$ by definition of the minimality of A, which is also in $\mathfrak{M}$, because each morphism with an observable source belongs to $\mathfrak{M}$. Hence r is an isomorphism showing that also A is observable.

Finally the equivalence of the notions "observable" and "state minimal" is proved in /Eh-Kr 73/ and the implication from "observable" to "reduced" in /Eh-K^3 74/, 7.8. ▭

Although the notions "reduced", "observable", "scoop minimal", "state minimal" and "strong observable" are considered in some papers (cf. all references), the questions, which nondeterministic automata are minimal or which have equivalent minimal automata, are not completely answered. We will give an answer for automata in pseudoclosed categories relative to the category of sets using an extended characterization of machine morphisms (cf. /Eh-K^3 74/, theorem 6.6).

3.2 LEMMA (Characterization of Machine Morphisms)

1) The machine morphism M(A) of an automaton A=(I, O, S, d, l) is the unique K'-morphism, such that the following diagrams commute:

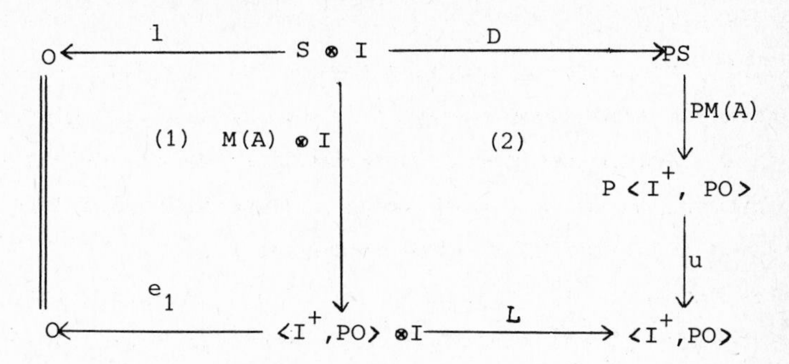

where the K'-morphism D is defined by $v_S D = d$ and e_1 is the restriction of $e_O = v_O\, ev_{PO}: \langle I^+, PO\rangle \otimes I^+ \longrightarrow O$ (cf. 1.1).

2) On the other hand a K'-morphism $f: S \longrightarrow \langle I^+, PO\rangle$ is the machine morphism of a suitable automaton A, if there is a K'-morphism $D: S \otimes I \longrightarrow PS$, such that (2) with M(A) replaced by f commutes.

Remark: This result can be extended to a corresponding behavior characterization provided that ε is included in all K-retractions (cf. /Eh-K[3] 74/, 6.7).

Proof: The commutativity of (1) is a direct consequence of the definitions of l^+, M(A) and e_1, and the commutativity of (2) is proved in /Eh-K[3] 74/, 6.6, where vice versa also is shown, that a K'-morphism $f: S \longrightarrow \langle I^+, PO\rangle$ satisfying the assumptions of 2) is machine morphism of the automaton $A_f = (I, O, S, d_f := v_S D, l_f := e_1(f \otimes I))$.

Thus given an automaton A and a morphism $f: S \longrightarrow \langle I^+, PO\rangle$, such that (1) and (2), with M(A) replaced by f, commute we obviously get: $A = A_f$, and hence: $M(A) = M(A_f) = f$, showing uniqueness of M(A). ▭

3.3 THEOREM (Characterization of Minimal Automata)

Let A=(I, O, S, d, l) be an automaton in (K, ⊗) pseudoclosed relative to the category of sets.

Then A is minimal, iff A is observable and only one K'-morphism $D: S \otimes I \longrightarrow PS$ exists, such that the following diagram is commutative:

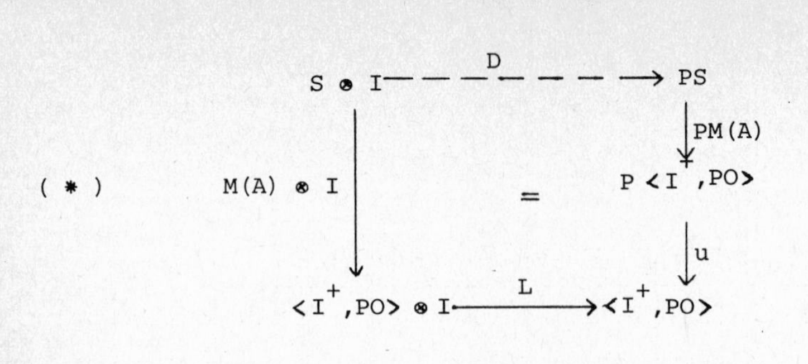

Proof: Let A be minimal, then A is also observable by 3.1. Assuming the existence of D_1 and D_2 with the above property we get two automata

$A_i = (I, O, S, d_i = v_S D_i, l_i = e_1(M(A) \otimes I))$ $(i=1,2)$

with M(A) as machine morphism according to lemma 3.2. Thus using the minimality of A there are two unique automata morphisms $f_i : A_i \longrightarrow A$ $(i=1,2)$ which are necessarily given by $f_i = 1_S : S \longrightarrow S$, because we have: $M(A) = M(A_i) = M(A) f_i$ and $M(A) \in \mathfrak{M}$. This implies $d_1 = d = d_2$ and hence $D_1 = D_2$. On the other hand given an arbitrary automata $A' = (I, O, S', d', l'$ with $E(A') \subseteq E(A)$ we will prove $f := c\ i\ e(A')$, where i is the inclusion of the behaviors and c is a coretraction of e(A), is an automata morphism from A' to A and hence the unique one, because A is observable:

i) Now we have:

$\underline{M(A) f} = M(A)\ c\ i\ e(A') = m(A) e(A) c\ i\ e(A') = m(A) i\ e(A') =$
$= m(A')\ e(A') = \underline{M(A')}$

which implies, that f is compatible with l and l'.

ii) Assuming $f\ d' \neq d\ (f \times I)$ we also have $P\ f\ D' \neq D\ (f \times I)$ with $v_{S'} D' = d'$ and $v_S d = D$ using the adjunction J P.

Thus there exist $s' \in S'$ and $x' \in I$, such that

$Pf \circ D'(s', x') \neq D\ (f \times I)\ (s', x') = D(f\ (s'), x')$.

Defining $D_o : S \times I \longrightarrow PS$ by

$D_o\ (f(s'), x') = Pf \circ D'(s', x')$ and $D_o(s,x) = D(s,x)$ otherwise

we have $D_o \neq D$, but

$\underline{u\ PM(A) D_o (f(s'), x')} = uPM(A) PfD'(s', x') = uPM(A') D'(s', x') =$
$= L(M(A') \times I)(s', x') = L(M(A) \times I)(f \times I)(s', x') =$
$= \underline{L(M(A) \times I)(f\ (s'), x')}$ and

$\underline{uPM(A)D_o(s,x)}$ = uPM(A)D(s,x) = $\underline{L(M(A)x\ I)(s,x)}$ otherwise, imply that D and D_o satisfy (*) and thus must be equal, which is a contradiction. Hence we have f d'=d(f x I) showing that f:A' ⟶ A is an automata morphism. ▭

4. SCOOP MINIMIZATION: ALGORITHM AND OPEN PROBLEMS

Now we are going to consider the minimization problems for initial automata (A, a), i.e. automata A=(I, O, S, d, l) in ($\underline{K}$, ⊗) together with an initial state morphism a:U ⟶ S in $\underline{K}$ (U unit object). The behavior E(A, a) of (A, a) is given by the $\underline{K}$-morphism

$$E(A, a) = (I^+ \xrightarrow{a \otimes I^+} S \otimes I^+ \xrightarrow{l^+} O).$$

In /Eh-Kr 74/ we have shown, that the constructions of equivalent observable, reduced and power automata can be extended to initial ones. Whereas in the case of automata in closed categories all results of section 2 can be carried over to initial automata using first a reachability construction (cf. /Eh-K^3 74/, 9.5), this does not remain true for the pseudoclosed case. Equivalent observable initial automata do not have isomorphic state objects in general, and for some pseudoclosed categories, e.g. $\underline{Rel}$ relative $\underline{Sets}$, there exists, in fact, no minimal initial automaton. In the hierarchy 3.1 the implications "strong observable" implies "minimal" and "observable" implies "state minimal"are no longer true for initial automata, even if we only regard reachable ones.
However, given an input-output morphism $f:I^+ \longrightarrow O$ there exists a "canonical" observable realization defined by the left shift closure of f in $\langle I^+, PO \rangle$, which is the smallest deterministic realization (i.e. the state transition belongs to $\underline{K}$'). Hence observability can be regarded only as a first step to minimize initial automata.

As far as we know up to now the scoop minimization, introduced in /Eh-Kr 74/ and /Eh-K^3 74/, is the best approximation for state minimization of initial automata. A slight modification of this construction is also considered in /Ma 75/. In this paper we will give an algorithm to scoop each nondeterministic automaton with finite state set to an equivalent scoop minimal one, based on the scoop construction in/Eh-K^3 74/, 10.5 and 10.7.

4.1 LEMMA

Given a nondeterministic automaton $(A = (I, O, S, d, l), a)$, a state $s_1 \in S$ and an equivalent subset of states $S_1 \subseteq S \setminus \{s_1\}$

1) Then $(m_1 : S \setminus \{s_1\} \longrightarrow S,\ n_1 : S \longrightarrow P(S \setminus \{s_1\}))$, where m_1 is the inclusion and n_1 is defined by $n_1(s_1) = S_1$ and $n_1(s) = \{s\}$ otherwise, is a scoop of A.
2) The initial automaton $(A(m_1, n_1), v_{S \setminus \{s_1\}} n\ a)$ - short $(A(s_1), a(s_1))$ -has state set $S \setminus \{s_1\}$ and is equivalent to (A, a).

Proof: The scoop property of (m_1, n_1) is a direct consequence of the equivalence of s_1 and S_1. Part 2 is proved in /Eh-K[3] 74/, 10.4, for arbitrary scoops. □

4.2 THEOREM (Algorithm for Scoop Minimization)

Given a nondeterministic initial automaton (A, a) with finite state set, which are numbered from 1 to $n \in \mathbb{N}$ without loss of generality.

Then the following iteration starting with (A, a) ends up with an equivalent scoop minimal automaton:

For i=1 step 1 until n apply lemma 4.1 to state i and an equivalent subset $S_i \subset S \setminus \{i\}$, if there is one.

The order of the states and of the tests with subsets is arbitrary and does not affect the state set of the scoop minimal automaton.

Proof: Subset of states are states of the deterministic power automaton and hence there are a well-known algrothim to decide the equivalence of a state with a subset.

Assume that the algorithm scooped the states $i_1 \leq \ldots . \leq i_m$ $(m \geq 0)$ and that the resulting automaton $(A(i_1) \ldots .(i_m), a(i_1) \ldots .(i_m))$ is not scoop minimal. Hence there exists a state $i \in S \setminus \{i_1, \ldots, i_m\}$ and a subset $S_i \subseteq S \setminus \{i_1, \ldots, i_m\}$ equivalent to i. But this means, that i is scooped in step i of the algorithm in contradiction to the assumption. Note that all scoop minimal automaton defined by scoops of A have the same state object $S \setminus \{i_1, \ldots ., i_m\}$. □

4.3 OPEN PROBLEMS

The above minimization does not lead to a state minimal automaton in general, a counterexample is given in /Eh-K[3] 74/, 10.6. Hence the following problems arise:

1) Characterize the class S of all those automata such that scoop minimization leads to state minimality.

2) Is it possible to construct for each automaton A, or at least for the deterministic observable realization of any behavior function $f: I^{+} \longrightarrow PO$, an equivalent automaton in the class S defined above?

If both problems could be solved we would have an algorithm for the minimization of nondeterministic initial automata and for the minimal realization of finitely realizable behavior functions. But note that even a suitable mathematical formulation of this problem "minimization of initial nondeterministic automata" is difficult to formulate because in any case we have the trivial solution to consider for each automaton A all automata A' with smaller state set, which is only a finite number up to isomorphism, and to prove the equivalence of A' and A.

REFERENCES

/Ar-Ma 73/ M.A.Arbib, E.G.Manes: Adjoint Machines. State-Behavior and Duality, Techn. Rep. 73B-1COINS, Univ. of Mass. at Amherst (1973), to appear in J. Pure Appl. Alg.

/Ar-Ma 74/ M.A. Arbib, E.G.Manes: Fuzzy morphisms in automata theory, in Lecture Notes in Comp.Sci. 25, 80-86. (1975)

/Ba 72/ E.S.Bainbridge: A Unified Minimal Realization Theory, with Duality for Machines in a Hyperdoctrine (Announcement of Results),Techn.Rep.,Comp. and Comm.Sci.Dept.,Univ. of Michigan (1972).

/Eh-K^3 74/ H.Ehrig, K.-D.Kiermeier,H.-J.Kreowski,W.Kühnel: Universal Theory of Automata, Teubner, Stuttgart 1974

/Eh-Kr 73/ H.Ehrig,H.-J.Kreowski: Systematic Approach of Reduction and Minimization in Automata and System Theory, Forschungsbericht 73-16 (1973), FB 20 der TU Berlin, to appear in revised version in: J. Comp.Syst. Sci.

/Eh-Kr 74/ H.Ehrig,H.-J.Kreowski: Power and initial automata in pseudoclosed categories, in Lecture Notes in Comp.Sci. 25, 144-150.

/Go 71/ J.A.Goguen: Discrete-Time-Machines in Closed Monoidal Categories I, Quarterly Rep. no. 30, Inst. f. Comp.Res., Univ. of Chicago (1971), condensed version in: Bull AMS 78 (1972), 777-783

/Ma 75/ E.G.Manes: Nondeterminism, preprint, Univ. of Mass. (1975).

LOGICAL AND ALGEBRAICAL MODELS OF THE NETWORKS OF ACTIVITIES

J. L. Kulikowski
Institute of Organization,Management and Control Sciences
Warsaw, Poland

I. INTRODUCTION

All, good and bad, achievements of human civilization can be considered as the results of less or more organized, sometimes mutually crossed, individual human activities. Reaching a desired aim as a result of a set of organized actions is one of the most important practical problems. By no doubt, good organization in the domain of political, military and economical affairs, in science, technology,public health, education, etc. if does not make a success directly, makes failing less probable. The problem of good organization of human activities seems also interesting from a theoretical point of view. It can be considered as a part of cybernetics related to several other disciplines: economy, technology, logics and applied mathematics, operations research, etc. Due to the uncertainty of external factors acting on the human beings and to the compositeness of the human reactions on them the problem seems also extremely difficult for investigations. Maybe, this difficulty explains why for many years the theory of organization avoided using more advanced mathematical tools. However, the backgrounds of this theory and a lot of valuable results we are owing to Ch. Babbage (1792-1871), F.W. Taylor (1856-1915), H. Le Chatelier (1850-1936), H.L. Gantt (1861-1919),F.B. Gilbreth (1868-1924), H. Fayol (1841-1925), K. Adamiecki (1866-1933),H. Ford (1863-1947), T. Kotarbiński (1886-) and others.

Even in the first period of the theory of organization development some attempts to the formalization of the problem were undertaken. However, significant results in this field has been reached since the years of the World War II. The jobs planning and scheduling methods based on graphs, activity networks, combinatorial and algebraical models, etc. become valuable only after having at our disposal modern computational tools for evaluation of numerical results of the composite problems stated by practice. The advantage of the well known PERT-diagram methods is the simplicity of their general philosophy. However, namely this limitates sometimes their practical applications. Real activities can not be reduced, without any loss of information, to "nodes" or "arrows" and the relations between them can not be considered as simple logical ones.The models thus needs in improvements.

II. GRAPHS AS MODELS OF THE NETWORKS OF ACTIVITIES

All over this course of lectures the capital letters:

- from the beginning of the latin alphabet (A, B, C, ...) will be used for some finite sets of elements;
- from the middle of the latin alphabet (I,J,K,...) will be used for the sets of integers enumerating the elements of the sets;
- from the end of the latin alphabet (U,V,W,...) will be used for some relations described on the sets of elements.

The corresponding lower-case letter will be used for the elements of the sets.

The logical symbols of negation - $\neg$, disjunction - $\vee$, conjunction - $\wedge$, implication - $\Rightarrow$, equivalence - $\equiv$ as well as the symbols of partial - $\exists$ and universal - $\forall$ quantors will be also used in a common sense.

The commonly used set-algebra symbols of an union - $\cup$, intersection - $\cap$, asymmetrical difference - $\setminus$, inclusions - $\in$, $\subset$ and cartesian products $\times$ will be also adopted here.

The figural brackets $\{\}$ containing the names of some elements will be used for unordered sets of the elements, while the straight ones will denote the ordered sets.Empty sets will be denoted by $\emptyset$.Formulae are numerated in the chapters independently; the references to them consist of two parts, first indicating the chapter-number, if concern to the formulae given in other chapters.

Among the formal models used for the description of the networks of activities graphs should be mentioned, first of all. The practical importance of graphs we owe to their simplicity as of two-arguments relations.

Let

$$A = \{a_i\},\ i \in I, \tag{1}$$

be a finite set of elements called nodes. Let us take into account a projection

$$\varphi:\ A \rightarrow A \tag{2}$$

(not obviously univalued) and a set

$$S = \{s\} = \{[a_i, a_j]\}, \quad i,j \in I \tag{3}$$

of all ordered pairs of elements satisfying to the condition

$$a_j = \varphi(a_i). \tag{4}$$

The element a_i will be called incident to a_j in this case and the pair s_{ij} will be called an arc from a_i to a_j.

The three elements

$$G = [A, S, \varphi] \tag{5}$$

will be called an oriented graph. A graph is called symmetrical (unoriented) if for any $i,j \in I$

$$(s_{ij} \in S) \Longrightarrow (s_{ji} \in S). \tag{6}$$

The pairs of arcs $\{s_{ij}, s_{ji}\}$ of symmetrical graphs are called edges of the graphs. However, we shall not deal with symmetrical graphs in this course of lectures. The term graph (without adjective) thus will be used for oriented graphs.

For the given graph G (defined by (5)) a graph

$$H = [B, T, \psi] \tag{7}$$

will be called:

1° a subgraph of G if

$$B \subset A \tag{8}$$

and ψ is a subprojection of φ defined as

$$\psi = \varphi \text{ for } a_i, a_j \in B, \tag{9}$$

and $\psi = \emptyset$ otherwise;

2° a partial subgraph of G if (8) holds and ψ is a partial subprojection of φ defined a

$$\psi = \varphi \text{ or } \emptyset \text{ for } a_i, a_j \in B, \tag{10}$$

and $\psi = \emptyset$ otherwise.

For the networks of activities description purposes the nodes of a graph can be interpreted as some "states"reached during a job realization and the arcs as "works" or "actions" performed in order to reach the consecutive states. A subgraph corresponds to a subjob consisting of a given subset of states and of all those actions of the network which transform the states within the given subset. A partial subgraph corresponds to a subjob containing only some of activities within the given subset of states.

It is clear that because of ordering of activities in time the graphs describing networks of activities should satisfy to some additional conditions.

A sequence of nodes of the graph G:

$$l(i_1, i_n) = [a_{i_1}, a_{i_2}, \ldots, a_{i_n}], \quad i_1, \ldots, i_n \in I, \tag{11}$$

for any natural n, such that

$$a_{i_k} = \varphi(a_{i_{k-1}}),\ 1 \leqslant k \leqslant n, \tag{12}$$

will be called a path from a_{i_1} to a_{i_n}. The integer n-1 will be called tle length of the path $l(i_1, i_n)$.

The graph G will be called acyclical one if for any a_i and for any n it does not exist any path from a_i to a_i.

If the states and the activities of the jobs are strongly determined in physical time, the graphs corresponding to real networks of activities should satisfy to the condition of acyclicity.

For any acyclical graph G a node a_i will be called foregoing the node a_j (a_j following the node a_i),

$$a_i \prec a_j, \tag{13}$$

if there exists a path $l(i,j)$ from a_i to a_j. The following properties of the relation are evident:

$$(a_i \prec a_j) \wedge (a_j \prec a_i) \Rightarrow (a_i = a_j), \tag{14}$$

$$(a_i \prec a_j) \wedge (a_i \neq a_j) \Rightarrow \neg(a_j \prec a_i), \tag{15}$$

$$(a_i \prec a_j) \wedge (a_j \prec a_k) \Rightarrow (a_i \prec a_k). \tag{16}$$

Therefore, the relation $\prec$ is a sort of semiordering.

A sequence of nodes:

$$h(i_1, i_n) = [a_{i_1}, a_{i_2}, \ldots, a_{i_n}] \tag{17}$$

for any natural n, such that

$$a_{i_k} = \varphi(a_{i_{k-1}}) \text{ or } a_{i_{k-1}} = \varphi(a_{i_k}),\ 1 \leqslant k \leqslant n, \tag{18}$$

will be called a chain linking a_{i_1} and a_{i_n}.

The graph G will be called compact if for any pair of its nodes there exists a chain linking the nodes; it will be called compact in strongsense if any given pair of nodes can be linked with a path, at least in one of two possible directions. For the networks of activities considerations it will be enough to take into account the compact graphs only, so as the graphs not satisfying to this condition correspond to some sets of mutually independent jobs being of less interest.

For a given acyclical compact graph G a node a_k will be called an upper bound of the nodes a_i, a_j if both

$$a_i \prec a_k \text{ and } a_j \prec a_k \tag{19}$$

hold. The upper bound a_k will be called a supremum of the given pair a_i, a_j if it does not exist any other upper bound a_l of this pair of nodes such that $a_l \prec a_k$. The lower bound and the infimum of the given pair a_i, a_j of nodes can be defined in a similar way.

The graph G becomes a lattice if any pair of its nodes has its lower and its upper bound.

The latticies are usually used as the graph-models of the networks of activities.

III. AN ALGEBRAIC APPROACH TO THE DESCRIPTION OF HUMAN ACTIVITIES

We are intended to investigate in a more detailed manner the relationships between the real networks of activities and their graphs--representants.

Let G (given by (II.5)) be a finite compact acyclical graph-lattice. It will be supposed that it contains only one "starting" node a_0 and only one "final" node a_f (in the sense of the above-introduced relation of semiordering). For any finite compact acyclical graph--lattice G representing a network of activities the condition can be easily satisfied by addition of fictional starting and final states and the corresponding fictional actions transforming the fictional starting state into the real starting ones and the real final states into the fictional final one. The set I of indices enumerating the nodes of the graph G will thus contain the elements 0 and f.

We shall denote by a_i^* the real state represented by the node a_i of the graph and by s_{ij}^* the real action represented by the arc s_{ij}, for $i,j \in I$.

The following simple propositions will be introduced:

$$\underline{P}_{ij}: \text{"}s_{ij}^* \text{ has been started"}, \tag{1}$$
$$\overline{P}_{ij}: \text{"}s_{ij}^* \text{ has been finished"},$$
$$\overline{Q}_i: \text{"}a_i^* \text{ has been reached"}, \tag{3}$$

for all $i,j \in I$. The set of the above-defined propositions will be denoted by L_0. Using L_0 and the classicalpropositional calculus operators: $\neg, \wedge, \vee, \Rightarrow, \equiv$ we can define an extended set of composite propositions based on the simple ones and on the rules of classical propositional calculus. The set of the composite propositions will be defined by L_1. The problem arises of a correspondence between the propositions from L_1 and some topological objects defined on the graph G.

Any composite sentence of the form:

$$\overline{Q}_i \wedge \underline{P}_{ij} \wedge \overline{P}_{ij} \wedge \overline{Q}_j \wedge \underline{P}_{jk} \wedge \overline{P}_{jk} \wedge \ldots \wedge \overline{P}_{mn} \wedge \overline{Q}_n \tag{4}$$

corresponds to a path $l(i,n)$ from a_i to a_n if and only if

$$s_{ij}, s_{jk}, \ldots, s_{mn} \in S. \tag{5}$$

Therefore, despite the fact that the expression (4) seems quite correct from the point of view of logical rules, it may be "semantically" nonadmissible, for example, if $s_{ij} \notin S$. However, the real world may impose also some "positive" conditions on the formal model. For example, no direct relationships between the logical values of the propositions of the type $\overline{Q}_i$ and $\underline{P}_{ij}$ follow from the logical rules. Nevertheless, in most part of graph-models describing the networks of activities the following is assumed:

$$\overline{Q}_i \Rightarrow \underline{P}_{ij}, \quad "Q_0" = \text{true}, \tag{6}$$

$$\underline{P}_{ij} \Rightarrow \overline{P}_{ji}, \tag{7}$$

$$\left(\bigwedge_{h \in H_i} \overline{P}_{hi}\right) \Rightarrow \overline{Q}_i, \tag{8}$$

for $i, j \in I$ and such that s_{ij}, s_{hi}, $h \in H_i$, are admitted as physically possible actions, H_i describing the subset of all states incident to the given state a_i. The above-given rules can be expresse as follows: 1° any state a_i reached initiates all actions going out of this state, 2° any action if started is also finished and vice-versa, 3° any state becomes reached iff all the input actions has been finished.

It is clear, that the above-formulated rules, in general, are not obligatory. For example, starting new actions, when a starting-state has been reached, may depend on some additional conditions.Any action started does not make it sure reaching the desired aim. Reaching a state may depend on oerforming the foregoing actions in more complicated manner. At last, reaching a state a_j going out of a given one a_i may be possible on different ways. All these additional conditions can not be taken into account when a simple graph-model of the network of activities is used. For example, there is no direct possibility to discriminate between the rule (6) and the one

$$\overline{Q}_i \Rightarrow (\underline{P}_{ij} \vee \underline{P}_{ik}), \tag{9}$$

supposing that a_i is incident both to a_j and a_k.

The alternative ways of reaching a_j^* from a_i^* can be described if multigraphs are used instead of the graphs.

For any given set of nodes A and a set of projections

$$\varphi^{(k)}: A \rightarrow A, \; k \in [1,2,3,\ldots] \tag{10}$$

such that

$$\left[(s_{ij}^{(k)} \in S^{(k)}) \wedge (k > 1)\right] \Rightarrow (s_{ij}^{(1)} \in S^{(1)}), \tag{11}$$

where

$$S^{(k)} = \left\{s_{ij}^{(k)}: a_j = \varphi^{(k)}(a_i)\right\}, \tag{12}$$

a multigraphs can be defined as a set

$$M = \left\{A, S^{(1)}, \varphi^{(1)}, S^{(2)}, \varphi^{(2)}, \ldots, S^{(k)}, \varphi^{(k)}, \ldots\right\}. \qquad (13)$$

The set of ordered pairs

$$S_{ij} = \left\{s_{ij}^{(1)}, s_{ij}^{(2)}, \ldots, s_{ij}^{(k^{*})}\right\} \qquad (14)$$

where $k^{*} = k^{*}(i,j)$, will be called a multiarc from a_i to a_j. For our purposes the multiarc S_{ij} can be interpreted as a set of alternative actions that can be undertaken in order to reach the state a_j^{*} starting out from the state a_i^{*}.

Therefore, for the multigraph used as a formal model of the network of activities the assumption (8) can be formulated:

$$\left(\bigwedge_{h \in H_i} \bigvee_{k} \overline{P}_{hi}^{(k)}\right) \Rightarrow \overline{Q}_i \qquad (15)$$

where

$$\overline{P}_{ij}^{(k)}: \text{"}s_{ij}^{(k)*} \text{ has been finished"} \qquad (16)$$

$s^{(k)*}$ denoting the k-th variant of the $[a_i^{*}, a_j^{*}]$ action performance. Multigraphs make thus the formal model more flexible.However,the real situations represented by the expression (9) can not be also imbedded into the multigraphs without additional elements of description. For certain applications (e.g. for information processing) the rules of choosing the actions starting from a given state a_i depend on the way the state a_i has been reached (this means that the number of the state does not give us full information for the following actions choosing). In this case the following approach is possible.

Let us define the sets of input and output actions for the given state a_i:

$$\overline{\Sigma}_i = \bigcup_{h \in I} S_{hi}^{*}, \quad i \in I \qquad (17)$$

$$\underline{\Sigma}_i = \bigcup_{j \in J} S_{ij}^{*}, \quad i \in I. \qquad (18)$$

Then the relation

$$R_i \subset \overline{\Sigma}_i \times \underline{\Sigma}_i, \quad i \in I \qquad (19)$$

will describe all possible ways of choosing the consecutive subsets of actions when the state a_i^{*} has been reached by the strongly definite manners.

Therefore, the algebra of relations seems the most adequate formal tool for the description of situation that can arise in the considered class of the networks of activities.

IV. THE LOGIC OF TIME-RELATIONS

The restrictions imposed on the networks of activities in the foregoing considerations were of logical type. It is clear that any logical relation of the type

$$\overline{P}_{ij} \Rightarrow \underline{P}_{jk}$$

where $\overline{P}_{ij}$ and $\underline{P}_{jk}$ have the meaning given by (III.2) and (III.1), correspondingly, imply also on the ordering of the activities s^{*}_{ij}, s^{*}_{jk} in time. However, in many cases the time-ordering of the activities, both in absolute and in relative sense, may be imposed on the network directly. Therefore, it seems desirable to investigate in more detailed form the exact meaning of the propositions containing some time-relation indications like:

"The state a_i has been reached at the time t",

"The operation s_{ij} starts not later then s_{mn}",

etc. The logical values of both above-given propositions are not constant in time. On the other hand,for operations planning technique we need some formal models in some sense independent on the course of events in the real world. Otherwise speaking, we are looking for a method of proving some time-relations concerning predicates.

The basic concepts of a time-relations logic were formulated by A. N. Prior in 1967; there will be given here some adaptation of those concepts to the jobs planning problems formulated by the author in 1972.

First of all, let us remark that there is a difference between the proposition:

"The state a_i has been reached"

which, if the state a_i has been really reached at the time t,is false for any time before t and is true for any time after t, and the proposition:

"There is such a time t that the proposition: ′The state a_i has been reached‵ becomes true for all $t' \geqslant t''$ ".

The last proposition is evidently true in any time.

The basic idea of the time-relations (tense) logics lies in some transformations of the first-type propositions into the second-type ones.

Let us denote by x any proposition and let us define the operator D_t acting on the proposition in such a way that we obtain:

$$D_t x \overset{def}{=} \text{"x at the time t"}. \tag{1}$$

According to the assumptions formulated by A.N. Prior, the "date-operator" D_t should satisfy to the conditions:

$$D_t(\neg x) \Rightarrow \neg D_t x, \tag{2}$$

$$D_t(x' \Rightarrow x'') \Rightarrow (D_t x' \Rightarrow D_t x''), \tag{3}$$

$$\forall t(D_t x) \Rightarrow x, \tag{4}$$

$$D_{t''}(D_{t'} x) \Rightarrow D_{t'+t''} x. \tag{5}$$

Now, the first example given in this chapter, according to (III.3), will be simply written as: $D_t \overline{Q}_i$. However, for our purposes there will be used some more general kind of operators:

1° a "weak time-interval operator" $d_{t',t''}$ transforming any proposition x into the one:

$$\exists (t' \leq t < t'')(D_t x); \tag{6}$$

2° a "strong time-interval operator" $D_{t',t''}$ transforming any proposition x into the one:

$$\forall (t' \leq t < t'')(D_t x), \tag{7}$$

where t, t′, t″ are some real values.

The Prior′s "date-operator" can be obtained in particular case:

$$d_{t',t''} x = D_{t',t''} x = D_t x. \tag{8}$$

The other kinds of so-called tense-operators can be also derived:

3° the "strong-past operator"

$$G_t x \overset{def}{=} D_{-\infty,t} x, \tag{9}$$

4° the "strong-future operator"

$$H_t x \overset{def}{=} D_{t,+\infty} x, \tag{10}$$

5° the "weak-past operator"

$$P_t x \overset{def}{=} d_{-\infty,t} x, \tag{11}$$

6° the "weak-future operator"

$$F_t x \overset{def}{=} d_{t,+\infty} x. \tag{12}$$

On the basis of:

a) the axioms of the classical propositional calculus,

b) the rules of classical logical inference,

c) the axioms of the set-algebra

the following properties od the tense-operator can be derived:

$$D_{t',t''} x \Rightarrow d_{t',t''} x, \quad t' < t'', \tag{13}$$

$$\neg d_{t',t''} x \Rightarrow \neg D_{t',t''} x, \tag{14}$$

$$[D_{t_1',t_1''} x \wedge D_{t_2',t_2''} y] \Rightarrow D_{t',t''} (x \wedge y) \tag{15}$$

for

$$t' = \max(t_1', t_2'), \quad t'' = \min(t_1'', t_2''), \tag{16}$$

$$[d_{t_1',t_1''} x \vee d_{t_2',t_2''} y] \Rightarrow d_{t',t''} (x \vee y) \tag{17}$$

for

$$t' = \min(t_1', t_2'),\ t'' = \max(t_1'', t_2''). \tag{18}$$

In addition to a)-c) the following will be assumed:

d) for any $A_{t',t''} \in \{D_{t',t''}, d_{t',t''}\}$ and for any

$$B_{t',t''} \in \{D_{t',t''}, d_{t',t''}\}$$

$$A_{t_1',t_1''} B_{t_2',t_2''} x = B_{t_2',t_2''} x \tag{19}$$

The mutual position of two time-intervals on the time-axis can be defined on the basis of the above-defined tense-operators. For example, let us take into account the mutual position of the beginning of the first and of the end of the second time-interval. If the propositions x and y are concerning the first and the second time-intervals, correspondingly, then the following operators can be defined:

$$D^{be}_{\tau,+}(x,y) = \exists(\tau' \geqslant \tau, t_1, t_2, t_3)(D_{t_1,t_2} x \wedge D_{t_3,t_1+\tau} y), \tag{20}$$

$$D^{be}_{\tau,-}(x,y) = \exists(\tau' \leqslant \tau, t_1, t_2, t_3)(D_{t_1,t_2} x \wedge D_{t_3,t_1+\tau} y). \tag{21}$$

The operator $D^{be}_{\tau,+}(x,y)$ can be read as "the beginning of the time -interval, where x is at least τ earlier then the end of the time-interval, where y". Otherwise speaking, it transforms the propositions x, y into the one being true if x is true always inside $[t_1,t_2)$ and y is true always inside $[t_3,t_4)$ where

$$t_4 \geqslant t_1 + \tau . \tag{22}$$

In a similar way, the operator $D^{be}_{\tau,-}(x,y)$ can be read as "the beginning of the time-interval, where x,is at most τ earlier then the end of the time-interval, where y". The parameter τ can reach any real value.

In similar way the other mutual positions of the time-intervals, where the propositions remain true, can be described by the operators of the type D^{bb} (beginning-to-beginning), D^{eb} (end-to-beginning) and D^{ee} (end-to-end). Also, the "time-interval duration" operators can be defined:

$$\vartheta^+_\tau x = \exists(t',t'')[d_{t',t''} x \wedge (t'' - t') > \tau], \tag{23}$$

$$\vartheta^-_\tau x = \exists(t',t'')[d_{t',t''} x \wedge (t'' - t') < \tau], \tag{24}$$

$$\Theta^+_\tau x = \exists(t',t'')[D_{t',t''} x \wedge (t'' - t') > \tau], \tag{25}$$

$$\Theta_{\tau}^{-}x = \exists(t',t'')[D_{t',t''}x \wedge (t'' - t') < \tau], \tag{26}$$

for any $\tau > 0$. Many formal properties of the above-defined operators and the rules of logical inference of the time-intervals logic can be derived. They may be used for formal proving the consistency of the logical and the time restrictions imposed on the networks of activities.

V. APPLICATION OF MODAL AND TOPOLOGICAL LOGICS

On the basis of the two-valued,classical logic no distonction between the assertions about the desired, possible and inevitable facts is possible. However, this distinction seems desirable for the sake of exactness of the language used to the description of the networks of activities. It becomes possible if some non-classical logical systems (multi-valued, inductive, probabilistic, fuzzy-sets, modal, topological etc.) are used instead of the classical propositional calculus or of the theory of two-valued predicates. For example, the three-valued logic admitts the following interpretation (the quotation marks " " denote the logical value of the proposition indicated within them):

"P" = 0 is read as "P is false",
"P" = 1 is read as "P is possible",
"P" = 2 is read as "P is true".

In a similar way there can be assigned to the propositions some logical values from the continuous interval $[0,1]$ interpreted as a "measure of probability" in the probabilistic logic or as a "measure of belonging to the set of true predicates" in the case of fuzzy-logic. However, both the above-mentioned logical systems seem unadequate to the practical situations, when the unique facts are dealt with and there is no direct method possible of evaluating the logical values of the assertions concerning some observed or planned facts to the reality. The modal logic originated by Aristotheles and developed in our times by G.H. von Wright, C.I. Lewis, R. Feys, H. Rasiowa and others seems more suitable to our needs. Despite the fact that modal logics can be derived on strongly axiomatic backgrounds, we shall prefer here a heuristic approach to it (described in details by R. Feys).

We shall denote by P,Q,... some predicates depending on a variable ζ (not indicated directly for the sake of simplicity) representing some "conditions of realization".

The rules of the classical propositional calculus will be admitted therefore, if P,Q,... are some admitted logical formulae, then also $\neg P$, $\neg Q$, $P \wedge Q$, $P \vee Q$, $P \Rightarrow Q$, and $P \equiv Q$ are admitted logical formulae.

The general and partial quantifiers will be used:

$\forall \xi$ denoted briefly as $\Box$,

$\exists \xi$ denoted briefly as $\Diamond$,

and the following interpretation will be given:

$\Box P$ is read as "necessarily P",

$\Diamond P$ is read as "possibly P".

The rules of classical logical inference (the rule of abstraction and of substitution) are assumed to be valid also in modal logic.

If $\vdash$ denotes the provability of a logicalformula then it is assumed that if

$$\vdash P, \text{ then also } \vdash \Box P. \tag{1}$$

The following axioms of modality are assumed:

$$\vdash \Box (P \Rightarrow Q) \Rightarrow (\Box P \Rightarrow \Box Q). \tag{2}$$

If T is a string of symbols of the type $\neg$ and $\Diamond$ ended by the symbol P of a formula, called a modality, then

$$\vdash T \Rightarrow \Box T. \tag{3}$$

For any proposition

$$\vdash \Box P \Rightarrow P. \tag{4}$$

The following denotations will be also used:

$$\Diamond P \overset{\text{den}}{=} \neg \Box \neg P, \tag{5}$$

$$P \Rightarrow Q \overset{\text{den}}{=} \Box (P \Rightarrow Q), \tag{6}$$

$$P \Leftrightarrow Q \overset{\text{den}}{=} \Box (P \equiv Q). \tag{7}$$

Now,the following properties of the modalities can be proved:

a) negation

$$\vdash \neg P \equiv \Diamond \neg P, \tag{8a}$$

$$\vdash \Box P \equiv \neg \Diamond \neg P, \tag{8b}$$

$$\vdash \neg \Diamond P \equiv \Box \neg P, \tag{8c}$$

$$\vdash \Diamond P \equiv \neg \Box \neg P; \tag{8d}$$

b) subordination

$$\vdash \Box P \Rightarrow P, \tag{9a}$$

$$\vdash P \Rightarrow \Diamond P, \tag{9b}$$

$$\vdash \Box P \Rightarrow \Diamond P; \tag{9c}$$

c) distributivity with respect to $\wedge$ and $\vee$:

$$\vdash \Box (P \wedge Q) \equiv (\Box P \wedge \Box Q), \tag{10a}$$

$$\vdash (\Box P \vee \Box Q) \Rightarrow \Box(P \vee Q), \tag{10b}$$
$$\vdash \Diamond(P \vee Q) \equiv (\Diamond P \vee \Diamond Q), \tag{10c}$$
$$\vdash \Diamond(P \wedge Q) \Rightarrow (\Diamond P \wedge \Diamond Q), \tag{10d}$$
$$\vdash \Diamond(P \wedge Q) \Rightarrow \Diamond P; \tag{10e}$$

d) distributivity with respect to $\Rightarrow$ and $\Leftrightarrow$:

$$\vdash (P \Rightarrow Q) \Rightarrow (\Box P \Rightarrow \Box Q), \tag{11a}$$
$$\vdash [(P \Rightarrow Q) \wedge \Box P] \Rightarrow \Box Q, \tag{11b}$$
$$\vdash (P \Rightarrow Q) \Rightarrow (\Diamond P \Rightarrow \Diamond Q), \tag{11c}$$
$$\vdash [(P \Rightarrow Q) \wedge \Diamond P] \Rightarrow \Diamond Q, \tag{11d}$$
$$\vdash (P \Rightarrow Q) \Rightarrow (\Box P \equiv \Box Q), \tag{11e}$$
$$\vdash (P \Leftrightarrow Q) \Rightarrow (\Diamond P \equiv \Diamond Q); \tag{11f}$$

e) deduction rules:

if $\vdash T \Rightarrow T'$, then also $\vdash \Box T \Rightarrow \Box T'$, (12a)

if $\vdash T \Rightarrow T'$, then also $\vdash \Diamond T \Rightarrow \Diamond T'$, (12b)

if $\vdash T \Leftrightarrow T'$, then also $\vdash \Box T \Leftrightarrow \Box T'$, (12c)

if $\vdash T \Leftrightarrow T'$, then also $\vdash \Diamond T \Leftrightarrow \Diamond T'$. (12d)

Now, the following way of application of the modal logic can be suggested. Let us suppose that for a certain network of activities a set of propositions of the type P_{ij}, Q_i, $i,j \in I$, describing the planned actions or works and reaching the corresponding states are defined. The propositions are considered, according to the heuristic interpretation of modal logic, as depending on some variable representing the real conditions of realization of the activities. There are also imposed some logical restrictions on the activities. Thus, the network is characterized by a set R of logical formulae expressing the relationships imposed on the network of activities under consideration. The set R is called consistent if there is no such a formula P that both P and $\neg P$ are provable on the basis of the rules of propositional calculus. Consistency is thus considered as a necessary condition of realizability of the set od activities leading to the desired result. Let us suppose, that during the realization of the works belonging to the network some additional information is reached in the form of modal assertions of the type $\Box P_{ij}$, $\Box Q_i$ for some $i,j \in I$. The corresponds to the works or states that has been already realized, the other ones still remaining doubtfull. The problem arises of inferring on the basis of the information reached about the realizability of the other states or works being of particular interest for the designer. This can be reached on the basis of the rules of modal logic inference. The results thus will be given in the form of some modalities: abc...mP, where a,b,c,... are the signs of the type $\neg$, $\Box$ or $\Diamond$.

The minimum possible number of the signs $\Diamond$ in the term T can be considered as a measure of "uncertainty" of the corresponding assertion.

Modal logic is only one of possible methods of expressing the uncertainty of assertions. The other one is based on the principle of evaluating the logical values of the propositions in a relative scale. This means that any logical value "P" can be evaluated only with respect to a logical value od some other proposition, "Q". Therefore, for any two propositions one of the following statements is possible:

$P \prec Q$ read as "P is less true as Q",

$P \succ Q$ read as "P is more true as Q",

$P \approx Q$ read as "P and Q are logically univalued",

$P \not\sim Q$ read as "P and Q are logically uncomparable".

The logical system based on this kind of assumptions is called a topological logic. The investigation of topological logics was originated by C.G. Hempel in 1937, the contributions has been made by H.A. Vessel. The applications of topological logics were originated by J. L. Kulikowski in 1972. In particular, it has been shown that the axioms of topological logic can be choosen in such a way that the evaluation of the logical values of the propositions becomes an operation equivalent to finding out some paths on graphs-lattices representing the logical relationships.

VI. LINGUISTIC APPROACH TO THE NETWORKS OF ACTIVITIES

We have formerly defined the sets of simple propositions $\overline{Q}_i$, $\underline{P}_{ij}$, $\overline{P}_{ij}$, $i,j \in I$, and we have used the propositional calculus for the description of composite works. The set of symbols:

$$\varphi = \{\overline{P}_{ij}, \underline{P}_{ij}, \overline{Q}, \neg, \vee, \wedge, \equiv, \Rightarrow\}, \quad i,j \in I, \tag{1}$$

can be considered as an alphabet and a set L_0 of all possible strings consisting of the symbols belonging to φ can be considered. Then, the subset $L_1 \subset L_0$ of the strings of symbols satisfying to the rules of propositional calculus can be taken into account. The three sets: φ, L_0 and L_1 define a sort of formal language describing the networks of activities. The language can be enriched by including to it the tense -logic or modal logic operators. In general, thus, arises a problem of construction and investigation of formal languages for the description of composite networks of activities. Some kinds of this sort of formal languages were, for example, considered by Yu.I. Klykov and M. Nowakowska.

Evolution of a logical superstructure of the formal languages under consideration is only one of two possible ways of their improvement and adaptation to the description of the networks of activities. The other one consists in a deepening of the internal structure of their basic semantical units, which in the former case were reduced to some statements about the "states" and "actions" in a real world.

There will be defined the following sets of symbols:

$\{D_m\}$, $m \in [1,2,3,\ldots]$ representing some "objects",

$\{R_n\}$, $n \in [1,2,3,\ldots]$ representing some "qualities of the objects"

$\{V_q\}$, $q \in [1,2,3,\ldots]$ representing some "relationships between the objects"

$\{W_r\}$, $r \in [1,2,3,\ldots]$ representing some "modes" of the relationships.

The symbols D_m, R_n, V_q and W_r correspond, approximately, to the nouns, adjectives, verbs and adverbs in the natural languages.

Next, there will be introduced the sets of symbols A_D, A_R, A_V and A_W such that:

$$\{D_m\} \subset A_D,\ \{R_n\} \subset A_R,\ \{V_q\} \subset A_V,\ \{W_r\} \subset A_W, \tag{2}$$

interpreted, correspondingly, as the set of composite objects, qualities, relationships and modes. If we denote by $*$ an operator of concatenation of symbols, then it will be assumed that:

$$R_{m'} * R_{m''} \in A_R \text{ for any } R_{m'}, R_{m''} \in A_R, \tag{3}$$

$$W_{r'} * W_{r''} \in A_W \text{ for any } W_{r'}, W_{r''} \in A_W, \tag{4}$$

$$R_n * D_m \in A_D \text{ for any } R_n \in A_R,\ D_m \in A_D, \tag{5}$$

$$W_r * V_q \in A_V \text{ for any } W_r \in A_W,\ V_q \in A_V. \tag{6}$$

In addition, the following rules will be introduced:

- absorption:

$$\ldots R_{m_\alpha} * R_{m_\beta} * R_{m_\gamma} * \ldots * R_{m_\delta} * R_{m_\beta} * R_{m_\varepsilon} * \ldots = \\ = \ldots R_{m_\alpha} * R_{m_\beta} * R_{m_\gamma} * \ldots * R_{m_\delta} * R_{m_\varepsilon} * \ldots, \tag{7}$$

$$\ldots W_{r_\alpha} * W_{r_\beta} * W_{r_\gamma} * \ldots * W_{r_\delta} * W_{r_\beta} * W_{r_\varepsilon} * \ldots = \\ = \ldots W_{r_\alpha} * W_{r_\beta} * W_{r_\gamma} * \ldots * W_{r_\delta} * W_{r_\varepsilon} * \ldots, \tag{8}$$

- symmetry

$$\ldots R_{m_\alpha} * R_{m_\beta} * R_{m_\gamma} * \ldots * R_{m_\delta} * R_{m_\varepsilon} * R_{m_\eta} * \ldots = \\ = \ldots R_{m_\alpha} * R_{m_\varepsilon} * R_{m_\gamma} * \ldots * R_{m_\delta} * R_{m_\beta} * R_{m_\eta} * \ldots, \tag{9}$$

$$\ldots W_{r_\alpha} * W_{r_\beta} * W_{r_\gamma} * \ldots * W_{r_\delta} * W_{r_\varepsilon} * W_{r_\eta} * \ldots = \\ = \ldots W_{r_\alpha} * W_{r_\varepsilon} * W_{r_\gamma} * \ldots * W_{r_\delta} * W_{r_\beta} * W_{r_\eta} * \ldots \tag{10}$$

for any m_α , m_β, ..., m_ϱ, r_α, r_β, ..., $r_\varrho \in [1,2,3,...]$.

Any concatenation of the symbols of the form

$$P = V_q * D_{m_1} * D_{m_2} * \ldots * D_{m_q}, \qquad (11)$$
$$V_q \in A_V,\ D_{m_1}, \ldots, D_{m_q} \in A_D,$$

will be called a proposition and will be read,in general, as: "A relationship V_q between $D_{m_1}, D_{m_2}, \ldots, D_{m_q}$ holds".We shall denote by P_{DRVW} the set of all possible propositions of this type based on the sets of simple propositions $\{\overline{Q}_i\}$, $\{\overline{P}_{ij}\}$, $\{\underline{P}_{ij}\}$, $i,j \in I$, can be considered as some subsets of the set P_{DRVW} for the corresponding sets $\{D_m\}$ and $\{V_q\}$ (the sets $\{R_m\}$ and $\{W_r\}$ are not used here).In fact, if we define $\{D_m\}$ as a set of all possible "states" a_i then the denotation $\overline{Q}_i, \underline{P}_{ij}$, and $\overline{P}_{ij}$ for $i,j \in I$ can be, correspondingly,changed into $V_1 * D_i$, $V_2 * D_i * D_j$, $V_3 * D_i * D_j$, which is nothing more but a simple changement of the code.

The language of "objects" and "verbs" can be used for the description of much more complicated relations. For example, an expression $V_q * D_{m_1} * D_{m_2} * D_{m_3} * D_{m_4}$ can be used for the description of a four-argument relation read, in general, as "D_{m_1} is reached by D_{m_2} starting from D_{m_3} and using D_{m_4}". In particular, this may mean: "Mr Smith (D_{m_2}) starting from Warsaw (D_{m_3}) arrives to Rome (D_{m_1}) by a plane (D_{m_4})" or "the detail D_{m_2} starting from the state D_{m_3} in subjected to the operation D_{m_1} using the tool D_{m_4}", etc. A question may arise, why not "Warsaw starting from a plane arrives to Mr Smith by Rome"? This shows us that the semantical aspects of the problem-oriented languages under consideration should be taken into account together with the syntactical ones. In the above-given case to any verb V_q there should be prescribed some additional restrictions of semantical rather then of syntactical kind. For example, if we define the subsets

$$A'_D,\ A''_D,\ A'''_D \subset A_D\,, \qquad (12)$$

A'_D meaning "towns", A''_D - "persons or goods" and A'''_D - means of transportation, then the area of the above given operator V_q will be restricted to the cartesian product of the subsets $A_D \times A_D \times A_D \times A_D$, which makes the full expression significant.However,the example shows us also, that the networks of activities description language does not belong to the context-free ones. Therefore,the theoretical approach to this kind of languages should be appropriate to their properties.

Till now, we did not use the sets of "qualities" and "modes". Their role lies in some extension of the "objects and verbs" languages. For example, if we define:

R_1 - important, R_2 - ill, R_3 - suspected,

W_1 - by schedule, W_2 - by charter, W_3 - illegally

etc. we can express the situation:

"ill and important Mr Smith starting from Warsaw arrives to Rome by a charter flight"

in the form

$$W_2 * V_q * D_{m_1} * R_2 * R_1 * D_{m_2} * D_{m_3} * D_{m_4}.$$

The sets of "qualities" and "modes" extend thus the sets of "objects" and "verbs", correspondingly, by cartesian products. Further extension is possible if we admit the following:

1° if $\{D_m\}$ is the set of simple objects, then an empty set $\emptyset$, the subsets of D_m and their combinations in the sense of the set-algebra are the elements of A_D;

2° if V_q is an s-argument relation in $(A_D)^s$, then any set of the form

$$D' = \{D_{m_\alpha} : "V_q \times D_{m_1} \times \ldots, \times D_{m_\alpha} \times \ldots \times D_{m_s}" = \text{true}\} \tag{13}$$

belong to A_D (an analogue of the -ing form in English).

VII. REFERENCES

Chapter II

1. Claude Berge. Théorie des graphes et ses applications. Dunod, Paris 1958.
2. Claude Berge. Graphes et hypergraphes. Dunod, Paris 1970.
3. Robert G. Busacker, Thomas L. Saaty. Finite graphs and networks. McGraw Hill, New York.
4. Frank Harary. Graph theory. Addison-Wesley, Reading, Mass. 1969.

Chapter III

1. Haskel B. Curry. Foundations of mathematical logic. Mc Graw Hill, New York 1963.
2. Juliusz L. Kulikowski. Algebraic methods in pattern recognition. Springer Verlag, Vienna 1972.

Chapter IV

1. A.A. Ivin. Logika vremeni (in Russian: The logic of time). In: "Ne-

klassicheskaya logika" (Non-classical logics).Nauka, Moscov 1970.
2. Juliusz L. Kulikowski. An application of a tense-logic system to formal planning. Control a. Cybernetics (Warszawa) vol. 2, 1973, No. 3/4.

Chapter V

1. Robert Feys. Modal logics. Paris 1965.
2. C.G. Hempel. A purely topological form of non-Aristotelian logic . Journal of Symbolic Logic vo. 2, 1937 No. 3.
3. Juliusz L. Kulikowski. Zastosowania nieklasycznych modeli logicznych w badaniach operacyjnych (in Polish: An application of non--classical logical models to operations research). In "Współczesne problemy zarządzania" (Actual problems of management). PWN, Warszawa 1974.
4. Helena Rasiowa. An algebraic approach to non-classical logics. North-Holland,PWN, Amsterdam, Warsaw 1974.
5. Ch. A. Vessel. O topologicheskoj logike (in Russian: About the topological logics). Nauka, Moscow 1970.

Chapter VI

1. Yu.I. Klykov. Situacjonnoe upravlenye bolshimi sistemami (in Russian: An situational control in large-scale systems).Energia, Moscow 1974.
2. Maria Nowakowska. Teoria działania. Próba formalizacji (in Polish: Theory of action. An attempt to formalization).PWN, Warszawa 1973.

GENERAL DYNAMICAL SYSTEMS: CONSTRUCTION AND REALIZATION

Franz R. Pichler
Lehrkanzel fur Systemtheorie
Johannes Kepler-University Linz
A-4045 Linz, Austria

(The author is presently with the School of Advanced Technology, State University of New York at Binghamton, Binghamton, New York 13901,U.S.A.)

ABSTRACT

The paper introduces some ideas of general systems theory which successfully establish a set theoretical and algebraic framework for construction and realization of dynamical systems. In chapter 4, these concepts are used to explain the linear-realization theory of Kalman for the discrete-time case. Chapter 5 is devoted to general ideas on the state concept and dynamical generation of processes.

1. Introductory Remarks

The aim of the paper is to introduce the concepts of construction and realization of dynamical systems in a very general way. We cannot expect to get exciting results here. Our considerations will be more or less well known to the specialists, but the uniformity of our representation can help to get a proper survey of the subject. We will not attempt to interpret the results on models of real life phenomena in this paper. We will, therefore, assume that the reader has some maturity in the field of dynamical systems and their applications in science and engineering. Furthermore, we shall not always give the proper citation to those scientists who first discovered the techniques which we are discussing. While the following is by no means a complete list, we feel it necessary to mention the contributions of Salovaara [1], Blomberg [2], Mesarovic [3], Macko [4], Windeknecht [5], Klir [6], Kalman [7] and Zadeh [8]. The "philosophy" used here is also used in my textbook on "Mathematische Systemtheorie" [9].

2. Construction of general dynamical systems

By construction, we mean the process of putting parts together in order to get a whole. In our case, this is a general dynamical system (gds). Since we shall stay on an abstract level, construction in the foregoing sense is in some respects identical to the process of definition. The "parts" from which we construct a general dynamical system are:

(1) two sets A and B, the sets of <u>input-states</u> and <u>output-states</u>, respectively.

(2) a set T together with a complete order $\leq$ and a <u>starting-time</u> t_0, which is the minimal element of T. We refer to T as the <u>time set</u>.

(3) a set X of functions of the kind $x:T\to A$. We call X the <u>input-set</u>; each function $x\varepsilon X$ is called an <u>input-function</u>.

(4) a set Q_0, the set of <u>initial-states</u>.

(5) a set Q with $Q_0 \subset Q$, the <u>state set</u>.

(6) a functional relation ϕ from $T\times Q_0\times X$ into Q, which we call the <u>state-relation</u>. $(t,q_0,x)\phi q$ will be referred to as "state q is reached by ϕ with x from q_0 at time t." The state-relation ϕ is assumed to have a "causality" property as follows:
Let $(t,q_0,x)\phi q$ and $(t,\overline{q}_0,\overline{x})\phi q$ for some $q_0,\overline{q}_0\varepsilon Q_0, q\varepsilon Q$, $t\varepsilon T$ and $x,\overline{x}\varepsilon X$. If $x|[t,t')=\overline{x}|[t,t')$ for some $t'\varepsilon T$ with $t\leq t'$ and

$(t',q_0,x)\phi q'$ and $(t',q_0,\overline{x})\phi\overline{q}'$ then $q'=\overline{q}'$. (Note: $[t,t')$ denotes as usual the interval of T which is given by $[t,t')$: $=\{t:t\varepsilon T\&t\leq t<t'\}$. $x|[t,t')$ is the restriction of x on $[t,t')$.)

(7) a functional relation β from $T\times Q\times A$ into B, the output-relation. $(t,q,a)\beta b$ has the meaning of "the pair (q,a), consisting of the state q and the input-state a determines at time t the output-state b."

We will now construct a dynamical system (with input and output) by the parts (1)-(7). The whole construction process is done by the definition and the subsequent interpretation:

Definition 1: The general dynamical system constructed by the parts (1)-(7) is given by the list $(A,B,X,Y,Q_0,Q,\phi,\beta,T,\leq,t_0)$ where Y denotes the set of all functional relations y from T to B which are determined by ϕ and β by $y=\{(t,b):(t,q_0,x)\phi q\&(t,q,x(t))\beta b\}$. We call y the output related to q and x. In short, we will denote a general dynamical system also by the pair (ϕ,β).

The interpretation of (ϕ,β) as a "machine" is as follows: If we have at the starting-time t_0 given the state q_0 and the input-function x is applied to (ϕ,β), then we reach at an arbitrary time t the state q for which $(t,q_0,x)\phi q$ is valid. In case of $(t,q,x(t))\beta b$, the output-state b will appear.

The causality property of ϕ which we assumed allows us to substitute ϕ by a functional relation ξ from $T\times T\times Q\times X$ into Q which is defined in the following way:

$$\xi:=\{(t',t,q,x)),q'):(\exists q_0\varepsilon Q_0)(t,q_0,x)\phi q\&(t',q_0,x)\phi q'\}$$

We interpret ξ as follows: In case of the validity of $(t',t,q,x)\xi q'$ we say "the state q at time t is mapped by the input x at time t' to the state q'." It is easily observed that the functional relation ξ has the following semi-group property:

In case of $(t',t,q,x)\xi q'$ and $(t'',t,q,x)\xi q''$ and $(t',t'',q'',x)\xi\overline{q}'$ where $t\leq t''\leq t'$ we have $q'=\overline{q}'$. The relation ξ is called the state-transition relation of the gds (ϕ,β). The state-transition relation ξ obviously implies the state relation ϕ. We can also describe a gds by the list $(A,B,X,Y,Q_0,Q,\xi,\beta,T,\leq,t_0)$ where we replaced ϕ by ξ. If this is done we also use the symbol (ξ,β) for it.

State transitions can be seen as part of the internal description of a system as given by a gds (ξ,β). Opposed to this stands the external description, the description of how the system looks from the outside.

The external description $S(\xi,\beta)$ of a gds (ξ,β) is defined by:

$$S(\xi,\beta):=\{(x,y):(\exists q_0\varepsilon Q)\ (\forall t\varepsilon T)(t,q_0,x)\phi q\ \&\ (t,q,x(t))\beta y(t)\}.$$

$S(\xi,\beta)$ is the set of all pairs (x,y) consisting of an input-function x together with the functional relation $y\subset T\times B$, the related output. Observe that y is in general not a function. In many cases it is possible to assume that the state relation ϕ and the output relation β are functions. Then ξ is also a function as is each output y.

3. Realization of general dynamical systems

Let $\bar{T}$ denote a completely ordered set and $t\varepsilon\bar{T}$. The intervals (t,∞), $[t,\infty)$, (∞,t) and $(\infty,t]$ are defined by:

$$(t,\infty):=\{\bar{t}:\bar{t}\varepsilon\bar{T}\&t<\bar{t}\},$$

$$[t,\infty):=(t,\infty)\cup\{t\},$$

$$(\infty,t):=\{\bar{t}:\bar{t}\varepsilon T\&\bar{t}<t\},$$

$$(\infty,t]:=(\infty,t)\cup\{t\}.$$

For a given $t_0\varepsilon\bar{T}$ we denote with T the interval $T:=[t_0,\infty)$. Let U and Y denote a set of functions of the kind $x:\bar{T}\to A$ and $y:T\to B$, respectively, and let f be an onto function $f:U\to Y$. We call $\bar{T}$ the <u>overall time-set</u>, t_0 again the <u>starting-time</u> of the output observations, T the (observed) <u>time-set</u>, U the <u>input-set</u>, Y the <u>output-set</u> and f the (observed) <u>process</u>. We can interpret f as an abstract result of a stimulus-response experiment, where the responses are observed from time t_0 on.

<u>Definition 2</u>: The process $f:U\to Y$ is called <u>nonanticipatory</u> if for all $u,\bar{u}\varepsilon U$ and for all $t\varepsilon T$ we have:

$$u|(\infty,t]=\bar{u}|(\infty,t]\Rightarrow f(u)(t)=f(\bar{u})(t).$$

Observe that in case of a nonanticipatory process $f:U\to Y$ and $u|(\infty,t]=\bar{u}|(\infty,t]$ we have also $f(u)|[t_0,t]=f(\bar{u})|[t_0,t]$. Let X denote the set of all functions which are restrictions of U onto T, that is $X:=\{x:(\exists u\varepsilon U)x=u|T\}$. Let S denote the relation from X to Y generated by $f:U\to Y$; $S:=\{(x,y):(\exists u\varepsilon U)x=u|T\&y\varepsilon Y\&y=f(u)\}$.

We call S the <u>general time system</u> (for the starting-time t_0) which is implied by f. Observe that S is in general not functional. We can interpret S as "the input-output data observed from time t_0 on."

<u>Definition 3</u>: Let f denote a nonanticipatory process $f:U\to Y$. By a <u>realization</u> of f we understand a general dynamical system (ξ,β) for which we have $S=S(\xi,\beta)$.

We can say that a realization (ξ,β) of $f:U\to Y$ <u>generates</u> f from time t_0 on. A given realization (ξ,β) of $f:U\to Y$ will in general not be unique,

because there will be many ways to construct the states and the state-transitions. A natural approach is to construct the states of a realization (ξ,β) by the input-functions $u\varepsilon U$ of f. The following construction of a realization demonstrates this.
The sets A,B,X,Y,T, the ordering $\leq$ and the starting-time t_0 for (ξ,β) can be read from the given process $f:U\to Y$. We know all parts of (ξ,β) if we know in addition the state sets Q_0 and Q, the state relation ϕ (or equivalently the state-transition relation ξ) and the output relation β. At first we construct Q_0 and Q. For an arbitrary given $t\varepsilon T$, let Q(t) denote the set $Q(t):=\{q:(\exists u\varepsilon U)q=u|(\infty,t)\}$. We define Q_0, Q of (ξ,β) by

$$Q_0:=Q(t_0) \text{ and } Q:=\bigcup_{t\varepsilon T} Q(t).$$

The state relation ϕ of (ξ,β) we define by $\phi\subset(T\times Q_0\times X)\times Q$ and $(t,q,x)\phi q'$ if $qx|(\infty,t)=q'$. (Note: qx denotes the usual concatenation of q and x which is given by $qx:\overline{T}\to A$ with $qx(t)=q(t)$ if $t<t_0$ and $qx(t)=x(t)$ if $t_0\leq t$.
It is obvious that the relation ϕ is functional and has the "causality" property. The output-relation β of (ξ,β) can now be defined by $\beta\subset(T\times Q\times A)\times B$ with $(t,q,a)\beta b$ if there exists an input $u\varepsilon U$ such that $u|(\infty,t)=q$ and $u(t)=a$ and $f(u)(t)=b$. Let us prove that the output relation β is functional: If $(t,q,a)\beta b$ and $(t,q,a)\beta\overline{b}$ then there exists functions $u,u\varepsilon U$ such that $u|(\infty,t)=q=\overline{u}|(\infty,t)$ and $u(t)=a=\overline{u}(t)$. Since f is assumed to be nonanticipatory it follows that $b=f(u)(t)=f(\overline{u})(t)=\overline{b}$.

Since the transition relation ξ follows uniquely from ϕ this finishes our construction of (ξ,β) for the given process. Also, we can already argue that the dynamical system (ξ,β) is a realization of f. Let us, however, present a more formal argument. We must show that $S=S(\xi,\beta)$ holds. Let $(x,y)\varepsilon S$, then there exists an input function $u\varepsilon U$ with $x=u|T$ and $f(u)=y$. Let $q:=u|(\infty,t_0)$ and let for an arbitrary given $t\varepsilon T$ define q' by $q':=u|(\infty,t)$. Then for all $t\varepsilon T$ we have $(t,q,x)\phi q'$ and $(t,q',u(t))\beta f(u)(t)$. Therefore, we have $(x,y)\varepsilon S(\xi,\beta)$. Conversely, if $(x,y)\varepsilon S(\xi,\beta)$ then there exists an initial state $q\varepsilon Q_0$ such that for all $t\varepsilon T$ we have $(t,q,x)\phi qx|(\infty,t)$ and $(t,qx|(\infty,t),x(t))\beta y(t)$. By definition of β $y(t)=f(qx)(t)$. Therefore $(x,y)\varepsilon S$.
Each state of the realization (ξ,β) consists of the "t-history" of a single input function. To get a state set of smaller cardinality it is interesting to introduce the following equivalence relation on the initial state set Q_0.
Definition 4: The Nerode-equivalence $\equiv$ on Q_0 is defined by the relation $\equiv\subset Q_0\times Q_0$ with $q\equiv\overline{q}$ if for all $x\varepsilon X$ we have (i) $qx\varepsilon U$ is equivalent

to $\bar{q}x\varepsilon U$, and (ii) $f(qx)=f(\bar{q}x)$.

To state this definition more informally:
Two initial states $q,\bar{q}$ are said to be Nerode-equivalent if they accept the same set of input functions and produce for each of these input functions the same output. $\equiv$ is obviously an equivalence relation. Therefore, we can construct the quotient set $Q_0/\equiv$. Similarly, we could define in an analogous way for each $t\varepsilon T$ a "t-Nerode equivalence $\underset{t}{\equiv}$" and construct $Q(t)/\underset{t}{\equiv}$. It is then straightforward how to construct a quotient $(\bar{\bar{\xi}},\bar{\bar{\beta}})$ using the Nerode-equivalences $\equiv$ and $\underset{t}{\equiv}$. The following chapter shows this construction for the linear discrete time-invariant case.

4. Realization of linear systems

In science, as well as in engineering, there is interest in generating processes of the kind $f:U\to Y$ by special classes of dynamical systems (ξ,β). A class of often used systems are the linear systems [A,B,C,D] which are described in the continuous-time case by equations of the form

$$\dot{z}(t)=Az(t)+Bx(t) \quad \text{and}$$

$$y(t)=Cz(t)+Dx(t),$$

[Note: I hope that no one is worried that I do not write (F,G,H)!]

or in the discrete-time case by

$$z(t+1)=Az(t)+Bx(t)$$
$$y(t) \;=Cz(t)+Dx(t).$$

A question which immediately arises at this point is: Under what conditions on f is it possible to generate f (exactly) by such a system [A,B,C,D] Furthermore: If f fulfills such conditions, how can we construct the linear transformations A,B,C and D from f (linear realization problem)? The algebraic theory of linear systems, as developed by R. E. Kalman [7] answers such questions. Our goal here is to consider the linear realization problem from our point of view. For the sake of simplicity, we will concentrate on the discrete-time case. We cannot expect to get new results here. But our representation should help to contribute to an easier understanding of the existing theory.

From the knowledge of the fundamental behavior of systems of the kind [A,B,C,D] we know that we have, in any case, to assume that

(i) the process $f:U\to Y$ has to be nonanticipatory,

(ii) the input set X should be closed by concatenation, that is, for each $x,\bar{x}\varepsilon X$ and each $t\varepsilon T$ we should also have $x\underset{t}{\circ}\bar{x}\varepsilon X$, where $x\underset{t}{\circ}\bar{x}:T\to A$ with $x\underset{t}{\circ}\bar{x}(t')$ for $t'<t$ and $x\underset{t}{\circ}\bar{x}(t'):=\bar{x}(t)$ for $t\leq t'$,

(iii) the overall time-set $\bar{T}$ has to be isomorphic to the set Z of integers, on which we assume to have the usual addition $+$ and the usual ordering $\leq$. For the starting-time we assume $t_0:=0$. Without loss of generality we can set $\bar{T}:=Z$ and $T=N_0=\{0,1,2,3,\ldots\}$,

(iv) the process $f:U\to Y$ has to be <u>time</u>-<u>invariant</u>, that is:

(a) For all $u\varepsilon U$ and all $t\varepsilon T$ we should also have $u\to t\varepsilon U$ ($f\to t$ denotes in general the <u>t-time shift</u> of u which is defined by $u\to t(t'):=u(t+t')$ for all $t'\varepsilon T$).

(b) For all $u\varepsilon U$ and all $t\varepsilon T$ we should have $f(u\to t)=f(u)\to t$.

(v) the set A of input-states and also the set B of output-states should be finite dimensional linear spaces over the same field $\mathbb{K}$. Therefore, without loss of generality, we can assume that $A=\mathbb{K}^m$ and $B=\mathbb{K}^p$.

(vi) U and Y are assumed to be linear spaces where the algebraic structure is, as usual, induced from A and B respectively.

(vii) $f:U\to Y$ is assumed to be <u>linear</u>, that is for all $u,\bar{u}\varepsilon U$ and all $k\varepsilon\mathbb{K}$ we should have $f(u+\bar{u})=f(u)+f(\bar{u})$ and $f(ku)=kf(u)$.

We want to see now what consequences the assumptions (i) to (vii) have for the realizations (ξ,β) and $(\overset{\equiv}{\xi},\overset{\equiv}{\beta})$ of chapter 3. Since $qx=q0+0x$ and since f is by (vii) assumed to be linear, we have $f(qx)=f(q0+0x)$ $=f(q0)+f(0x)$ and consequently in case that $f(qx)=f(\bar{q}x)$ the equivalent equality $f(q0)=f(\bar{q}0)$ or by linearity $f((q-\bar{q})0)=0$. Since also $f(00)(0)$ $=0$ we have $(q-\bar{q})\equiv 0$ and consequently $[q]=q+[0]$, where $[0]$ denotes the subspace of Q_0 which is generated by $0\varepsilon Q_0$. Let $\overset{\equiv}{Q}_0$ denote the quotient set $\overset{\equiv}{Q}_0:=Q_0/\equiv$. From the foregoing consideration we see that we have $\overset{\equiv}{Q}_0=Q_0/[0]$, that is, that $\overset{\equiv}{Q}_0$ is identical to the quotient-space of Q_0 by $[0]$.

Since f is time-invariant we have $Q(t)\subseteq Q_0$ for all $t\varepsilon\mathbb{N}_0$ and therefore $Q=Q_0$. This suggests we can define $\overset{\equiv}{Q}:=\overset{\equiv}{Q}_0$. For $\phi\subset(\mathbb{N}_0\times\overset{\equiv}{Q}_0\times X)\times\overset{\equiv}{Q}_0$, we now have $(t,[q],x)\overset{\equiv}{\phi}[\bar{q}]\Leftrightarrow[\bar{q}]=[(qx\to t)\mid(\infty,0)]$. Since $qx\to 1=(q0\to 1)+(0x\to 1)$, we get for $t=1$ $[\bar{q}]=[(q0\to 1)\mid(\infty,t)]+[(0x\to 1)\mid(\infty,t)]$.
It is quite obvious that the first element $[(q0\to 1)\mid(\infty,t)]$ of this sum depends only on $[q]$. It represents the autonomous part of the "dynamic" of the system. We will describe this part of the dynamic by the operation $\alpha:\overset{\equiv}{Q}_0\to\overset{\equiv}{Q}_0$ with $[q]\mapsto\alpha([q]):=[(q0\to 1)\mid(\infty,t)]$. On the other

hand, the second element, $[(0x\to 1)\,|\,(\infty,t)]$ depends only on the value $x(t)$ of x at time t. It can therefore be described by an operation $[\;]:\mathbb{K}^m\to\overline{\overline{Q}}_0$ with $a\mapsto[a]:=[(0x\to 1)\,|\,(\infty,t)]$. Both α and $[\;]$ are linear. The "1-step" state transformation $(1,[q],x)\mapsto[\overline{q}]$ can now be represented in the form

$$[\overline{q}]=\alpha([q])+[x(0)].$$

For the output relation $\beta\subset(\mathbb{N}_0\times\overline{\overline{Q}}_0\times\mathbb{K}^m)\times\mathbb{K}^p$ we have $(t,[q],a)\beta b\iff(\exists x\varepsilon X)$ $x(t)=a$ and $f(qx)(t)=b$. Since f is assumed to be time-invariant β is also time-invariant because we have

$f(qx)(t)=f((qx\to t)\,|\,(\infty,t))(0)+f(0u)(0)$ where $u(0)=x(t)$.

Therefore, the output state b can be computed by an equation of the form

$$b=c([q])+d(a)$$

where the functions $c:\overline{\overline{Q}}_0\to\mathbb{K}^p$ and $d:\mathbb{K}^m\to\mathbb{K}^p$ are defined by

$c([\overline{q}]:=f(qx\to t)\,|\,(\infty,t))(0)$ and

$d(a):=f(0u)(0)$ **with $u(0)=a$ respectively.**

It is easy to observe that both c and d are also linear functions.

The form of the foregoing equations for state-transition and output-computation is already very similar to the equations of a system $[A,B,C,D]$. An isomorphism can be established if we are able to find a map $\gamma:\overline{\overline{Q}}_0\to\mathbb{K}^n$ where n is the dimension of the state space of $[A,B,C,D]$ which has the following two properties:

(1) γ is a vector space isomorphism between $\overline{\overline{Q}}_0$ and $\mathbb{K}^n$,

(2) γ makes the following diagram commutative

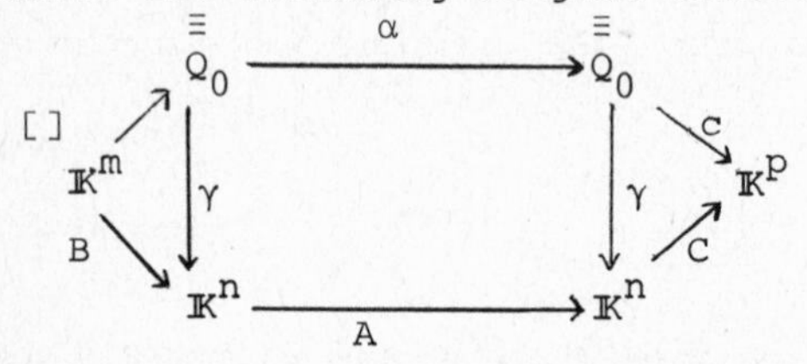

In the case that we have been successful in establishing an isomorphism $\gamma:\overline{\overline{Q}}_0\to\mathbb{K}^n$, the dynamical systems $(\overline{\overline{\xi}},\overline{\overline{\beta}})$ and $[A,B,C,D]$ will be isomorphic and the linear transformations A,B,C,D and $\alpha,[\;],c,d$ are related by

$$A=\gamma\alpha\gamma^{-1},\; B=\gamma[\;],\; C=c\gamma^{-1} \text{ and } D=d.$$

In order to find such a map γ we have to assume that $\overline{\overline{Q}}_0$ and α have certain additional properties. Kalman [10] first investigated that for $\overline{\overline{Q}}_0$ the structure of a finitely generated R-torsion module would fit this situation. Let us explore this investigation in the setting of our framework. The first step is to identify Q_0 as a $\mathbb{K}^m[[\zeta]]$-

module by the assignment $\nu:Q_0\to\mathbb{K}^m[[\zeta]]$ with $q\mapsto\nu(q):=\sum_{i=1}^{\infty} q(-i)\zeta^{i-1}$. Obviously ν establishes an isomorphism between Q_0 and the image set $\mathrm{Im}(\nu)\subset\mathbb{K}^m[[\zeta]]$ of ν. Let $[0(\zeta)]$ denote the subspace of $\mathbb{K}^m[[\zeta]]$ which corresponds to $[0]\subset Q_0$. Then $\mathrm{Im}(\nu)/[0(\zeta)]$ is isomorphic to $\overline{\overline{Q}}_0=Q_0/[0]$ by the isomorphism $\nu_0:=\nu/[0]$. With M we denote in the following the quotient space $M:=\mathrm{Im}(\nu)/[0(\zeta)]$. We want to investigate the operation $\alpha(\zeta):M\to M$ which corresponds to $\alpha:\overline{\overline{Q}}_0\to\overline{\overline{Q}}_0$. It is easy to observe that we have $\nu((q0\to1)|(\infty,0))=\zeta\nu(q)$ and also $\nu_0([(q0\to1)|(\infty,0)])=\zeta\nu_0([q])$. This result gives us the motivation to extend the scalar multiplication on $\mathbb{K}^m[[\zeta]]$ and $\mathbb{K}^m[[\zeta]]/[0(\zeta)]$ from the field $\mathbb{K}$ to the ring $\mathbb{K}[\zeta]$ of formal polynomials in the following way:

For $p(\zeta)\in\mathbb{K}[\zeta]$ and $q(\zeta)\in\mathbb{K}^m[[\zeta]]$ the product $p(\zeta)q(\zeta)$ is defined by the multiplication of the components of $q(\zeta)$ by $p(\zeta)$. (Observe that $\mathbb{K}^m[[\zeta]]\cong(\mathbb{K}[[\zeta]])^m$, therefore, the components of $q(\zeta)$ can be identified as power-series with scalar coefficients). Then $\mathbb{K}^m[[\zeta]]$ and, in the usual way, also $\mathbb{K}^m[[\zeta]]/[0(\zeta)]$ and the subspace M of it, receives the structure of a $\mathbb{K}[\zeta]$-module.

In the context of $\mathbb{K}[\zeta]$-modules the operation $\alpha(\zeta):M\to M$ now has a definite and simple meaning: It is the special $\mathbb{K}[\zeta]$-module homomorphism on M which is given by the assignment $[q(\zeta)]\mapsto\alpha(\zeta)[q(\zeta)]=\zeta[q(\zeta)]$. Our problem of finding an isomorphism $\gamma:\overline{\overline{Q}}_0\to\mathbb{K}^n$ has now the following equivalent form:

Given a $\mathbb{K}[\zeta]$-module homomorphism $\alpha(\zeta):M\to M$, find a vector space isomorphism $\gamma:M\to\mathbb{K}^n$ such that the following diagram is commutative:

$$\begin{array}{ccc} M & \xrightarrow{\alpha(\zeta)} & M \\ \downarrow\gamma & & \downarrow\gamma \\ \mathbb{K}^n & \xrightarrow{A} & \mathbb{K}^n \end{array}$$

This problem is well known in Linear Algebra. There is shown that such an isomorphism can be established if and only if M is a finitely generated torsion-module. We cannot expect that the $\mathbb{K}[\zeta]$-module M derived from $f:U\to Y$ has, in general, these properties. It is, therefore, necessary to find out under what conditions on the process $f:U\to Y$ these properties of M are guaranteed.

Definition 5: A linear process $f:U\to Y$ is called past-finite if each Nerode-equivalence class $[q]$ contains at least one element $\hat{q}$ with finite support. That means that the set $\mathrm{supp}(\hat{q}):=\{t:t\in(\infty,0)\,\&\,q(t)\neq0\}$ is finite.

Let us now assume that the discrete-time process $f:U\to Y$ is past-finite and let $\hat{Q}_0$ denote the set of all functions $q\varepsilon Q_0$ with finite support. For each $\hat{q}\varepsilon Q_0$ let $\hat{t}$ denote the minimal element of supp$(\hat{q})$. Then $\nu(\hat{q})$ is a polynomial of degree $-\hat{t}$. Each equivalence class $[q]\varepsilon Q_0$ can now be generated by a function $\hat{q}\varepsilon Q_0$ with finite support or, to state it equivalently, each equivalence class $[\nu(q)]=[q(\zeta)]$ of the $\mathbb{K}[\zeta]$-module M can now be generated by polynomials $\hat{q}(\zeta)\varepsilon\mathbb{K}[\zeta]$. Therefore, in the case of a past-finite process $f:U\to X$ the $\mathbb{K}[[\zeta]]$-module M (the state space of the realization $(\underline{\underline{\xi}},\underline{\underline{\beta}})$ of f) is isomorphic to a submodule of $\mathbb{K}^m[\zeta]/[0(\zeta)]$. Since $\mathbb{K}^m[\zeta]$ can be finitely generated by the constant polynomials $e_1(\zeta):=(1,0,...,0)$, $e_2(\zeta):=(0,1,...,0),...,e_m(\zeta):=(0,0,...,1)$ so can now the module $M\subset\mathbb{K}^m[\zeta]/[0(\zeta)]$.

It is interesting to derive this "finiteness" property of the module M, which we found to be necessary for the solution of the realization problem, from a more general property of the process $f:U\to Y$. For a given $t\varepsilon\overline{T}$ and $u\varepsilon U$ let $U\underset{t}{\geq}u$ denote the set

$$U\underset{t}{\geq}u:=\{\overline{u}:\overline{u}\varepsilon U \;\&\; \overline{u}|[t,\infty)=u\}.$$

We have then the following definition:

<u>Definition 6</u>: A process $f:U\to Y$ is said to have <u>finite memory</u> if for any $u\varepsilon U$ there exists a $t\varepsilon(\infty,t_0)$ such that $U\underset{t}{\geq}u\subset[u]$.

In the case of a process f with finite memory, there exists for each $u\varepsilon U$ a $\hat{t}\varepsilon(\infty,t_0)$ such that the values $u(t)$ of u for $t<\hat{t}$ have no influence on the output $f(u)$. It is clear that a discrete-time linear process $f:U\to Y$ which has finite memory is past-finite in the strong sense, that for <u>each</u> function $q\varepsilon Q$ there exists a $\hat{t}$ such that $[0\overset{\circ}{_{\hat{t}}}q]=[q]$ ($0\overset{\circ}{_{t}}q$ denotes here the concatenation $0\overset{\circ}{_{\hat{t}}}q\varepsilon Q$ with $0\overset{\circ}{_{\hat{t}}}q(t)=0$ for $t<\hat{t}$ and $0\overset{\circ}{_{\hat{t}}}q(t)=q(t)$ for $t\geq\hat{t}$).

For the following we will assume that our process $f:U\to Y$ is past-finite. Then, we know already, the $\mathbb{K}[\zeta]$-module M is a submodule of $\mathbb{K}^m[\zeta]/[0(\zeta)]$ and therefore finitely generated. We explore the properties which we have to assume on the process $f:U\to Y$ further in order that M becomes a torsion-module.

By definition M is a torsion module if for any $[\hat{q}(\zeta)]\varepsilon M$ there exists a polynomial $p(\zeta)\varepsilon\mathbb{K}[\zeta]$ such that $p(\zeta)[\hat{q}(\zeta)]=[0(\zeta)]$ or equivalently $p(\zeta)\hat{q}(\zeta)\varepsilon[0(\zeta)]$. It is sufficient to show that for each "generator" $[e_i(\zeta)]$, $i\varepsilon\{1,2,...,m\}$, such a polynomial $p_i(\zeta)$ exists:

Let $\{m_1,m_2,...,m_s\}$ denote a smallest subset of the set $\{[e,(\zeta)],[e_2(\zeta)],...,[e_m(\zeta)]\}$ which generates M. Let

$p_1(\zeta), p_2(\zeta), \ldots, p_s(\zeta)$ denote polynomials (of minimal degree) with the property that $p_i(\zeta)m_i = [0_s(\zeta)]$ for $i=1,2,\ldots,s$. The polynomial $p(\zeta)$ may be defined by $p(\zeta) := \prod_{i=1}^{s} p_i(\zeta)$. Then, for any $m \in M$ we have $p(\zeta)m = [0(\zeta)]$ which can be seen as follows: If $m = \sum_{i=1}^{s} m_i(\zeta)m_i$, then

$$p(\zeta)m = \left(\prod_{j=1}^{s} p_j(\zeta)\right)\left(\sum_{i=1}^{s} m_i(\zeta)m_i\right)$$

$$= \sum_{i=1}^{s} \left(\prod_{\substack{j=1 \\ j\neq i}}^{s} p_j(\zeta)\right) m_i(\zeta)p_i(\zeta)m_i$$

$$= \sum_{i=1}^{s} \left(\prod_{\substack{j=1 \\ j\neq i}}^{s} p_j(\zeta)\right) [0(\zeta)]$$

$$= [0(\zeta)].$$

The property $p_i(\zeta)m_i = [0(\zeta)]$ for $i=1,2,\ldots,s$ which guarantees that M becomes a torsion module, has an intrinsical system-theoretical interpretation:

Let $p_i(\zeta) = a_{i0} + a_{i1}\zeta + \ldots + a_{in_i-1}\zeta^{n_i-1} + \zeta^{n_i}$ for $i=1,2,\ldots,s$.

We have then

$$p_i(\zeta)m_i = a_{i0}m_i + a_{i1}\zeta m_i + \ldots + a_{in_i-1}\zeta^{n_i-1}m_i + \zeta^{n_i}m_i = [0(\zeta)]$$

or equivalently

$$\zeta^{n_i}m_i = -a_{i0}m_i - a_{i1}\zeta m_i - \ldots - a_{in_i-1}\zeta^{n_i-1}m_i.$$

Since $m_i = [e_{k_i}(\zeta)]$ for a certain $k_i \in \{1,2,\ldots,n\}$ we see that this property of the generators m_i has in the context of the process $f: U \to Y$ the following meaning: For all states $q_i \in Q_0$ which correspond to generators m_i in M, that is $\nu[q_i] = m_i$, there must exist a time $n_i \in \{0,1,2,\ldots\}$ such that $[(q_i 0 \to n_i) | (\infty, 0)]$ is linearly dependent from the set $\{[q_i], [(q_i 0 \to 1) | (\infty,0)], \ldots, [(q_i 0 \to n_i - 1) | (\infty, 0)]\}$. From the foregoing disucssion, we see that the "unit-pulse"-inputs $e_1, e_2, \ldots, e_m \in U$ as given by $e_1(-1) = (1,0,0,\ldots,0,0)$, $e_2(-1) = (0,1,\ldots,0,0), \ldots,$ $e_m(-1) = (0,0,\ldots,0,1)$ and $e_1(t) = e_2(t) = \ldots = e_m(t) = (0,0,\ldots,0,0)$ for all other $t \in \overline{T}$, can be used for this experiment. This motivates the following definition

Definition 7: A linear discrete time process $f: U \to Y$ (which has, in addition, all properties defined so far) will be called finite-dimensional realizable if each input e_i generates by translation a finite dimensional linear subspace E_i of $\bar{\bar{Q}}_0$. From the foregoing discussion it is clear that the property "finite-dimensional realizable" is a sufficient condition for the process $f:U\to Y$ that the related $\mathbb{K}[\zeta]$-module M is a torsion-module. Let us comment on the "experiments" to determine this property on a given process. From the definition of $\equiv$ on Q_0 and since f is linear, we know that there is a 1-1 correspondence between the elements of $\overset{\equiv}{Q}_o$ and Y by the assignment $[q]\mapsto f(g0)$. (Assume that $f(g0)=f(\bar{g}0)$. Then by linearity we have for any $x\varepsilon X$ $f(qx)=f(q0+0x)=f(q0)+f(0x)=f(\bar{q}0)+f(0x)=f(\bar{q}x)$, therefore, $q\equiv\bar{q}$).

Observe further that this assignment is also linear. Therefore, it extends to an isomorphism between $\overset{\equiv}{Q}_0$ and the set

$\{y:y\varepsilon Y \text{ and } (\exists q\varepsilon Q_0) f(q0)=y\}\subset Y.$

Furthermore, by time-invariance of f, we also have $f(q0)\to t=f(q0\to t)$.The states $[e_i|(\infty,0)],[(e_i\to 1)|(\infty,0)],\ldots,[(e_i\to t)|(\infty,0)],\ldots$have therefore in Y the counterparts $f(e_i),f(e_i\to 1),\ldots,f(e_i\to t),\ldots$. These outputs can be observed by "experiments." A more natural approach to determine if a process is "finite-dimensional realizable" is therefore to prove that for all $i=1,2,\ldots,m$ there exists a n_i such that for $t\geq n_i$ the outputs depend linearly on the outputs $f(e_i),f(e_i\to 1),\ldots,f(e_i\to n_i-1)$. It is convenient to assure this property by demanding that the rank of all matrices E_i which are given for $i=1,2,\ldots,m$ by the infinite Hankel-matrices

$$E_i := \left[\begin{array}{c|c|c}
f(e_i)(0) & f(e_i)(1) & f(e_i)(2)\ldots \\ \hline
f(e_i)(1) & f(e_i)(2) & f(e_i)(3)\ldots \\ \hline
f(e_i)(2) & f(e_i)(3) & f(e_i)(4)\ldots \\ \hline
\vdots & \vdots & \vdots
\end{array}\right]$$

is finite. An algorithm to get a minimal set $\{m_1,m_2,\ldots,m_s\}$ of generators of M and the corresponding minimal dimensions $n_1,n_2,\ldots,n_s$ can be easily established (n_1:=rank E_1; n_k:=rank E_{i_k} if rank $[E_{i_1}|E_{i_2}|\ldots E_{i_{k-1}}]<$rank $[E_{i_1}|E_{i_2}|\ldots|E_{i_{k-1}}|E_{i_k}]$ where $i_1:=1$ and $i_k<i_{k+1}\leq m$ for $k=2,3,\ldots$). The dimension n of the state space $\mathbb{K}^n$ of [A,B,C,D] is then given by $n=n_1+n_2+\ldots+n_s$.

We have now established all properties of $f:U\to Y$ in order that the corresponding general dynamical system $(\overset{\equiv}{\zeta},\overset{\equiv}{\beta})$ can be isomorphically simulated by a discrete-time linear system [A,B,C,D]. We do not want to go further into computational aspects of such a realization. The reader who is interested in such questions is advised to consult the existing literature, for instance Kalman [7] or Rissanen [11].

5. Realization of general dynamical processes by sub-processes.

In this last chapter, we present some fundamental ideas on the realization problem of general dynamical systems. We will not go too deeply into detail here. The main point is to interest the reader in the approaches of general dynamical systems realization as suggested by the work of Klir [6], Windeknecht [5], and Zadeh [9]. For a more comprehensive discussion of the concepts presented in this chapter, we refer the reader to Pichler [9].

Let $\overline{T}$ denote, as before, a set with a complete order $\leq$ on it, M a set and P a set of functions $p:\overline{T}\to M$. We refer to P now as an M-process. $\overline{T}$ is the overall time-set, M the value-set of the M-process P. Any "part" of P could be called a state of P. The parts which are usually used as states in system theory are either restrictions of P concerning the time-set $\overline{T}$ or the value-set M or subsets of P. To start with, we take the latter concept of states of P and present the following definition:

Definition 8: Any subset q of the process P will be called a state of P.

For a fixed time-point $t_0 \varepsilon \overline{T}$, we define T as before by $T:=[t_0,\infty)$. We call T the (observed) time-set of P. For a given set Q of states of P each function $z:T\to Q$ will be called a trajectory of P.

Definition 9: A state-description of P is given by any pair (P,Z) where Z is a set of trajectories which cover at every time point $t\varepsilon T$ the process P, that is for all $t\varepsilon T$ we must have $\bigcup_{z\varepsilon Z} z(t)=P$.

Z will be called the state-process of (P,Z). The state-set Q, which is associated with (P,Z), is given by the set $Q:=\{q: (\exists z\varepsilon Z)(\exists t\varepsilon T) q=z(t)\}$. The set $Q(t):=\{q:(\exists z\varepsilon Z) q=z(t)\}$ will be called the set of states of (P,Z) which are reached at time t.

In the following, we try to develop some properties of state-descriptions so that the state process can be generated by a general dynamical system.

Definition: A state-description (P,Z) will be called concatenation-closed if for any time $t\varepsilon T$ and any state $q\varepsilon Q(t)$ we have $q\overset{o}{_t}q=q$ where $q\overset{o}{_t}q$ is defined by:

$q\overset{o}{_t}q:=\{p:(\exists\overline{p},\overline{\overline{p}}\varepsilon q)\, p|(\infty,t)=\overline{p} \text{ and } p|[t,\infty)=\overline{\overline{p}}\}$.

In the case that a state-description (P,Z) is concatenation-closed each "history" $p|(\infty,t)$ of the state $q\varepsilon Q(t)$ where $p\varepsilon q$, can be combined with any "prediction" $\bar{p}|[t,\infty)$ of q and vice versa.

Definition 11: A state-description (P,Z) is called past-extending if for each time $t\varepsilon T$ and each $q\varepsilon Q(t)$ the knowledge of $p\varepsilon q$ and $p\underset{t}{\leq} q:=\{\bar{p}:\bar{p}\varepsilon q\&\bar{p}|(\infty,t)=p|(\infty,t)\}$ determines uniquely the corresponding state q.

In the case of a past extending state description, a state $q\varepsilon Q(t)$ is already determined if we know a "history" $p|(\infty,t)$ of it and if in addition we know the set $p\underset{t}{\leq} q$ which gives the predictions of q which are related to p.

Definition 12: We call (P,Z) weakly-transitional, if for any set $z\varepsilon Z$ and $t,t'\varepsilon T$ with $t\leq t'$ the relation $\phi(t,t')$ from $P\times Q$ to $P(Q)$, the power set of Q, which is given by $(p,q)\phi(t,t')U:\Longleftrightarrow(\exists z\varepsilon Z)q=z(t)\&p\varepsilon z(t)\&U=p\underset{t}{<}z(t')$ is functional.

It is quite obvious that a state-description which is concatenation-closed, past-extending and weakly-transitional has all the properties which one expects for a dynamical system. We define therefore:

Definition 10: A state-description (P,Z) which is concatenation-closed, past-extending and weakly-transitional is called dynamical.

The construction of a state-description (P,Z) for a given M-process P is already a part of the model building process in the sense that we should be able to interpret (P,Z) on a model of a real object. The states of (P,Z) should reflect important properties of the model so that they help to create a problem-solution relation. The following general examples of state-descriptions (P,Z) are mainly for the purpose of demonstration and to assist in classification.

Let P denote a given M-process. We define then:

(0) the finest state-description (P,Z_0) of P by
$Z_0:=\{z:(\exists p\varepsilon P)(\forall t\varepsilon T)z(t)=\{P\}\}$,

(1) the coarsest state-description (P,Z_1) of P by
$Z_1:=\{z:(\forall t\varepsilon T)z(t)=P\}$,

(~) the natural state-description $(P,\tilde{Z})$ of P by
$\tilde{Z}:=\{z:(\exists p\varepsilon P)(\forall t\varepsilon T)z(t)=p\underset{t}{<}P\}$,

(^) the weakly autonomous state-description of $(P,\hat{Z})$ of P by
$\hat{Z}:=\{z:(\exists p\varepsilon P)(\forall t\varepsilon T)z(t)=P\underset{t}{\geq}p\}$,

($\equiv$) the <u>state</u>-<u>reduced</u> state description $(P,\overset{\equiv}{Z})$ of P by

$$\overset{\equiv}{Z} :=\{z:(\exists p\varepsilon P)(\forall t\varepsilon T)\, z(t)=\{\bar{p}:\bar{p}\varepsilon P \,\&\, (\bar{p} \underset{t}{\leq} P)\,|[t,\infty)=(p \underset{t}{\leq} P\ |[t,\infty)\}\}$$

$(U|[t,\infty)$ denotes the restrictions of all functions $u\varepsilon U$ to $[t,\infty)$).
The following theorem is then easy to observe:

<u>Theorem</u>: For each M-process P, the state descriptions (0)-($\equiv$), are dynamical.

At this point let us introduce two further properties of state description (P,Z) which are of practical use.
<u>Definition 13</u>: A state description (P,Z) is called <u>constructable</u> (<u>strongly</u> <u>constructable</u>) if for all $t\varepsilon T$ and $q\varepsilon Q(t)$ by $q|(\infty,t)$ (by $p|(\infty,t)$ with $p\varepsilon q$) the state q is uniquely determined. (P,Z) is called <u>reconstructable</u> (<u>strongly</u> <u>reconstructable</u>) if for all $t\varepsilon T$ and $q\varepsilon Q(t)$ by $q|[t,\infty)$ (by $p|[t,\infty)$ with $p\varepsilon q$) the state q is uniquely determined.
The following table (the proof of which is easily composed) shows to what extent the foregoing examples (P,Z_0) to $(P,\overset{\equiv}{Z})$ of state-descriptions are constructable and reconstructable respectively.

	constr.	str. constr.	reconstr.	str. reconstr.
(P,Z_0)				
(P,Z_1)	X	X	X	X
$(P,\widetilde{Z})$	X	X		
$(P,\hat{Z})$			X	X
$(P,\overset{\equiv}{Z})$	X	X	X	

We see that only (P,Z_1) is strongly constructable <u>and</u> strongly reconstructable. This property, of being strongly constructable and strongly reconstructable, is usually very welcome in model-building. However, it offers here little help since (P,Z_1) is trivial. We can furthermore read from the table that $(P,\overset{\equiv}{Z})$ is strongly constructable and reconstructable. But in general it is not strongly reconstructable. Let us therefore introduce a property of the M-process P which assures that $(P,\overset{\equiv}{Z})$ also becomes strongly reconstructable.
<u>Definition 14</u>: An M-process P is called <u>homogeneous</u> if for all $p\varepsilon P$ and all $t\varepsilon T$ we have $(P \underset{t}{\geq} p) \underset{t}{\rho} (p \underset{t}{\leq} P) \subset P$.

It is not too difficult to observe that in the case of an homogeneous M-process P it is true that $(P,\overset{\equiv}{Z})$ is strongly reconstructable.

Finally, it should be mentioned that in the case that the M-process P is an input-output-process (that is if M is the cartesian product $M=A\times B$ of a set A of the input-states and a set B of output-states) which is nonanticipatory, a dynamical state description (P,Z) implies a general dynamical system (ξ,β) as introduced in chapter 2.

Acknowledgement

I am indebted to Michael McGoff from the School of Advanced Technology for assisting me in the translation of my paper from the German and to Helen Tarbell who typed the manuscript.

....

References

[1] Salovaara, S.: On Set Theoretical Foundations of System Theory. Acta Polytechnica Scandinavia, Mathematics and Computing Machinery Series No. 15. Helsinki 1967.

[2] Blomberg, H.: On set theorectical and algebraic systems theory-Part 1. In: Advances in Cybernectics and Systems Research (ed. F. Pichler and R. Trappl). Proceedings of the European Meeting. Vienna 1972. Hemisphere Publishing Corporation, 1025 Vermont Avenue, N. W., Washington, D. C. 20005. 1972.

[3] Mesarovic, M.D. and Y. Takahara: General Systems Theory: Mathematical Foundations. Academic Press., New York 1975.

[4] Macko, D.: Natural States and Past-Determinacy in General Time Systems. In: Information Sciences, Vol. 3, No. 1 (1971) pp. 1-16

[5] Windeknecht, T.G.: General Dynamical Processes II. The State Space Approach Bookmanuscript, received 1972.

[6] Klir,G.: An Approach to General Systems Theory. Van Nostrand Reinhold Company. New York 1969.

[7] Kalman, R.E., P.L. Falb, M.A. Arbib: Topics in Mathematical System Theory. Mc-Graw-Hill Book Company. New York 1969. Chapter 10.

[8] Zadeh, L.A.: The concept of state in System Theory. In: Views on General Systems Theory (ed. M.D. Mesarovic) John Wiley. New York 1965. pp. 3-42.

[9] Pichler, F.: Mathematische Systemtheorie: Dynamische Konstruktionen Walter de Gruyter. Berlin-New York 1975.

[10] Kalman, R.E.: Algebraic structure of linear dynamical systems. I. The module of Σ. Proc. Nat. Acad. Sci.(USA). Vol 54, pp. 1503-1508.

[11] Rissanen, J.: Recursive Identification of Linear Systems. SIAM J. Control Vol. 9, No. 3, 1971, pp. 420-430.

Vol. 59: J. A. Hanson, Growth in Open Economies. V, 128 pages. 1971.

Vol. 60: H. Hauptmann, Schätz- und Kontrolltheorie in stetigen dynamischen Wirtschaftsmodellen. V, 104 Seiten. 1971.

Vol. 61: K. H. F. Meyer, Wartesysteme mit variabler Bearbeitungsrate. VII, 314 Seiten. 1971.

Vol. 62: W. Krelle u. G. Gabisch unter Mitarbeit von J. Burgermeister, Wachstumstheorie. VII, 223 Seiten. 1972.

Vol. 63: J. Kohlas, Monte Carlo Simulation im Operations Research. VI, 162 Seiten. 1972.

Vol. 64: P. Gessner u. K. Spremann, Optimierung in Funktionenräumen. IV, 120 Seiten. 1972.

Vol. 65: W. Everling, Exercises in Computer Systems Analysis. VIII, 184 pages. 1972.

Vol. 66: F. Bauer, P. Garabedian and D. Korn, Supercritical Wing Sections. V, 211 pages. 1972.

Vol. 67: I. V. Girsanov, Lectures on Mathematical Theory of Extremum Problems. V, 136 pages. 1972.

Vol. 68: J. Loeckx, Computability and Decidability. An Introduction for Students of Computer Science. VI, 76 pages. 1972.

Vol. 69: S. Ashour, Sequencing Theory. V, 133 pages. 1972.

Vol. 70: J. P. Brown, The Economic Effects of Floods. Investigations of a Stochastic Model of Rational Investment. Behavior in the Face of Floods. V, 87 pages. 1972.

Vol. 71: R. Henn und O. Opitz, Konsum- und Produktionstheorie II. V, 134 Seiten. 1972.

Vol. 72: T. P. Bagchi and J. G. C. Templeton, Numerical Methods in Markov Chains and Bulk Queues. XI, 89 pages. 1972.

Vol. 73: H. Kiendl, Suboptimale Regler mit abschnittweise linearer Struktur. VI, 146 Seiten. 1972.

Vol. 74: F. Pokropp, Aggregation von Produktionsfunktionen. VI, 107 Seiten. 1972.

Vol. 75: GI-Gesellschaft für Informatik e.V. Bericht Nr. 3. 1. Fachtagung über Programmiersprachen · München, 9.–11. März 1971. Herausgegeben im Auftrag der Gesellschaft für Informatik von H. Langmaack und M. Paul. VII, 280 Seiten. 1972.

Vol. 76: G. Fandel, Optimale Entscheidung bei mehrfacher Zielsetzung. II, 121 Seiten. 1972.

Vol. 77: A. Auslender, Problèmes de Minimax via l'Analyse Convexe et les Inégalités Variationelles: Théorie et Algorithmes. VII, 132 pages. 1972.

Vol. 78: GI-Gesellschaft für Informatik e.V. 2. Jahrestagung, Karlsruhe, 2.–4. Oktober 1972. Herausgegeben im Auftrag der Gesellschaft für Informatik von P. Deussen. XI, 576 Seiten. 1973.

Vol. 79: A. Berman, Cones, Matrices and Mathematical Programming. V, 96 pages. 1973.

Vol. 80: International Seminar on Trends in Mathematical Modelling, Venice, 13–18 December 1971. Edited by N. Hawkes. VI, 288 pages. 1973.

Vol. 81: Advanced Course on Software Engineering. Edited by F. L. Bauer. XII, 545 pages. 1973.

Vol. 82: R. Saeks, Resolution Space, Operators and Systems. X, 267 pages. 1973.

Vol. 83: NTG/GI-Gesellschaft für Informatik, Nachrichtentechnische Gesellschaft. Fachtagung „Cognitive Verfahren und Systeme", Hamburg, 11.–13. April 1973. Herausgegeben im Auftrag der NTG/GI von Th. Einsele, W. Giloi und H.-H. Nagel. VIII, 373 Seiten. 1973.

Vol. 84: A. V. Balakrishnan, Stochastic Differential Systems I. Filtering and Control. A Function Space Approach. V, 252 pages. 1973.

Vol. 85: T. Page, Economics of Involuntary Transfers: A Unified Approach to Pollution and Congestion Externalities. XI, 159 pages. 1973.

Vol. 86: Symposium on the Theory of Scheduling and its Applications. Edited by S. E. Elmaghraby. VIII, 437 pages. 1973.

Vol. 87: G. F. Newell, Approximate Stochastic Behavior of n-Server Service Systems with Large n. VII, 118 pages. 1973.

Vol. 88: H. Steckhan, Güterströme in Netzen. VII, 134 Seiten. 1973.

Vol. 89: J. P. Wallace and A. Sherret, Estimation of Product. Attributes and Their Importances. V, 94 pages. 1973.

Vol. 90: J.-F. Richard, Posterior and Predictive Densities for Simultaneous Equation Models. VI, 226 pages. 1973.

Vol. 91: Th. Marschak and R. Selten, General Equilibrium with Price-Making Firms. XI, 246 pages. 1974.

Vol. 92: E. Dierker, Topological Methods in Walrasian Economics. IV, 130 pages. 1974.

Vol. 93: 4th IFAC/IFIP International Conference on Digital Computer Applications to Process Control, Part I. Zürich/Switzerland, March 19–22, 1974. Edited by M. Mansour and W. Schaufelberger. XVIII, 544 pages. 1974.

Vol. 94: 4th IFAC/IFIP International Conference on Digital Computer Applications to Process Control, Part II. Zürich/Switzerland, March 19–22, 1974. Edited by M. Mansour and W. Schaufelberger. XVIII, 546 pages. 1974.

Vol. 95: M. Zeleny, Linear Multiobjective Programming. X, 220 pages. 1974.

Vol. 96: O. Moeschlin, Zur Theorie von Neumannscher Wachstumsmodelle. XI, 115 Seiten. 1974.

Vol. 97: G. Schmidt, Über die Stabilität des einfachen Bedienungskanals. VII, 147 Seiten. 1974.

Vol. 98: Mathematical Methods in Queueing Theory. Proceedings 1973. Edited by A. B. Clarke. VII, 374 pages. 1974.

Vol. 99: Production Theory. Edited by W. Eichhorn, R. Henn, O. Opitz, and R. W. Shephard. VIII, 386 pages. 1974.

Vol. 100: B. S. Duran and P. L. Odell, Cluster Analysis. A Survey. VI, 137 pages. 1974.

Vol. 101: W. M. Wonham, Linear Multivariable Control. A Geometric Approach. X, 344 pages. 1974.

Vol. 102: Analyse Convexe et Ses Applications. Comptes Rendus, Janvier 1974. Edited by J.-P. Aubin. IV, 244 pages. 1974.

Vol. 103: D. E. Boyce, A. Farhi, R. Weischedel, Optimal Subset Selection. Multiple Regression, Interdependence and Optimal Network Algorithms. XIII, 187 pages. 1974.

Vol. 104: S. Fujino, A Neo-Keynesian Theory of Inflation and Economic Growth. V, 96 pages. 1974.

Vol. 105: Optimal Control Theory and its Applications. Part I. Proceedings 1973. Edited by B. J. Kirby. VI, 425 pages. 1974.

Vol. 106: Optimal Control Theory and its Applications. Part II. Proceedings 1973. Edited by B. J. Kirby. VI, 403 pages. 1974.

Vol. 107: Control Theory, Numerical Methods and Computer Systems Modeling. International Symposium, Rocquencourt, June 17–21, 1974. Edited by A. Bensoussan and J. L. Lions. VIII, 757 pages. 1975.

Vol. 108: F. Bauer et al., Supercritical Wing Sections II. A Handbook. V, 296 pages. 1975.

Vol. 109: R. von Randow, Introduction to the Theory of Matroids. IX, 102 pages. 1975.

Vol. 110: C. Striebel, Optimal Control of Discrete Time Stochastic Systems. III. 208 pages. 1975.

Vol. 111: Variable Structure Systems with Application to Economics and Biology. Proceedings 1974. Edited by A. Ruberti and R. R. Mohler. VI, 321 pages. 1975.

Vol. 112: J. Wilhlem, Objectives and Multi-Objective Decision Making Under Uncertainty. IV, 111 pages. 1975.

Vol. 113: G. A. Aschinger, Stabilitätsaussagen über Klassen von Matrizen mit verschwindenden Zeilensummen. V, 102 Seiten. 1975.

Vol. 114: G. Uebe, Produktionstheorie. XVII, 301 Seiten. 1976.

Vol. 115: Anderson et al., Foundations of System Theory: Finitary and Infinitary Conditions. VII, 93 pages. 1976

Vol. 116: K. Miyazawa, Input-Output Analysis and the Structure of Income Distribution. IX, 135 pages. 1976.

Vol. 117: Optimization and Operations Research. Proceedings 1975. Edited by W. Oettli and K. Ritter. IV, 316 pages. 1976.

Vol. 118: Traffic Equilibrium Methods, Proceedings 1974. Edited by M. A. Florian. XXIII, 432 pages. 1976.

Vol. 119: Inflation in Small Countries. Proceedings 1974. Edited by H. Frisch. VI, 356 pages. 1976.

Vol. 120: G. Hasenkamp, Specification and Estimation of Multiple-Output Production Functions. VII, 151 pages. 1976.

Vol. 121: J. W. Cohen, On Regenerative Processes in Queueing Theory. IX, 93 pages. 1976.

Vol. 122: M. S. Bazaraa, and C. M. Shetty,Foundations of Optimization VI. 193 pages. 1976

Vol. 123: Multiple Criteria Decision Making. Kyoto 1975. Edited by M. Zeleny. XXVII, 345 pages. 1976.

Vol. 124: M. J. Todd. The Computation of Fixed Points and Applications. VII, 129 pages. 1976.

Vol. 125: Karl C. Mosler. Optimale Transportnetze. Zur Bestimmung ihres kostengünstigsten Standorts bei gegebener Nachfrage. VI, 142 Seiten. 1976.

Vol. 126: Energy, Regional Science and Public Policy. Energy and Environment I. Proceedings 1975. Edited by M. Chatterji and P. Van Rompuy. VIII, 316 pages. 1976.

Vol. 127: Environment, Regional Science and Interregional Modeling. Energy and Environment II. Proceedings 1975. Edited by M. Chatterji and P. Van Rompuy. IX, 211 pages. 1976.

Vol. 128: Integer Programming and Related Areas. A Classified Bibliography. Edited by C. Kastning. XII, 495 pages. 1976.

Vol. 129: H.-J. Lüthi, Komplementaritäts- und Fixpunktalgorithmen in der mathematischen Programmierung. Spieltheorie und Ökonomie. VII, 145 Seiten. 1976.

Vol. 130: Multiple Criteria Decision Making, Jouy-en-Josas, France. Proceedings 1975. Edited by H. Thiriez and S. Zionts. VI, 409 pages. 1976.

Vol. 131: Mathematical Systems Theory. Proceedings 1975. Edited by G. Marchesini and S. K. Mitter. X, 408 pages. 1976.